Optically Polarized Atoms

Optically Polarized Atoms

Understanding Light–Atom Interactions

Marcis Auzinsh
University of Latvia, Latvia

Dmitry Budker
University of California at Berkeley, California, USA

Simon M. Rochester
University of California at Berkeley, California, USA

Great Clarendon Street, Oxford, OX2 6DP,
United Kingdom

Oxford University Press is a department of the University of Oxford.
It furthers the University's objective of excellence in research, scholarship,
and education by publishing worldwide. Oxford is a registered trade mark of
Oxford University Press in the UK and in certain other countries

© M. Auzinsh, D. Budker, and S. Rochester 2010

The moral rights of the authors have been asserted

First published 2010
First published in paperback 2014

Impression: 1

All rights reserved. No part of this publication may be reproduced, stored in
a retrieval system, or transmitted, in any form or by any means, without the
prior permission in writing of Oxford University Press, or as expressly permitted
by law, by licence or under terms agreed with the appropriate reprographics
rights organization. Enquiries concerning reproduction outside the scope of the
above should be sent to the Rights Department, Oxford University Press, at the
address above

You must not circulate this work in any other form
and you must impose this same condition on any acquirer

Published in the United States of America by Oxford University Press
198 Madison Avenue, New York, NY 10016, United States of America

British Library Cataloguing in Publication Data
Data available

ISBN 978–0–19–956512–2 (hbk.)
ISBN 978–0–19–870502–4 (pbk.)

Printed and bound by
Clays Ltd, St Ives plc

Links to third party websites are provided by Oxford in good faith and
for information only. Oxford disclaims any responsibility for the materials
contained in any third party website referenced in this work.

Acknowledgments

We are greatly indebted to Drs. Mikhail G. Kozlov, Alexander O. Sushkov, Sam Heifets, Vyacheslavs Kashcheyevs, Andrey Jarmola, and Kyle Beloy, Profs. Tanya Zelevinsky, Arlene Wilson-Gordon, Eugene D. Commins, James Higbie, Angom Dilip Kumar Singh, and Oleg P. Sushkov, and Mr. Victor M. Acosta, Mr. Byung Kyu (Andrew) Park, and Mr. Afrooz Family for carefully reading parts of the manuscript and providing us with critical feedback.

We have also benefitted enormously from "trying out" the material on our colleagues, and most importantly, students in the courses we taught at the University of Latvia and the University of California at Berkeley. Among the students who have provided invaluable feedback are Brendan Abolins, TzuCheng Chuang, Ilya Feschenko, Dylan Gorman, Peter Grisins, Ahram Kim, Patrick Lii, Jian-Long Liu, James McBride, Guil Miranda, Shawn Leighton, Etsuko Meida, Guillermina Ramirez-San Juan, Henning Schröder, Darwin Windes, and Alan Wu.

Professor Derek F. Jackson-Kimball deserves special mention for supplying key ideas and pedagogical ways of presenting them; we have also borrowed ideas from our research collaborators, in particular, Profs. Yevgeniy Alexandrov, Ruvin Ferber, Wojciech Gawlik, Alexander Pines, Jason E. Stalnaker, Antoine Weis, and Max Zolotorev, Drs. Damon English, Micah P. Ledbetter, Yuri P. Malakyan, Chih-Hao Li, Szymon Pustelny, Jeff Urban, and Valeriy V. Yashchuk, and Mr. Nathan Leefer. Some parts of the presentation are greatly influenced by the lectures on quantum mechanics of, and discussions with, Prof. Robert Littlejohn.

The material in this book is closely related to our research. Therefore, we also acknowledge the support of our research sponsors: the Latvian Science Foundation, the University of Latvia, the US National Science Foundation, the Office of Naval Research, NASA, NATO, the US-Israel Bi-National Science Foundation, the Foundational Questions Institute, and the US Department of Energy. Two of us (M.A. and D.B.) were supported by the Miller Institute for Basic Research in Science during the time some of the ideas presented here were formed.

Preface

This book is written for students who intend to practice atomic physics professionally. It reviews a number of topics including atomic structure, light polarization, interaction of atoms with light, etc., from an "experimental perspective," in which qualitative understanding of the phenomena and the ability to perform practical calculations are emphasized over theoretical rigor. In addition to the traditional topics, the book also describes several more specialized topics (dear to the authors' hearts) such as polarization moments for atoms and light, angular-momentum probability distributions, and nonlinear magneto-optical phenomena. Given this selection of topics, we hope that the book will also be useful to practicing researchers working in atomic, molecular, and optical physics, as well as to those interested in familiarizing themselves with this exciting field.

The book originated in part from the lectures and seminar courses the authors have taught over the years at the University of Latvia (M.A.), the University of California at Berkeley (D.B.), and a mini-course that two of us (D.B. and S.M.R.) delivered at the Bar-Ilan University in January 2008. We therefore envisage that the book can be used as a text for various courses suitable for advanced undergraduate students who are already familiar with basic electrodynamics and quantum mechanics and are beginning to explore more specialized topics, and also for graduate students.

In order to assist readers in using the present text, we have set up a web page (http://ukcatalogue.oup.com/product/9780199565122.do, which is linked to from the Oxford web page http://www.oup.co.uk/) containing *Mathematica*® software for atomic-physics calculations, animated examples, as well as *PowerPoint*® presentations for selected chapters that may be useful for instructors using the book for courses.

While we tried our best to avoid errors, we know from prior experience that some are bound to creep into the text. Once such errors are found, we will maintain a list of errata that will also be accessible from the book's web page. We would greatly appreciate feedback from readers, in particular, reports (via e-mail provided at the book's web page) of any errors discovered in the text.

With these preliminary remarks, we humbly submit the book to the readers' kind attention.

Marcis Auzinsh (marcis.auzins@lu.lv),
Dmitry Budker (budker@berkeley.edu),
and Simon Rochester (simonr@berkeley.edu)

Riga, Latvia and Berkeley, California, July 2009

Contents

List of acronyms   xi

    PART I   INTRODUCTION TO LIGHT–ATOM INTERACTIONS

1 Introduction   3
    1.1   Why was this book written?   3
    1.2   How to use this book   3
    1.3   Relation to other texts   4
    1.4   Formalism of quantum mechanics   5

2 Atomic states   8
    2.1   Energy states of the hydrogen atom   8
    2.2   Angular momentum of the electron in the hydrogen atom   9
    2.3   Multi-electron atoms   19
    2.4   Hyperfine interactions and hyperfine structure of atomic states   26
    2.5   Parity of atomic states   27

3 A bit of angular-momentum theory   30
    3.1   Classical rotations   30
    3.2   Quantum-mechanical rotations   36
    3.3   The angular-momentum operator   39
    3.4   Rotations in the Zeeman basis   43
    3.5   Addition of angular momenta; Clebsch–Gordan coefficients   45
    3.6 $3j$ and $6j$ symbols   46
    3.7   Irreducible tensors and tensor products   49
    3.8   The Wigner–Eckart theorem   51

4 Atoms in external electric and magnetic fields   55
    4.1   Linear Zeeman effect   55
    4.2   Zeeman effect in the manifold of hyperfine levels, Breit–Rabi diagrams   58
    4.3   Atoms in an electric field: the Stark effect   62
    4.4   Combined effect of electric and magnetic fields   72
    4.5   Atoms in oscillating fields   75

5 Polarized atoms   82
    5.1   The density matrix   82
    5.2   Rotation of density matrices   88

		5.3 Angular-momentum probability surfaces	89
		5.4 Angular-momentum probability surfaces and the density matrix: equivalence and symmetries	90
		5.5 Temporal evolution of the density matrix: the Liouville equations	92
		5.6 Example: alignment-to-orientation conversion	94
		5.7 Multipole moments	96
6	**Polarized light**		109
	6.1	The light polarization ellipse	110
	6.2	Partially polarized light and unpolarized light	114
	6.3	Spin angular momentum of polarized light	116
	6.4	Spherical basis for light polarization	118
	6.5	The polarization density matrix	121
	6.6	Angular-momentum probability surfaces for light	122
	6.7	Stokes parameters for partially polarized light	123
7	**Atomic transitions**		126
	7.1	Two-level system under the action of a periodic perturbation	126
	7.2	Selection rules for electric-dipole transitions	129
	7.3	Probability calculation for electric-dipole transitions	133
	7.4	Line strength	141
	7.5	Higher-multipole radiative transitions	144
	7.6	Multipole expansion	148
	7.7	Two-photon and multi-photon transitions	152
	7.8	Visualization of atomic transitions	153
8	**Coherence in atomic systems**		159
	8.1	Dark and bright states	159
	8.2	Quantum beats	164
	8.3	The Hanle effect	166
9	**Optical pumping**		169
	9.1	Linear and nonlinear processes; saturation parameters	169
	9.2	Optical pumping on closed transitions	173
	9.3	Optical pumping on open transitions	183
10	**Light–atom interaction observed in transmitted light**		186
	10.1	Effect of atoms on transmitted light	186
	10.2	Magneto-optical effects with linearly polarized light	190
	10.3	Perturbative approach	213

PART II   ADVANCED TOPICS

11	**Nonlinear magneto-optical rotation**		219
	11.1	Nested nonlinear magneto-optical rotation features	219
	11.2	Bennett-structure effects	220
	11.3	The role of alignment-to-orientation conversion in nonlinear magneto-optical rotation	221

11.4	Buffer-gas vapor cells	224
11.5	Antirelaxation-coated cells	225
11.6	Optically thick media	228
11.7	Nonlinear magneto-optical rotation with modulated light	234

12 Perturbative and approximate methods for light–atom interactions — 239
- 12.1 Polarization transfer in spontaneous decay — 239
- 12.2 Perturbative solution of the steady-state density matrix — 243
- 12.3 The optical-field case — 244
- 12.4 Repopulation and depopulation — 246
- 12.5 Optical excitation — 248
- 12.6 Absorption and optical rotation signals — 248
- 12.7 What kind of atomic polarization can influence the absorption and emission of light? — 252
- 12.8 The broad-line approximation — 252

13 Polarization effects in transitions with partially resolved hyperfine structure — 257
- 13.1 Depopulation pumping — 259
- 13.2 Excited state and repopulation pumping — 263
- 13.3 Absorption — 266
- 13.4 Fluorescence — 269
- 13.5 Comparison of different cases — 269

14 The effect of hyperfine splitting on nonlinear magneto-optical rotation — 271
- 14.1 Doppler-free transit effect — 272
- 14.2 Doppler-broadened transit effect — 277
- 14.3 Wall effect — 280
- 14.4 Higher nuclear spin and the $D2$ line — 283
- 14.5 Comparison of quantitative results for different cases — 287

15 Coherence effects revisited — 289
- 15.1 Dark and bright states — 289
- 15.2 Quantum beats — 292
- 15.3 The Hanle effect — 296

16 Collapse and revival in quantum beats — 303

17 Nuclear quadrupole resonance and alignment-to-orientation conversion — 309

18 Selective addressing of high-rank polarization moments — 314
- 18.1 General technique and production and detection of the $\kappa = 2$ and $\kappa = 4$ moments — 314
- 18.2 Production and observation of the $\kappa = 6$ hexacontatetrapole moment — 319
- 18.3 Production and detection of the hexadecapole moment in the Earth's magnetic field — 322

19 Tensor structure of the DC- and AC-Stark polarizabilities — 329

20 Photoionization of polarized atoms with polarized light — 333
- 20.1 Photoionization cross-section — 334

x   Contents

    20.2 Formulas for $\sigma_{0,1,2}$ 336

Appendix A **Constants, units, and notations**   339

Appendix B **Units of energy, frequency, and wavelength**   342

Appendix C **Reference data for hydrogen and the alkali atoms**   343

Appendix D **Classical rotations**   344
    D.1 Rotations in the Cartesian basis   344
    D.2 The spherical basis   347

Appendix E **Nonlinear magneto-optical rotation with hyperfine structure**   352
    E.1 Perturbation theory with polarization moments   352
    E.2 Doppler-free transit effect   354
    E.3 Doppler-broadened transit effect   356
    E.4 Wall effect   357

Appendix F **The Atomic Density Matrix software package**   358

Bibliography   360

Index   367

List of acronyms

Acronym	Meaning
AC	literally "alternating current," often used for any oscillating quantity
AMO	atomic, molecular, and optical (physics)
AMPS	angular-momentum probability surface
AOC	alignment-to-orientation conversion
BLA	broad-line approximation
CGS	centimeter-gram-second (system of units)
CPT	coherent population trapping, or combined C, P, and T transformation
cw	literally "continuous wave," often used as antonym to pulsed
DC	literally "direct current," often used for any non-oscillating quantity
EDM	(permanent) electric-dipole moment
EIA	electromagnetically induced absorption
EIT	electromagnetically induced transparency
EFG	electric field gradient
FM NMOR	nonlinear magneto-optical rotation with frequency modulated light
FWHM	full width at half maximum
hfs	hyperfine structure
HWHM	half width at half maximum
NLZ	nonlinear Zeeman (effect)
NMOR	nonlinear magneto-optical rotation
NMR	nuclear magnetic resonance
NQR	nuclear quadrupole resonance
PM	polarization moment
QED	quantum electrodynamics
RF	radio frequency
RWA	rotating-wave approximation
STIRAP	stimulated Raman adiabatic passage

Part I

Introduction to light–atom interactions

1 Introduction

1.1 Why was this book written?

The original idea for writing this book came about during a visit by one of us (M.A.) to Berkeley. Our joint research project necessitated understanding the peculiarities of the interactions of alkali atoms with narrow-band laser light under conditions in which the hyperfine structure of the transition is only partially resolved (this problem is of importance in atomic magnetometry, and more generally, in the nonlinear optics of alkali atoms). As we gained an understanding of the problem, we thought it would be instructive to share this understanding in the form of a tutorial write-up. As we contemplated what preliminary material would be needed to understand the problem at hand, we came to a realization that our write-up could take a more useful form as a book emphasizing the angular-momentum and symmetry aspects of light–atom interactions (including a detailed description of polarized light and atomic magnetic sublevels), a subject that is very important for practicing atomic, molecular, and optical physicists.

1.2 How to use this book

The book is divided into two parts: *Introduction to light–atom interactions* and *Advanced topics*. When used as a text for a one-semester course, the instructor might wish to cover the first part in its entirety, and then select one or several topics from the second part that will illustrate the use of the concepts and techniques developed in the first part.

Our intent was to make the book self-contained (for example, by including reference material in the appendices). Nevertheless, throughout the text, and especially in the second part of the book, we provide copious references to earlier work. It is not necessary for the reader to look up these references in order to understand the material. However, we believe that it may be useful to have these references on hand, both to give readers a feel for this actively developing field of science, and to guide them towards a more detailed exploration of a given topic should they have interest and/or need.

Students who used a draft version of the book as the primary text for our courses at Riga and Berkeley have expressed a desire to do exercises as they work through the text. Although we do not specifically delineate exercises, we have sprinkled numerous (worked) examples throughout the text. We, therefore, encourage the reader to go through the text with paper and pencil in hand (or a laptop, as may be these days). In addition, *Atomic Physics. An Exploration Through Problems and Solutions*, Second Edition, by Budker *et al.* (2008) can be used in conjunction with the present text. We frequently provide references to specific relevant problems in that book for the reader's convenience.

As in the bulk of our previous published work, in this text we primarily use the Gaussian (CGS) system of units. We prefer this system to the SI system, which is becoming gradually more prevalent, because equations written in Gaussian units are not polluted with coefficients like $4\pi\epsilon_0$ (where ϵ_0 is the *vacuum permittivity*, equal to unity in Gaussian units), and because electric and magnetic fields have the same dimensions in Gaussian units, reflecting the deep symmetry between these fields inherent in the Maxwell equations. Appendix A provides a list of fundamental constants in Gaussian units. The reader interested in converting between the systems of units should encounter no difficulty doing so using numerous web resources; see also the in-depth discussion of the relation between various units in the appendix of the book by Jackson (1975).

> Finally, some examples, auxiliary remarks, and supplementary notes appear in the text, indicated like this. We feel that this material enriches the text; however, it can be skipped without breaking the "story line."

1.3 Relation to other texts

A sizable fraction of the material presented in this book can be found in other books and review papers. However, we hope that this text will address a considerable gap in the literature: in our own experience and that of our colleagues and students, most of the textbooks are either too elementary and at times superficial for someone who wishes to work in atomic physics professionally, or too advanced, with mathematical details obscuring intuitive physical pictures.

Nevertheless, there are many useful books on related subjects. Some that we ourselves use in our work are

- *Atomic and Laser Spectroscopy* by A. Corney (1988). This excellent text was written in the mid-1970s, and was recently reprinted as an Oxford Classic Text in the Physical Sciences. It covers a broad range of topics, from reviews of electrodynamics and quantum mechanics to more specialized ones such as optical double resonance and optical pumping. There is some overlap of topics (for example, the chapter on the Hanle effect and resonance fluorescence) with the material of this book; however, Corney's discussion is often briefer than ours, and we feel we have a lot to add to it, even for readers already familiar with Corney's book.
- *Atomic Spectra and Radiative Transitions* by I. I. Sobelman (1992). This is an excellent reference on atomic structure, angular-momentum theory, and theory of atomic transitions. While this is probably the most frequently used book in our laboratories, it is too technical and "theoretical" to be an introductory textbook.
- *Quantum Theory of Angular Momentum* by D. A. Varshalovich *et al.* (1988) is the most complete reference known to us on irreducible tensors, spherical harmonics, angular-momentum coupling coefficients, etc. It is by no means a textbook, and is useful to the reader already proficient in the subject matter.
- *Angular Momentum in Quantum Mechanics* by A. R. Edmonds (1974) is a classic reference on angular-momentum theory bridging the gap between formal mathematical literature and "experimental" references. It does not discuss light–atom interactions.
- *Angular Momentum* by R. N. Zare (1988) is an excellent textbook covering angular-momentum theory, spherical tensors, and their applications. As most examples in this book are from molecular physics, they are predominantly, by necessity, more complex than those arising from the simpler atomic systems considered in the present book.

- *Polarization of Light* by S. Huard (1997) is an excellent text and reference on various methods of representing polarization states of light and analyzing propagation of polarized light through anisotropic media in the context of linear optics.
- *Quantum Mechanics. Nonrelativistic Theory* by L. D. Landau and E. M. Lifshitz (Landau and Lifshitz 1977) is, arguably, the best book on quantum mechanics ever written. There is not much overlap with the content of the present book.
- *Optical Polarization of Molecules* by M. Auzinsh (one of the authors of the present book) and R. Ferber (Auzinsh and Ferber 1995) discusses a variety of processes related to the interaction of molecules with light. We recommend this text to the readers of the present book as a follow-up source on applications of much of the material presented here in molecular spectroscopy.
- *Interference of Atomic States* by E. B. Alexandrov, M. P. Chaika, and G. I. Khvostenko (Aleksandrov et al. 1993) discusses various phenomena related to the angular distribution of fluorescence from excited atomic states, the Hanle effect, quantum beats, etc. We recommend this book as complementary reading to the present text.
- *Foundations of Laser Spectroscopy* by S. Stenholm (2005) gives an introduction to the density-matrix formalism, discusses how to describe relaxation processes, and, in general, provides a very nice introduction to theoretical spectroscopy. For one of the authors (D.B.) and some of his colleagues, this book was the first encounter with density matrices.
- *The Quantum Theory of Light* by R. Loudon (2000) is an introductory textbook on quantum optics (the subject we touch upon only peripherally in this book), full of deep insights and qualitative explanations along with clear mathematical formulations. We always recommend this book to our students.
- *The Theory of Coherent Atomic Excitation* by B. Shore (1990) is an encyclopedia of dynamic light–atom interactions, including such processes as Rabi oscillations, excitation of two- and multi-level atoms, etc. A unique source of unabbreviated insight.
- *The Physics of Laser–Atom Interactions* by D. Suter (1997) is an excellent modern textbook covering a wide range of topics of current relevance. The book is admirable in its success in explaining complex phenomena such as energy-sublevel dynamics in the presence of modulated laser excitation using the simplest possible models. Suter's book, however, does not emphasize the angular, kinematic aspects of laser–atom interactions, a principal focus of the present text.
- *Optically Pumped Atoms* by W. Happer et al. (2010), despite the similar title, has only minimal overlap with the present text in terms of the content. While extremely useful for professionals with a firm grasp of the basics who wish to perform quantitative computer modeling of optical-pumping processes, it may be challenging for others, whom we, on the other hand, see as our primary audience.

1.4 Formalism of quantum mechanics

We will assume that the reader has some familiarity with undergraduate quantum mechanics; here we briefly recall some basic facts and establish notations to be used in the book.

The central concept in quantum mechanics is the *wave function* usually designated as ψ. The meaning of the wave function is that it is the (generally complex) *probability amplitude* for the system to be in a particular state. The recipe for finding a corresponding probability is to evaluate the absolute value squared of the wave function:

$$P = |\psi|^2. \tag{1.1}$$

This assumes that the quantum system can be in a set of discrete states. In some cases, for example, when the wave function describes the spatial location of the particle, the modulus squared of the wave function actually denotes *probability density*.

Another central concept in quantum mechanics is that of an *operator*—a certain procedure that, when applied to a quantum-mechanical state, transforms it into another (or possibly the same) state. If the operator is designated as $\mathcal{O}$, its action on a wave function can be written

$$\psi' = \mathcal{O}\psi. \tag{1.2}$$

If the resultant wave function ψ' happens to be the same as the initial wave function ψ up to a complex numerical constant c

$$\psi' = \mathcal{O}\psi = c\psi, \tag{1.3}$$

then the wave function ψ is said to be the *eigenfunction* of the operator $\mathcal{O}$, and the constant c is the corresponding *eigenvalue*.

A particularly important quantum mechanical operator is the energy operator, a.k.a. the *Hamiltonian*. The eigenfunctions of a time-independent Hamiltonian are called *stationary states*, for which probabilities do not depend on time. A set of these states $\{\psi_k\}$ forms a complete *basis* that can be chosen to be *orthonormal*, which means that an arbitrary wave function can be represented as a sum of the basis states with complex coefficients c_k:

$$\psi = \sum_k c_k \psi_k, \tag{1.4}$$

and that the *inner product* of two basis functions is

$$\psi_l \cdot \psi_k = \delta_{kl}. \tag{1.5}$$

Taking a inner product (also called a *scalar product*) of two wave function means integrating the product of the complex conjugate ψ_l^* with ψ_k over the range of variables, the set of which we schematically designate as x, on which the wave functions depend:

$$\psi_l \cdot \psi_k = \int \psi_l^*(x) \psi_k(x) dx. \tag{1.6}$$

Any complete basis can be used to write operators and wave functions in *matrix notation*. A wave function can be represented by a vector of coefficients (Eq. 1.4). A linear operator takes one such vector and transforms it to another. It can be written as a matrix operating on the vector by matrix multiplication. To find the matrix elements of the operator represented in a particular basis, one needs to know how the matrix acts on the basis states.

It can be shown using the methods of linear algebra that a basis of states that are simultaneously eigenstates of two operators $\mathcal{A}$ and $\mathcal{B}$ can be found if and only if $\mathcal{A}$ and $\mathcal{B}$ commute, i.e., if

$$[\mathcal{A}, \mathcal{B}] = \mathcal{AB} - \mathcal{BA} = 0. \tag{1.7}$$

There is an alternative and very convenient way to write such expressions, the *Dirac notation*. In Dirac notation, a wave function ψ_ξ (where ξ is a label used to identify the particular wave function) is identified with a *ket*

$$|\xi\rangle, \tag{1.8}$$

while the complex conjugate of the wave function is identified with a *bra*

$$\langle\xi|. \tag{1.9}$$

The scalar product of two wave functions ψ_ξ and ψ_ζ is written

$$\langle\zeta|\xi\rangle. \tag{1.10}$$

Suppose we start with some wave function ψ_ξ and act on it with an operator $\mathcal{O}$. We might be interested in the "degree of the overlap" of the new wave function $\mathcal{O}|\xi\rangle$ with some other wave function $|\zeta\rangle$. This degree of overlap is given by the quantum mechanical *amplitude*, which is just the scalar product

$$\langle\zeta|\mathcal{O}|\xi\rangle, \tag{1.11}$$

which also goes by the name of the *matrix element* of the operator $\mathcal{O}$ between the states $|\xi\rangle$ and $|\zeta\rangle$.

The eigenstates of a system described by a time-independent Hamiltonian H can be found by solving the *Schrödinger equation*

$$H|\xi\rangle = E_\xi|\xi\rangle. \tag{1.12}$$

While, following tradition, the discussion of quantum mechanics usually begins with wave functions, a more general description is, in fact, in terms of *density matrices*, which we will discuss beginning in Chapter 5.

Throughout the discussion in the subsequent chapters, we will encounter a variety of quantum-mechanical operators. A common procedure by which a quantum-mechanical operator is introduced follows the *correspondence principle*, which stipulates that quantum-mechanical expressions should be equivalent to the classical ones in the appropriate limits. In this procedure one takes a classical expression for a physical quantity and replaces the variables in the expression with the corresponding quantum-mechanical operators. For example, the classical expression for kinetic energy

$$K = \frac{\mathbf{p}^2}{2m}, \tag{1.13}$$

where $\mathbf{p}$ is the particle's momentum and m is its mass, in quantum mechanics becomes

$$\hat{K} = \frac{\hat{\mathbf{p}}^2}{2m} = \frac{(-i\hbar\nabla)^2}{2m}, \tag{1.14}$$

where we have substituted the quantum-mechanical expression for momentum $\hat{\mathbf{p}} = -i\hbar\nabla$. Note that we will generally omit the "hats," except for the places where we need to emphasize the operator character of the expressions. Another example is the *electric-dipole moment* of a system of particles. According to the rule that we have just introduced, the quantum-mechanical expression will be essentially identical to its classical counterpart:

$$\hat{\mathbf{d}} = \sum q_n \hat{\mathbf{r}}_n, \tag{1.15}$$

where the summation is over the particles in the system, and q_n and $\hat{\mathbf{r}}_n$ are the charge and position operator of the n-th particle.

2
Atomic states

The goal of this book is to discuss the "practical theory" of light–atom interactions. We begin by recalling the relevant features of atomic structure. In this chapter, we examine the energy levels of the simplest atom, hydrogen, recalling the gross, fine, and hyperfine structure, and then move on to a discussion of energy levels in complex atoms, paying particular attention to the various atomic angular momenta, which will be found to play a central role in light–atom interactions.

More detailed descriptions of atomic structure can be found in many excellent textbooks, such as those by Foot (2005) or Bransden and Joachain (2003), which contain material about atomic states at a level and in a style that correlates well with this book.

2.1 Energy states of the hydrogen atom

Generally, in quantum mechanics, systems of bound particles have discrete energy states. In the nonrelativistic approximation, and neglecting the electron's intrinsic angular momentum (spin), these energy states can be found as solutions of the time-independent Schrödinger equation (Eq. 1.12). In the simplest case of the hydrogen atom the equation is written

$$\left(-\frac{\hbar^2}{2m_e}\triangle - \frac{e^2}{r}\right)\psi_{nlm}(\mathbf{r}) = E_n\psi_{nlm}(\mathbf{r}), \tag{2.1}$$

where $\hbar$ is Planck's constant h divided by 2π, m_e and e are the mass of the electron and the magnitude of the electron's charge, and $\mathbf{r}$ is its position vector of the electron relative to the nucleus (we have assumed that the nucleus is infinitely heavy). The Laplace operator $\triangle = \nabla^2 = \sum_i \partial^2/\partial r_i^2$ is here expressed as the sum of second derivatives with respect to the electron's Cartesian coordinates r_i. The eigenvalues E_n correspond to allowed energies for the atom, and the eigenfunctions $\psi_{nlm}(\mathbf{r})$ are the electronic wave functions. The electron density at a given location $\mathbf{r}$ in an atom is given by $|\psi_{nlm}(\mathbf{r})|^2$. Under the approximation in which we have neglected the effects of the electron spin and relativity, the energy depends only on one quantum number n, called the *principal quantum number*. The energy levels obtained from Eq. (2.1) are

$$E_n = -\frac{1}{2}\frac{m_e e^4}{\hbar^2}\frac{1}{n^2}, \tag{2.2}$$

where the principal quantum number can assume any value $n = 1, 2, 3, \ldots$. The dependence on the combination $m_e e^4/\hbar^2$ can be derived dimensionally, since it is the only combination of the fundamental constants m_e, e, and $\hbar$ appearing in Eq. (2.1) that has the dimensions of energy. As such it is known as the atomic unit of energy, or the *Hartree*. One-half of the Hartree, or $m_e e^4/(2\hbar^2)$, is a more commonly used unit and is known as the *Rydberg*. The wave functions

$\psi_{nlm}(\mathbf{r})$ are labeled by two additional quantum numbers: the *orbital quantum number* l and the *magnetic quantum number* m, both related to the orbital angular momentum of the electron. Thus, in general, each energy value of the hydrogen atom corresponds to several different wave functions, i.e., the atomic states are *degenerate*. Ignoring the electron spin, we will see that, given the values that can be assumed by the quantum numbers l and m, the number of different wave functions that correspond to a given energy state in the hydrogen atom is equal to n^2.

2.2 Angular momentum of the electron in the hydrogen atom

2.2.1 The angular momentum quantum numbers

The orbital angular momentum quantum number l (a nonnegative integer) is related to the modulus of the angular momentum $|\mathbf{l}|$.[1] The modulus assumes only the discrete values

$$|\mathbf{l}| = \sqrt{l(l+1)}\hbar. \qquad (2.3)$$

This can be obtained by solving the Schrödinger equation, or by considering the commutation relations for the components of the angular-momentum operator, as will be done in Sec. 3.3. As we can see from Eq. (2.3), Planck's constant has the dimensions of angular momentum. This fact motivated Niels Bohr at the beginning of the 20th century to introduce a model for the atom. The *Bohr model* treated the motion of the electron as classical, but quantized by requiring orbits to have angular momentum equal to an integer multiple of $\hbar$. This occurred not many years after Max Planck introduced the constant in order to explain a great mystery of physics at the very end of the 19th century—the spectrum of black body radiation. Later it was discovered that the Bohr model does not give a completely correct picture of the hydrogen atom; nevertheless, it played an important role in the development of the theory of atoms as we know it today. The Bohr model is also closely related to the *vector model*, a useful tool that we will use in this book to visualize the coupling of angular momenta.

Another result that is obtained from the solution of the Schrödinger equation is that a state with a given energy must have an upper bound on the orbital angular momentum it is allowed to have.

> This particular result is also true in classical physics and is often discussed in the context of Kepler's problem; see, for example, Landau and Lifshitz (1976), Chapter 3 §15. The energy of an elliptical orbit depends only on the length of the major axis of the ellipse, but not on the angular momentum. On the other hand, for orbits with a given energy (or major axis), the *eccentricity* decreases with an increase in angular momentum, and reaches zero for a circular orbit, which, therefore, corresponds to an orbit of the highest angular momentum for a given energy. The connection between the Kepler orbits and the quantum theory is discussed in detail in the classic text by Sommerfeld and Brose (1934).

For a state with energy E_n, where n is the principal quantum number, the angular momentum quantum number l must be less than n, i.e. it can assume the values $l = 0, 1, 2, \ldots, n-1$. In classical physics, once the magnitude of the angular momentum is specified, there are two more parameters (e.g., two angles) that can be measured to specify the direction, thus specifying the angular-momentum vector completely. In quantum mechanics, however, it is in principle impossible to know the direction in space of the angular momentum. The maximal information that we can have is one projection of the electron's angular momentum on an axis,

---

[1] In quantum mechanics $\mathbf{l} = l_x\hat{\mathbf{x}} + l_y\hat{\mathbf{y}} + l_z\hat{\mathbf{z}}$ is a vector operator, not a classical vector. By $|\mathbf{l}|$ we are referring, strictly speaking, to the square root of the expectation value of the $\mathbf{l} \cdot \mathbf{l}$ operator with respect to an eigenstate with quantum number l.

usually called the *quantization axis*. The reference frame is traditionally chosen so that the quantization axis is along the z-axis. It can be shown that if the quantization axis is singled out, the projection l_z on this axis can only take on the values

$$l_z = m\hbar, \qquad m = -l, -l+1, \ldots, l-1, l, \qquad (2.4)$$

which is reasonable considering that one would expect the magnitude of the projection on an axis to be less than the total magnitude $\sqrt{l(l+1)}\hbar$ of the angular momentum. The quantum number m is called the magnetic quantum number because, experimentally, a magnetic field is often used as the means to single out a direction in space. The direction of the quantization axis can also be defined by a static or oscillating (for example, optical) electric field, gravitational field, or by some other means. For a theoretical description of an atom we do not even need to use physical means (i.e., an external field) to define this direction. We can simply choose any convenient direction and place the quantization axis along it. Since the total number of possible m values is $2l+1$, the total number of possible combinations of l and m values for a given n is easily seen to be n^2, as was stated in the previous section.

The degeneracy with respect to m can be expected just on the grounds that all directions in space are equivalent, since different values of m correspond to different arrangements of the electronic angular momentum in space (specifically, the angle between the angular momentum and the quantization axis). The degeneracy with respect to l, which indexes the magnitude of the angular momentum, on the other hand, is a special feature of the $1/r$ Coulomb potential. Thus when corrections to the Coulomb potential are introduced, we expect this degeneracy to be broken.

In classical physics, for a system with a definite projection of angular momentum on a chosen axis (for example, z), it is possible to simultaneously determine the projections of angular momentum on the other two coordinate axes (x and y) to arbitrary precision. In quantum physics, there are *uncertainty relations* that put restrictions on one's ability to determine more than one angular-momentum projection at a time, for example,

$$\Delta l_x \Delta l_y \geq \frac{\hbar |\langle l_z \rangle|}{2}, \qquad (2.5)$$

where l_i are the projections of the angular momentum on the corresponding axes. Here Δl_x and Δl_y signify standard deviations; the standard deviation in the measurement of an observable A, is defined by

$$\Delta A = \sqrt{\langle A^2 \rangle - \langle A \rangle^2}. \qquad (2.6)$$

If a state has definite quantum numbers l and m then the directions transverse to the quantization axis (z) are equivalent, i.e., the state is symmetric with respect to rotations about the quantization axis. This means, for example, that $\langle l_x \rangle = \langle l_y \rangle = 0$, and $\Delta l_x = \Delta l_y$. We can use this to evaluate the standard deviations for l_x and l_y for this eigenstate. We have, writing the $\mathbf{l}^2$ operator in terms of components,

$$\mathbf{l}^2 = l_x^2 + l_y^2 + l_z^2. \qquad (2.7)$$

Taking expectation values,

$$l(l+1)\hbar^2 = 2\langle l_x^2 \rangle + m^2\hbar^2, \qquad (2.8)$$

so $\Delta l_x = \Delta l_y = \hbar\sqrt{[l(l+1) - m^2]/2}$. We can check that this agrees with the uncertainty relation by substituting into Eq. (2.5):

$$\Delta l_x \Delta l_y = \frac{l(l+1) - m^2}{2}\hbar^2 \geq \frac{|m|\hbar^2}{2}, \tag{2.9}$$

or $l(l+1) \geq m^2 + |m| = |m|(|m| + 1)$, which is always true. Interestingly, the standard deviation for transverse components of angular momentum is the largest for the smallest magnitude of the quantum number m, for which the uncertainty condition (2.5) is the least restrictive.

> There are also generalized momentum-coordinate uncertainty relations. The very first uncertainty relation usually written in quantum mechanics courses is that for linear momentum in a certain direction and the position coordinate for the same direction. For the orbital angular momentum case considered here, one generalized relation reads (see, for example, Atkins and Friedman 1996, Sec. 3.3):
>
> $$\Delta l_z \Delta \sin\varphi \geq \frac{\hbar|\langle\cos\varphi\rangle|}{2}. \tag{2.10}$$
>
> Here φ is the azimuthal angle for the electron. The uncertainty relation of Eq. (2.10) is a bit more complicated than that for the linear momentum and the corresponding coordinate, $\Delta p \Delta x \geq \hbar/2$. This is because φ, in contrast to x, is a cyclic variable, so that the values φ and $\varphi + 2\pi N$, where N is an integer, are physically equivalent. A detailed discussion of uncertainty relations in this case can be found in Carruthers and Nieto (1968).

2.2.2 Wave functions of the hydrogen atom

As we see from Eq. (2.1), each combination of quantum numbers n, l, and m labels a specific hydrogen atom wave function $\psi_{nlm}(\mathbf{r})$. For a hydrogenic atom, one can separate the radial (r) and angular (θ, φ) variables so that the wave functions can be written in polar coordinates in the form

$$\psi_{nlm}(\mathbf{r}) = R_{nl}(r)Y_{lm}(\theta, \varphi), \tag{2.11}$$

where the radial part of the wave function $R_{nl}(r)$ depends only on the quantum numbers n and l, and the angular part of the wave function $Y_{lm}(\theta, \varphi)$ depends only on the quantum numbers l and m. The functions $Y_{lm}(\theta, \varphi)$ are called *spherical functions* or *spherical harmonics*. They can be further separated as $Y_{lm}(\theta, \varphi) = \Theta_{lm}(\theta)\Phi_m(\varphi)$, where $\Phi_m(\varphi) \propto e^{im\varphi}$ and $\Theta_{lm}(\theta)$ is given in terms of an *associated Legendre polynomial* in $\cos\theta$. The explicit form of the spherical harmonics for several of the lowest values of the quantum numbers l and m is presented in Table 2.1. Some explicit radial functions for hydrogen are shown in Table 2.2. The radial functions are defined in terms of the *Bohr radius*

$$a_0 = \frac{\hbar^2}{m_e e^2}. \tag{2.12}$$

From the form of the functions in Table 2.2, we can see that the Bohr radius sets the scale for the size of the hydrogen atom.

It is an important point that this separation of the wave function into radial and angular parts is possible not only for the hydrogen atom, but for any atom—as long as we can assume that each electron in the atom is subject to a *central field*, i.e., one whose potential depends only on the distance from the nucleus. (For a multi-electron atom, the combined field produced

12 Atomic states

Table 2.1 Angular parts of the wave function of an atom for the lowest values of quantum numbers l and m.

Spherical functions $Y_{lm}(\theta, \varphi)$
$Y_{00}(\theta, \varphi) = \sqrt{\frac{1}{4\pi}}$
$Y_{10}(\theta, \varphi) = \sqrt{\frac{3}{4\pi}} \cos\theta$
$Y_{1\pm 1}(\theta, \varphi) = \mp\sqrt{\frac{3}{8\pi}} \sin\theta\, e^{\pm i\varphi}$
$Y_{20}(\theta, \varphi) = \sqrt{\frac{5}{16\pi}} \left(3\cos^2\theta - 1\right)$
$Y_{2\pm 1}(\theta, \varphi) = \mp\sqrt{\frac{15}{8\pi}} \sin\theta \cos\theta\, e^{\pm i\varphi}$
$Y_{2\pm 2}(\theta, \varphi) = \sqrt{\frac{15}{32\pi}} \sin^2\theta\, e^{\pm 2i\varphi}$

Table 2.2 Radial parts of the wave function of a hydrogen atom for the lowest values of quantum numbers n and l.

Hydrogen radial functions $R_{nl}(r)$
$R_{10}(r) = 2 a_0^{-3/2} e^{-r/a_0}$
$R_{20}(r) = \frac{1}{\sqrt{2}} a_0^{-3/2} e^{-\frac{r}{2a_0}} \left(1 - \frac{r}{2a_0}\right)$
$R_{21}(r) = \frac{1}{2\sqrt{6}} a_0^{-3/2} e^{-\frac{r}{2a_0}} \frac{r}{a_0}$

by the atomic nucleus and the rest of the electrons may be only approximately central, as discussed in Sec. 2.3.) The radial function depends on the particular electrostatic potential created by the nucleus and electrons in the particular atom. Thus, for atoms of different species, the radial parts $R_{nl}(r)$ of the wave function are different for the same set of quantum numbers. However, the angular parts $Y_{lm}(\theta, \varphi)$, which are determined by rotational symmetry, are the same for two states with the same l and m. The fact that the part of the wave function that is related to the angular momentum of the electron can be found in a rather simple analytical form for an arbitrary atom is very important for our discussion. It simplifies the quantum theory of angular momentum and allows one to solve many problems in atomic physics involving angular momentum precisely, without employing any approximations.

2.2.3 Electron spin and the fine structure interaction

Experimentally it is found that, in addition to orbital angular momentum, each electron possesses an intrinsic angular momentum. This *spin angular momentum* can be characterized by the *spin quantum number s*.

> It is tempting to think about spin classically in terms of a small spinning object and its corresponding angular momentum. As many classical pictures of this sort, this can be useful, but only to a certain point. For example, the classical picture implies that if the object is spinning at a finite rate—the only possibility in a classical world—the object has to have a finite spatial extent in order to have finite angular momentum. Suppose the spatial extent is r. The dimensional estimate of the moment of inertia is then $I = m_e r^2$. Since the energy of rotation has to be less than $m_e c^2$, it follows that $m_e r^2 \omega^2 < m_e c^2$. Thus the angular momentum is less than $I\omega \sim m_e r c$, where ω is the angular velocity of the rotation, and we set all numerical factors to unity. For an angular momentum $\sim \hbar$, this implies that the spatial extent has to be on the order of the *Compton wavelength of the electron* $\lambda = \hbar/(m_e c) \approx 3.9 \times 10^{-11}$ cm.

Yet it has been checked with high-energy colliders that the electron is a point-like particle without any structure down to $\sim 10^{-18}$ cm (Particle Data Group 2008).

An alternate amusing classical picture is that the spin angular momentum is just the angular momentum of the combination of the Coulomb electric and dipole magnetic field produced by the electron:

$$\mathbf{s} = \mathbf{L}_{EM} = \frac{1}{4\pi c} \int \mathbf{r} \times (\mathcal{E} \times \mathbf{B}) \, d^3r \quad (2.13)$$

[see, for example, Jackson (1975) for a derivation of Eq. (2.13)]. This picture gives an estimate of the spatial extent of the electron's internal structure on the order of the *classical radius of the electron*, $r_e = e^2/(mc^2) \approx 2.8 \times 10^{-13}$ cm, also ruled out by the collider experiments.

Another conceptual problem with the idea of the electron being a spinning object is the fact that the electron has half-integer angular momentum, which means that it would need to rotate in space by an angle 4π in order to come back to its original state (as opposed to 2π for conventional systems; see Sec. 3.3.4), a property that is difficult (but not impossible!—see Richard Feynman's lecture in the book by Feynman and Weinberg, 1987) to associate with a classical object.

In contrast to the orbital angular momentum quantum number l, which can take on different values, the spin quantum number s for a single electron has a single half-integer value of $1/2$. This means that the modulus of the spin angular momentum of the electron is always

$$|\mathbf{s}| = \sqrt{s(s+1)}\hbar = \frac{\sqrt{3}}{2}\hbar. \quad (2.14)$$

The projection of the spin on the quantization axis can only have one of the two values

$$s_z = m_s \hbar = \pm \frac{1}{2}\hbar. \quad (2.15)$$

The often used expression that the electron spin is "up or down" (i.e., along or opposite to the quantization direction) has to be used with a certain degree of caution. The modulus of the spin is greater than its projection. This is true for any quantum angular momentum.

In Eq. (2.15) we use the subscript s with the magnetic quantum number m. We will use such subscripts whenever we refer to magnetic quantum numbers that characterize different kinds of angular momenta to show explicitly which momentum we mean.

The fact that the electron has an intrinsic angular momentum has direct consequences for the energy levels and atomic states of the hydrogen atom. Both the orbital and spin angular momentum have associated magnetic moments (μ_l and μ_s, respectively).

The magnitude of the magnetic moment associated with orbital motion of an electron can be estimated from classical considerations. Indeed, we know that the magnitude of the magnetic moment of a current loop is $\mu = |\mu| = IA/c$, where I is the current, and A is the area of the loop. Thinking in terms of a classical orbit of radius r, the electron is moving with speed p/m_e (**p** is the momentum), so the revolution rate is $(2\pi r m_e/p)^{-1}$, and $I = ep/(2\pi r m_e)$, so that

$$\boldsymbol{\mu} = -\frac{e}{2m_e c}\mathbf{l} = -\gamma \mathbf{l}, \quad (2.16)$$

where $\gamma = e/(2m_e c)$ is the gyromagnetic ratio.

The classical picture suggests that the magnetic moment associated with orbital motion is collinear with the orbital angular momentum. In fact, because the charge of the electron is negative ($-e$), the magnetic moment is directed oppositely to the orbital angular momentum.

14 Atomic states

Because in quantum mechanics the modulus of angular momentum is quantized (see Eq. 2.3), the modulus of the magnetic moment must also be quantized:

$$\mu_l = -\frac{e\hbar}{2m_e c}\sqrt{l(l+1)} = -\mu_B\sqrt{l(l+1)}, \tag{2.17}$$

where $\mu_B = e\hbar/(2m_e c)$ is a quantum unit of magnetic dipole moment called the *Bohr magneton*. The value of the Bohr magneton is $\approx 0.93 \times 10^{-20}$ erg G^{-1} (see Table A.1). For practical calculations, it is often convenient to express the Bohr magneton as $\mu_B/h \approx 1.40$ MHz G^{-1}.

The projection of the magnetic dipole moment on the quantization axis is quantized as well:

$$(\mu_l)_z = -\frac{e}{2m_e c}l_z = -\frac{e\hbar}{2m_e c}m = -\mu_B m. \tag{2.18}$$

It turns out that there is also a magnetic moment associated with the electron's spin, and the relation between this magnetic moment and the spin is given by a relation similar to Eq. (2.17):

$$\mu_s \approx -2\frac{e\hbar}{2m_e c}\sqrt{s(s+1)} = -2\mu_B\sqrt{s(s+1)}. \tag{2.19}$$

An important difference here is the factor $g_s \approx 2$ for the case of the electron spin, multiplying the Bohr magneton. This factor is called the *Landé g-factor*. (Comparing with Eq. (2.17) we would say that the g-factor for orbital angular momentum is $g_l = 1$.) The quantum-mechanical *Dirac equation* for spin-1/2 particles predicts $g_s = 2$ here, but *quantum-electrodynamical (QED)* corrections that take into account perpetual creation and annihilation of virtual particles introduce small corrections to the numerical factor. We will ignore this *anomalous magnetic moment* throughout this book, and will assume that the total magnetic moment associated with an electron is

$$\boldsymbol{\mu} = -\mu_B\left(\mathbf{l} + 2\mathbf{s}\right)/\hbar. \tag{2.20}$$

> The magnetic moment of the electron has been measured to better then one part in a trillion making it one of the most precisely known physical constants. The "2" is really $2 \times 1.00115965218073(28)$ (Hanneke *et al.* 2008).
>
> We note also that Hanneke *et al.* (2008) use the same sign convention for the g-factors as we use here. Unfortunately, this is yet another case where there is considerable "diversity" in the conventions used by various authors. For example, the authoritative Committee on Data for Science and Technology (CODATA) adopts a convention where the g-factor for the electron is negative (Mohr *et al.* 2008).

In general, when both spin and orbital angular momenta are present, neither of the two are conserved. Classically, if we have angular momenta associated with different kinds of motion (for example, orbital motion of a planet, and its spin around its axis), it is only the total angular momentum of the system that is conserved, but not the individual components (although on a short time scale, conservation of individual angular momenta may be an excellent approximation, as we are used to in the planetary case).

The mechanism for coupling the orbital and spin angular momentum of an atomic electron is known as the *spin–orbit interaction*. One way to think about it is the following. Due to orbital motion, the electric field of the atom produces a magnetic field in the rest frame of the electron. This magnetic field interacts with the electron's magnetic moment associated with the spin (Eq. 2.19), producing an energy shift depending on the relative orientation of $\mathbf{l}$ and $\mathbf{s}$, i.e. on the value of the *total angular momentum* $\mathbf{j} = \mathbf{l} + \mathbf{s}$.

While states with particular values of m_l and m_s can be used to describe the system, they are no longer eigenstates when the spin–orbit interaction is considered. On the other hand, states with definite values of j, m_j are eigenstates, as **j** commutes with the total Hamiltonian. Classically, we would describe this by saying that the total angular momentum is a constant of the motion for any isolated system.

What are the properties of the total angular momentum **j**? As with all quantum-mechanical angular momenta, it is characterized by two quantum numbers: j and m_j parametrize the modulus $|\mathbf{j}|$ and projection on the quantization axis, j_z, according to the general rule

$$|\mathbf{j}| = \sqrt{j(j+1)}\hbar, \qquad j_z = m_j \hbar. \tag{2.21}$$

As usual, $|m_j| \leq j$. In order to determine the allowed values of j, it can be helpful (though not strictly accurate, as discussed below) to consider **l** and **s** as adding like classical vectors. Then j must have a value between $|l - s|$ and $l + s$, i.e., $j = |l - s|, |l - s| + 1, \ldots, l + s$; since $s = 1/2$, there are two allowed values, $j = l \pm 1/2$, for $l > 0$, and only the value $j = 1/2$ for $l = 0$.

The interaction between the orbital motion of an electron and its spin (the *spin–orbit interaction*), together with other relativistic corrections to the Schrödinger equation (2.1), cause the energies of the states with the same n but different j to differ slightly. The hydrogen atom energy including the relativistic effects can be found as a perturbative correction to the energy (Eq. 2.2) obtained from the nonrelativistic Schrödinger equation (Sobelman 1992). The total energy is given by the *Dirac formula*

$$E_n = \left[-\frac{1}{n^2} + \alpha^2 \left(\frac{3}{4n} - \frac{1}{j + 1/2} \right) \frac{1}{n^3} \right] \frac{m_e e^4}{2\hbar^2}, \tag{2.22}$$

where the dimensionless constant

$$\alpha = \frac{e^2}{\hbar c} \tag{2.23}$$

is the *fine-structure constant* whose numerical value is $\alpha \approx 1/137$ (see Table A.1). It is interesting to note that according to the Bohr model the velocity of the electron in the first hydrogen orbit is equal to αc, as one might guess from dimensional analysis, since $e^2/\hbar$ is the only combination of the fundamental constants e, $\hbar$, and m that appear in the Schrödinger equation (2.1) that has the dimensions of velocity.

> The possibility that α and other fundamental constants may actually be changing with time was raised at the dawn of modern physics (see, for example, a most informative and entertaining account in the book by Barrow, 2002), even though such a variation, if discovered, would call for a thorough revision of our most basic understanding of the laws of nature. This has recently become a particularly "hot" area of experimental research following evidence from some astrophysical observations that α may have been about one part in 10^5 smaller ten billion years ago than it is now (see the book by Karshenboim and Peik, 2004, for a detailed account of the background and experimental details of the searches for variation of fundamental constants). Laboratory experiments that measure frequencies of atomic transitions have reached a sensitivity to the present-day rate of variation of α at a level of a part in 10^{15} per year, but so far have found no variation.

This motivates the notion that the fine-structure constant is a measure of the magnitude of relativistic effects in the atom. For example, the relativistic energy corrections in Eq. (2.22) are proportional to α^2. Since $\alpha \approx 1/137$, these effects are fairly small compared to the gross

16 *Atomic states*

structure of the atom, which is the reason for the name "fine structure". From Eq. (2.22), the fine-structure splitting between two atomic states with $j = l - 1/2$ and $j = l + 1/2$ is given by

$$\Delta E_{FS} = \frac{1}{l(l+1)} \alpha^2 \frac{1}{n^3} \frac{m_e e^4}{2\hbar^2}. \tag{2.24}$$

As we will see in Chapter 4, the fact that the splitting decreases as $1/n^3$ with increasing principal quantum number n has important consequences for atoms in external magnetic and electric fields. As a numerical example, two states belonging to the same level in hydrogen with $l = 1$ and $n = 2$ but having different total angular momenta $j = 1/2$ and $j = 3/2$ (denoted $2P_{1/2}$ and $2P_{3/2}$ in "spectroscopic notation"[2]) are according to Eq. (2.24) split by the energy interval $\Delta E \approx 0.366$ cm^{-1} $\times hc \approx 10.9$ GHz $\times h$ (see Appendix B for a discussion of energy units). If n is changed to 3, with the remaining quantum numbers the same, the fine-structure splitting is only $\Delta E \approx 0.108$ cm^{-1} $\times hc \approx 3.24$ GHz $\times h$.

It is interesting to note that, according to the Dirac formula (2.22), states of the hydrogen atom that have the same values of quantum numbers n and j, but different l, have the same energy—the spin–orbit interaction does not split these states. In the real hydrogen atom, however, these states, for example, the state with quantum numbers $n = 2$, $l = 0$, $j = 1/2$ and the state with $n = 2$, $l = 1$, $j = 1/2$, denoted $2S_{1/2}$ and $2P_{1/2}$, have slightly different energies (the *Lamb shift*). The magnitude of the Lamb shift can be calculated using the methods of quantum electrodynamics. The mechanism responsible for the shift is the interaction of the atomic electron with fluctuations in the vacuum. Each energy level is shifted by a different amount, resulting in the splitting of otherwise degenerate levels. The Lamb-shift splitting between the $2S_{1/2}$ and $2P_{1/2}$ states is $\Delta E \approx 0.035$ cm^{-1} $\times hc \approx 1057$ MHz $\times h$, which is an order of magnitude smaller than the fine-structure splitting between $2P_{1/2}$ and $2P_{3/2}$.

2.2.4 Vector model of the atom

Pictures can often be very helpful in analyzing physical phenomena, and angular-momentum coupling is no exception. Unfortunately, as is also common, there is a tradeoff between accuracy and intelligibility.[3] Later in the book, we will properly represent the angular-momentum states as probability distributions. However, when trying to visualize the coupling of angular momenta, one would need to combine two or more distributions with possibly arbitrary phases between them. The mental gymnastics involved in this are most likely more difficult than the corresponding quantum-mechanical calculation. Instead, a simpler—though less accurate—approach is commonly used, one which still illustrates the main properties of quantum-mechanical angular momentum.

One picture that we can draw represents a quantum-mechanical angular momentum as a vector using the expectation values of its angular-momentum projections. For example, to represent the angular momentum **j**, we would draw the vector $(\langle j_x \rangle, \langle j_y \rangle, \langle j_z \rangle)$. Such a representation is simple and unambiguous (and we use it below, see, for example, Fig. 2.3), and has a well-defined and quantum-mechanically justifiable meaning. However, this picture describes only the average over many measurements of the angular momentum, and omits significant information about the angular-momentum state itself. To construct a more

---
[2]For details on this notation see Sec. 2.3.

[3]According to A. Pais (1991), in a 1949 interview Niels Bohr discussed "jest and seriousness in science." When asked what is complementary to truth, he replied: "clarity..."

Angular momentum of the electron in the hydrogen atom   17

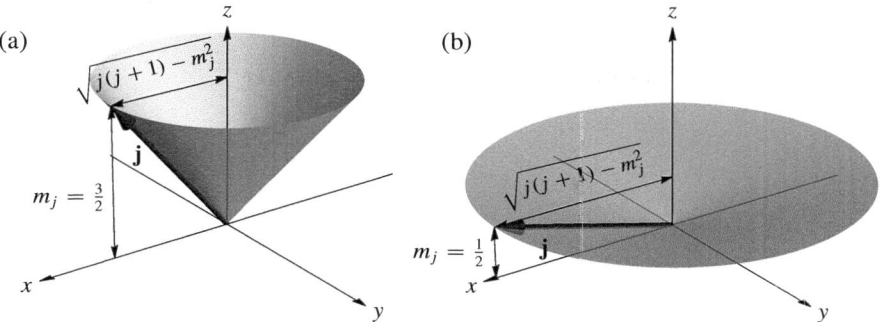

Fig. 2.1 A somewhat flawed (see text) depiction of an angular-momentum-j state with a fixed projection on the z-axis. The figures are drawn for $j = 3/2$ and $m_j = 3/2$ (a) and $m_j = 1/2$ (b). The figures accurately represent the fact that for these states, only one z-projection can be measured, and that the expectation values of the angular-momentum projections in the transverse directions are zero. On the other hand, these pictures may give a false impression about the range of values of the transverse projections that can be measured for such states (see text).

sophisticated picture, we consider some aspects of an angular-momentum state that we would like to represent.

The basic quantum-mechanical result concerning an angular momentum, say $\mathbf{j}$, is that the maximum possible projection on any given axis (say, z) that can be measured in an experiment[4] is $(j_z)_{\max} = j$. Let us, for the moment, consider just this maximal-projection state. Before we attempt to draw this state, let us recall some more results from quantum mechanics. First, the expectation value of the operator $\mathbf{j}^2$ is equal to $j(j + 1)$. This means that any vector that we use to represent the state should have length $\sqrt{j(j + 1)}$. Clearly, even for a state with maximum z-projection, the vector cannot point along the z-axis, because its z-projection must be j. As a consequence, the vector will have a component along x or y (or both). Physically, a measurement of j_x or j_y can produce a nonzero result. This is one of the consequences of the *uncertainty relations*. However, the expectation values $\langle j_x \rangle$ and $\langle j_y \rangle$ are both equal to zero (in any state with a definite projection j_z). This means that we cannot choose a fixed vector to represent the state, since at least one of $\langle j_x \rangle$ and $\langle j_y \rangle$ would be nonzero.

These considerations suggest that we should use a collection of vectors to represent the state, namely, all of the vectors with length $\sqrt{j(j + 1)}$ and z-projection equal to j. An example of this is shown in Fig. 2.1(a). Here the angular-momentum state under discussion is represented with a cone with its vertex at the origin and the axis coinciding with z. The height of the cone is j, and the length of the generatrix of the cone is $\sqrt{j(j + 1)}$. The cone can be thought of as representing a vector with fixed length whose end position is undetermined, i.e., distributed on a circle, while the z-projection is always fixed. (Sometimes the cone may be omitted, or represented by a circle, in order to simplify a diagram.)

We might declare this a success, especially since this pictorial representation works equally well for a state with any allowed definite value of j_z ($-j, -j + 1, ..., j$), as shown in Fig. 2.1(b), and all of the properties we set out to illustrate are indeed represented in the picture. Using this representation, the addition of $\mathbf{l}$ and $\mathbf{s}$ to form $\mathbf{j}$ appears as shown in Fig. 2.2.

---

[4] Here we introduce a common and convenient practice of measuring angular momentum in units of $\hbar$.

18 Atomic states

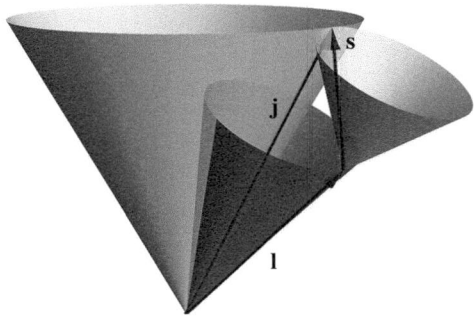

Fig. 2.2 A vector-model depiction of orbital angular momentum **l** and spin **s** coupling into the total angular momentum **j**.

Unfortunately, this representation, known as the *vector model*, presents some problems. Suppose we ask the following, seemingly innocent, question: OK, we know that the expectation value of j_x is zero (as seen in the picture), and that if we attempt to measure j_x, we will generally get a nonzero value (the average of the measured values will, of course, be zero). What is the *maximal* possible value of j_x that can be measured?

Looking at our picture, we would say that whatever value we measure should satisfy $j_x \leq \sqrt{j(j+1) - j^2} = \sqrt{j}$. This is, in fact, *not* true. A straightforward quantum mechanical calculation would show that, indeed, the maximal value of a transverse projection that we can measure is $j_x = j$.

An explanation of this "paradox" lies in a fundamental principle of quantum mechanics that says that a measurement perturbs the system in a way that, generally, increases the measurement uncertainty. In this particular case, after the measurement of a transverse projection, the system is no longer in a state with a fixed z-projection, which indicates that it has been perturbed by the measurement, and the limitations that have followed from the system being in such a state no longer hold.

Whatever the reason for the flaw of the picture, we need to recognize this, and avoid drawing wrong conclusions based on it.

Another question arises from the common interpretation that the vector represents a state that "precesses" around its cone. For example, if we say that **l** and **s** in Fig. 2.2 precess around **j**, then what is the frequency of this precession?

An important idea emphasized in this book is that the quantum-mechanical analog of a classical oscillating (or rotating) system is a coherent superposition of stationary states. In the situation we are discussing, states characterized by a set of quantum numbers $|lsjm_j\rangle$ are eigenstates, (or stationary states). If a system is prepared in a stationary state, the quantum numbers do not change with time. This means that if at $t = 0$ we prepare a $|lsjm_j\rangle$ state, the probabilities of then measuring certain values of m_l and m_s will not depend on t. Thus no "precession" is observed for a case such as shown in Fig. 2.2: the angular-momentum vectors are spread out around the cones, but do not precess around them.

On the other hand, suppose at some point we prepare a state $|lm_lsm_s\rangle$ with definite values of m_l and m_s. Such a state is *not* a stationary state, and so at a later time, we would, generally, measure some different values of m_l and m_s. A way to understand the time evolution of the state is to write it in the basis of stationary states. The resulting expression is a coherent superposition of eigenstates $|lsjm_j\rangle$ with different values of j. The time dependence for the amplitude of a given eigenstate in the superposition is given by $\exp[-i(E_j/\hbar)t]$, where E_j is

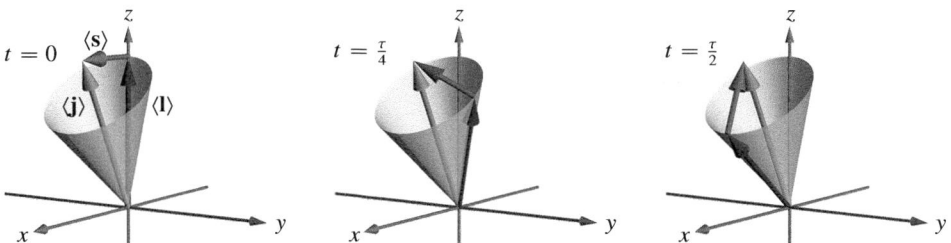

Fig. 2.3 Illustration of quantum-beat evolution of the $\langle \mathbf{l} \rangle$ and $\langle \mathbf{s} \rangle$ vectors, resembling precession around $\langle \mathbf{j} \rangle$. The plots represent the results of a quantum-mechanical calculation for $l = 1$ and $s = 1/2$. At $t = 0$, the system is prepared with $l_z = 1$ and $s_x = 1/2$. The initial state is not an energy eigenstate, and so it undergoes quantum beats with a frequency corresponding to the energy splitting between eigenstates with $j = 3/2$ and $1/2$. The expectation vectors at one quarter and one half of the quantum-beat period τ are shown. Note that $\langle \mathbf{j} \rangle$ remains constant.

the energy of a state with a given j. The differences between various eigenenergies define a set of *beat frequencies* associated with the system. We note that, in this case, these beats are not actually precession—it is the magnitudes, not the directions, of the expectation vectors $\langle \mathbf{l} \rangle$ and $\langle \mathbf{s} \rangle$ that change in time.

Quantum beats that resemble precession can be obtained if at $t = 0$ we prepare an initial state for which, for example, the $\mathbf{l}$ vector has a definite projection on $\hat{\mathbf{z}}$ and the $\mathbf{s}$ vector has a definite projection on $\hat{\mathbf{x}}$ (Fig. 2.3). In this case of $l = 1$ and $s = 1/2$, there is only one beat frequency corresponding to the energy difference between the $j = 3/2$ and $j = 1/2$ states, i.e., the fine-structure splitting. Figure 2.3 illustrates the time evolution of the expectation vectors of the angular momenta in this case, obtained from an exact quantum-mechanical calculation.

2.3 Multi-electron atoms

Even in the nonrelativistic approximation, the Schrödinger equation for an atom with more than one electron is not analytically solvable. This is not too surprising, because the three-body problem in classical mechanics does not generally have an analytical solution either. There exist various techniques for approximate analytical and numerical solution of the Schrödinger equation for multi-particle systems. A discussion of these techniques is beyond our present scope, although some aspects of the quantum-mechanical description of multi-electron systems will be of great importance to us. In this and the following sections, we discuss the different schemes of angular-momentum coupling in complex atoms as well as the quantum numbers that characterize these complex atoms.

In a multi-electron atom, each electron interacts not only with the atomic nucleus but also with the other electrons. The nonrelativistic Schrödinger equation in this case is

$$\left(-\frac{\hbar^2}{2m_e} \sum_i^Z \Delta_i - \sum_i^Z \frac{Ze^2}{r_i} + \frac{1}{2} \sum_{i \neq j}^Z \frac{e^2}{r_{ij}} \right) \psi(\mathbf{r}_1 \ldots \mathbf{r}_Z) = E\psi(\mathbf{r}_1 \ldots \mathbf{r}_Z), \quad (2.25)$$

where i, j are the individual electron labels, $r_{ij} = |\mathbf{r}_i - \mathbf{r}_j|$ is the distance between two electrons, and Z is the number of protons (equal to the number of electrons for a neutral atom). This means that each electron in a multi-electron atom experiences a more complicated electric field than the simple $\propto 1/r$ Coulomb potential of the hydrogen atom. Due to this more complex

20 Atomic states

potential, the energies of the atomic states depend on the orbital angular momentum of the electrons l in addition to the principal quantum number n, in contrast to the solutions of the hydrogenic Schrödinger equation (2.1).

2.3.1 Gross structure, *LS* coupling

Despite the fact that in a multi-electron atom each electron interacts not only with the nucleus but also with the rest of the electrons, very often we can assume that each electron "sees" a total electric field that is approximately central. Then each electron can be described by its own Schrödinger equation with a central potential, the solution of which, as discussed above for the case of hydrogen, can be separated into radial and angular parts. The angular part is characterized by the orbital angular-momentum quantum numbers for this electron. This allows a description of multi-electron atoms in which each electron is characterized by a corresponding orbital angular momentum $\mathbf{l}_i$ (and spin $\mathbf{s}_i$). Let us consider the radial Schrödinger equation for the i-th electron with a given l_i, and make a further approximation that the potential it "sees" does not depend on the state of either this electron or the any of the other electrons (we will need to refine this approximation in the subsequent analysis). The solution of the radial Schrödinger equation will then generally include a set of bound states and energy levels. The standard practice is to label these states with a *principal quantum number* $n_i = l_i + 1, l_i + 2, ...$ starting from the lowest-energy bound-state for this l_i.[5] This is done in analogy with hydrogen, where the value of the principal quantum number can be no smaller than the orbital angular momentum plus one. A set of the numbers n_i, l_i for all electrons is the *electronic configuration* of a state in a multi-electron atom. The energy of the configuration in this approximation is the sum of the energies of the individual electrons, and is generally different for different configurations. Because the configuration makes no reference to projections of $\mathbf{l}_i$ or $\mathbf{s}_i$, the energy levels obtained in this approximation are generally highly degenerate.

Let us examine in greater detail the approximation of neglecting the non-central part of the potential for each electron. The approximate potential includes a part of the mutual electronic repulsion (the term proportional to $1/r_{ij}$ in Eq. 2.25) that is centrally symmetric. The neglected part of the potential is called the *residual Coulomb interaction*. This interaction is responsible for lifting some of the energy degeneracies within a given configuration. When this interaction is included, the energy of a state becomes dependent on the way that the individual $\mathbf{l}_i$ and $\mathbf{s}_i$ combine.

In addition to the residual Coulomb interaction, the spin–orbit ($\mathbf{l} \cdot \mathbf{s}$) interaction (Sec. 2.2.3) also lifts degeneracies within a configuration. Generally both of these interactions have to be taken into account, but particularly for light atoms, the former interaction is more important, and we initially neglect the spin–orbit interaction. This approximation is known as *LS coupling* or *Russell–Saunders coupling*. The term *LS* coupling is potentially confusing, in view of the fact that in this scheme it is the $\mathbf{l} \cdot \mathbf{s}$ interaction that is neglected. In the following, we describe the reason for this terminology.

Within the central-field approximation, not only can we ascribe a definite value l_i to each electron, but the projection of the orbital angular momentum of each electron on the quantization axis is a *good quantum number*, i.e., the projection does not change in time (it is conserved). Noncentral-field corrections change this situation. Classically, if one imagines an orbit of an electron, a perturbation from another electron can cause a torque that would

---

[5]There is a beautiful *oscillation theorem* (see, for example, Landau and Lifshitz 1977) that states that the number of zeros of the radial wave function with $r \neq 0$ is $n_i - l_i - 1$.

force the plane of the orbit to precess, corresponding to a change in the direction of the orbital angular momentum. Quantum-mechanically, we can say that the noncentral forces lead to mixing of the states of an electron with different projections of the orbital angular momentum.[6] Either picture leads us to the conclusion that the projections of the individual orbital momenta are not conserved due to the noncentral-field interactions among the electrons. Not everything is lost, however, as the orbital angular momenta form an overall collective system whose total angular momentum, denoted L, is conserved, along with its projection on the quantization axis.[7]

In our discussion of multi-electron states so far, we have largely ignored atomic spins. In fact, from what we have said, one might get the (wrong) impression that the energies of atomic states do not depend on the electron spins, at least, as long as we neglect spin–orbit interactions. This is emphatically *not* true, and we now bring an additional all-important piece of physics into our discussion to see why this is the case.

First of all, when we introduced electronic configurations, we did not discuss whether all configurations we can think of are indeed possible. Let us take neutral uranium as an example. There are 92 electrons, and the simplest configuration one can think of is where each of the electrons is in the lowest $l = 0$ state. The notation for such a configuration would be $1s^{92}$ (here "1" stands for $n = 1$ and "92" indicates the number of electrons with these values of n and l—this is a standard notation for electronic configurations). Of course, such a "boring" configuration is impossible. A crucially important factor that determines possible configurations is that electrons are *fermions*, i.e., particles that obey *Fermi–Dirac statistics*, meaning that the wave function of a system of electrons must be antisymmetric with respect to permutation of any two of them. An immediate consequence of this is the famous *Pauli exclusion principle* that says that no two electrons in an atom can have an identical set of quantum numbers. This property, in fact, determines the richness of the *periodic table* by ensuring, for example, that all electrons in the ground configuration of an atom do not reside in the same 1s orbital.

> The possibility of small violations of the Pauli principle and other deviations from the "proper" quantum behavior for identical particles has been discussed theoretically, and experiments have been mounted to look for such effects, so far, only putting upper limits on such violations (Hilborn and Tino 2000). A discovery of an effect of this sort would shake the foundations of our present understanding of Nature.

The fact that electrons are fermions and all allowed wave functions should be properly antisymmetric leads to a dependence of the energy of atomic states on the *total* electronic spin, S. The mechanism for this is the so-called *exchange interaction*, and it works roughly like this. The value of the total spin (resulting from quantum-mechanical addition of the spins of individual electrons; see Sec. 3.5) affects the symmetry properties of the spin part of the total multi-electron wave function. Because the *total* wave function, which is a product of the spin and the spatial parts, has to be antisymmetric with respect to permutation of any two electrons, the symmetry of the spatial wave function is thus affected. This symmetry, in turn, has a direct effect on the energy of the state because it affects the energy of the Coulomb

---

[6]There is also mixing of states with different values of l_i. However, since, in the central-field approximation, such states are, generally, nondegenerate, we neglect this mixing, in contrast to the mixing between states with the same l_i but different projections which are degenerate in the central-field approximation.

[7]Let us remind the reader that at this point we have gone beyond the central-field approximation, but have not yet included spin–orbit interactions.

repulsion between electrons. If a wave function is such that the spatial part is symmetric with respect to a permutation of two electrons, the electrons tend to be closer to each other (stronger repulsion) compared to the antisymmetric case where these electrons can never be found at the same location, and tend to be further apart (weaker repulsion). Note that symmetrization of the spatial wave function would not affect the energy for a purely central potential; therefore, the effect we are discussing is explicitly beyond the central-field approximation.

We have now accounted for the effect of the noncentral Coulomb interactions between electrons, including the effect of the exchange interaction, but we are still neglecting spin–orbit interactions. In this approximation, the energies of the electronic states generally depend on the electronic configuration, and also on the values of the total electronic orbital angular momentum L and spin S. The values of L and S for a given configuration determine the *electronic term* of the state. The usual notation for a term is ^{2S+1}L, where the quantity $2S+1$ is called the *multiplicity*. As an example, the configuration and term of the ground state of helium are $1s^2 \, ^1S$.

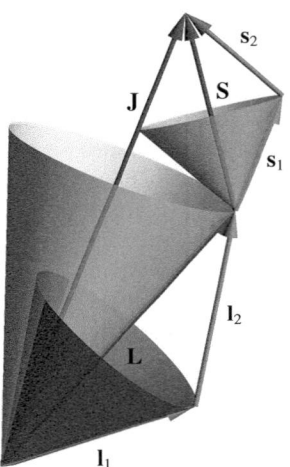

Fig. 2.4 Vector model diagram of the LS angular-momentum coupling scheme.

Under the present approximation, the energy of atomic states for a given configuration and term is independent of the (individually conserved) projections of the total orbital and spin angular momenta. This degeneracy is reduced by inclusion of the spin–orbit interaction,[8] which leads to a splitting within a term according to the value of the total angular momentum of the atom J. This splitting, however, is classified as *fine structure*, as opposed to the *gross structure* of the electronic states that we have discussed so far. An interesting situation leading to a breakdown of the LS coupling scheme (where L and S are good quantum numbers) occurs when spin–orbit interactions exceed the noncentral Coulomb interactions (for example, in heavy atoms). We discuss this in more detail below.

The LS coupling scheme is illustrated for the case of two electrons (the helium atom or another atom with two valence electrons) using the vector model in Fig. 2.4. The orbital momenta $\mathbf{l}_1$ and $\mathbf{l}_2$ of two individual electrons add to produce the total orbital angular momentum $\mathbf{L}$, and similarly, the spins $\mathbf{s}_1$ and $\mathbf{s}_2$ add into $\mathbf{S}$. These, in turn, add to produce $\mathbf{J}$, whose quantum numbers J and m_J are "good" for an isolated atom, meaning that eigenstates of $\mathbf{J}^2$ and J_z are energy eigenstates.[9] The quantum numbers m_l, m_s for each electron and m_L, m_S are all "not good" meaning that eigenstates of $\mathbf{J}^2$ and J_z are superpositions of states with different energies. Using the hierarchy of the energy splittings discussed above and the relation between the energy splitting of eigenstates and the "precession" frequencies in the vector model (end of Sec. 2.2.4), we can introduce a hierarchy of various "precessions" shown in Fig. 2.4. According to our preceding discussion, relative reorientation of $\mathbf{l}_1$ and $\mathbf{l}_2$ or $\mathbf{s}_1$ and $\mathbf{s}_2$ corresponds to the

---

[8]In multi-electron atoms, in addition to the spin–orbit interaction, spin–spin and spin–other-orbit interactions between electrons also exist. For all but the lightest multi-electron atoms, they are relatively unimportant (Bethe and Salpeter 1977).

[9]We write $\mathbf{J}^2$ for the dot product of the $\mathbf{J}$ operator with itself, $\mathbf{J}^2 = \mathbf{J} \cdot \mathbf{J}$, in order to distinguish it from J^2, the square of the quantum number.

energy of the residual Coulomb interaction. On the other hand, relative reorientation of **L** and **S** corresponds to the fine-structure energy. Since in *LS* coupling the former is much larger, we arrive at a picture (often presented but less often properly explained in textbooks) in which there is a hierarchy of precessions. First, we have fast precession of **l**$_1$ and **l**$_2$ around **L** and a similarly fast precession of **s**$_1$ and **s**$_2$ around **S**. On the time scale of the fast precession (which is just $\hbar$ divided by the corresponding energy-difference scale), the vectors **L** and **S** are stationary. Looking at longer time scales, the vectors **L** and **S** (slowly) precess around **J**, which is stationary without approximation for an isolated atom (as long as we do not consider the effect of the nuclear spin, see Sec. 2.4).

Once again, we remind the reader that that this needs to be taken with a grain of salt because there is no "precession" in a stationary state, as discussed in Sec. 2.2.4.

2.3.2 jj and intermediate coupling schemes

In some cases, such as in heavy atoms, the energy of the spin–orbit interactions can exceed the residual Coulomb energy. In this case, the *LS* coupling scheme no longer holds. To find a good alternative, let us step back to the central-field approximation, i.e., neglect both the residual Coulomb and spin–orbit interactions. As we discussed before, in this approximation, the energy depends only on the electron configuration, and eigenstates are generally highly degenerate. Now we remove our approximations one by one. As the spin–orbit interaction is the larger perturbation, we take it into consideration first. At a single-electron level, this means that we need to add **l**$_i$ and **s**$_i$ into **j**$_i$, the total angular momenta for individual electrons. Introducing the residual Coulomb interactions next, we add individual **j**$_i$ into **J**. This coupling scheme is called the *jj coupling scheme*. While the j_i are good quantum numbers in this scheme, the total orbital- and spin-angular-momentum quantum numbers L and S are not, in contrast to the *LS* coupling scheme. The total J is good in both coupling schemes.

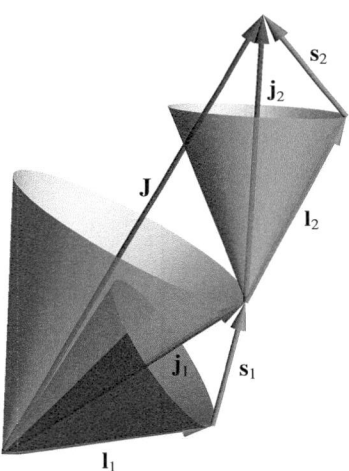

Fig. 2.5 A vector-model illustration of jj coupling in a two-electron atom.

The jj coupling scheme can be illustrated using the vector model (Fig. 2.5). In this case the faster precessions are those of **l**$_i$ and **s**$_i$ around **j**$_i$. The slower precession is that of vectors **j**$_i$ around the total angular momentum **J**.

For the case of a two-electron atom, the relation between the two coupling schemes we have discussed so far is diagrammed in Fig. 2.6. Although this illustration involves only two electrons it can be easily extended to atoms with more than two electrons just by adding rows to the graph.

In practice, there are many atomic states that do not fully conform to either *LS* or jj coupling schemes; sometimes even within one and the same atom, some states are better described with *LS*

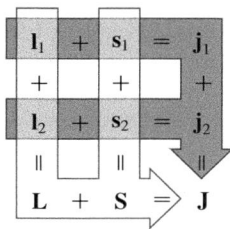

Fig. 2.6 Schematic illustration of the relation between the jj and *LS* coupling schemes.

24 *Atomic states*

coupling and others with jj coupling. Depending on the specific hierarchy of noncentral interactions, various versions of *intermediate coupling* may be used.

The coupling scheme has clear consequences for the atomic-transition selection rules that will be discussed in Chapter 7. The rules related to the total orbital and spin angular momenta, for example, are only obeyed insofar as L and S are good quantum numbers. The selection rules for the total angular momentum **J**, on the other hand, are much more strict, because J and m_J are good quantum numbers under any coupling scheme, in the absence of hyperfine interactions (Sec. 2.4).

2.3.3 Notation of states in multi-electron atoms

From the above discussion of the angular momentum coupling schemes, we see that an atomic state is characterized by, in addition to its energy, a certain set of angular-momentum quantum numbers. There is a conventional labeling scheme, called *spectroscopic notation*, for representing these quantum numbers. We have already mentioned most of the elements of spectroscopic notation, but it may be worthwhile to summarize it here. Note that an excellent reference on notation, basic ideas, data, and formulas of atomic spectroscopy (Martin and Wiese 2007) is available online;[10] see also Drake (2006).

First, the *configuration*, or set of electronic orbitals, is given. This is a list of principal (n_i) and orbital (l_i) quantum numbers for each electron. The n_i are represented with integers, but the values of the l_i are designated with code letters as given in the following table.

$$l = \begin{array}{cccccccccccccccc} 0 & 1 & 2 & 3 & 4 & 5 & 6 & 7 & 8 & 9 & 10 & 11 & 12 & 13 & 14 & 15 & \cdots \\ \updownarrow & \updownarrow & \updownarrow & \updownarrow & \updownarrow & \updownarrow & \updownarrow & \updownarrow & \updownarrow & \updownarrow & \updownarrow & \updownarrow & \updownarrow & \updownarrow & \updownarrow & \updownarrow & \\ s & p & d & f & g & h & i & k & l & m & n & o & q & r & t & u & \cdots \end{array} \qquad (2.26)$$

This assignment of letters seems illogical at first sight. It is alphabetical after the first three letters (with the omission of j to avoid confusion with the total angular momentum quantum number and of s and p that appear in the beginning of the sequence), but what motivates the initial sequence s, p, d?

The notation dates from the early days of spectroscopy; in the 19th century spectroscopists observed that some spectral lines were strong (and thus were called "principal"), some were very narrow ("sharp") and some were broad ("diffuse"). It turns out that there is a strong relation between these features and the orbital angular momenta of the states connected by the transition that forms a particular spectral line (Condon and Shortley 1951).

If there are multiple electrons with the same n and l, the number of such electrons is shown with a superscript. Thus the ground-state configuration of sodium, with 11 electrons, is $1s^2 2s^2 2p^6 3s$. A configuration can be abbreviated using the symbol for a noble gas to stand for its ground-state configuration; the Na configuration can then be written [Ne]3s, where [Ne] stands for the neon ground-state configuration $1s^2 2s^2 2p^6$. Sometimes the noble-gas configuration is considered to be understood and simply dropped.

To specify more precisely the atomic state in the LS coupling scheme (the most common situation), we give the *term*, consisting of the total electronic orbital angular momentum L and the total electronic spin S. The total electronic angular momentum J must also be given to specify a state. The quantum number L is represented by the same alphabetic code used above, but with an uppercase letter. A numerical subscript gives J, and a preceding superscript

---

[10] http://physics.nist.gov/Pubs/AtSpec/index.html

Table 2.3 Fine-structure splitting of the lowest $^2P_{1/2}$–$^2P_{3/2}$ doublets of alkali atoms.

	n	Z	ΔE (cm^{-1} × hc)	ΔE (GHz × h)
Li	2	3	0.338	10.1
Na	3	11	17.22	516.6
K	4	19	57.7	1731
Rb	5	37	237.6	7128
Cs	6	55	554.0	16620

Table 2.4 Fine-structure splitting $^2P_{1/2}$–$^2P_{3/2}$ in the sodium atom for states with different principal quantum number.

Na	n	ΔE (cm^{-1} × hc)	ΔE (GHz × h)
	3	17.22	516.6
	4	5.49	164.7
	5	2.49	74.7
	6	1.50	45.0

denotes the total spin through the numerical quantity $2S + 1$, called the *multiplicity*, for reasons described below. We thus arrive at the notation $^{2S+1}L_J$. For example, some typical states are $^1S_0$, $^2S_{1/2}$, and $^2P_{3/2}$. The multiplicity is said to be "singlet" for 1, "doublet" for 2, and "triplet" for 3, so that, for example, $^2P_{3/2}$ is read as "doublet P three-halves".

The origin of the term "multiplicity" is the following. If the total spin of an atom for a particular term is smaller than or equal to the orbital angular momentum, then the number of fine-structure components (possible values of J) is $2S + 1$, i.e., it is equal to the multiplicity. If the spin is larger than the orbital angular momentum, however, the number of components is equal to $2L + 1$. Thus, for example, for a ^2S term, the spin S is $1/2$ and the orbital angular momentum L is zero, so the number of components is $2L + 1 = 1$ (only $J = S = 1/2$ is possible).

The complete designation of configuration plus term plus the total angular momentum (which is written as a subscript to the right of the term) is then given for the ground state of barium, for example, as [Xe]6s^2 $^1S_0$. If the configuration is understood, an atomic state may be referred to using just its term and the total angular momentum: "singlet S zero" in the case of the Ba ground state.

2.3.4 Fine structure in multi-electron atoms

The energies of states with different J belonging to the same LS coupling term are split by the *fine-structure* (spin–orbit) interactions. Typical magnitudes of fine-structure splittings exemplified by the intervals between the lowest $P_{1/2}$ and $P_{3/2}$ states in alkali atoms are listed in Table 2.3. As can be seen from the table, the magnitude of the fine-structure splitting increases significantly for the heavier alkali atoms. It can be shown that this splitting scales approximately as Z^2. At the same time, if we look at the dependence of the $P_{1/2}$–$P_{3/2}$ splitting on n for a particular atom, we see that it decreases for higher n states. This is clearly illustrated by the example of sodium, given in Table 2.4.

2.4 Hyperfine interactions and hyperfine structure of atomic states

Up to this point we have only analyzed the electronic angular momenta, and neglected the angular momentum of the atom's nucleus. The angular momentum of the nucleus is called the *nuclear spin* and is usually denoted **I**. As for an electron, there is generally a magnetic moment associated with a nuclear spin

$$\boldsymbol{\mu}_I = g_I \mu_N \mathbf{I}/\hbar, \tag{2.27}$$

where g_I is the (dimensionless) *nuclear Landé factor* specific to the particular nucleus and $\mu_N = e\hbar/(2m_p c)$ is the *nuclear magneton*.[11] The nuclear magneton differs from the Bohr magneton in that the proton mass is substituted for the electron mass. The value of the nuclear magneton is $\approx 5.05 \times 10^{-24}$ erg G^{-1} (see Table A.1). For practical calculations, it is often convenient to express the nuclear magneton as $\mu_N/h \approx 762$ Hz G^{-1}. As we see, the nuclear magneton is approximately 2000 times smaller than the Bohr magneton due to the ratio of proton and electron masses. This means that the interaction of the nuclear spin with the electrons is very weak compared to the electron–electron interactions. Thus this interaction causes small splittings within each fine-structure level to form hyperfine components.

The total angular momentum formed when the nuclear spin **I** is coupled to the electronic angular momentum **J** is usually denoted **F**. This coupling can again be depicted in a vector model with the components **I** and **J** precessing around the total angular momentum **F** (Fig. 2.7).

The number of possible hyperfine structure states can be determined in a manner similar to the case of fine-structure splitting: the minimal value of the quantum number F is $|J - I|$ and the maximal value is $J + I$. If the nuclear spin is smaller than or equal to the electron angular momentum, then the number of possible values of F is $2I + 1$. If the nuclear spin is larger, the number of possible values of F is $2J + 1$.

The degeneracy between levels with different F is removed by *hyperfine interactions*, which can be thought of as the interaction of various electric and/or magnetic moments associated with the nucleus with electric and magnetic fields and their gradients created by the electrons at the nucleus. Accordingly, the hyperfine interaction can be expanded in a *multipole* series. The lowest terms of the series are dominant; they represent the magnetic dipole and the electric quadrupole interactions between the nucleus and the electrons, leading to hyperfine shifts given by

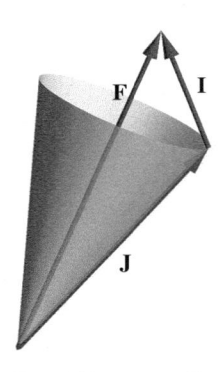

Fig. 2.7 Vector diagram of the coupling of nuclear (**I**) and electronic (**J**) angular momentum.

$$\Delta E_F = \frac{1}{2} A K + B \frac{\frac{3}{2} K(K+1) - 2I(I+1)J(J+1)}{2I(2I-1)2J(2J-1)}, \tag{2.28}$$

where $K = F(F+1) - I(I+1) - J(J+1)$. Here the constants A and B characterize the strengths of the magnetic-dipole and the electric-quadrupole interaction, respectively. B is zero unless I and J are both greater than 1/2. This is because, as discussed further in Sec. 5.7, a nucleus with angular momentum I can only support multipole moments of rank κ with

---

[11]Note also the sign difference in the conventional definition of Eq. (2.27) with that for the case of the electron (Eq. 2.20).

$\kappa \leq 2I$, and similarly, a field multipole associated with a system of electrons with angular momentum J cannot exceed $2J$. Since the hyperfine interaction is between the nuclear moment and the field multipole of the same rank, there is no quadrupole interaction for atoms with I or atomic states with J equal to 0 or $1/2$.

It turns out that *parity conservation* and *time-reversal invariance* dictate that nuclear moments are limited to even-rank electric moments and odd-rank magnetic moments. This is why a term for the electric-dipole moment ($\kappa = 1$), for example, does not appear in Eq. (2.28). The next higher-order allowed terms, the *magnetic octupole* ($\kappa = 3$) or *electric hexadecapole* ($\kappa = 4$) hyperfine interactions, can also be considered (Childs 1992) but the shifts due to these interactions are usually negligible.

To discuss a typical magnitude of the hyperfine splitting in multi-electron atoms, let us once again, consider alkali atoms. Different isotopes of the same atomic species can have different nuclear spins. As a specific example, for the sodium isotope ^{23}Na with nuclear spin $I = 3/2$, the ground-state fine-structure level 3s $^2S_{1/2}$ is split into two hyperfine-structure components with $F = 1$ and 2. These components are separated by 1.771 GHz $\times h$ (the $F = 2$ state has higher energy). The first excited state, 3s $^2P_{1/2}$ (the lower of the two fine-structure components of the ^2P term), also has two hyperfine structure components with $F = 1$ and 2, but the separation between them is only 188.9 MHz $\times h$. This is a typical situation in which the higher-lying excited states in multi-electron atoms have smaller hyperfine splitting.

Finally, let us look at how strongly the hyperfine splitting can vary between different isotopes of the same atom. An example is the two stable isotopes of Rb, namely ^{85}Rb ($I = 5/2$) and ^{87}Rb ($I = 3/2$). The ground state 5s $^2S_{1/2}$ of both these isotopes is split into two hyperfine levels. For ^{85}Rb these levels are $F = 2, 3$ and for ^{87}Rb the levels are $F = 1, 2$. The splitting of these levels is $\Delta E \approx 3.035$ GHz $\times h$ for ^{85}Rb and $\Delta E \approx 6.835$ GHz $\times h$ for ^{87}Rb, differing by more than a factor of 2.

In all of the above examples, the hyperfine splitting is due to the magnetic-dipole interaction only (because $J = 1/2$ for all the states we have mentioned). When both magnetic-dipole and electric-quadrupole interactions are allowed, their contributions to the shifts are typically of the same order of magnitude.

2.5 Parity of atomic states

Let us assume for a moment that each electron in an atom moves in a central field created by the atomic nucleus and the rest of the electrons. As we have discussed above, under this assumption, the state of each electron can be characterized by a wave function that can be separated into a part that depends on the radial coordinate and a part that depends on the angular coordinates (Eq. 2.11)

$$\psi_{nlm}(r, \theta, \varphi) = R_{nl}(r) Y_{lm}(\theta, \varphi).$$

This separation is allowed as a direct result of angular-momentum conservation in a central potential.

Let us examine what happens with the wave function if we invert the coordinates with respect to the origin of the reference frame. This means that we let $x \to -x$, $y \to -y$, and $z \to -z$. For spherical coordinates, the transformation becomes $r \to r$, $\theta \to \pi - \theta$, and $\varphi \to \varphi + \pi$.

Because the potential is radial, it is invariant with respect to spatial inversion. Thus, the problem possesses inversion symmetry. Formally, it is said that the operator of spatial inversion commutes with the Hamiltonian of the electron. Since the problem has a certain symmetry, this means that the solution (i.e., the electron eigenfunctions) can also be chosen to have this symmetry. Again, in formal language, the stationary solutions have to be simultaneous eigenstates of the Hamiltonian and the inversion operator. For the special case of degenerate states, we can always choose a basis of states satisfying this property.

Now, what about the eigenvalues of the inversion operator? Notice that if we perform spatial inversion twice, we return the coordinate frame back to its original state, and we can demand that the wave function also remains invariant. If the eigenvalue of the inversion operator is P, we conclude that $P^2 = 1$, so that P can take on one of the two possible values, $+1$ or -1. From this argument, we see that the density of the probability to find an electron at a certain point in space, being proportional to the modulus squared of the wave function, should not change under spatial inversion. Wave functions that remain unchanged under spatial inversion are said to be of positive or *even parity*. Wave functions that change sign under spatial inversion are said to be of negative or *odd parity*.

Looking at the decomposition of Eq. (2.5), it is clear that under coordinate-system inversion, the radial part of the wave function $R_{nl}(r)$ remains unchanged, because inversion does not change the modulus of the radius vector r. Thus, the parity of the wave function is determined by the angular part.

As one can see by examining the properties of the spherical functions $Y_{lm}(\theta, \varphi)$ (see Sec. 2.2.2), the effect of coordinate inversion on the angular part of the wave function is

$$Y_{lm}(\pi - \theta, \varphi + \pi) = (-1)^l Y_{lm}(\theta, \varphi). \tag{2.29}$$

Since a multi-electron wave function is a (properly antisymmetrized) product of wave functions for each electron, the parity of a multi-electron state is a product of parities for each electron:

$$P = (-1)^{l_1}(-1)^{l_2} \cdots (-1)^{l_n} = (-1)^{\sum_i l_i}. \tag{2.30}$$

The reader is well advised not to confuse $\sum_i l_i$ in the expression (2.30) with L, which is generally not the same.

Finally, for a multi-electron atom, it is only an approximation that each electron "sees" a central potential. What effect do noncentral forces have on our consideration of parity? The answer is that parity for individual electrons, in fact, becomes ill-defined; however, the electron cloud as a whole is in the central field of the nucleus, and thus, the overall parity of an electronic state remains a perfectly good quantum number.

Spatial inversion is an example of a *discrete transformation*. Such transformations differ from *continuous transformations*, for example, rotations or translations, in that the discrete transformations cannot be reduced to a sequence of infinitesimally small constituent transformations. For example, spatial inversion (P) takes a right-handed object and transforms it into a left-handed object. There is no way you can rotate your right glove to look the same as your left glove! In order to transform between the two types of gloves, a mirror reflection is needed, and there is no such thing as an infinitesimal reflection.

In addition to spatial inversion, other discrete transformations are *time reversal* (T), which can be thought as "running the movie backwards," and *charge conjugation* (C) that reverses the charges of all particles. Then there are combined transformations, for example, CP, meaning the combination of spatial inversion and charge reversal.

Conservation laws are related to the invariance of the laws of nature with respect to various transformations. For example, translational invariance leads to linear-momentum conservation, while rotational invariance leads to angular-momentum conservation. This general result is known as *Noether's theorem* (Emmy Noether, 1915), and is nicely discussed, for example, by Landau and Lifshitz (1976).

Prior to 1956, physicists, for the most part, held a sacred belief that the laws of nature are also invariant with respect to discrete transformations. In the case of spatial inversion, the conservation law corresponding to this invariance is the law of *parity conservation*. This is, in fact, the basis for assigning a definite parity to stationary atomic states.

It came as a shock to the physics community when in the late 1950s and 1960s, *parity (P) violation* and then also *CP violation* were experimentally discovered. In fact, remarkably, *all* discrete symmetries have now been experimentally shown to be violated, with the sole exception of the combination of all three symmetries, *CPT*. This symmetry is currently being checked by more and more precise experiments.

Atomic and molecular experiments are at the forefront of experimental studies of discrete-symmetry violations; however, the manifestations of these violations in atomic and molecular systems typically appear as quite small effects that can usually be neglected, except in the specific context of the study of discrete-symmetry violations.

Since this fascinating subject is outside of the scope of this book, we will generally assume that there are no symmetry violations, although we will refer to some of the experiments in this field.

There are several books and review papers available to the reader interested in further exploring discrete-symmetry violations in atoms and molecules (Khriplovich 1991; Khriplovich and Lamoreaux 1997; Frois and Bouchiat 1999; Ginges and Flambaum 2004; Budker *et al.* 2008).

3
A bit of angular-momentum theory

The subjects of light–atom interactions and atomic polarization are intimately connected to the properties of the angular momentum of the atoms (and also of the light). Therefore, in this chapter, we provide a brief survey of the results in the theory of angular momentum that will be used in our subsequent discussion.

We begin by examining rotations, which are central to the theory of angular momentum. We present the concepts of active and passive rotations, and co- and contravariance. We then discuss the quantum-mechanical angular-momentum operator, and several topics in quantum-mechanical angular momentum, including angular-momentum coupling and the Wigner-Eckart theorem.

3.1 Classical rotations

We first present results concerning classical rotations in the Cartesian and spherical bases. A more detailed exposition can be found in Appendix D.

3.1.1 Rotations in the Cartesian basis

A rotation of a vector $\mathbf{v}$ to become a new vector $\mathbf{v}'$ is written $\mathbf{v}' = R\mathbf{v}$, where R is the *classical rotation operator*. The rotation can be specified in the *axis-angle* parameterization as a rotation by an angle θ about a direction $\hat{\mathbf{n}}$, notated $R_{\hat{\mathbf{n}}}(\theta)$, where the right-hand rule is used to fix the rotation direction: if the thumb of the right hand points along $\hat{\mathbf{n}}$, the fingers curl in the direction of a positive rotation.

A rotation can be considered to act on the Cartesian components v_i of $\mathbf{v}$ as follows:

$$\mathbf{v}' = R\mathbf{v} = R\hat{\mathbf{x}}_i v_i = \hat{\mathbf{x}}_j \mathsf{R}_{ji} v_i, \tag{3.1}$$

where $\mathsf{R}_{ji} = \hat{\mathbf{x}}_j \cdot R\hat{\mathbf{x}}_i$ is the rotation matrix corresponding to R, and we use the summation convention in which summation over repeated indices is assumed. For example, the matrix for a rotation by an angle θ about $\hat{\mathbf{z}}$ is given by

$$\mathsf{R}_{\hat{\mathbf{z}}}(\theta) = \begin{pmatrix} \cos\theta & -\sin\theta & 0 \\ \sin\theta & \cos\theta & 0 \\ 0 & 0 & 1 \end{pmatrix}. \tag{3.2}$$

A rotation thought of in this sense—in which the basis vectors are held fixed, while the vector components are rotated as $v'_i = \mathsf{R}_{ij} v_j$—is known as an *active rotation*.

There is another type of rotation, called a *passive rotation*, in which we think of the basis (rather than the vector) as being rotated. In a passive rotation the basis vectors and the vector components must transform in a complementary way so that the vector itself does not rotate. This relationship is expressed by saying that the components and the basis vectors are *relatively contravariant*. We can write

$$\mathbf{v} = R^{-1}R\mathbf{v} = \hat{\mathbf{x}}_k R_{kj}^{-1} R_{ji} v_i = R_{jk} \hat{\mathbf{x}}_k R_{ji} v_i. \tag{3.3}$$

Here we have used the fact that R is *orthogonal*, meaning that its transpose equals its inverse (see Appendix D). We can therefore write a passive rotation as $\mathbf{v} = \hat{\mathbf{x}}_j'' v_j''$, with $\hat{\mathbf{x}}_j'' = R_{jk}\hat{\mathbf{x}}_k$ and $v_j'' = R_{ji} v_i = v_j'$. Evidently, applying R to the basis vectors has the opposite effect as applying it to the vector components, so that applying it to each has the desired effect of leaving $\mathbf{v}$ constant. Thus, even though they are relatively contravariant, in the Cartesian basis the vector components and the basis vectors transform according to the same rule.

We can draw an equivalence between active and passive rotations using the fact that a simultaneous rotation of the vector and the basis vectors—while keeping the components fixed—has no effect from a practical point of view. Thus, two rotations that have the same effect on the vector components can be thought of as equivalent. The active and passive rotations that we have written are equivalent in the sense that the vector components transform in the same way. However, passive rotations are usually described in terms of the rotation of the basis vectors. Since the basis vectors in Eq. (3.3) rotate in the opposite direction from the vector in Eq. (3.1), we say that a passive rotation with matrix $R^{(p)}$ is equivalent to the opposite active rotation with matrix $R^{(a)}$:

$$R_{\hat{\mathbf{n}}}^{(p)}(\theta) = R_{\hat{\mathbf{n}}}^{(a)}(-\theta) = \left(R_{\hat{\mathbf{n}}}^{(a)}(\theta)\right)^{-1}. \tag{3.4}$$

Then an active rotation of $\mathbf{v}$ by an angle θ is given by

$$\mathbf{v}' = \hat{\mathbf{x}}_i R_{ij}^{(a)}(\theta) v_j, \tag{3.5}$$

a passive rotation of the basis vectors by an angle θ is described by

$$\mathbf{v} = R_{jk}^{(p)}(\theta) \hat{\mathbf{x}}_k R_{ji}^{(p)}(\theta) v_i, \tag{3.6}$$

and the active rotation equivalent to this passive rotation is given by

$$\mathbf{v}'' = \hat{\mathbf{x}}_j R_{ji}^{(p)}(\theta) v_i = \hat{\mathbf{x}}_j R_{ji}^{(a)}(-\theta) v_i. \tag{3.7}$$

When we omit the superscript label on the rotation operator, we will continue to mean an active rotation.

3.1.2 The spherical basis

Let us introduce a new set of basis vectors, defined by

$$\hat{\boldsymbol{\epsilon}}_1 = -\frac{1}{\sqrt{2}}(\hat{\mathbf{x}} + i\hat{\mathbf{y}}), \tag{3.8a}$$

$$\hat{\boldsymbol{\epsilon}}_0 = \hat{\mathbf{z}}, \tag{3.8b}$$

$$\hat{\boldsymbol{\epsilon}}_{-1} = \frac{1}{\sqrt{2}}(\hat{\mathbf{x}} - i\hat{\mathbf{y}}). \tag{3.8c}$$

The fact that these unit vectors—describing the *spherical basis*—are complex has consequences for the algebra of rotations.

32 A bit of angular-momentum theory

The transformation between the Cartesian and spherical bases can be written in terms of a matrix U, given by

$$U = \begin{pmatrix} -\frac{1}{\sqrt{2}} & -\frac{i}{\sqrt{2}} & 0 \\ 0 & 0 & 1 \\ \frac{1}{\sqrt{2}} & -\frac{i}{\sqrt{2}} & 0 \end{pmatrix}, \qquad (3.9)$$

so that

$$\begin{pmatrix} \hat{\boldsymbol{\epsilon}}_1 \\ \hat{\boldsymbol{\epsilon}}_0 \\ \hat{\boldsymbol{\epsilon}}_{-1} \end{pmatrix} = U \begin{pmatrix} \hat{\mathbf{x}}_1 \\ \hat{\mathbf{x}}_2 \\ \hat{\mathbf{x}}_3 \end{pmatrix}. \qquad (3.10)$$

This can also be notated as

$$\hat{\boldsymbol{\epsilon}}_q = U_{qj} \hat{\mathbf{x}}_j, \qquad (3.11)$$

keeping track of the fact that q runs from 1 to -1, while j runs from 1 to 3. Here the matrix U is *unitary*, meaning that its inverse is equal to its conjugate transpose:

$$U^{-1} = (U^*)^t = U^\dagger. \qquad (3.12)$$

This means that the inverse transformation is given by

$$\hat{\mathbf{x}}_j = U^\dagger_{jq} \hat{\boldsymbol{\epsilon}}_q. \qquad (3.13)$$

We now consider the components of $\mathbf{v}$ in the spherical basis, v^q, defined by

$$\mathbf{v} = \hat{\boldsymbol{\epsilon}}_q v^q, \quad \text{with} \quad v^q = \hat{\boldsymbol{\epsilon}}_q^* \cdot \mathbf{v}. \qquad (3.14)$$

The components are given in terms of the Cartesian components of $\mathbf{v}$ by

$$v^q = U^*_{qj} v_j \qquad (3.15)$$

or, explicitly,

$$v^1 = -\frac{1}{\sqrt{2}}(v_x - i v_y), \qquad (3.16a)$$

$$v^0 = v_z, \qquad (3.16b)$$

$$v^{-1} = \frac{1}{\sqrt{2}}(v_x + i v_y). \qquad (3.16c)$$

The inverse transformation is given by

$$v_j = U^t_{jq} v^q. \qquad (3.17)$$

We see that the transformation that takes the vector components into the spherical basis is the complex conjugate of the one given by Eq. (3.11) that acts on the basis vectors.

Classical rotations

Now let us consider rotations in the spherical basis. We can write a passive rotation as

$$\mathbf{v} = R^{-1}R\mathbf{v} = \tilde{R}^*_{q_1 q_2}\hat{\boldsymbol{\epsilon}}_{q_2}\tilde{R}_{q_1 q}v^q = \hat{\boldsymbol{\epsilon}}'_{q_1}v'^{q_1}, \tag{3.18}$$

where $\tilde{R} = (URU^\dagger)^*$ is the rotation matrix in the spherical basis, and the transformations of the basis vectors and vector components are given by

$$\hat{\boldsymbol{\epsilon}}'_{q_1} = \tilde{R}^*_{q_1 q_2}\hat{\boldsymbol{\epsilon}}_{q_2}, \tag{3.19a}$$

$$v'^{q_1} = \tilde{R}_{q_1 q}v^q, \tag{3.19b}$$

respectively. The spherical basis vectors and vector coefficients must be relatively contravariant in order that $\mathbf{v}$ remain unchanged under a passive rotation. In contrast to the Cartesian case, here the relatively contravariant quantities have different transformation properties—the rotation matrix for the basis vectors, $\tilde{R}^*$, is the complex conjugate of that for the vector components, $\tilde{R}$. To keep track of these different types of quantities, we will call sets of quantities that transform like $\hat{\boldsymbol{\epsilon}}_q$ in Eq. (3.19a) *covariant*, and sets that transform like v^q in Eq. (3.19b) *contravariant*. As may be already clear, this distinction is indicated in the notation by writing the index on contravariant quantities as a superscript. We will discuss the relationship between covariant and contravariant quantities further below.

An active rotation in the spherical basis is given by

$$(v^q)' = \tilde{R}_{qq_1}v^{q_1}, \tag{3.20}$$

with $\tilde{R} = (URU^\dagger)^*$. For example, for a rotation by an angle θ about $\hat{\mathbf{z}}$, with Cartesian rotation matrix given by Eq. (3.2), the corresponding rotation matrix for contravariant quantities in the spherical basis is

$$\tilde{R} = \begin{pmatrix} -\frac{1}{\sqrt{2}} & \frac{i}{\sqrt{2}} & 0 \\ 0 & 0 & 1 \\ \frac{1}{\sqrt{2}} & \frac{i}{\sqrt{2}} & 0 \end{pmatrix} \begin{pmatrix} \cos\theta & -\sin\theta & 0 \\ \sin\theta & \cos\theta & 0 \\ 0 & 0 & 1 \end{pmatrix} \begin{pmatrix} -\frac{1}{\sqrt{2}} & 0 & \frac{1}{\sqrt{2}} \\ -\frac{i}{\sqrt{2}} & 0 & -\frac{i}{\sqrt{2}} \\ 0 & 1 & 0 \end{pmatrix}$$

$$= \begin{pmatrix} e^{-i\theta} & 0 & 0 \\ 0 & 1 & 0 \\ 0 & 0 & e^{i\theta} \end{pmatrix}. \tag{3.21}$$

This shows that a rotation about the z-axis induces phase shifts in the spherical components, but does not mix them. This is an important feature of the spherical basis.

The matrix $\tilde{R}$ applies to active rotations of the contravariant components, and can be notated in full as $\tilde{R}^{(a)}_{\hat{\mathbf{n}}}(\theta)$. In a passive rotation of the basis vectors by an angle θ, the components undergo a transformation equivalent to an active rotation of $-\theta$, as discussed in the previous section. The matrix that describes this equivalent rotation is then given by $\tilde{R}^{(p)}_{\hat{\mathbf{n}}}(\theta) = \tilde{R}^{(a)}_{\hat{\mathbf{n}}}(-\theta) = [\tilde{R}^{(a)}_{\hat{\mathbf{n}}}(\theta)]^{-1}$.

It can be seen that taking the complex conjugate converts covariant quantities into contravariant and vice versa. Therefore we can define the set of contravariant basis vectors

$$\hat{\boldsymbol{\epsilon}}^q = \hat{\boldsymbol{\epsilon}}^*_q = U^*_{qj}\hat{\mathbf{x}}_j, \tag{3.22}$$

which transform under passive rotations as $(\hat{\boldsymbol{\epsilon}}^q)' = \tilde{R}^{(p)}_{q'q}\hat{\boldsymbol{\epsilon}}^q$. This definition, however, depends on the fact that the basis vectors $\hat{\mathbf{x}}_i$ are real.

Table 3.1 Transformation properties under rotation for various sets of quantities in the spherical basis.

Quantity	Type	Rotation type	Transformation
v^q	contravariant components	active	$\tilde{R}^{(a)}_{q'q} v^q$
		passive	$\tilde{R}^{(p)}_{q'q} v^q$
$\hat{\epsilon}_q$	covariant basis vectors	passive	$\tilde{R}^{(p)*}_{q'q} \hat{\epsilon}_q$
v_q	covariant components	active	$\tilde{R}^{(a)*}_{q'q} v_q$
		passive	$\tilde{R}^{(p)*}_{q'q} v_q$
$\hat{\epsilon}^q$	contravariant basis vectors	passive	$\tilde{R}^{(p)}_{q'q} \hat{\epsilon}^q$

There is another, more general, way to convert between covariant and contravariant components. It can be shown (see Appendix D) that if y^q is a set of contravariant quantities, the set $z_q = (-1)^q y^{-q}$ transforms as a covariant set. Likewise, if y_q is a covariant set of quantities, $z^q = (-1)^q y_{-q}$ is contravariant. For the spherical basis vectors, the contravariant set obtained in this way coincides with the above definition of $\hat{\epsilon}^q$. For the spherical components, this set comprises the covariant components corresponding to $\hat{\epsilon}^q$. Therefore, we make the definition $v_q = (-1)^q v^{-q}$. The transformation rules for various quantities in the spherical basis are summarized in Table 3.1.

Note that it is also necessary to keep track of covariance and contravariance when considering scalar products of vectors expressed in the spherical basis. A scalar product $\mathbf{v} \cdot \mathbf{w}$ must remain invariant under rotations. This means that if the scalar product is written out in terms of components, the components of $\mathbf{v}$ must be relatively contravariant to those of $\mathbf{w}$.

To write the scalar product in terms of the spherical components of $\mathbf{v}$ and $\mathbf{w}$, we convert from the Cartesian components:

$$\mathbf{v} \cdot \mathbf{w} = v_j w_j = v_j U^t_{jq} w^q = U_{qj} v_j w^q = v_q w^q. \tag{3.23}$$

Employing the transformation between covariant and contravariant components, the scalar product can be written in the alternate forms:

$$\mathbf{v} \cdot \mathbf{w} = v_q w^q = (-1)^q v_q w_{-q} = v^q w_q = (-1)^q v^q w^{-q}. \tag{3.24}$$

3.1.3 Euler angles

The Cartesian rotation matrices for rotations about $\hat{\mathbf{x}}$, $\hat{\mathbf{y}}$, and $\hat{\mathbf{z}}$ are given by

$$R_{\hat{\mathbf{x}}}(\theta) = \begin{pmatrix} 1 & 0 & 0 \\ 0 & \cos\theta & -\sin\theta \\ 0 & \sin\theta & \cos\theta \end{pmatrix}, \tag{3.25a}$$

$$R_{\hat{\mathbf{y}}}(\theta) = \begin{pmatrix} \cos\theta & 0 & \sin\theta \\ 0 & 1 & 0 \\ -\sin\theta & 0 & \cos\theta \end{pmatrix}, \tag{3.25b}$$

$$R_{\hat{\mathbf{z}}}(\theta) = \begin{pmatrix} \cos\theta & -\sin\theta & 0 \\ \sin\theta & \cos\theta & 0 \\ 0 & 0 & 1 \end{pmatrix}. \tag{3.25c}$$

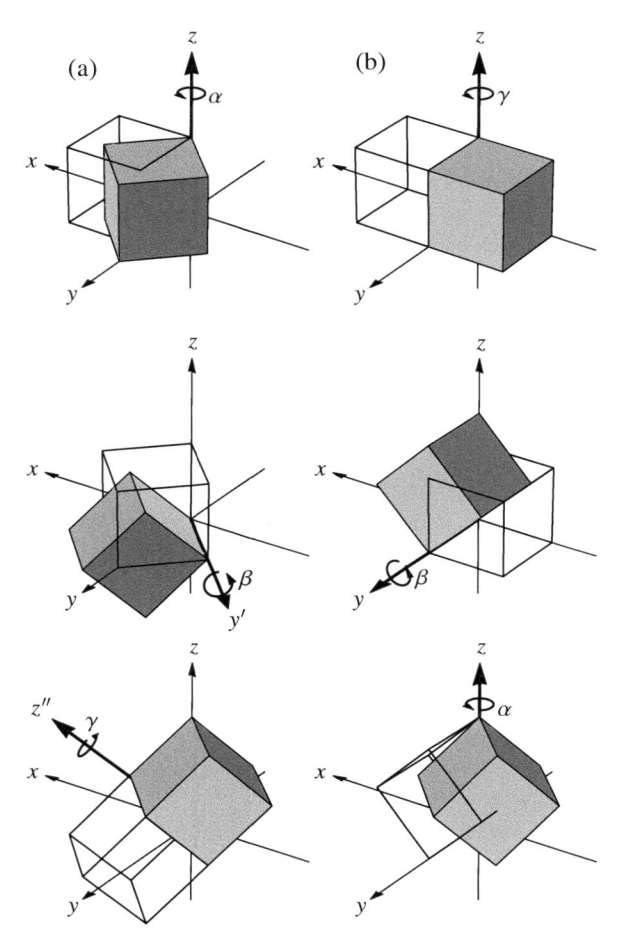

Fig. 3.1 Rotation of a body by (a) Euler angles α, β, γ with respect to the body-fixed z, y', and z'' axes is equivalent to rotation by (b) the same angles but in the opposite order with respect to the original space-fixed axes z, y, and z. For each step, the rotation direction is indicated by a curved arrow; the initial position of the body is shown as a cubic frame, and the final position is shown as a filled cube.

Rather than attempting to write out the matrix for rotation about an arbitrary axis, we can decompose an arbitrary rotation into a series of rotations about the Cartesian axes. An arbitrary rotation can be described by three successive rotations by the *Euler angles* α, β, and γ. To describe these rotations in the active sense, we make use of *body-fixed* axes, which undergo the same active rotation as the rotating object, so as to remain fixed with respect to it, as opposed to the original set of *space-fixed* axes, which do not change during an active rotation. The sequence of Euler rotations can then be described as (Fig. 3.1a)

1. A rotation by angle α ($0 \leq \alpha \leq 2\pi$) around the z-axis. (Prior to this rotation, the space-fixed and body-fixed axes coincide.)
2. A rotation by angle β ($0 \leq \beta \leq \pi$) around the body-fixed y-axis (denoted as the y'-axis in the space-fixed coordinate system).
3. A rotation by angle γ ($0 \leq \gamma \leq 2\pi$) around the body-fixed z-axis (the z''-axis in the space-fixed system).

This combination of rotations can produce any desired resultant rotation. In other words, an axis-angle rotation $R_{\hat{\mathbf{n}}}(\phi)$ can be written as

36 A bit of angular-momentum theory

$$R_{\hat{n}}(\phi) = R_{\hat{z}''}(\gamma) R_{\hat{y}'}(\beta) R_{\hat{z}}(\alpha) \tag{3.26}$$

for some combination of $\alpha, \beta,$ and γ.

It is much more convenient, however, to describe rotations with respect to the space-fixed coordinate system. To do this, we use a remarkable relation between space-fixed rotations and body-fixed rotations. Namely, the combination of three active Euler rotations with respect to the body-fixed axes $\hat{z}, \hat{y}', \hat{z}''$ is equivalent to the same three active rotations with respect to the space-fixed axes $\hat{z}, \hat{y}, \hat{z}$, but performed in the opposite order (Fig. 3.1b). Denoting a rotation by Euler angles as $R(\alpha, \beta, \gamma)$, we have

$$\begin{aligned} R(\alpha, \beta, \gamma) &= R_{\hat{z}''}(\gamma) R_{\hat{y}'}(\beta) R_{\hat{z}}(\alpha) \\ &= R_{\hat{z}}(\alpha) R_{\hat{y}}(\beta) R_{\hat{z}}(\gamma). \end{aligned} \tag{3.27}$$

For a proof of this property, see the book by Sakurai and Tuan (1994).

Using this property and the rotation matrices given above, we can write a general active Euler-angle rotation as

$$R(\alpha, \beta, \gamma) = \begin{pmatrix} \cos\alpha & -\sin\alpha & 0 \\ \sin\alpha & \cos\alpha & 0 \\ 0 & 0 & 1 \end{pmatrix} \begin{pmatrix} \cos\beta & 0 & \sin\beta \\ 0 & 1 & 0 \\ -\sin\beta & 0 & \cos\beta \end{pmatrix} \begin{pmatrix} \cos\gamma & -\sin\gamma & 0 \\ \sin\gamma & \cos\gamma & 0 \\ 0 & 0 & 1 \end{pmatrix}. \tag{3.28}$$

3.1.4 Commutation relations

Rotations in three-dimensional space, in contrast to the case of two-dimensional rotations, have an important property that the result of a composite transformation consisting of consecutive rotations around different axes generally depends on the order in which these rotations are performed. This *noncommutativity* property for sequential rotations about different axes is illustrated in Figs. 3.2 and 3.3.

In Fig. 3.2 we show the result of two consecutive $\pi/2$ rotations around two orthogonal axes, and the two rotations in the opposite order. It is apparent that the results of these two sequences of rotations are completely different. In order to quantify this difference, however, it is helpful to consider the case of small rotations.

The case of rotations by small angles is shown in Fig. 3.3, and can be described by the *commutation relations*

$$[R_i(\delta), R_j(\delta)] = \epsilon_{ijk} \left(R_k(\delta^2) - 1 \right), \tag{3.29}$$

to second order in a small angle δ. Here $R_i \equiv R_{\hat{x}_i}$ ($\hat{x}_{1,2,3}$ are the Cartesian unit vectors), the square brackets designate the *commutator* $[A, B] = AB - BA$, and ϵ_{ijk} is the totally antisymmetric tensor.

3.2 Quantum-mechanical rotations

In quantum mechanics, instead of rotating vectors, we wish to rotate wave functions. Under a classical rotation R, a state will transform according to a quantum-mechanical rotation operator $\mathscr{D}(R)$ corresponding to R. In other words, for a given classical rotation that takes a vector $\mathbf{v}$ into a new vector $\mathbf{v}' = R\mathbf{v}$, there is a quantum-mechanical rotation operator $\mathscr{D}(R)$ that takes a wave function $|\psi\rangle$ into a new wave function $|\psi'\rangle = \mathscr{D}(R)|\psi\rangle$. We will evaluate $\mathscr{D}$ explicitly later in this chapter; in this section we examine some of its general properties.

We will assume that the quantum-mechanical rotation operators share some features of their classical counterparts. In particular, as the length of a classical vector is preserved upon

Quantum-mechanical rotations 37

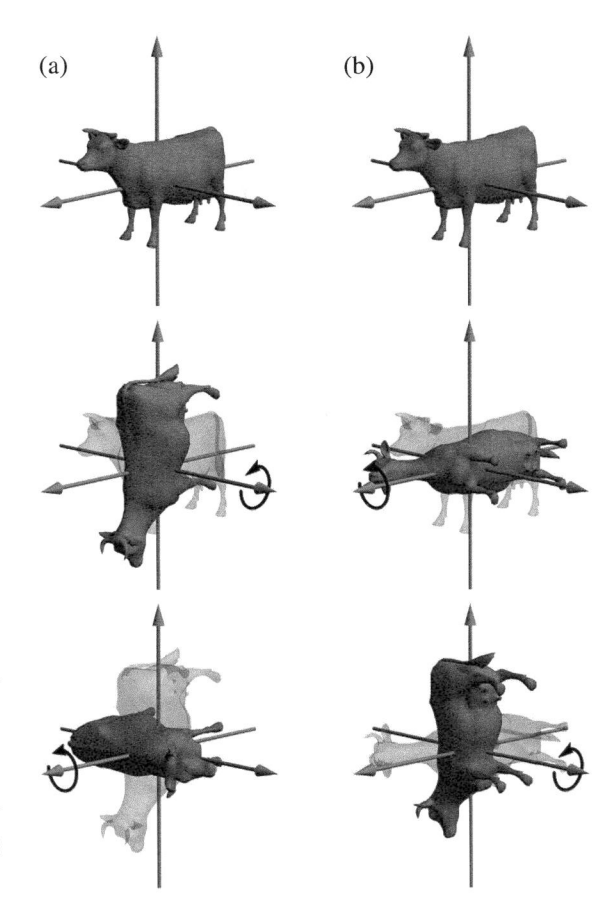

Fig. 3.2 Two consecutive rotations performed in one order in the left column (a) and the opposite order in the right column (b). The steps proceed from top to bottom—the transparent objects show the starting position for each step, the circular arrows indicate the rotation applied, and the solid objects show the results of the rotation.

rotation, the norm of a wave function should also be preserved. Since wave functions are not real-valued, $\mathscr{D}$ is not necessarily orthogonal, but it is unitary, i.e.,

$$\mathscr{D}^\dagger \mathscr{D} = 1. \tag{3.30}$$

In addition, we assume that the quantum-mechanical rotation operators satisfy the same commutation relations as the classical operators. In other words, for active rotations,

$$[\mathscr{D}_i(\delta), \mathscr{D}_j(\delta)] = \epsilon_{ijk}(\mathscr{D}_k(\delta^2) - 1), \tag{3.31}$$

where $\mathscr{D}_i(\delta) = \mathscr{D}[R_i(\delta)]$.

These two requirements can be taken as the definition of the quantum-mechanical rotation operators. In order to find explicit formulas for them, it is helpful to write them in terms of the angular momentum operator, discussed in the next section.

Note that quantum-mechanical rotations can also be viewed as applying to operators, rather than wave functions. To see this, consider the expectation value of an operator A with respect to a state $|\psi\rangle$: $\langle A \rangle = \langle \psi | A | \psi \rangle$. We now apply a rotation $\mathscr{D}(R)$ to the state, to obtain the rotated

38   *A bit of angular-momentum theory*

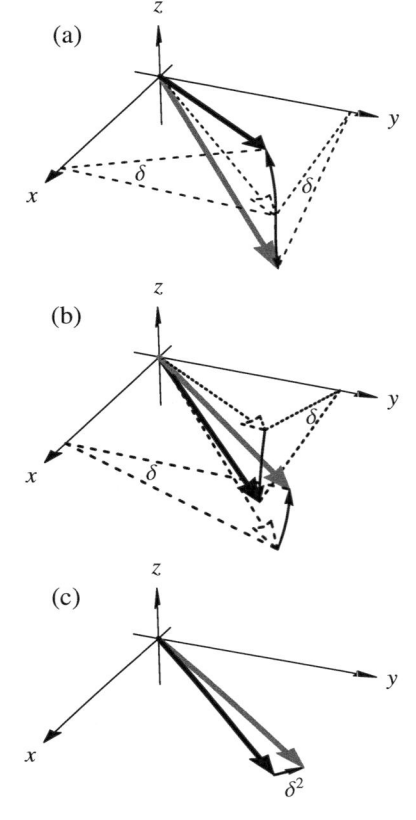

Fig. 3.3 Classical commutation relation for rotations about the x and y axes (Eq. 3.29). We start with two identical copies of a vector along $\hat{\mathbf{x}} + \hat{\mathbf{y}}$. One copy (black) is rotated by (a) an angle δ about $\hat{\mathbf{x}}$ and then (b) the same angle about $\hat{\mathbf{y}}$, while the other copy (gray) undergoes the same rotations but in the opposite order. The difference in the resultant vectors is approximately the same as the difference between the original vector and a copy rotated by δ^2 about $\hat{\mathbf{z}}$ as shown in (c).

state $|\psi'\rangle = \mathscr{D}|\psi\rangle$ (and the Hermitian conjugate $\langle\psi'| = \langle\psi|\mathscr{D}^\dagger$). The expectation value of the operator with respect to the rotated state is then given by

$$\langle A'\rangle = \langle\psi'|A|\psi'\rangle = \langle\psi|\mathscr{D}^\dagger A \mathscr{D}|\psi\rangle = \langle\psi|A'|\psi\rangle. \tag{3.32}$$

This shows that, rather than rotating the state, we can obtain an equivalent result by instead rotating the *operator*, using the formula

$$A' = \mathscr{D}^\dagger A \mathscr{D}. \tag{3.33}$$

Furthermore, we can also relate $\mathscr{D}$ to R by this method. If $\mathbf{A}$ is a vector observable, its expectation value $\langle\mathbf{A}\rangle = \langle\psi|\mathbf{A}|\psi\rangle$ should transform as a classical vector. In other words, under the active rotation R,

$$\langle\mathbf{A}'\rangle = R\langle\mathbf{A}\rangle. \tag{3.34}$$

The expectation value after the rotation can also be written

$$\langle\mathbf{A}'\rangle = \langle\psi'|\mathbf{A}|\psi'\rangle = \langle\psi|\mathscr{D}^\dagger \mathbf{A} \mathscr{D}|\psi\rangle; \tag{3.35}$$

combining Eqs. (3.34) and (3.35), we have

$$\langle\psi|\mathscr{D}^\dagger \mathbf{A} \mathscr{D}|\psi\rangle = R\langle\psi|\mathbf{A}|\psi\rangle. \tag{3.36}$$

Because this equation is true for any state $|\psi\rangle$, we can drop $|\psi\rangle$ and write the operator equation

$$\mathscr{D}^\dagger \mathbf{A} \mathscr{D} = R\mathbf{A}. \tag{3.37}$$

This equation can be viewed as a constraint on the form of $\mathscr{D}$ if the form of $\mathbf{A}$ is known, or vice versa.

3.3 The angular-momentum operator

3.3.1 Definition

In classical physics, the angular momentum of an object can be given an intuitive definition in terms of the physical quantities describing its angular velocity and moment of inertia. This sort of definition can be translated into quantum mechanics to describe orbital angular momentum, but it cannot describe spin angular momentum, which has no direct classical analog (see Sec. 2.2.3). However, a more abstract definition can be made in terms of the rotation operators; this definition can be used to describe all forms of angular momentum.

The angular-momentum operator $\mathbf{J}$ is defined to be the *generator of infinitesimal rotations*. This means that the rotation operator for a rotation $R_{\hat{\mathbf{n}}}(d\phi)$ by a small angle $d\phi$ about an axis $\hat{\mathbf{n}}$ can be written

$$\mathscr{D}[R_{\hat{\mathbf{n}}}(d\phi)] = 1 - i\mathbf{J} \cdot \hat{\mathbf{n}}\, d\phi/\hbar, \tag{3.38}$$

i.e., $\mathbf{J}$ is related to the difference between the original state and the rotated state. (This relation can also be made in the classical case.) The factor of i is required to make the components of $\mathbf{J}$ Hermitian. [This follows from the unitarity of $\mathscr{D}(R)$, Eq. (3.30).] The minus sign is a universally adopted convention.

Finite rotations can be found in terms of the angular momentum operator by compounding many infinitesimal rotations:

$$\begin{aligned}\mathscr{D}[R_{\hat{\mathbf{n}}}(\phi)] &= \lim_{N \to \infty} [1 - i\mathbf{J} \cdot \hat{\mathbf{n}}\, \phi/(\hbar N)]^N \\ &= \exp(-i\mathbf{J} \cdot \hat{\mathbf{n}}\, \phi/\hbar).\end{aligned} \tag{3.39}$$

As for classical rotations, an arbitrary quantum-mechanical rotation is conveniently specified in terms of Euler angles, so that it takes the form

$$\begin{aligned}\mathscr{D}(\alpha, \beta, \gamma) &= \mathscr{D}_{\hat{\mathbf{z}}}(\alpha)\, \mathscr{D}_{\hat{\mathbf{y}}}(\beta)\, \mathscr{D}_{\hat{\mathbf{z}}}(\gamma) \\ &= \exp(-iJ_z \alpha/\hbar) \exp(-iJ_y \beta/\hbar) \exp(-iJ_z \gamma/\hbar).\end{aligned} \tag{3.40}$$

Thus, once we determine the properties of the angular momentum operator, in particular its eigenstates and eigenvalues, we can find explicit matrix representations of the rotation operators.[1]

We cannot find a complete set of states that are simultaneously eigenstates of J_x, J_y, and J_z. This is because the components of $\mathbf{J}$ do not commute with each other; this is directly related to the fact that rotations about different axes do not commute. On the other hand, if we look at the eigenstates of just one component of $\mathbf{J}$, the angular momentum of each of these eigenstates is not completely specified. In the vector model (which we discussed in

---

[1] The reader should be cautioned that, unfortunately, there are a variety of conventions and definitions in use related to angular momentum. For example, there is a difference between the definitions of the $\mathscr{D}$ operators and the order in which the Euler rotations are applied between Landau and Lifshitz (1977) and Edmonds (1974).

some detail in Chapter 2), we would say that an eigenstate of J_z has a particular projection of angular momentum on the z-axis, but that vectors of many different lengths could have this same projection. To differentiate between angular momentum vectors of different lengths, the operator $\mathbf{J}^2 = J_x^2 + J_y^2 + J_z^2$ is formed that describes the magnitude of the angular momentum. It commutes with each of the J_i. That means that we can find a complete set of states that are eigenstates of both $\mathbf{J}^2$ and any one of the components, for example, J_z. These eigenstates describe the angular momentum of a particle to the maximum extent allowed by quantum mechanics.

3.3.2 Raising and lowering operators

As tools to find the eigenvalues of $\mathbf{J}^2/\hbar^2$ and $J_z/\hbar$, denoted by λ and μ, respectively, and their corresponding eigenstates $|\lambda\mu\rangle$, one can use the raising and lowering operators defined as $J_\pm = J_x \pm i J_y$. Using Eq. (3.31), we find the commutation relations

$$[\mathbf{J}^2, J_\pm] = [\mathbf{J}^2, J_i] = 0 \qquad (3.41\text{a})$$
$$[J_x, J_\pm] = \mp \hbar J_z \qquad (3.41\text{b})$$
$$[J_y, J_\pm] = -i\hbar J_z \qquad (3.41\text{c})$$
$$[J_z, J_\pm] = \pm \hbar J_\pm \qquad (3.41\text{d})$$
$$[J_+, J_-] = 2\hbar J_z. \qquad (3.41\text{e})$$

These relations will help us find the matrix elements of $J_\pm$. From $[J_z, J_\pm] = \pm \hbar J_\pm$ we have, on one hand,

$$\langle \lambda \mu' | [J_z, J_\pm] | \lambda \mu \rangle = \pm \hbar \langle \lambda \mu' | J_\pm | \lambda \mu \rangle, \qquad (3.42)$$

and on the other hand,

$$\langle \lambda \mu' | [J_z, J_\pm] | \lambda \mu \rangle = \langle \lambda \mu' | J_z J_\pm - J_\pm J_z | \lambda \mu \rangle$$
$$= \hbar (\mu' - \mu) \langle \lambda \mu' | J_\pm | \lambda \mu \rangle. \qquad (3.43)$$

Comparing Eqs. (3.42) and (3.43), we have

$$(\mu' - \mu \mp 1) \langle \lambda \mu' | J_\pm | \lambda \mu \rangle = 0, \qquad (3.44)$$

i.e., the matrix element of $J_\pm$ is zero unless $\Delta \mu = \pm 1$. Thus we can write

$$J_\pm |\lambda\mu\rangle = \hbar C^\pm (\lambda\mu) |\lambda, \mu \pm 1\rangle, \qquad (3.45)$$

where $C^\pm(\lambda\mu)$ are constants, and applying the Hermitian conjugate of this equation to itself on the left, we have

$$\hbar^2 \left| C^\pm (\lambda\mu) \right|^2 = \langle \lambda\mu | J_\pm^\dagger J_\pm | \lambda\mu \rangle. \qquad (3.46)$$

To determine the constant $C^\pm$, we expand $\mathbf{J}^2$ as

$$\mathbf{J}^2 = J_z (J_z \pm \hbar) + J_\mp J_\pm, \qquad (3.47)$$

which is rewritten to be

$$J_\pm^\dagger J_\pm = \mathbf{J}^2 - J_z(J_z \pm \hbar), \tag{3.48}$$

since $J_\mp = J_\pm^\dagger$. Thus

$$\begin{aligned}\langle\lambda\mu|J_\pm^\dagger J_\pm|\lambda\mu\rangle &= \langle\lambda\mu|\mathbf{J}^2 - J_z(J_z \pm \hbar)|\lambda\mu\rangle \\ &= \lambda - \mu(\mu \pm 1),\end{aligned} \tag{3.49}$$

so

$$\left|C^\pm(\lambda\mu)\right|^2 = \lambda - \mu(\mu \pm 1). \tag{3.50}$$

From Eq. (3.45) we have

$$J_\pm|\lambda\mu\rangle = e^{i\varphi}\hbar\sqrt{\lambda - \mu(\mu \pm 1)}|\lambda, \mu \pm 1\rangle, \tag{3.51}$$

where the phase φ is conventionally taken to be zero, making $C^\pm$ real. The names "raising" and "lowering" come from the effect of the operators on the state $|\lambda\mu\rangle$.

3.3.3 Angular momentum eigenvalues

The allowed range for μ can be found from the relation

$$\mathbf{J}^2 - J_z^2 = J_x^2 + J_y^2, \tag{3.52}$$

which gives

$$\begin{aligned}\langle\lambda\mu|\mathbf{J}^2 - J_z^2|\lambda\mu\rangle &= \hbar^2(\lambda - \mu^2) \\ &= \langle\lambda\mu|J_x^2|\lambda\mu\rangle + \langle\lambda\mu|J_y^2|\lambda\mu\rangle \\ &\geq 0,\end{aligned} \tag{3.53}$$

since J_x^2 and J_y^2 are nonnegative. Thus $\lambda \geq 0$ and $-\sqrt{\lambda} \leq \mu \leq \sqrt{\lambda}$. Repeatedly applying J_+ or J_- to a state $|\lambda\mu\rangle$ will violate the second condition unless there are some maximum and minimum values $\mu^\pm$ of μ such that $J_\pm|\lambda\mu^\pm\rangle = 0$. From Eq. (3.49) we have

$$\langle\lambda\mu^\pm|J_\pm^\dagger J_\pm|\lambda\mu^\pm\rangle = \hbar^2\left[\lambda - \mu^\pm(\mu^\pm \pm 1)\right] = 0, \tag{3.54}$$

or

$$\lambda = \mu^+(\mu^+ + 1) = \mu^-(\mu^- - 1), \tag{3.55}$$

which is only satisfied (assuming $\mu^- \leq \mu^+$) if $\mu^- = -\mu^+$. Designating μ by m and μ^+ by J, the eigenvalue for $\mathbf{J}^2/\hbar^2$ is given by $\lambda = J(J+1)$, and we now notate the eigenstates by $|Jm\rangle$, so that

$$\mathbf{J}^2|Jm\rangle = \hbar^2 J(J+1)|Jm\rangle. \tag{3.56}$$

Equation (3.51) becomes

$$J_\pm|Jm\rangle = \hbar\sqrt{J(J+1) - m(m \pm 1)}|J, m \pm 1\rangle, \tag{3.57}$$

and the components of $\mathbf{J}$ acting on the basis states are given by

42 *A bit of angular-momentum theory*

$$J_x|Jm\rangle = \frac{1}{2}(J_+ + J_-)|Jm\rangle$$
$$= \frac{\hbar}{2}\sqrt{J(J+1) - m(m+1)}|J, m+1\rangle \quad (3.58\text{a})$$
$$+ \frac{\hbar}{2}\sqrt{J(J+1) - m(m-1)}|J, m-1\rangle$$

$$J_y|Jm\rangle = \frac{-i}{2}(J_+ - J_-)|Jm\rangle$$
$$= -\frac{i\hbar}{2}\sqrt{J(J+1) - m(m+1)}|J, m+1\rangle \quad (3.58\text{b})$$
$$+ \frac{i\hbar}{2}\sqrt{J(J+1) - m(m-1)}|J, m-1\rangle$$

$$J_z|Jm\rangle = \hbar m|Jm\rangle. \quad (3.58\text{c})$$

Since repeated application of J_+ takes $|J, -J\rangle$ to $|JJ\rangle$ in $2J$ integer steps, J must be integer or half-integer.

3.3.4 Spin-$\frac{1}{2}$ eigenstates and operators

The $2J+1$ eigenstates of the J_z operator corresponding to the eigenvalues $m = -J, -J+1, \ldots, J$ form a complete basis for the subspace with total angular momentum quantum number J. If there were any other eigenstates with eigenvalues not in this list, repeated application of J_+ or J_- on such a state would run afoul of $-\sqrt{J(J+1)} \leq m \leq \sqrt{J(J+1)}$, and the series could not terminate, because Eq. (3.55) has no more allowed solutions. Thus for a given subspace with angular momentum J, we can write the components of $\mathbf{J}$ as $(2J+1) \times (2J+1)$ matrices acting on column vectors written in the standard form as

$$\begin{pmatrix} |JJ\rangle \\ |J, J-1\rangle \\ \vdots \\ |J, -J\rangle \end{pmatrix}. \quad (3.59)$$

For $J = 1/2$, the matrices can be read off from Eq. (3.58) as

$$J_x = \frac{\hbar}{2}\begin{pmatrix} 0 & 1 \\ 1 & 0 \end{pmatrix}, \quad J_y = \frac{\hbar}{2}\begin{pmatrix} 0 & -i \\ i & 0 \end{pmatrix}, \quad J_z = \frac{\hbar}{2}\begin{pmatrix} 1 & 0 \\ 0 & -1 \end{pmatrix}, \quad (3.60)$$

or $\mathbf{J} = \frac{\hbar}{2}\boldsymbol{\sigma}$, where $\boldsymbol{\sigma}$ is the *Pauli spin operator*. We can use this expression to find $\mathscr{D}_{\hat{\mathbf{n}}}(\phi)_{1/2} \equiv \mathscr{D}[R_{\hat{\mathbf{n}}}(\phi)]_{1/2}$ (Eq. 3.39) explicitly.

In Cartesian coordinates, we have $\hat{\mathbf{n}} = (n_x, n_y, n_z) = (\sin\theta\cos\phi, \sin\theta\sin\phi, \cos\theta)$. Thus the component of $\boldsymbol{\sigma}$ along $\hat{\mathbf{n}}$ is given by

$$(\boldsymbol{\sigma} \cdot \hat{\mathbf{n}}) = \begin{pmatrix} n_z & n_x - in_y \\ n_x + in_y & -n_z \end{pmatrix}$$
$$= \begin{pmatrix} \cos\theta & e^{-i\phi}\sin\theta \\ e^{i\phi}\sin\theta & -\cos\theta \end{pmatrix}. \quad (3.61)$$

Expanding Eq. (3.39) in a power series gives

$$\mathscr{D}_{\hat{\mathbf{n}}}(\eta)_{1/2} = \exp(-i\boldsymbol{\sigma}\cdot\hat{\mathbf{n}}\,\eta/2)$$
$$= \sum_{k=0}^{\infty} \frac{(-\eta/2)^k}{k!} (\boldsymbol{\sigma}\cdot\hat{\mathbf{n}})^k \qquad (3.62)$$
$$= I\cos\frac{\eta}{2} - i\begin{pmatrix} \cos\theta & e^{-i\phi}\sin\theta \\ e^{i\phi}\sin\theta & -\cos\theta \end{pmatrix} \sin\frac{\eta}{2},$$

since, using the properties of the Pauli matrices, it can be shown that

$$(\boldsymbol{\sigma}\cdot\hat{\mathbf{n}})^k = \begin{cases} I, & k \text{ even;} \\ \boldsymbol{\sigma}\cdot\hat{\mathbf{n}}, & k \text{ odd.} \end{cases} \qquad (3.63)$$

In particular

$$\mathscr{D}_{\hat{\mathbf{z}}}(\gamma)_{1/2} = \begin{pmatrix} e^{-i\gamma/2} & 0 \\ 0 & e^{i\gamma/2} \end{pmatrix} \qquad (3.64a)$$

$$\mathscr{D}_{\hat{\mathbf{y}}}(\beta)_{1/2} = \begin{pmatrix} \cos\frac{\beta}{2} & -\sin\frac{\beta}{2} \\ \sin\frac{\beta}{2} & \cos\frac{\beta}{2} \end{pmatrix}, \qquad (3.64b)$$

resulting in the matrix for rotation by Euler angles

$$\mathscr{D}(\alpha,\beta,\gamma)_{1/2} = \mathscr{D}_{\hat{\mathbf{z}}}(\alpha)_{1/2}\,\mathscr{D}_{\hat{\mathbf{y}}}(\beta)_{1/2}\,\mathscr{D}_{\hat{\mathbf{z}}}(\gamma)_{1/2}$$
$$= \begin{pmatrix} e^{-i(\alpha+\gamma)/2}\cos\frac{\beta}{2} & -e^{-i(\alpha-\gamma)/2}\sin\frac{\beta}{2} \\ e^{i(\alpha-\gamma)/2}\sin\frac{\beta}{2} & e^{i(\alpha+\gamma)/2}\cos\frac{\beta}{2} \end{pmatrix}. \qquad (3.65)$$

This matrix gives the *Wigner D-functions* for angular momentum $J = 1/2$. They are functions of the Euler angles that describe the rotation of a state written in the basis of angular momentum eigenstates. We discuss the D-functions for arbitrary angular momenta in the next section.

> Equation (3.65) explicitly shows that a rotation of a spin-1/2 system described by an arbitrary wave function by an angle of 2π results in the change of the overall sign of the wave function. It thus takes two full rotations to bring the wave function back to its original value. This is the famous and completely "nonclassical" result that was actually put to rigorous experimental tests using *neutron interferometry* as discussed in the book by Rauch and Werner (2000).

3.4 Rotations in the Zeeman basis

3.4.1 General properties

The issues of active vs. passive rotations and co- vs. contravariant components discussed in Sec. 3.1 also arise for quantum-mechanical rotations. Suppose we have a state ket $|\psi\rangle$ representing the angular degrees of freedom of a state with angular momentum J. We can expand the ket in the basis of eigenstates $|m\rangle$ of the J_z operator:

$$|\psi\rangle = |m\rangle\langle m|\psi\rangle = |m\rangle\psi^m, \qquad (3.66)$$

where $\psi^m = \langle m|\psi\rangle$ are the Zeeman components of $|\psi\rangle$.

44   A bit of angular-momentum theory

Consider a rotation applied to $|\psi\rangle$:

$$|\psi'\rangle = \mathscr{D}|\psi\rangle. \tag{3.67}$$

This rotation can be written as an active rotation:

$$|\psi'\rangle = |m_1\rangle\langle m_1|\mathscr{D}|m_2\rangle\langle m_2|\psi\rangle = |m_1\rangle\mathscr{D}_{m_1 m_2}\psi^{m_2} = |m_1\rangle\psi'^{m_1}, \tag{3.68}$$

where

$$\psi'^{m_1} = \mathscr{D}_{m_1 m_2}\psi^{m_2}. \tag{3.69}$$

We can also write a passive rotation:

$$|\psi\rangle = \mathscr{D}^{-1}|\psi'\rangle = \mathscr{D}^{-1}|m_1\rangle\mathscr{D}_{m_1 m}\psi^{m} = |m_2\rangle\mathscr{D}^{-1}_{m_2 m_1}\mathscr{D}_{m_1 m}\psi^{m} = |m_1\rangle'\psi'^{m_1}, \tag{3.70}$$

with

$$|m_1\rangle' = \mathscr{D}^{*}_{m_1 m_2}|m_2\rangle. \tag{3.71}$$

Comparing with Eq. (3.18), we see that this is quite similar to the case of vectors in the spherical basis: relatively contravariant quantities transform with rotation operators that are complex conjugates of one another. The parallel goes further, however. Consider the operator $\mathscr{D}_{\hat{z}}(\theta)$ for rotations about $\hat{z}$ by an angle θ. In matrix form for a $J = 1$ state, we have in the Zeeman basis

$$\mathscr{D}_{\hat{z}}(\theta) = e^{-iJ_z\theta} = \begin{pmatrix} e^{-i\theta} & 0 & 0 \\ 0 & 1 & 0 \\ 0 & 0 & e^{i\theta} \end{pmatrix}. \tag{3.72}$$

This matrix is identical to the spherical-basis rotation matrix $\tilde{R}_{\hat{z}}(\theta)$ given by Eq. (3.21). In fact, the components ψ^m of a $J = 1$ state $|\psi\rangle$ transform just like the contravariant spherical components v^q of a vector $\mathbf{v}$. Likewise, the basis states $|m\rangle$ transform like the covariant spherical basis vectors $\hat{\mathbf{e}}_q$. Furthermore, taking the Hermitian conjugate of the above expressions, we see that the dual basis states $\langle m|$ transform like contravariant spherical basis vectors $\hat{\mathbf{e}}^q$, and the components ψ_m of $\langle\psi|$ expanded over $\langle m|$ transform like covariant spherical basis components v_q.

The $2J + 1$ Zeeman components and basis states for states with angular momenta besides $J = 1$ also transform according to the rules described in this section, but with different matrix representations of the rotation operators. We therefore generalize the definition of the spherical basis, and call a basis that has the transformation properties of the angular-momentum-J Zeeman basis $|m\rangle$ the spherical basis of rank J. A rank-J basis has $2J + 1$ basis vectors. Thus the spherical basis described in Sec. 3.1.2 has rank 1. Similarly to the definition made for the rank-1 spherical basis, quantities that transform like the Zeeman basis states $|m\rangle$ are defined to be covariant, while those that transform like the basis $\langle m|$ (or the coefficients ψ^m) are defined to be contravariant.

3.4.2   Wigner D-functions

The Wigner D-functions are the explicit matrix elements of the quantum-mechanical rotation operators in the Zeeman (spherical) basis. They are defined in terms of the Euler angles (Sec. 3.1.3) by the formula

$$D^{(J)}_{m'm}(\alpha, \beta, \gamma) = \langle m'|\mathscr{D}_{\hat{z}}(\alpha)\,\mathscr{D}_{\hat{y}}(\beta)\,\mathscr{D}_{\hat{z}}(\gamma)|m\rangle. \tag{3.73}$$

They are useful whenever the effect of a specific rotation on a quantum state needs to be found. There are various methods for obtaining a general expression for the Wigner D-functions. The matrix for rotations about the z-axis follows directly from the fact that the basis states are eigenstates of J_z. We have $\langle J m' | J_z | J m \rangle = m \hbar \delta_{m'm}$, so

$$[\mathscr{D}_{\hat{z}}(\phi)]_{m'm} = \langle J m' | e^{-i J_z \phi/\hbar} | J m \rangle = e^{-im\phi} \delta_{m'm}. \tag{3.74}$$

Thus the D-function can be written

$$\begin{aligned} D^{(J)}_{m'm}(\alpha, \beta, \gamma) &= \langle m' | \mathscr{D}_{\hat{z}}(\alpha) \, \mathscr{D}_{\hat{y}}(\beta) \, \mathscr{D}_{\hat{z}}(\gamma) | m \rangle \\ &= \sum_{m'',m'''} \delta_{m'm'''} e^{-im'\alpha} \langle m''' | \mathscr{D}_{\hat{y}}(\beta) | m'' \rangle e^{-im\gamma} \delta_{mm''} \\ &= e^{-i(\alpha m' + \gamma m)} d^{(J)}_{m'm}(\beta), \end{aligned} \tag{3.75}$$

where $d^{(J)}_{m'm}(\beta)$ is the matrix for rotations about the y-axis. One technique for finding a formula for this matrix involves relating angular-momentum eigenstates to states of a system consisting of two uncoupled harmonic oscillators (see Sakurai and Tuan 1994 for details). This approach results in *Wigner's formula* for $d^{(J)}_{m'm}(\beta)$:

$$\begin{aligned} d^{(J)}_{m'm}(\beta) = \sum_k (-1)^{k-m+m'} & \frac{\sqrt{(J+m)!\,(J-m)!\,(J+m')!\,(J-m')!}}{(J+m-k)!\,k!\,(J-k-m')!\,(k-m+m')!} \\ & \times \left(\cos\frac{\beta}{2}\right)^{2J-2k+m-m'} \left(\sin\frac{\beta}{2}\right)^{2k-m+m'}, \end{aligned} \tag{3.76}$$

where the sum runs over values of k for which none of the arguments of the factorials are negative. For the $J = 1$ case, for example, this becomes

$$d^{(1)}_{m'm}(\beta) = \begin{pmatrix} \cos^2\frac{\beta}{2} & -\frac{\sin\beta}{\sqrt{2}} & \sin^2\frac{\beta}{2} \\ \frac{\sin\beta}{\sqrt{2}} & \cos\beta & -\frac{\sin\beta}{\sqrt{2}} \\ \sin^2\frac{\beta}{2} & \frac{\sin\beta}{\sqrt{2}} & \cos^2\frac{\beta}{2} \end{pmatrix}. \tag{3.77}$$

3.5 Addition of angular momenta; Clebsch–Gordan coefficients

In classical physics, if we add two angular momenta with given magnitudes, the resulting vector sum (the *total angular momentum*) has a magnitude that lies between the sum of the magnitudes of the constituent angular momenta and the magnitude of their difference (the *triangle rule*).

In quantum mechanics, a similar result holds, with the exception of the fact that the total angular-momentum quantum number should be integer (if the sum of the constituent angular-momenta quantum numbers is integer) or half-integer (if the sum is half-integer). For example, if we add **J** and **I** to form **F**, we have:

$$|J - I| \leq F \leq J + I. \tag{3.78}$$

The projections along the quantization axis simply add:

$$F_z = J_z + I_z, \tag{3.79}$$

or

46   *A bit of angular-momentum theory*

$$m_F = m_J + m_I. \tag{3.80}$$

Suppose we have a total-angular-momentum state $|F, m_F\rangle$, which originated from the addition of **J** and **I**. In general, there are several combinations $|J, m_J\rangle|I, m_I\rangle$ that can contribute to $|F, m_F\rangle$:

$$\begin{aligned}|Fm_F\rangle &= \left(\sum_{m_J,m_I}|Jm_J\rangle|Im_I\rangle\langle Jm_J|\langle Im_I|\right)|Fm_F\rangle \\ &= \sum\langle Jm_JIm_I|Fm_F\rangle|Jm_J\rangle|Im_I\rangle.\end{aligned} \tag{3.81}$$

The inverse transformation can be formed the same way:

$$\begin{aligned}|Jm_J\rangle|Im_I\rangle &= \left(\sum_{F,m_F}|Fm_F\rangle\langle Fm_F|\right)|Jm_J\rangle|Im_I\rangle \\ &= \sum\langle Fm_F|Jm_JIm_I\rangle|Fm_F\rangle.\end{aligned} \tag{3.82}$$

The coefficients $\langle Jm_JIm_I|Fm_F\rangle$ are called the *vector-addition* or *Clebsch–Gordan* coefficients. In the standard convention, Clebsch–Gordan coefficients are real, so

$$\langle Fm_F|Jm_JIm_I\rangle = \langle Jm_JIm_I|Fm_F\rangle. \tag{3.83}$$

The values of the Clebsch–Gordan coefficients are derived in numerous books on quantum mechanics starting from the fact that there is only one possibility to obtain a *stretched state* in which $m_F = J + I$ or $-(J + I)$, leading to

$$\langle JJII|I+J, I+J\rangle = \langle J,-J,I,-I|I+J, -(I+J)\rangle = 1, \tag{3.84}$$

followed by application of raising or lowering operators, and making use of orthogonality of the wave functions corresponding to different total angular momenta. We will not reproduce this derivation here. General formulas, as well as simplified formulas for special cases, are given, for example, in the book by Sobelman (1992). Numerical values can also be obtained from built-in functions in software packages such as *Mathematica*.

Various symmetry and sum rules can be formulated for the Clebsch–Gordan coefficients, as discussed, for instance, in the books by Sobelman (1992); Varshalovich *et al.* (1988); and Judd (1998). For example, here is the rule for permutation of the arguments:

$$\langle J_2m_2J_1m_1|Jm\rangle = (-1)^{J_1+J_2-J}\langle J_1m_1J_2m_2|Jm\rangle. \tag{3.85}$$

3.6   3j and 6j symbols

In the literature, a set of coefficients called the *Wigner 3j symbols* are commonly used instead of the Clebsch–Gordan coefficients. The $3j$ symbols, written as 2×3 matrices in parentheses, are related to the Clebsch–Gordan coefficients in the following way (Sobelman 1992; Varshalovich *et al.* 1988; Judd 1998):

$$\begin{pmatrix}J_1 & J_2 & J \\ m_1 & m_2 & m\end{pmatrix} = (-1)^{J_1-J_2-m}\frac{1}{\sqrt{2J+1}}\langle J_1m_1J_2m_2|J,-m\rangle. \tag{3.86}$$

Since the usual phase convention for the Clebsch–Gordan coefficients is chosen so that they are all real, the $3j$ symbols are real as well. Furthermore, the triangle and the projection rules for the Clebsch–Gordan coefficients must necessarily carry over to the $3j$ symbols, so that a $3j$ symbol is zero unless it satisfies

$$|J_1 - J_2| \leq J \leq J_1 + J_2 \tag{3.87}$$

and

$$m_1 + m_2 + m = 0. \tag{3.88}$$

The advantage of using $3j$ symbols is that they are more symmetric than the Clebsch–Gordan coefficients. For example, for the Clebsch–Gordan coefficients, we have the relation of Eq. (3.85). This relation can be expressed using $3j$ symbols by noting that, according to Eqs. (3.86) and (3.85), odd permutations of the columns of the $3j$ symbol obey

$$\begin{pmatrix} J_1 & J_2 & J \\ m_1 & m_2 & m \end{pmatrix} = (-1)^{J_1+J_2+J} \begin{pmatrix} J_2 & J_1 & J \\ m_2 & m_1 & m \end{pmatrix} = (-1)^{J_1+J_2+J} \begin{pmatrix} J_1 & J & J_2 \\ m_1 & m & m_2 \end{pmatrix} = \ldots. \tag{3.89}$$

Furthermore, even permutations of the columns leave the value of a $3j$ symbol unchanged:

$$\begin{pmatrix} J_1 & J_2 & J \\ m_1 & m_2 & m \end{pmatrix} = \begin{pmatrix} J & J_1 & J_2 \\ m & m_1 & m_2 \end{pmatrix} = \begin{pmatrix} J_2 & J & J_1 \\ m_2 & m & m_1 \end{pmatrix}. \tag{3.90}$$

One of the consequences of Eq. (3.89) is that all $3j$ symbols with two identical columns are zero if $J_1 + J_2 + J$ is odd, since if, for example, $J_1 = J_2$ and $m_1 = m_2$, then Eq. (3.89) implies

$$\begin{pmatrix} J_1 & J_1 & J \\ m_1 & m_1 & m \end{pmatrix} = (-1)^{2J_1+J} \begin{pmatrix} J_1 & J_1 & J \\ m_1 & m_1 & m \end{pmatrix}, \tag{3.91}$$

which can only be satisfied if $2J_1 + J$ is even or if

$$\begin{pmatrix} J_1 & J_1 & J \\ m_1 & m_1 & m \end{pmatrix} = 0. \tag{3.92}$$

There are other simply stated symmetry relations involving $3j$ symbols, such as

$$\begin{pmatrix} J_1 & J_2 & J \\ m_1 & m_2 & m \end{pmatrix} = (-1)^{J_1+J_2+J} \begin{pmatrix} J_1 & J_2 & J \\ -m_1 & -m_2 & -m \end{pmatrix}, \tag{3.93}$$

as well as useful *sum rules*

$$\sum_{J,m} (2J+1) \begin{pmatrix} J_1 & J_2 & J \\ m_1 & m_2 & m \end{pmatrix} \begin{pmatrix} J_1 & J_2 & J \\ m'_1 & m'_2 & m \end{pmatrix} = \delta_{m_1,m'_1} \delta_{m_2,m'_2}, \tag{3.94}$$

$$\sum_{m_1,m_2} \begin{pmatrix} J_1 & J_2 & J \\ m_1 & m_2 & m \end{pmatrix} \begin{pmatrix} J_1 & J_2 & J' \\ m_1 & m_2 & m' \end{pmatrix} = \frac{1}{2J+1} \delta_{J,J'} \delta_{m,m'}. \tag{3.95}$$

These sum rules translate directly into the properties of atomic transitions as will be discussed in Chapter 7.

We have seen that with two angular momenta J_1 and J_2 either the Clebsch–Gordan coefficients or the $3j$ symbols can be employed to relate the uncoupled basis $|J_1 m_1\rangle |J_2 m_2\rangle$ to the coupled basis $|(J_1 J_2) J m\rangle$. However, if we are interested in adding three angular momenta J_1, J_2, and J_3 to get the resultant J, the situation becomes somewhat more complicated. For instance, there is generally no unique expansion of $|(J_1 J_2 J_3) J m\rangle$ in terms of $|J_1 m_1\rangle |J_2 m_2\rangle |J_3 m_3\rangle$: the expansion coefficients depend on the order in which the angular momenta are added and on the value of the intermediate angular momenta resulting from the addition of the first two angular momenta (Judd 1998). However, if we specify the intermediate vector $\mathbf{J}_a$ that results from adding, for example, $\mathbf{J}_1 + \mathbf{J}_2 = \mathbf{J}_a$, then we are in fact able to unambiguously express the state $|(J_1 J_2) J_a J_3 J m\rangle$ as a linear combination of the states $|J_1 m_1\rangle |J_2 m_2\rangle |J_3 m_3\rangle$. But we could just as easily add $\mathbf{J}_2 + \mathbf{J}_3 = \mathbf{J}_b$ to form an intermediate vector J_b and obtain the states $|(J_2 J_3) J_b J_1 J m\rangle$.

Thus an important issue becomes how to relate the basis $|(J_1 J_2) J_a J_3 J m\rangle$ and the basis $|(J_2 J_3) J_b J_1 J m\rangle$. A useful tool for this task is the *Wigner 6j symbol*.

The transformation between the two bases can be described in terms of the coefficients $\langle J_a J_3 J m | J_b J_1 J m \rangle$,

$$|J_b J_1 J m\rangle = \sum_{J_a} |J_a J_3 J m\rangle \langle J_a J_3 J m | J_b J_1 J m\rangle. \tag{3.96}$$

Immediately we can note that the coefficients are independent of m by applying, for example, the raising operator J_+ to both sides of Eq. (3.96):

$$J_+ |J_b J_1 J m\rangle = \sum_{J_a} J_+ |J_a J_3 J m\rangle \langle J_a J_3 J m | J_b J_1 J m\rangle, \tag{3.97}$$

or explicitly,

$$\sqrt{J(J+1) - m(m+1)} |J_b J_1 J, m+1\rangle$$
$$= \sum_{J_a} \sqrt{J(J+1) - m(m+1)} |J_a J_3 J, m+1\rangle \langle J_a J_3 J m | J_b J_1 J m\rangle, \tag{3.98}$$

so that

$$|J_b J_1 J, m+1\rangle = \sum_{J_a} |J_a J_3 J, m+1\rangle \langle J_a J_3 J m | J_b J_1 J m\rangle. \tag{3.99}$$

If we begin with $m = -J$, this shows by iteration that the coefficients are all equal, so they can be denoted simply as $\langle J_a J_3 J | J_b J_1 J \rangle$.

The $6j$ symbols, which are written as 2×3 matrices in curly brackets, are directly related to these coefficients:

$$\begin{Bmatrix} J_3 & J & J_a \\ J_1 & J_2 & J_b \end{Bmatrix} = \frac{(-1)^{J_1+J_2+J_3+J}}{\sqrt{(2J_a+1)(2J_b+1)}} \langle J_a J_3 J | J_b J_1 J \rangle. \tag{3.100}$$

Similar to $3j$ symbols, $6j$ symbols are real numbers, and they have several important symmetry properties. The $6j$ symbols are zero unless they satisfy triangular conditions for the entries in the positions denoted by $\diamond$'s:

$$\begin{Bmatrix} \cdot & \diamond & \cdot \\ \diamond & \cdot & \diamond \end{Bmatrix}, \quad \begin{Bmatrix} \diamond & \diamond & \diamond \\ \cdot & \cdot & \cdot \end{Bmatrix}, \quad \begin{Bmatrix} \cdot & \cdot & \diamond \\ \diamond & \diamond & \cdot \end{Bmatrix}, \quad \begin{Bmatrix} \diamond & \cdot & \cdot \\ \cdot & \diamond & \diamond \end{Bmatrix}. \tag{3.101}$$

The triangular conditions for these entries are expressions of the triangular conditions for the addition of pairs of angular momenta, for example, the addition of $\mathbf{J}_1$ and $\mathbf{J}_2$ to form the intermediate angular momentum $\mathbf{J}_a$.

The $6j$ symbols are invariant under any permutation of the columns, for example

$$\begin{Bmatrix} J_1 & J_2 & J_3 \\ L_1 & L_2 & L_3 \end{Bmatrix} = \begin{Bmatrix} J_2 & J_1 & J_3 \\ L_2 & L_1 & L_3 \end{Bmatrix}, \tag{3.102}$$

and are also invariant under an interchange of the upper and lower arguments in each of any two columns,

$$\begin{Bmatrix} J_1 & J_2 & J_3 \\ L_1 & L_2 & L_3 \end{Bmatrix} = \begin{Bmatrix} L_1 & L_2 & J_3 \\ J_1 & J_2 & L_3 \end{Bmatrix}. \tag{3.103}$$

An important context in atomic physics where the $6j$ symbols appear and can be quite useful is when there is a need to relate matrix elements of an operator in the coupled angular momentum basis to those in the uncoupled basis. This is discussed in Sec. 3.8.

Much more information about $3j$ and $6j$ symbols can be found in the books by Sobelman (1992); Varshalovich et al. (1988); and Judd (1998), and symbolic and numerical values can easily be evaluated using programs such as *Mathematica*.

3.7 Irreducible tensors and tensor products

We can refer to a set of objects that transform into each other under rotations via a linear transformation a *tensor*. So far, we have seen vector components (in the spherical or Cartesian bases) and components of wave functions, and their associated basis sets, as examples of tensors. Another trivial example of a tensor is a scalar, which has only one component, and whose rotation transform is the identity.

Physically, it is clear that any vector can be rotated to point along any other vector, so that for a general rotation R of a vector $\mathbf{v}$, each component v'_i of the rotated vector depends on all three components v_j of the original vector. This means that, for example, there is no basis $(\hat{\mathbf{a}}, \hat{\mathbf{b}}, \hat{\mathbf{c}})$ such that the component of the vector in the $(\hat{\mathbf{a}})$ subspace, $v_a\hat{\mathbf{a}}$, remains in this subspace under all possible rotations, while the component $v_b\hat{\mathbf{b}} + v_c\hat{\mathbf{c}}$ remains in the $(\hat{\mathbf{b}}, \hat{\mathbf{c}})$ subspace. A tensor with this property is called *irreducible*—there is no basis in which it can be written as the sum of two or more tensors, the components of each of which transform only into themselves. All of the tensors that we have discussed so far have this property.

An irreducible tensor defined on a spherical basis of rank κ is called a rank-κ irreducible spherical tensor and has $2\kappa + 1$ components, commonly indexed with the label q. Thus, for example, the three spherical components of a vector comprise an irreducible spherical tensor of rank 1. As another important example, the spherical harmonics $Y_{lm}(\theta, \varphi)$ with $m = -l, \ldots, l$ form an irreducible spherical tensor of rank l. Thus an alternate definition of an irreducible spherical tensor is a collection of $2l+1$ quantities that rotate like (have the same rotation matrix as) the spherical harmonics of rank l. The standard notation for the covariant components of a general irreducible tensor A of rank κ is A^κ_q and for the contravariant components $A^{\kappa q}$.

To find an example of a *reducible* tensor, we consider a rank-2 Cartesian tensor $T = T_{ij}\hat{\mathbf{x}}_i\hat{\mathbf{x}}_j$. (The rank of a Cartesian tensor is given by the number of its indices, and is related to the rank of an irreducible tensor in a manner described below.) Such a tensor has nine components T_{ij} and nine basis tensors $\hat{\mathbf{x}}_i\hat{\mathbf{x}}_j$, and transforms under a spatial rotation by applying the rotation operator R twice:

$$T' = T_{ij} R\hat{\mathbf{x}}_i R\hat{\mathbf{x}}_j = T_{ij}\hat{\mathbf{x}}_m R_{mi}\hat{\mathbf{x}}_n R_{nj} = R_{mi}R_{nj}T_{ij}\hat{\mathbf{x}}_m\hat{\mathbf{x}}_n. \tag{3.104}$$

It is easy to see that T is reducible. We construct a new basis element

$$\frac{1}{3}\hat{\mathbf{x}}_i\hat{\mathbf{x}}_i = \frac{1}{3}\left(\hat{\mathbf{x}}\hat{\mathbf{x}} + \hat{\mathbf{y}}\hat{\mathbf{y}} + \hat{\mathbf{z}}\hat{\mathbf{z}}\right). \tag{3.105}$$

The component of T with respect to this basis element is $T_{ii}/3$, i.e., one-third of the trace of T. But the trace of a matrix is invariant under an orthogonal transformation.[2] Thus this component is invariant under rotations and so forms an irreducible subspace of rank zero. Similarly, one can construct three more basis elements that form an irreducible subspace of rank one; the remaining five elements form an irreducible subspace of rank two.

An important case of a rank-two Cartesian tensor is a *dyadic* $T = \mathbf{ab}$ formed out of the direct product of two vectors $\mathbf{a}$ and $\mathbf{b}$. The components $T_{ij} = a_i b_j$ can then be decomposed into terms for each irreducible subspace as

$$T_{ij} = \frac{\mathbf{a}\cdot\mathbf{b}}{3}\delta_{ij} + \frac{a_ib_j - a_jb_i}{2} + \left(\frac{a_ib_j + a_jb_i}{2} - \frac{\mathbf{a}\cdot\mathbf{b}}{3}\delta_{ij}\right). \tag{3.106}$$

The first term is a scalar (invariant under rotations), the second term is directly related to the vector product $\mathbf{a}\times\mathbf{b}$, which behaves as a vector under rotations, and the final term is a symmetric traceless tensor of rank two.

Note that each of the terms, denoted T_{ij}^κ, in our decomposition of the dyadic has $2\kappa + 1$ independent components, where κ is the rank of the tensor:

$$T_{ij}^0 = \frac{\mathbf{a}\cdot\mathbf{b}}{3}\delta_{ij} \tag{3.107}$$

has only one independent component,

$$T_{ij}^1 = \frac{(a_ib_j - a_jb_i)}{2} \tag{3.108}$$

has three independent components,[3] corresponding to the components of $\mathbf{a}\times\mathbf{b}$, and

$$T_{ij}^2 = \frac{(a_ib_j + a_jb_i)}{2} - \frac{\mathbf{a}\cdot\mathbf{b}}{3}\delta_{ij} \tag{3.109}$$

has five independent components, since it is both symmetric and traceless. That gives $1 + 3 + 5 = 9$ independent components, so we have recovered the original number of independent components of the dyadic.

In order to see what the decomposition of T into irreducible tensors looks like in the spherical basis, we first discuss various types of products of irreducible spherical tensors. There are several different kinds of tensor products that are encountered in practical calculations. We have already considered the scalar product of vectors expressed in the spherical basis:

$$\mathbf{U}\cdot\mathbf{V} = \sum_{i=1}^{3} U_i V_i = \sum_{q=-1}^{1} U^{1q} V_q^1 = \sum_{q=-1}^{1} (-1)^q U_q^1 V_{-q}^1. \tag{3.110}$$

---

[2] Using the invariance of the trace under cyclic permutation: $\text{Tr}\left(\mathsf{R}^t T \mathsf{R}\right) = \text{Tr}\left(T\mathsf{R}\mathsf{R}^t\right) = \text{Tr}\,T$.

[3] This can be seen by noting that this formula is antisymmetric in i and j. Thus the diagonal elements of T_{ij}^1 are zero, and the matrix elements below the diagonal can be found from those above the diagonal. Thus, specifying the three matrix elements above the diagonal determines the entire matrix.

In general, the scalar product of any two tensors A^κ and B^κ of the same rank is given by

$$A^\kappa \cdot B^\kappa = \sum_{q=-\kappa}^{\kappa} A^{\kappa q} B_q^\kappa = \sum_{q=-\kappa}^{\kappa} (-1)^q A_{-q}^\kappa B_q^\kappa. \tag{3.111}$$

A generalization of the scalar product is the *tensor product*, which combines two tensors A^{κ_1} and B^{κ_2} into a third tensor of a specific rank κ and index q:

$$(A^{\kappa_1} B^{\kappa_2})_q^\kappa = \sum_{q_1, q_2} \langle \kappa_1 q_1 \kappa_2 q_2 | \kappa q \rangle A_{q_1}^{\kappa_1} B_{q_2}^{\kappa_2}. \tag{3.112}$$

From the Clebsch–Gordan-coefficient triangle rule we can see that the possible values of κ are given by $|\kappa_1 - \kappa_2| < \kappa < \kappa_1 + \kappa_2$. Thus only two tensors of the same rank can combine to form a rank-zero tensor. (It is important to note, however, that the tensor product of rank zero is actually different from the scalar product by a numerical factor dependent on the rank of the tensors being combined.)

As for the Cartesian case, the *direct product* of two tensors, $A_{q_1}^{\kappa_1} B_{q_2}^{\kappa_2}$, is a (generally reducible) tensor of rank $\kappa_1 + \kappa_2$. Using the tensor product, it can be expanded into a sum of irreducible tensors:

$$A_{q_1}^{\kappa_1} B_{q_2}^{\kappa_2} = \sum_{\kappa=|\kappa_1-\kappa_2|}^{\kappa_1+\kappa_2} \sum_{q=-\kappa}^{\kappa} \langle \kappa_1 q_1 \kappa_2 q_2 | \kappa q \rangle (A^{\kappa_1} B^{\kappa_2})_q^\kappa. \tag{3.113}$$

Thus, we can find the irreducible spherical tensor components of the tensor T by performing this expansion on the direct product of the vectors **a**, **b** in the spherical basis, $a_{q_1} b_{q_2}$.

Using Eq. (3.112), we find the relationship between the Cartesian tensor components T_{ij}^1, T_{ij}^2 and the irreducible tensor components T_q^1, T_q^2:

$$T_0^1 = i\sqrt{2} T_{xy}^1, \tag{3.114a}$$

$$T_{\pm 1}^1 = -T_{xz}^1 \mp i T_{yz}^1, \tag{3.114b}$$

$$T_0^2 = \sqrt{\frac{3}{2}} T_{zz}^2, \tag{3.114c}$$

$$T_{\pm 1}^2 = \mp T_{xz}^2 - i T_{yz}^2, \tag{3.114d}$$

$$T_{\pm 2}^2 = \frac{1}{2} \left(T_{xx}^2 - T_{yy}^2 \pm 2i T_{xy}^2 \right). \tag{3.114e}$$

Note that since these quantities are strictly defined as coefficients of irreducible sets of basis tensors, they depend on a choice of normalization of the basis elements. As a consequence, different versions of these formulas may appear (see, for example, Varshalovich *et al.* 1988).

The decomposition of higher-rank tensors becomes quite complicated: for instance, from the 27 independent Cartesian components of a third-rank tensor, one can construct seven irreducible tensors (one zero rank, three first rank, two second rank, and one third rank)! Not to mention that the decomposition is not unique.

3.8 The Wigner–Eckart theorem

The formalism of irreducible spherical tensors can be applied directly to quantum-mechanical operators, in particular to any set of operator components that satisfies the general definition of

52   A bit of angular-momentum theory

an irreducible spherical tensor. For example, the spherical-basis components J_q of the vector operator **J** form a rank-one irreducible spherical tensor.

Because of the close connection between the angular momentum operator and quantum-mechanical rotations, it can be shown that an irreducible spherical tensor operator also satisfies the commutation relations

$$[J_z, T_q^\kappa] = \hbar q T_q^\kappa, \tag{3.115a}$$

$$[J_\pm, T_q^\kappa] = \hbar\sqrt{\kappa(\kappa+1) - q(q\pm 1)}\, T_{q\pm 1}^\kappa, \tag{3.115b}$$

where $J_\pm$ are the raising and lowering operators.[4] This can be regarded as an alternate definition of an irreducible tensor operator.

Note that Eqs. (3.115a) and (3.115b) for irreducible tensor operators are analogous to the last of Eqs. (3.58) and Eq. (3.57), where instead of operating with J_z or $J_\pm$ on a basis state, we form the commutator with a tensor operator. The index q varies from $-\kappa$ to $+\kappa$, and the T_q^κ's are the $2\kappa + 1$ components of the rank-κ irreducible tensor operator. From Eqs. (3.115a) and (3.115b), one can also derive

$$[J_i, [J_i, T_q^\kappa]] = \hbar^2 \kappa(\kappa+1)\, T_q^\kappa, \tag{3.116}$$

where summation over the repeated index i is implied. From the above relations we see that κ is analogous to j, and q is analogous to m.

A ubiquitous feature of atomic physics problems is the necessity to calculate matrix elements of operators between various atomic states. An essential tool for performing such calculations is the *Wigner–Eckart theorem*, which states that the matrix elements of the covariant components of an irreducible tensor operator T_q^κ between states of a standard (Zeeman) angular-momentum basis are given by the product of a constant independent of magnetic quantum numbers (m, m', q) and an appropriate Clebsch–Gordan coefficient:

$$\langle \xi' J' m' | T_q^\kappa | \xi J m \rangle = \frac{\langle \xi' J' \| T^\kappa \| \xi J \rangle}{\sqrt{2J'+1}} \langle J m \kappa q | J' m' \rangle, \tag{3.117}$$

where the quantity

$$\langle \xi' J' \| T^\kappa \| \xi J \rangle \tag{3.118}$$

is the so-called *reduced matrix element*, and we employ the standard general angular momentum basis $|\xi J m\rangle$, with

$$\mathbf{J}^2 |\xi J m\rangle = \hbar^2 J(J+1) |\xi J m\rangle, \tag{3.119a}$$

$$J_z |\xi J m\rangle = \hbar m |\xi J m\rangle, \tag{3.119b}$$

where ξ accounts for all other quantum numbers. It can be shown that for a Hermitian operator T_q^κ,

$$\langle \xi J \| T^\kappa \| \xi' J' \rangle = (-1)^{J-J'} \langle \xi' J' \| T^\kappa \| \xi J \rangle^*. \tag{3.120}$$

This will be useful for relating amplitudes of direct and inverse processes.

---

[4] Note that the spherical components J_1 and J_{-1} differ from the raising and lowering operators J_+ and J_- by numerical factors.

The meaning of the Wigner–Eckart theorem can be understood by realizing that the definition of an irreducible tensor T_q^κ says that it has properties similar to those of an eigenstate of angular momentum $|J = \kappa, m = q\rangle$. In particular, they rotate in the same manner. From this perspective, taking the matrix element $\langle \xi' J' m' | T_q^\kappa | \xi J m \rangle$ looks a lot like coupling two angular momentum states $|\kappa q\rangle$ and $|J m\rangle$ into the final state $|J' m'\rangle$. Thus it is natural to see that the result is proportional to the corresponding Clebsch–Gordan coefficient $\langle J m \kappa q | J' m' \rangle$.

An important example of the application of the Wigner–Eckart theorem is when we calculate the expectation value of the components of a vector operator T_q^1 for a given state $|J m\rangle$. In this case, the initial and final states coincide, and using the Wigner–Eckart theorem and the fact that $\mathbf{J}$ is a vector operator, we can write $\langle T_q^1 \rangle = \langle J m | T_q^1 | J m \rangle \propto \langle J m | J_q | J m \rangle = \langle J_q \rangle$. In other words, the expectation value of any vector observable lies along the expectation value of the total angular momentum. For instance, for a state with a given J, the expectation values of $\mathbf{L}$ and $\mathbf{S}$ (assuming LS-coupling), as well as the magnetic moments μ_L, μ_S, and the total magnetic moment μ, all lie along the expectation value of $\mathbf{J}$. This is consistent with our discussion of the vector model in Chapter 2.

Expressing the Clebsch–Gordan coefficient in Eq. (3.117) with a $3j$ symbol, one arrives at the following form of the Wigner–Eckart theorem:

$$\langle \xi' J' m' | T_q^\kappa | \xi J m \rangle = (-1)^{J'-m'} \langle \xi' J' \| T^\kappa \| \xi J \rangle \begin{pmatrix} J' & \kappa & J \\ -m' & q & m \end{pmatrix}. \quad (3.121)$$

We give the Wigner–Eckart theorem without proof here; proofs can be found in most advanced quantum-mechanics texts, for example, that by Sakurai and Tuan (1994).

> The definition of reduced matrix elements that we employ throughout this book is the definition assumed, for example, by Edmonds (1974); Fano and Racah (1959); Sobelman (1992); Varshalovich et al. (1988); Budker et al. (2008); Zare (1988); and Auzinsh and Ferber (1995). The reader should be forewarned that there is another convention in the literature, e.g., Brink and Satchler (1993), for the reduced matrix element, although it is less commonly used, in which $\langle \xi' J' \| T^\kappa \| \xi J \rangle$ absorbs the factor $\sqrt{2J'+1}$, so the Wigner–Eckart theorem reads
>
> $$\langle \xi' J' m' | T_q^\kappa | \xi J m \rangle = \langle \xi' J' \| T^\kappa \| \xi J \rangle \langle J m \kappa q | J' m' \rangle.$$
>
> We consistently use the form of Eqs. (3.117) and (3.121) throughout this book.

The significance of the Wigner–Eckart theorem lies in its explicit separation of the matrix element into two factors: the reduced matrix element $\langle \xi' J' \| T^\kappa \| \xi J \rangle$, which is a property of the particular physical observable being considered, and the angular coefficient (Clebsch–Gordan coefficient or a $3j$ symbol), which depends only on the geometry of the problem, i.e., the orientation of the physical observables with respect to the quantization axis. What makes the theorem so useful is that all the dependence of the matrix element on the magnetic quantum numbers is contained in the angular coefficient. This allows one to find ratios between different matrix elements, and to determine matrix elements for all values of q, m, and m' once this has been done for one particular case.

In some problems, the need arises to relate the reduced matrix element for some tensor operator T^κ found in an uncoupled angular momentum basis, for example $|J m_J\rangle |I m_I\rangle$, where J represents the total electronic angular momentum and I represents the nuclear spin, with the reduced matrix element in a coupled basis, for example $|F m_F\rangle$, where F is the total angular momentum. If T^κ commutes with I, then the formula relating the reduced matrix elements (derived, for example, in the books by Sobelman 1992; Judd 1998) is found to be

$$\langle J'IF'\|T^\kappa\|JIF\rangle = (-1)^{J'+I+F+\kappa}\sqrt{(2F+1)(2F'+1)}\begin{Bmatrix} J' & F' & I \\ F & J & \kappa \end{Bmatrix}\langle J'\|T^\kappa\|J\rangle. \quad (3.122)$$

Further chapters will provide a plethora of examples of the utility of irreducible tensors and the Wigner–Eckart theorem.

For more detailed discussions of tensors and tensor operators, see texts such as those by Fano and Racah (1959), Zare (1988), and Sakurai and Tuan (1994).

4
Atoms in external electric and magnetic fields

In Chapter 2, we discussed the general properties of the states of free atoms and the quantum numbers used to characterize these states. In this chapter, we look at how the states of atoms change in the presence of external fields: homogeneous, static magnetic and electric fields, as well as AC electromagnetic fields, including some aspects of the influence of a light field. We will return to the discussion of the latter subject in Chapter 7.

Following a quantum-mechanical approach, we introduce various perturbations of the atomic system by writing out the corresponding contributions to the Hamiltonian. The energies of the atomic states and their wave functions can then be found by solving the Schrödinger equation. In general, there are infinitely many such states. Frequently, however, it may be possible to restrict the consideration to a small subset of atomic states, and thus work with a Hamiltonian matrix of low dimension. For example, when considering the effect of a weak static magnetic field on an isolated atomic state of total angular momentum F, it is sufficient to restrict the consideration to a subset or even one of the Zeeman sublevels.

Another significant simplification can be brought about by a judicious choice of the basis. In particular, the selection of either coupled or uncoupled basis states and the direction of the quantization axis can be made to partially or totally diagonalize the Hamiltonian due to an external field.

4.1 Linear Zeeman effect

Historically, the study of the influence of external magnetic and electric fields on the properties of atoms played an important role in the development of our understanding of the structure of atoms. In 1845, Michael Faraday discovered the rotation of the polarization of light propagating in glass subject to a magnetic field—the *Faraday effect*. Later on, Faraday tried but did not succeed in seeing the influence of the magnetic field on the spectrum of radiating atoms. This influence was first seen in 1896 by Pieter Zeeman who studied the spectrum of sodium and observed broadening of spectral lines in a magnetic field. The reason for the broadening was what became known as the *anomalous Zeeman effect*—splitting of a spectral line into more than three components upon application of a magnetic field (see below). In 1897, Zeeman repeated his experiment with cadmium and saw magnetic-field-induced splitting of a spectral line into three components—the *normal Zeeman effect*. The normal Zeeman effect was successfully explained by Hendrik Lorentz. The explanation published in 1897 was based on the classical-physics picture of a radiating atom being an oscillator. It is described in many atomic physics textbooks, for example, the book by White (1934). (See also the discussion in Sec. 7.8.)

A modern-day description of the Zeeman effect can proceed as follows. If an atom is placed in an external magnetic field, it acquires an energy

$$\Delta E = -\boldsymbol{\mu} \cdot \mathbf{B}. \tag{4.1}$$

For now, let us assume that the atom is in a state with zero total spin ($S = 0$). For such a state the total angular momentum **J** coincides with the orbital angular momentum **L**. The relation between the orbital angular momentum **L** and the associated orbital magnetic moment $\boldsymbol{\mu}_L$ is represented by the gyromagnetic ratio $\gamma = e/(2m_e c) = \mu_B/\hbar$ as discussed in Sec. 2.2.3:

$$\boldsymbol{\mu}_L = -\gamma \mathbf{L} = -\frac{\mu_B}{\hbar}\mathbf{L}. \tag{4.2}$$

Here we use a subscript to indicate that this magnetic moment is associated with the orbital angular momentum. As we will see, there is a magnetic moment associated with each angular momentum of an atom. For a magnetic field along z, and using $L_z = m_L \hbar$, we obtain

$$(\mu_L)_z = -\mu_B m_L, \tag{4.3}$$

and an energy shift

$$\Delta E_{m_L} = -(\mu_L)_z B = \mu_B B m_L. \tag{4.4}$$

The magnetic splitting of the Zeeman sublevels can be observed through a corresponding splitting of spectral lines. If both the initial and the final state of the transition have $S = 0$, they are both split according to Eq. (4.4), generally, into a different number of magnetic sublevels. The shift of the transition energy is given by

$$\Delta E^f - \Delta E^i = \mu_B B (m_L^f - m_L^i). \tag{4.5}$$

As we see in Chapter 7, there is a selection rule for the magnetic quantum number determining which sublevels can be coupled by an electric-dipole transition: $\Delta m_L = m_L^f - m_L^i = 0, \pm 1$. From this we see that there are three possible values for $\Delta E^f - \Delta E^i$, so that the spectral line is generally split into three components. Historically, this is known as the *normal Zeeman effect*. Such splitting coincides with the predictions of the Lorentz theory based on classical physics, leading to the description of the effect as "normal."

If the atom possesses nonzero spin, the total angular momentum **J** of the atom in the *LS*-coupling scheme is composed of the orbital angular momentum **L** and the spin angular momentum **S**, each with an associated magnetic moment, $\boldsymbol{\mu}_L$ and $\boldsymbol{\mu}_S$, respectively. The Landé factors for the orbital and spin angular momentum of the electron were discussed for single-electron atoms in Sec. 2.2.3. To find the magnetic moments $\boldsymbol{\mu}_L$ and $\boldsymbol{\mu}_S$ for a multi-electron atom in the *LS*-coupling scheme, we add the magnetic moments of the individual electrons vectorially. Since the single-electron g-factors are all fixed ($g_l = 1$, $g_s \approx 2$), it follows that $g_L = g_l$ and $g_S = g_s$. The magnetic moments $\boldsymbol{\mu}_L$ and $\boldsymbol{\mu}_S$ add to form the total magnetic moment $\boldsymbol{\mu}$ (Fig. 4.1):

$$\boldsymbol{\mu} = \boldsymbol{\mu}_L + \boldsymbol{\mu}_S = -\mu_B(g_L \mathbf{L} + g_S \mathbf{S})/\hbar. \tag{4.6}$$

Since $g_L \neq g_S$, the total magnetic moment of an atom in the vector model is not collinear (proportional) to its total angular momentum $\mathbf{J} = \mathbf{L} + \mathbf{S}$ (see Fig. 4.1). Nevertheless, due to the fine-structure interaction, **L** and **S** "precess" around **J** (see the discussion of the meaning of

Fig. 4.1 A vector-model picture of the formation of the magnetic moment of a fine-structure state of an atom. The vectors **S** and **L** have associated magnetic moments μ_S, μ_L respectively. Because the charge of the electron is negative, μ_S and μ_L are directed oppositely to **S** and **L**. Because g_L and g_S are not equal the resultant magnetic moment $\mu = \mu_L + \mu_S$ associated with **J** is not collinear with **J**. However, μ, as with all other vectors in the system, "precesses" about **J** so that the perpendicular component $\mu_\perp$ averages to zero, leaving only the parallel component μ_J.

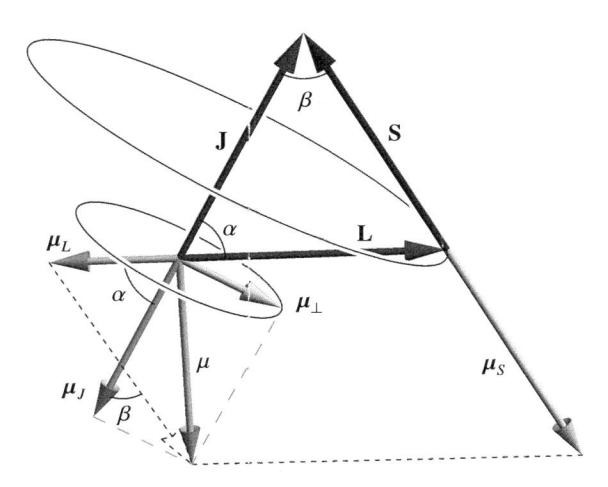

"precession" in Sec. 2.2.4). As a result, the component of the magnetic moment $\mu_\perp$ averages to zero, and only the part of μ_J that is collinear with **J** is left. The requirement that the average total magnetic moment of an atom must be collinear with the average total angular momentum also follows from symmetry (the Wigner–Eckart theorem; see the detailed discussion of this point in Appendix F.1 in the book by Budker et al., 2008).

From geometrical considerations, the total magnetic moment parallel to the angular momentum of an atom can be calculated as (see Fig. 4.1)

$$\mu_J = \mu_L \cos\alpha + \mu_S \cos\beta$$
$$= -\mu_B \sqrt{L(L+1)} \cos\alpha - 2\mu_B \sqrt{S(S+1)} \cos\beta. \quad (4.7)$$

The quantities $\cos\alpha$ and $\cos\beta$ can be found from the vector relation

$$\mathbf{J} = \mathbf{L} + \mathbf{S} \quad (4.8)$$

and the law of cosines, yielding

$$\cos\alpha = \frac{\mathbf{J}^2 + \mathbf{L}^2 - \mathbf{S}^2}{2|\mathbf{J}||\mathbf{L}|}, \quad (4.9a)$$

$$\cos\beta = \frac{\mathbf{J}^2 + \mathbf{S}^2 - \mathbf{L}^2}{2|\mathbf{J}||\mathbf{S}|}. \quad (4.9b)$$

This allows us to derive the Landé factor for the fine-structure state of an atom with arbitrary values of the quantum numbers J, L and S:

$$g_J = -\frac{\mu_J}{\mu_B} \frac{\hbar}{|\mathbf{J}|} = 1 + \frac{J(J+1) + S(S+1) - L(L+1)}{2J(J+1)}. \quad (4.10)$$

Finally, we can calculate the magnetic energy shift,

$$\Delta E_{m_J} = -\mu_{J_z} B = g_J \mu_B B m_J, \quad (4.11)$$

for a magnetic sublevel m_J of a fine-structure state characterized by quantum numbers J, L, and S. This equation can be regarded as a generalization of Eq. (4.4). From Eqs. (4.10) and

(4.11) we see that different fine-structure states will generally have different Landé factors and, at a given magnetic field strength B, different magnetic-sublevel splittings. Consequently, within the transitions determined by the $\Delta m_J = 0, \pm 1$ selection rule (Chapter 7), there will be multiple spectral components that depend on the particular fine-structure states involved in the transition. Because the Zeeman shifts of these components are in general different, the number of Zeeman-split components of a spectral line in this case is greater than three. For this reason, the Zeeman effect for states with $S \neq 0$ is historically called the *anomalous Zeeman effect*, to differentiate it from the normal Zeeman effect, in which three components are observed. However, the usage of the terminology "normal" and "anomalous" is diminishing. This is welcome, as, from the contemporary perspective, there is nothing "anomalous" about quantum mechanics, or the fact that electrons have spin.

4.2 Zeeman effect in the manifold of hyperfine levels, Breit–Rabi diagrams

Now that we have discussed the general case of the Zeeman effect for an LS-coupled J-state, let us consider the Zeeman effect for hyperfine F-states. We will calculate the Landé factors for these states, including the effect of the nuclear spin I.

If the nuclear spin $\mathbf{I}$ is included in the total angular momentum $\mathbf{F} = \mathbf{I} + \mathbf{J}$, the average magnetic moment now points along $\mathbf{F}$ rather than $\mathbf{J}$. The g-factor for a given state is therefore multiplied by a factor corresponding to the projection of $\mathbf{J}$ along $\mathbf{F}$. (There is also an effect due to the direct interaction between the nuclear magnetic moment and the magnetic field, but we neglect it as the nuclear magnetic moment is considerably smaller than the Bohr magneton.) The calculation is similar to that in the previous section and leads to

$$\Delta E_{m_F} = -(\mu_F)_z B = g_F \mu_B B m_F, \qquad (4.12)$$

where

$$g_F = g_J \left[\frac{F(F+1) + J(J+1) - I(I+1)}{2F(F+1)} \right]. \qquad (4.13)$$

Here is the derivation of Eq. (4.13). We are interested in finding the projection of $\mathbf{J}$ on $\mathbf{F}$, so we begin by finding the expectation value of the scalar product $\mathbf{J} \cdot \mathbf{F}$. We can conveniently find this product from

$$\mathbf{F} - \mathbf{J} = \mathbf{I}. \qquad (4.14)$$

Squaring both sides of Eq. (4.14) and taking the expectation values, we find

$$\langle \mathbf{J} \cdot \mathbf{F} \rangle = \frac{F(F+1) + J(J+1) - I(I+1)}{2} \hbar^2. \qquad (4.15)$$

The expectation value of $\mathbf{J}$ is directed along that of $\mathbf{F}$ and has magnitude proportional to $\langle \mathbf{J} \cdot \mathbf{F} \rangle$. This can be written as

$$\langle \mathbf{J} \rangle = \frac{\langle \mathbf{J} \cdot \mathbf{F} \rangle}{\langle \mathbf{F}^2 \rangle} \langle \mathbf{F} \rangle = \frac{F(F+1) + J(J+1) - I(I+1)}{2F(F+1)} \langle \mathbf{F} \rangle. \qquad (4.16)$$

Finally, since for an eigenstate of F_z, $\langle \mathbf{F} \rangle = m_F \hbar \hat{\mathbf{z}}$, and the expectation value of the magnetic moment is

$$\langle \boldsymbol{\mu} \rangle = -g_F \mu_B m_F \hat{\mathbf{z}} = -g_J \mu_B \langle \mathbf{J} \rangle / \hbar, \qquad (4.17)$$

we arrive at the result of Eq. (4.13).

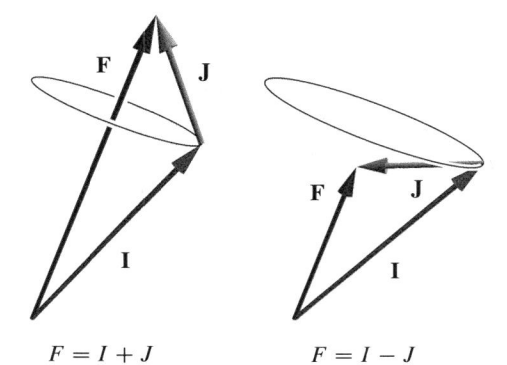

Fig. 4.2 Illustration of the orientation of **J** relative to the total angular momentum **F** for $F = I \pm J$. Angular momentum magnitudes correspond to the case of the ground state of ^{87}Rb, with $J = 1/2$ and $I = 3/2$.

Let us now look more closely at $^2S_{1/2}$ atomic states. Important examples of such states include the ground electronic states of hydrogen, alkali atoms, and elements of the I B group of the periodic table (copper, silver, gold). For $^2S_{1/2}$ states $S = 1/2$, $L = 0$, $J = 1/2$, and $g_J = 2$. The possible values of the total angular momentum in this case are $F = I \pm 1/2$. Thus, from Eqs. (4.10) and (4.13), the Landé factors for two hyperfine components of the $^2S_{1/2}$ atomic states are given by:

$$g_F = \pm \frac{2}{2I + 1}. \tag{4.18}$$

The relative signs of the Landé factors and the magnitudes of the Larmor frequencies can be understood as follows. Since we neglect the magnetic moment of the nucleus, the magnetic moment of an atomic hyperfine state is entirely due to the magnetic moment of the electron (μ_e). The Landé factor is a result of the relationship between the electronic angular momentum **J** (responsible for μ_e) and the total angular momentum **F**. As seen from Fig. 4.2, when $F = I + 1/2$, the electronic angular momentum points in the direction of the total angular momentum, so the g-factor is positive. However, when $F = I - 1/2$, the electronic angular momentum is directed oppositely to the total angular momentum and the g-factor is negative. The magnitudes of the Landé factors decrease as the nuclear spin gets larger. This is because as the nuclear spin increases, the total angular momentum increases faster than the magnetic moment. The latter increases only slightly due to a change in the angle between **J** and **F** (Fig. 4.2).

If the effect of the magnetic moment of the nucleus is not neglected, the formula for g_F becomes

$$g_F = g_J \frac{F(F+1) + J(J+1) - I(I+1)}{2F(F+1)} - g_I \frac{\mu_N}{\mu_B} \frac{F(F+1) + I(I+1) - J(J+1)}{2F(F+1)}. \tag{4.19}$$

Because $\mathbf{F} = \mathbf{I} + \mathbf{J}$, the additional term describing the effect of the nuclear magnetic moment has the same dependence on angular momentum as the electronic magnetic moment term with $I \leftrightarrow J$. The sign and the additional ratio of magnetic moments follow from the definitions of the g-factors, including the fact that the Bohr magneton is defined in terms of a positive charge e, the magnitude of the electron charge.

So far we have discussed the Zeeman effect in relatively small magnetic fields where Zeeman splittings are much smaller than any other energy intervals. In this regime, energies

60 *Atoms in external electric and magnetic fields*

of the Zeeman sublevels are linear in the applied field. When the magnetic field becomes sufficiently strong, so that the Zeeman shifts become comparable to other energy intervals in the atom, the character of the Zeeman shifts changes. In the case where the relevant intervals are those due to fine structure, the magnetic-field level splitting is called the *Paschen–Back* effect. In the strong-field regime, some of the eigenstates of the unperturbed Hamiltonian are strongly mixed, and a new set of quantum numbers characterizes the eigenstates of the system. We will not discuss the Paschen–Back effect further; however, we will analyze in some detail a similar case of the *magnetic decoupling* of hyperfine structure, which is important in many applications.

We now consider the effect of a uniform magnetic field $\mathbf{B} = B\hat{z}$ on the hyperfine levels of the $^2S_{1/2}$ ground state of hydrogen. Initially, we will neglect the effect of the nuclear (proton) magnetic moment. The energy eigenstates for the Hamiltonian describing the hyperfine interaction are also eigenstates of the operators $\{F^2, F_z, I^2, S^2\}$. Therefore if we write out a matrix for the hyperfine Hamiltonian H_{hf} in the coupled basis $|Fm_F\rangle$, it is diagonal. However, the Hamiltonian H_B for the interaction of the magnetic moment of the electron with the external magnetic field,

$$H_B = -\boldsymbol{\mu}_e \cdot \mathbf{B} = 2\mu_B B S_z/\hbar, \tag{4.20}$$

is diagonal in the uncoupled basis $|(SI)m_S, m_I\rangle$, made up of eigenstates of the operators $\{I^2, I_z, S^2, S_z\}$. We can write the matrix elements of the Hamiltonian in the coupled basis by relating the uncoupled to the coupled basis. (We could also carry out the analysis in the uncoupled basis, if we so chose.)

The relationship between the coupled $|Fm_F\rangle$ and uncoupled $|(SI)m_Sm_I\rangle$ bases (see the discussion of the Clebsch–Gordan expansions in Chapter 3) is

$$|1,1\rangle = |(\tfrac{1}{2}\tfrac{1}{2})\tfrac{1}{2}\tfrac{1}{2}\rangle, \tag{4.21a}$$

$$|1,0\rangle = \frac{1}{\sqrt{2}}\left(|(\tfrac{1}{2}\tfrac{1}{2})\tfrac{1}{2},-\tfrac{1}{2}\rangle + |(\tfrac{1}{2}\tfrac{1}{2}),-\tfrac{1}{2}\tfrac{1}{2}\rangle\right), \tag{4.21b}$$

$$|1,-1\rangle = |(\tfrac{1}{2}\tfrac{1}{2}),-\tfrac{1}{2},-\tfrac{1}{2}\rangle, \tag{4.21c}$$

$$|0,0\rangle = \frac{1}{\sqrt{2}}\left(|(\tfrac{1}{2}\tfrac{1}{2})\tfrac{1}{2},-\tfrac{1}{2}\rangle - |(\tfrac{1}{2}\tfrac{1}{2}),-\tfrac{1}{2}\tfrac{1}{2}\rangle\right). \tag{4.21d}$$

Employing the hyperfine energy shift formula (2.28) and Eq. (4.20), one finds for the matrix of the overall Hamiltonian $H = H_{\text{hf}} + H_B$ in the coupled basis

$$H = \begin{pmatrix} \tfrac{A}{4}+\mu_B B & 0 & 0 & 0 \\ 0 & \tfrac{A}{4}-\mu_B B & 0 & 0 \\ 0 & 0 & \tfrac{A}{4} & \mu_B B \\ 0 & 0 & \mu_B B & -\tfrac{3A}{4} \end{pmatrix}, \tag{4.22}$$

where we order the states $(|1,1\rangle, |1,-1\rangle, |1,0\rangle, |0,0\rangle)$.

We can use this matrix to solve for the energies of the states as a function of B by employing the Schrödinger equation

$$H|\psi\rangle = E|\psi\rangle \tag{4.23}$$

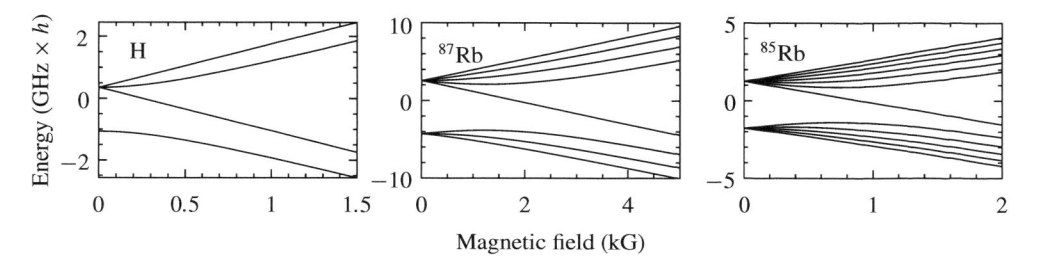

Fig. 4.3 Energies of the ground-state $^2S_{1/2}$ hyperfine manifold of hydrogen ($I = 1/2$), ^{87}Rb ($I = 3/2$), and ^{85}Rb ($I = 5/2$) as a function of applied magnetic field. Such a plot is known as the *Breit–Rabi diagram*. At low fields, the system is well described in the coupled basis, while at high fields the energy eigenstates are best approximated by the uncoupled basis.

which implies that

$$(H - E)|\psi\rangle = 0. \tag{4.24}$$

In order to have a nontrivial solution, the determinant of the matrix should be zero:

$$\begin{vmatrix} \frac{A}{4} + \mu_B B - E & 0 & 0 & 0 \\ 0 & \frac{A}{4} - \mu_B B - E & 0 & 0 \\ 0 & 0 & \frac{A}{4} - E & \mu_B B \\ 0 & 0 & \mu_B B & -\frac{3A}{4} - E \end{vmatrix} = 0. \tag{4.25}$$

The above expression is known as the *secular equation*.[1] The matrix is block diagonal, which makes the solution easy, yielding the following energies

$$E_1 = \frac{A}{4} + \mu_B B, \tag{4.26a}$$

$$E_2 = \frac{A}{4} - \mu_B B, \tag{4.26b}$$

$$E_3 = -\frac{A}{4} + \frac{A}{2}\sqrt{1 + 4\frac{\mu_B^2 B^2}{A^2}}, \tag{4.26c}$$

$$E_4 = -\frac{A}{4} - \frac{A}{2}\sqrt{1 + 4\frac{\mu_B^2 B^2}{A^2}}, \tag{4.26d}$$

which are plotted as a function of B in Fig. 4.3. Also shown in this figure are Breit–Rabi diagrams for the ground states of the two isotopes of rubidium with nuclear spins $I = 3/2$ and $5/2$, which have more sublevels, but show qualitatively the same behavior as hydrogen.

Returning to hydrogen, from Eq. (4.24), in addition to the eigenvalues given by Eqs. (4.26a)–(4.26d), we can also find the eigenfunctions of the full Hamiltonian, which yields

---

[1] "Secular equation" is a synonym for "characteristic equation;" the term has obscure historical origins in astronomy.

$$|\psi_1\rangle = |11\rangle, \tag{4.27a}$$

$$|\psi_2\rangle = |1, -1\rangle, \tag{4.27b}$$

$$|\psi_3\rangle = \frac{\left(A + \sqrt{4\mu_B^2 B^2 + A^2}\right)|10\rangle + 2\mu_B B|00\rangle}{\sqrt{4\mu_B^2 B^2 + \left(A + \sqrt{4\mu_B^2 B^2 + A^2}\right)^2}}, \tag{4.27c}$$

$$|\psi_4\rangle = \frac{\left(A - \sqrt{4\mu_B^2 B^2 + A^2}\right)|10\rangle + 2\mu_B B|00\rangle}{\sqrt{4\mu_B^2 B^2 + \left(A - \sqrt{4\mu_B^2 B^2 + A^2}\right)^2}}. \tag{4.27d}$$

Note that the "stretched" components with maximal angular-momentum projection do not mix with other states and have energies that depend linearly on the magnetic field, while the other components are mixed by the magnetic field, causing nonlinear shifts. The high-field eigenstates are, in fact, equal-amplitude superpositions of the two coupled-basis states.

If we include the effect of the proton's magnetic moment, we have

$$\boldsymbol{\mu} = \boldsymbol{\mu}_e + \boldsymbol{\mu}_p, \tag{4.28}$$

so

$$H_B = -\boldsymbol{\mu} \cdot \mathbf{B} = g_e \mu_B B S_z/\hbar - g_p \mu_N B I_z/\hbar. \tag{4.29}$$

The minus sign in the second term on the right-hand side of Eq. (4.29) is related to the sign difference in the definition of the electron and nuclear magnetic moments (compare Eqs. (2.27) and (2.20)). In the high-field limit we expect that the highest energy state should be the $|(\frac{1}{2}\frac{1}{2})\frac{1}{2}, -\frac{1}{2}\rangle$ state. In the low-field limit, the $|11\rangle = |(\frac{1}{2}\frac{1}{2})\frac{1}{2}\frac{1}{2}\rangle$ state is the highest energy state, so these two levels must cross at some magnetic field.

From Eqs. (4.26a) and (4.26c), where we neglected the proton magnetic moment, for sufficiently high fields ($2\mu_B B/A \gg 1$), the difference in energy between the two highest lying energy levels is

$$E_1 - E_3 \approx \frac{A}{2}. \tag{4.30}$$

The levels will cross when the difference in energy between $|(\frac{1}{2}\frac{1}{2})\frac{1}{2}\frac{1}{2}\rangle$ and $|(\frac{1}{2}\frac{1}{2})\frac{1}{2}, -\frac{1}{2}\rangle$ due to the interaction of the proton's magnetic moment with the magnetic field is equal to this energy difference. This occurs for the magnetic field

$$B \approx \frac{A}{2 \times 5.58\mu_N} \approx 167 \text{ kG}. \tag{4.31}$$

Such a field can be obtained in the laboratory, for example, using a superconducting magnet.

4.3 Atoms in an electric field: the Stark effect

We now turn to the discussion of the influence of an external electric field on the atomic spectra (the *Stark effect*[2]). Similarly to the case of the magnetic field, in which the Hamiltonian is

---

[2] The discoverer of the electric-field effect on spectral lines, Johannes Stark (1874–1957), is infamous for his active membership in the Nazi party. Apparently, almost simultaneously with Stark, in 1913, the effect of the electric field on spectral lines was also independently discovered by an Italian physicist Antonino Lo Surdo (1880–1949). Later, Lo Surdo also became a staunch supporter of Mussolini and the racial programs of the fascist government. The story of Stark, Lo Surdo and their discoveries is related by Leone *et al.* (2004).

Atoms in an electric field: the Stark effect

given by $H_B = -\boldsymbol{\mu} \cdot \mathbf{B}$, here the Hamiltonian is $H_{\mathcal{E}} = -\mathbf{d} \cdot \boldsymbol{\mathcal{E}}$, where $\mathbf{d}$ is the electric-dipole operator. Despite the similarities in the Hamiltonians, there are substantial differences between the two cases. For example, any atomic state with nonzero total angular momentum can have a magnetic moment, but a state must be degenerate with another state of opposite parity to have an electric-dipole moment. This can be seen from the selection rule that a nonzero electric-dipole matrix element must be between states of opposite parity (see Chapter 7), and, therefore, the diagonal matrix elements must be zero in the absence of parity violation (Sec. 2.5). If that were not enough, such a *permanent electric-dipole moment* (EDM) would violate *time-reversal invariance*. The consequence of the lack of a permanent EDM for nondegenerate states is that the effect of the electric field comes in at second order (the shifts are quadratic in the electric field).

In this section, we discuss both this *quadratic Stark effect* for atoms, conveniently characterized in terms of *electric polarizabilities*, as well as what happens when the energy shifts approach or exceed the unperturbed level separations, at which point the energy levels are effectively degenerate and the quadratic dependence may turn into the *linear Stark effect*.

4.3.1 Polarizability of a conducting sphere

We begin the discussion of the effect of an electric field on an atom by performing an estimate using classical physics. An appropriate model of the atom from the point of view of classical electrostatics is a conducting sphere of radius a_0, the Bohr radius of the atom (Chapter 2). If we put a conducting sphere in an electric field, the field inside the sphere will be zero. This results from the redistribution of charge on the surface of the sphere, which produces a new field that compensates the original field in the interior of the sphere. The new charge distribution of the sphere can be described by a *induced electric-dipole moment* $\mathbf{d}(\mathcal{E})$, related to the applied electric field $\mathcal{E}$ by the polarizability α (not to be confused with the fine-structure constant):

$$\mathbf{d} = \alpha \boldsymbol{\mathcal{E}}. \tag{4.32}$$

The energy of the sphere in the electric field is

$$E = -\int_0^{\mathcal{E}} \mathbf{d} \cdot d\boldsymbol{\mathcal{E}}' = -\int_0^{\mathcal{E}} \alpha \mathcal{E}' d\mathcal{E}' = -\frac{1}{2}\alpha \mathcal{E}^2. \tag{4.33}$$

The electrons on the conductor will arrange themselves so that the total electric field on the sphere is perpendicular to the surface (otherwise there would be a tangential force on the free charges, causing them to move until equilibrium is reached). To determine the external field generated by the sphere, we must find a solution of the Maxwell equations for free space that satisfies this boundary condition and goes to zero at infinity. By uniqueness theorems for electrostatics, this will be the correct solution for our situation (see, for example, the book by Jackson, 1975). Because the conductor is spherically symmetric, there is only one preferred direction specified in the problem: that of the applied electric field. Thus we can guess that a dipole field, which is specified by one vector, is the field configuration that we seek. To verify this we note that the electric field $\mathcal{E}_d$ from a dipole $\mathbf{d}$ is

$$\boldsymbol{\mathcal{E}}_d = \frac{1}{r^3}\left[3\left(\mathbf{d} \cdot \hat{\mathbf{r}}\right)\hat{\mathbf{r}} - \mathbf{d}\right]. \tag{4.34}$$

At the surface of the sphere ($r = a_0$) the component of the total field $\mathcal{E} + \mathcal{E}_d$ not along $\hat{\mathbf{r}}$ must vanish, i.e., we require

$$\hat{\mathbf{r}} \times (\mathcal{E} + \mathcal{E}_d) = \hat{\mathbf{r}} \times \left(\mathcal{E} + \frac{1}{a_0^3} [3(\mathbf{d} \cdot \hat{\mathbf{r}})\hat{\mathbf{r}} - \mathbf{d}] \right)$$
$$= \hat{\mathbf{r}} \times \mathcal{E} - \frac{1}{a_0^3} \hat{\mathbf{r}} \times \mathbf{d} \qquad (4.35)$$
$$= 0.$$

This is satisfied if

$$\mathbf{d} = a_0^3 \mathcal{E}. \qquad (4.36)$$

This means that the dipole field satisfies the boundary conditions, and that, from Eq. (4.32), the polarizability of the conducting sphere is

$$\alpha = a_0^3. \qquad (4.37)$$

Armed with this analogy, we crudely estimate that the natural scale for atomic polarizability is the cube of the Bohr radius. The quantity $a_0^3 \approx 0.148 \times 10^{-24}$ cm^3 is also the *atomic unit of polarizability*. A more practical unit of atomic polarizability that is sometimes used is $a_0^3/h \approx 248$ Hz/(kV/cm)^2.

4.3.2 Polarizability of the hydrogen ground state

We now check our educated guess by looking at the ground state of atomic hydrogen. Actually, we will assume a "simplified hydrogen" in which we neglect not only the hyperfine structure, but also the electron spin. This is an excellent approximation for our purposes. Orienting the quantization axis (z) along the electric field gives for the Hamiltonian

$$H_\mathcal{E} = -\mathbf{d} \cdot \mathcal{E} = e\mathcal{E}z = e\mathcal{E}r\cos\theta. \qquad (4.38)$$

Since $H_\mathcal{E}$ only connects states of opposite parity,[3] there is no first-order shift ($\Delta E^{(1)}$),

$$\Delta E_1^{(1)} = \langle 100|H_\mathcal{E}|100\rangle = 0, \qquad (4.39)$$

where $|nlm\rangle$ denotes the unperturbed state with principal quantum number n, orbital angular momentum quantum number l, and magnetic quantum number m. Therefore we must use second-order perturbation theory (discussed in any quantum mechanics textbook), in which the energy shift of a state $|i\rangle$ is given by

$$\Delta E_i^{(2)} = \sum_{k \neq i} \frac{|\langle k|H_\mathcal{E}|i\rangle|^2}{E_i^{(0)} - E_k^{(0)}}. \qquad (4.40)$$

Or, for our case,

$$\Delta E_1^{(2)} = \sum_{n \neq 1, l, m} \frac{|\langle nlm|H_\mathcal{E}|100\rangle|^2}{E_1^{(0)} - E_n^{(0)}}. \qquad (4.41)$$

The polarizability α can be found from $\Delta E_1^{(2)} = -\alpha \mathcal{E}^2/2$. To simplify the sum, we use the approximation that $E_1^{(0)} - E_n^{(0)} \approx E_1^{(0)} - E_2^{(0)}$, which is justified by the eigenenergies for hydrogen given by Eq. (2.2). We then have

---

[3]The reader unfamiliar with the electric-dipole selection rules is referred to Sec. 7.2, where they are discussed in detail.

$$\frac{1}{2}\alpha\mathcal{E}^2 \approx \frac{1}{E_2^{(0)} - E_1^{(0)}} \sum_{nlm} |\langle nlm|H_\mathcal{E}|100\rangle|^2, \quad (4.42)$$

where we are allowed to return $n = 1$ to the sum because that term is zero (Eq. 4.39). Now consider the sum

$$\sum_{nlm} |\langle nlm|H_\mathcal{E}|100\rangle|^2 = \sum_{nlm} \langle 100|H_\mathcal{E}|nlm\rangle\langle nlm|H_\mathcal{E}|100\rangle$$

$$= \langle 100|H_\mathcal{E} \left(\sum_{nlm} |nlm\rangle\langle nlm| \right) H_\mathcal{E}|100\rangle \quad (4.43)$$

$$= \langle 100|H_\mathcal{E}^2|100\rangle,$$

where we have made use of the completeness relation

$$\sum_{nlm} |nlm\rangle\langle nlm| = 1. \quad (4.44)$$

Based on Eqs. (4.42) and (4.43), we have

$$\frac{1}{2}\alpha\mathcal{E}^2 \approx \frac{\langle 100|H_\mathcal{E}^2|100\rangle}{E_2^{(0)} - E_1^{(0)}}. \quad (4.45)$$

The ground state wave function for hydrogen is

$$\psi_{100}(r) = \frac{1}{\sqrt{\pi}a_0^{3/2}} e^{-r/a_0}, \quad (4.46)$$

so the matrix element in (4.45) is given by the integral

$$\langle 100|H_\mathcal{E}^2|100\rangle = \frac{e^2\mathcal{E}^2}{\pi a_0^3} \int_0^\infty r^4 e^{-2r/a_0} dr \int_0^\pi \cos^2\theta \sin\theta d\theta \int_0^{2\pi} d\phi \quad (4.47)$$

$$= e^2\mathcal{E}^2 a_0^2,$$

where we use Eq. (4.38) and evaluate the integral using polar coordinates. The difference in energy between the ground and first excited states of hydrogen is

$$E_2^{(0)} - E_1^{(0)} = -\frac{e^2}{2a_0}\left(\frac{1}{4} - 1\right) = \frac{3e^2}{8a_0}. \quad (4.48)$$

Employing Eqs. (4.47) and (4.48) in (4.45), we find an estimate for the polarizability of the ground state of hydrogen:

$$\alpha \approx \frac{16}{3}a_0^3 \approx 0.79 \times 10^{-24} \text{ cm}^3. \quad (4.49)$$

Because of the approximation made in Eq. (4.42), this is in fact an upper limit on the polarizability of the hydrogen ground state. It turns out that it is close to the exact value (see, for example, Bethe and Salpeter 1977):

$$\alpha = \frac{9}{2}a_0^3 \approx 0.67 \times 10^{-24} \text{ cm}^3. \quad (4.50)$$

As we see, this calculation for hydrogen confirms our guess for the order of magnitude of the polarizability.

4.3.3 Polarizabilities of Rydberg states

Rydberg states are atomic states with a large value of the principal quantum number n. These states have very large polarizabilities, as we can show with simple scaling arguments.

We write Eq. (4.40) for a $\hat{z}$-directed electric field:

$$\Delta E_i^{(2)} = \sum_k \frac{|\langle k|d_z|i\rangle|^2 \mathcal{E}^2}{E_i^{(0)} - E_k^{(0)}}. \tag{4.51}$$

One can expect that the sum is dominated by terms corresponding to levels with close values of the principal quantum number n: $n_k \approx n_i$, since for these levels the energy denominators are smallest and dipole matrix elements are large because of the overlap between wave functions. The dipole matrix elements scale as the radius of the electron's orbits which go as n^2. The scaling of the energy denominators can be estimated from the fact that $E_n \propto n^{-2}$, so the distance between nearby states goes as $\propto n^{-3}$.

> Note that while the number of states with a given n goes as n^2 (Sec. 2.2.1), only those states allowed by the electric-dipole selection rules actually contribute to the sum (4.51). (An electric field along $\hat{z}$ can only couple levels that have the same value of m and values of l that differ by ± 1; see Chapter 7.) Thus this dependence of the "density of states" does not contribute an additional factor of n^2.

Combining the scaling factors ($|\langle k|d_z|i\rangle| \propto n^2$ and $E_i - E_k \propto n^{-3}$) in Eq. (4.51), we find that polarizabilities α of highly-excited states scale as

$$\alpha \propto n^7, \tag{4.52}$$

which is confirmed by more elaborate calculations (Bethe and Salpeter 1977).

4.3.4 Linear Stark effect

As we have seen above, the energy shifts for a fixed value of the electric field increase rapidly with the principal quantum number n, while the unperturbed separations between levels of opposite parity fall as $1/n^3$. Thus, for high enough values of n, we expect that easily achievable laboratory fields (≤ 100 kV/cm) will be able to produce Stark shifts larger than the unperturbed level separations. In this case, the nondegenerate perturbation theory that we have been using is no longer applicable, and the method of *degenerate perturbation theory*, which we have already applied to the calculation of the Breit–Rabi diagrams in Sec. 4.2, must be used.[4] This situation also occurs for molecules, even in low-lying states, because opposite parity states are separated by at most the rotational intervals, which are typically three orders of magnitude smaller than the electronic energy intervals. There are also some special cases of non-Rydberg nearly degenerate opposite-parity atomic states, for example, those in atomic dysprosium. (These have found use in tests of the Standard Model, such as a search for possible time variation of the fine-structure constant; see Cingöz *et al.* 2007, and references therein.) Another interesting example is atomic barium. There are several levels in Ba with $n < 10$ (low enough not to be considered Rydberg states) that have polarizabilities of several million atomic units (Li *et al.* 2004). This occurs because there are close-lying levels of opposite parity with large dipole matrix elements between them. In all of these examples, the large polarizabilities and

---

[4]We would like to emphasize that the method of degenerate perturbation theory is in fact applicable in all cases. Nondegenerate perturbation theory is a simplification applicable to cases in which the energy shifts due to the perturbation are smaller than the relevant unperturbed energy separations.

small unperturbed level separations lead to significant deviations from quadratic behavior of the Stark shift even for laboratory-scale electric fields.

This same behavior occurs for the $2s_{1/2}$ and $2p_{1/2}$ states of atomic hydrogen, which are separated only by the Lamb shift $\hbar\omega_{sp}$ (see Sec. 2.2.3). An electric field $\mathcal{E}$ along $\hat{z}$ applied to a hydrogen atom mixes the 2s and 2p levels, and because of the electric-dipole selection rules (Chapter 7), $\mathcal{E}$ only mixes states with $m = m'$, i.e., $|2, 0, 0\rangle$ and $|2, 1, 0\rangle$. Solving the secular equation, as in Sec. 4.2, for the Hamiltonian H describing this system,

$$H = \begin{pmatrix} 0 & -d_{sp}\mathcal{E} \\ -d_{sp}\mathcal{E} & \hbar\omega_{sp} \end{pmatrix}, \qquad (4.53)$$

we obtain for the eigenenergies

$$E_1 = \frac{\hbar\omega_{sp}}{2}\left(1 - \sqrt{1 + 4\frac{d_{sp}^2\mathcal{E}^2}{\hbar^2\omega_{sp}^2}}\right), \qquad (4.54a)$$

$$E_2 = \frac{\hbar\omega_{sp}}{2}\left(1 + \sqrt{1 + 4\frac{d_{sp}^2\mathcal{E}^2}{\hbar^2\omega_{sp}^2}}\right). \qquad (4.54b)$$

Here we have neglected relaxation of the levels. For small electric fields, $|d_{sp}\mathcal{E}| \ll \hbar\omega_{sp}$, we recover the usual quadratic dependence of energies on the electric field:

$$E_1 \approx -\frac{d_{sp}^2\mathcal{E}^2}{\hbar\omega_{sp}}, \qquad (4.55a)$$

$$E_2 \approx \hbar\omega_{sp} + \frac{d_{sp}^2\mathcal{E}^2}{\hbar\omega_{sp}}, \qquad (4.55b)$$

while for large fields, $|d_{sp}\mathcal{E}| \gg \hbar\omega_{sp}$, we find

$$E_1 \approx -|d_{sp}\mathcal{E}|, \qquad (4.56a)$$
$$E_2 \approx +|d_{sp}\mathcal{E}|. \qquad (4.56b)$$

For large electric fields we observe a linear Stark shift. The atom appears to have a permanent dipole moment d_{sp}, but this is because the states are now completely mixed by the electric field. The atom did not possess an intrinsic dipole moment prior to the application of the field.

The dipole matrix element

$$d_{sp} = -e\langle 210|z|200\rangle \qquad (4.57)$$

can be evaluated by putting in the appropriate hydrogen wave functions and integrating. The result is

$$d_{sp} = -\sqrt{3}\, ea_0. \qquad (4.58)$$

Since $ea_0/h \approx 1.28$ MHz/(V/cm), linear Stark shifts for the $n = 2$ levels in hydrogen occur for

$$|\mathcal{E}| \gg \left|\frac{\hbar\omega_{sp}}{2d_{sp}}\right| \approx 250 \text{ V/cm}. \qquad (4.59)$$

4.3.5 Formalism of static polarizability

Let us now go back to the case of the quadratic Stark effect. We neglect the effect of hyperfine structure, which we will discuss later in Sec. 4.3.6. In the absence of other external fields, the direction of the applied electric field is a natural choice for the quantization axis. In fact, application of the electric field does not break the axial symmetry around this axis, and therefore m_J remains a good quantum number. This can also be seen from the fact that the matrix elements of the dipole interaction are all diagonal in m_J. The m_J states are shifted, and this shift in general depends on m_J. Because the shift is quadratic in electric field, reversing the direction of the electric field, or equivalently the direction of the angular momentum projection, should have no effect. It follows that the shifts of the m_J and $-m_J$ states are the same.

We now derive the most general form of the quadratic energy shifts due to an electric field. (An alternative derivation based on tensor formalism is found in Chapter 19.) We start with Eq. (4.51), which we can write more explicitly for a particular state $|\xi J m\rangle$, where ξ denotes additional quantum numbers besides J and m, as

$$\Delta E^{(2)}_{\xi J m} = \mathcal{E}^2 \sum_{\xi' J' m'} \frac{|\langle \xi J m | d_z | \xi' J' m' \rangle|^2}{E_{\xi J} - E_{\xi' J'}}. \tag{4.60}$$

To calculate the m-dependence of the matrix element in Eq. (4.60) we use the Wigner–Eckart theorem (Sec. 3.8):

$$\langle \xi J m | T^\kappa_q | \xi' J' m' \rangle = (-1)^{J-m} \langle \xi J \| T_\kappa \| \xi' J' \rangle \begin{pmatrix} J & k & J' \\ -m & q & m' \end{pmatrix}. \tag{4.61}$$

Here T^κ_q is the q-th spherical component of an irreducible tensor operator of rank κ, and $\langle \xi J \| T_\kappa \| \xi' J' \rangle$ is the reduced matrix element. The operator $\mathbf{d}$ is a vector, so it has tensor rank 1, and d_z expressed in spherical tensor components is just d_0, so evaluating the $3j$ symbols explicitly (Sobelman 1992, Sec. 4.2.2) we have

$$\langle \xi J m | d_z | \xi' J' m' \rangle = \langle \xi J m | d_0 | \xi' J' m' \rangle$$

$$= (-1)^{J-m} \langle \xi J \| d \| \xi' J' \rangle \begin{pmatrix} J & 1 & J' \\ -m & 0 & m' \end{pmatrix}$$

$$= \delta_{m'm} \langle \xi J \| d \| \xi' J' \rangle \times \begin{cases} \frac{\sqrt{J^2-m^2}}{\sqrt{J(2J+1)(2J-1)}} & \text{for } J' = J-1 \\ \frac{m}{\sqrt{J(J+1)(2J+1)}} & \text{for } J' = J \\ -\frac{\sqrt{(J+1)^2-m^2}}{\sqrt{(J+1)(2J+1)(2J+3)}} & \text{for } J' = J+1 \\ 0 & \text{for } |J'-J| > 1 \text{ or } J = J' = 0. \end{cases} \tag{4.62}$$

In each of these cases, we have terms contributing to $\Delta E_{\xi J m}$ that are either independent of m or depend on m^2. Thus we can write the general form of $\Delta E_{\xi J m}$ as

$$\Delta E_{\xi J m} = \mathcal{E}^2 \left(A_{\xi J} + B_{\xi J} m^2 \right), \tag{4.63}$$

where $A_{\xi J}$ and $B_{\xi J}$ are constants not depending on m. Note that each of the $2J+1$ sublevels, except for the $m = 0$ level, is doubly degenerate with respect to the sign of m.

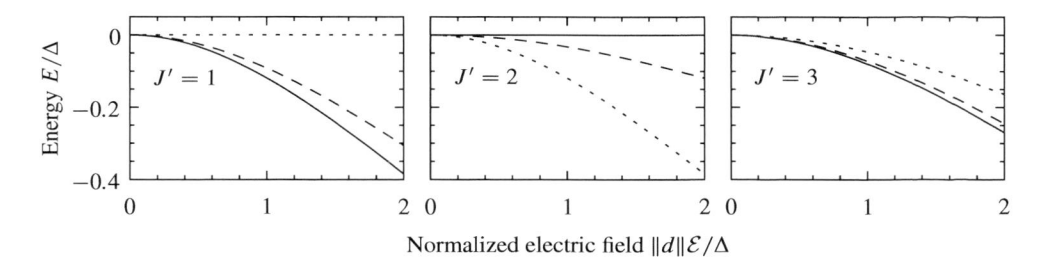

Fig. 4.4 Electric-field splitting of a $J = 2$ state for the cases in which the effect is dominated by a single partner state with angular momentum J' that is higher in energy than the $J = 2$ state by $\Delta = E_{J'} - E_{J=2}$. Solid lines—$m = 0$, dashed lines—$m = \pm 1$, dotted lines—$m = \pm 2$.

If we define the *scalar and tensor polarizabilities*, α_0 and α_2, by

$$A_{\xi J} = -\frac{\alpha_0}{2} + \frac{\alpha_2(J+1)}{2(2J-1)}, \tag{4.64a}$$

$$B_{\xi J} = -\frac{3\alpha_2}{2J(2J-1)}, \tag{4.64b}$$

then we have

$$\Delta E_{\xi J m} = -\frac{\mathcal{E}^2}{2}\left(\alpha_0 + \alpha_2 \frac{3m^2 - J(J+1)}{J(2J-1)}\right). \tag{4.65}$$

Since

$$\sum_{m=-J}^{J} m_J^2 = \frac{1}{3}J(1+J)(1+2J), \tag{4.66}$$

we see that the second term averaged over all Zeeman sublevels is zero. Thus the scalar polarizability is proportional to the shift of the level as a whole and the tensor polarizability characterizes the splitting. As shown in Chapter 19, α_2 is only nonzero for states with $J > 1/2$. Note also that α_2 is normalized so that for the sublevels $m = \pm J$, the tensor shift is $-\alpha_2 \mathcal{E}^2/2$.

It sometimes happens that the electric field splitting of a given state is dominated by the interaction with a single opposite-parity "partner" state. In this case the sublevel splitting takes a characteristic form depending on the angular momentum of the partner state. This is illustrated in Fig. 4.4 for a $J = 2$ state interacting with $J' = 1, 2, 3$ states in the presence of an electric field. Thus, measuring the scalar and tensor polarizabilities of a given state provides information about the dominantly mixed opposite-parity partner (Rochester *et al.* 1999; Li *et al.* 2004). For example, it follows from Eqs. (4.62) and (4.65) that the angular momentum of the partner state can be determined from the ratio of the scalar and tensor polarizabilities according to

$$\frac{\alpha_2}{\alpha_0} = \begin{cases} -1 & \text{for } J' = J - 1 \\ \frac{2J-1}{J+1} & \text{for } J' = J \\ \frac{-J(2J-1)}{(J+1)(2J+3)} & \text{for } J' = J + 1. \end{cases} \tag{4.67}$$

Equation (4.65) is written under the assumption that J and m are good quantum numbers. This is strictly true for m (as long as the electric field is applied along z). However, it is important to realize that the eigenfunctions in the presence of an electric field contain admixtures of opposite-parity states with total angular momenta that, in general, can equal J or $J \pm 1$. Under conditions in which the quadratic Stark effect is observed, these admixtures are normally small; however, there are situations where they are very important. An example is *Stark-induced transitions* between levels of the same parity that are forbidden in the absence of an electric field.

4.3.6 Stark effect in the manifold of hyperfine levels

In the case of a hyperfine level, we expect the Stark shifts in the form analogous to Eq. (4.65):

$$\Delta E_S = -\frac{1}{2}\alpha_0 \mathcal{E}^2 - \frac{1}{2}\alpha_2^F \mathcal{E}^2 \frac{3m_F^2 - F(F+1)}{F(2F-1)}, \qquad (4.68)$$

where the scalar polarizability is the same as in the absence of hyperfine splittings, and α_2^F is the tensor polarizability specific to the given hyperfine sublevel F.

In order to calculate α_2^F, we can expand the hyperfine-structure levels in the $|J m_J I m_I\rangle$ basis and write the shift of each of the levels in this basis using Eq. (4.65). Then the shift of the hyperfine level is the sum over these shifts, weighted by the squares of the expansion coefficients. A detailed derivation yielding

$$\alpha_2^F = \alpha_2 \frac{3X(X-1) - 4F(F+1)J(J+1)}{(2F+3)(2F+2)J(2J-1)}, \qquad (4.69)$$

where

$$X = F(F+1) + J(J+1) - I(I+1), \qquad (4.70)$$

can be found in the book by Armstrong (1971).

According to Eq. (4.69), the tensor polarizability of a hyperfine state is proportional to the polarizability of the corresponding J state. This would mean that a $J = 1/2$ state, such as the ground state of an alkali atom, which has no tensor splitting in the absence of hyperfine structure, should not have any splitting even when hyperfine structure is taken into account. However, this is not strictly true. The effect of the hyperfine energy splittings Δ_hf lead to nonzero tensor polarizabilities that scale as $\Delta_\text{hf}/\Delta E$, where ΔE is the splitting between the relevant J states. For cesium the effect gives a ground-state tensor polarizability that is $\sim 10^{-7}$ that of the scalar polarizability, which is not suppressed for $J = 1/2$. The effect is of interest in cesium because of its relevance to *atomic clocks*. A discussion and references to the theory and experiments (both, at times, controversial) related to this effect are given by Weis and Ulzega (2007).

For sufficiently strong electric fields, the Stark shifts may become comparable to or larger than the hyperfine-structure energy splitting. In this case, as with the Zeeman effect (Sec. 4.2), the hyperfine structure "uncouples"—states with the same values of m_F but different values of F become mixed.

In order to analyze the situation, we will construct an *effective Hamiltonian* for the Stark effect

$$H_S = -\frac{1}{2}\alpha_0 \mathcal{E}^2 - \frac{1}{2}\alpha_2 \mathcal{E}^2 \frac{3J_z^2 - \mathbf{J}^2}{\hbar^2 J(2J-1)}, \qquad (4.71)$$

that takes into account the energy shifts of the $|J m\rangle$ states according to Eq. (4.65).

Let us illustrate this with a specific example of a hyperfine manifold for $J = 3/2$ and $I = 3/2$ and focus on the states with $m_F = 2$, namely $|F = 3, m_F = 2\rangle$ and $|F = 2, m_F = 2\rangle$,

To find the Hamiltonian for these states, we expand them in terms of the uncoupled states $|(JI)m_J m_I\rangle \equiv |Jm_J\rangle|Im_I\rangle$. This gives

$$|F = 3, m_F = 2\rangle = \frac{1}{\sqrt{2}}|J = \tfrac{3}{2}, m_J = \tfrac{3}{2}\rangle|I = \tfrac{3}{2}, m_I = \tfrac{1}{2}\rangle$$
$$+ \frac{1}{\sqrt{2}}|J = \tfrac{3}{2}, m_J = \tfrac{1}{2}\rangle|I = \tfrac{3}{2}, m_I = \tfrac{3}{2}\rangle, \quad (4.72a)$$

$$|F = 2, m_F = 2\rangle = \frac{1}{\sqrt{2}}|J = \tfrac{3}{2}, m_J = \tfrac{3}{2}\rangle|I = \tfrac{3}{2}, m_I = \tfrac{1}{2}\rangle$$
$$- \frac{1}{\sqrt{2}}|J = \tfrac{3}{2}, m_J = \tfrac{1}{2}\rangle|I = \tfrac{3}{2}, m_I = \tfrac{3}{2}\rangle, \quad (4.72b)$$

where the appropriate Clebsch–Gordan coefficients $\langle Jm_J Im_I|Fm_F\rangle$ have been explicitly evaluated (see Sec. 3.5). In the uncoupled basis, the effective Hamiltonian is diagonal with the matrix elements given by Eq. (4.65) in terms of the scalar and tensor polarizabilities. Thus, we can use the expansions (4.72) and find, for example, that the off-diagonal matrix element $\langle F = 3, m_F = 2|H_S|F = 2, m_F = 2\rangle$ is given by

$$\langle F = 3, m_F = 2|H_S|F = 2, m_F = 2\rangle$$
$$= \frac{1}{2}\langle J = \tfrac{3}{2}, m_J = \tfrac{3}{2}|\langle I = \tfrac{3}{2}, m_I = \tfrac{1}{2}|H_S|J = \tfrac{3}{2}, m_J = \tfrac{3}{2}\rangle|I = \tfrac{3}{2}, m_I = \tfrac{1}{2}\rangle$$
$$- \frac{1}{2}\langle J = \tfrac{3}{2}, m_J = \tfrac{1}{2}|\langle I = \tfrac{3}{2}, m_I = \tfrac{3}{2}|H_S|J = \tfrac{3}{2}, m_J = \tfrac{1}{2}\rangle|I = \tfrac{3}{2}, m_I = \tfrac{3}{2}\rangle \quad (4.73)$$
$$= \frac{1}{2}\left(-\frac{1}{2}\alpha_0\mathcal{E}^2 - \frac{1}{2}\alpha_2\mathcal{E}^2\right) - \frac{1}{2}\left(-\frac{1}{2}\alpha_0\mathcal{E}^2 + \frac{1}{2}\alpha_2\mathcal{E}^2\right)$$
$$= -\frac{1}{2}\alpha_2\mathcal{E}^2.$$

The entire Hamiltonian, including the hyperfine shifts, which are given by Eq. (2.28), is

$$H = \begin{pmatrix} -\tfrac{3}{4}A - \tfrac{3}{4}B - \tfrac{1}{2}\alpha_0\mathcal{E}^2 & -\tfrac{1}{2}\alpha_2\mathcal{E}^2 \\ -\tfrac{1}{2}\alpha_2\mathcal{E}^2 & \tfrac{9}{4}A + \tfrac{1}{4}B - \tfrac{1}{2}\alpha_0\mathcal{E}^2 \end{pmatrix}, \quad (4.74)$$

with the basis states written in the order ($|F = 2, m_F = 2\rangle$, $|F = 3, m_F = 2\rangle$). Diagonalizing this Hamiltonian (i.e., solving the secular equation), we find the energy eigenvalues

$$E_{1,2} = \frac{1}{4}\left(3A - B - 2\alpha_0\mathcal{E}^2 \mp 2\sqrt{9A^2 + 6AB + B^2 + \alpha_2^2\mathcal{E}^4}\right). \quad (4.75)$$

This result agrees with Eq. (4.69) for electric fields small enough that the Stark shifts are much smaller than the hyperfine splitting. However, it is valid for a wider range of electric fields. The Stark shifts can exceed the hyperfine intervals, but must still be small enough that the Stark shifts can be described by the scalar and tensor polarizabilities in the $|Jm_J\rangle$ basis. This means that the shifts must be much smaller than both the fine-structure intervals and the energy separation from the levels of opposite parity that are mixed by the electric-dipole interaction.

In this example, we diagonalized the Hamiltonian in the coupled basis. Naturally, exactly the same result is also obtained in the uncoupled ($|Jm_J\rangle|Im_I\rangle$) basis, in which the effective Stark Hamiltonian is diagonal, but the hyperfine Hamiltonian has nonzero off-diagonal terms.

4.4 Combined effect of electric and magnetic fields

So far we have discussed atoms subject to either an external static magnetic or electric field. The case when electric and magnetic fields are applied simultaneously is also interesting. This occurs in practice, for example, in a class of atomic parity violation experiments, when the application of both fields is required to observe the parity-violation effect. Another example is in experiments with magneto-optical effects, where nominally only a magnetic field is applied. The electric field of the laser light, while not static, produces effects similar to those of a static field, and the combined effect of the magnetic and electric shifts must be taken into account in analyzing the experiment. This situation can result in the phenomenon of *alignment-to-orientation conversion*, discussed in Secs. 5.6 and 11.3, in which an atomic state with a preferred direction is generated, despite optical pumping with linearly polarized light, which has only a preferred axis and no preferred direction.

4.4.1 Atom in orthogonal electric and magnetic fields

We have discussed above the unperturbed Hamiltonian and the effect of external magnetic and electric fields separately. Now, the time has come for us to consider situations where the fields are applied simultaneously. Formally, all we have to do is to add the unperturbed Hamiltonian and the Hamiltonians describing the effect of each of the fields into one combined Hamiltonian, and then solve the eigenvalue/eigenvector problem. A small problem is that some of the fields may be directed along "inconvenient axes." Below, we use rotation matrices to manipulate the Hamiltonian. Before we use this general technique, however, we discuss an example where we do the basis change by hand.

Until now we always chose the quantization axis along the direction of the external field. This is the most natural choice, and it simplifies the calculations. The simplification comes from the fact that the magnetic quantum number m is a good quantum number even in the presence of the external field. But if the electric and magnetic fields are not collinear, then we have a new situation. We can assume that one of the fields, for example, the magnetic field, defines the quantization axis. The electric field will cause the magnetic sublevels to be mixed in the basis defined by the magnetic field.

Let us see how this works for an atomic state with angular momentum J. The magnetic-field Hamiltonian can be written as before:

$$H_{B_z} = -\boldsymbol{\mu} \cdot \mathbf{B} = g_J \mu_B B_z J_z / \hbar. \qquad (4.76)$$

If the electric field is directed along $\hat{\mathbf{x}}$, the effective electric-field Hamiltonian, in analogy to Eq. (4.65), is

$$H_{\mathcal{E}_x} = -\frac{1}{2}\alpha_0 \mathcal{E}^2 - \frac{1}{2}\alpha_2 \mathcal{E}_x^2 \frac{3J_x^2 - \mathbf{J}^2}{\hbar^2 J(2J-1)}. \qquad (4.77)$$

The matrix elements of J_x are given in Eq. (3.58). The α_0 term gives a common shift of all the levels. If we are only interested in splittings, we can write just the contribution of the α_2 term and the magnetic field Hamiltonian as

$$H = \begin{pmatrix} g_J\mu_B B_z + \frac{1}{4}\alpha_2\mathcal{E}_x^2 & 0 & -\frac{3}{4}\alpha_2\mathcal{E}_x^2 \\ 0 & -\frac{1}{2}\alpha_2\mathcal{E}_x^2 & 0 \\ -\frac{3}{4}\alpha_2\mathcal{E}_x^2 & 0 & -g_J\mu_B B_z + \frac{1}{4}\alpha_2\mathcal{E}_x^2 \end{pmatrix}. \qquad (4.78)$$

The energies of the atomic states in the combined field are the eigenvalues of this Hamiltonian:

$$E_\pm = \frac{1}{4}\alpha_2\mathcal{E}_x^2 \pm \sqrt{\frac{9}{16}\alpha_2^2\mathcal{E}_x^4 + g_J^2\mu_B^2 B_z^2}, \tag{4.79a}$$

$$E_0 = -\frac{1}{2}\alpha_2\mathcal{E}_x^2. \tag{4.79b}$$

The respective eigenfunctions, giving the atomic wave functions, are

$$|E_+\rangle = A_+|J=1, m=1\rangle - A_-|J=1, m=-1\rangle, \tag{4.80a}$$
$$|E_-\rangle = A_-|J=1, m=1\rangle + A_+|J=1, m=-1\rangle, \tag{4.80b}$$
$$|E_0\rangle = |J=1, m=0\rangle, \tag{4.80c}$$

where

$$A_\pm = \sqrt{\frac{\sqrt{\frac{9}{16}\alpha_2^2\mathcal{E}_x^4 + g_J^2\mu_B^2 B_z^2} \pm g_J\mu_B B_z}{\sqrt{\frac{9}{4}\alpha_2^2\mathcal{E}_x^4 + 4g_J^2\mu_B^2 B_z^2}}}. \tag{4.81}$$

Note that the x-directed electric field mixes the $m = \pm 1$ states with each other, but there is no mixing involving the $m = 0$ state. This is because J_x is decomposed into J_+ and J_-. These change m by ± 1. Application of J_x^2, therefore, either returns m to its initial value, or changes m by ± 2. In this situation, $m = 0$ can only mix with $m = \pm 2$, which do not exist in a $J = 1$ state. Thus the $m = 0$ state remains unmixed.

It is straightforward to verify that in the limit when one of the fields vanishes, the formulas above give the correct asymptotics—the pure Zeeman or the pure Stark effect. It is also instructive to plot out the evolution of the eigenenergies as a function of one of the fields with the value of the other field fixed at a finite value (Fig. 4.5). Here we see a continuous evolution of the level-splitting pattern between that characteristic of the Zeeman effect and that characteristic of the Stark effect.

> The dependence of the energy levels on the magnetic field shown in Fig. 4.5(a) is an example of an *avoided crossing*: in the absence of the electric field, the $m = \pm 1$ sublevels "cross" at $B = 0$; however, when electric field is applied, these sublevels mix, the eigenstates are pushed apart, and the crossing is avoided.
>
> Avoided crossings, for which we have presented a quantum mechanical picture but which can equally well be introduced classically, play a very important role in various phenomena in classical mechanics, optics, solid-state physics, and, of course, in atomic and molecular physics. We will encounter some further examples in Chapter 8.

4.4.2 Electric field shifts of magnetically split Zeeman levels

We now wish to further analyze the effect of an electric field on a state whose Zeeman sublevels are split by a strong magnetic field. Let us determine the additional level shifts of the Zeeman sublevels arising when a weak electric field $\mathcal{E}$ is applied at an arbitrary angle θ to the magnetic field. We neglect the effect of the scalar polarizability (α_0), which shifts all the sublevels by the same amount, but include the effect of the tensor polarizability (α_2), which in the frame in which the z-axis is along the electric field is described by a diagonal Hamiltonian with matrix elements (see Eq. 4.65)

$$H_{mm} = -\alpha_2 \frac{\mathcal{E}^2}{2} \frac{3m^2 - F(F+1)}{F(2F-1)} = C\left[3m^2 - F(F+1)\right]. \tag{4.82}$$

74 Atoms in external electric and magnetic fields

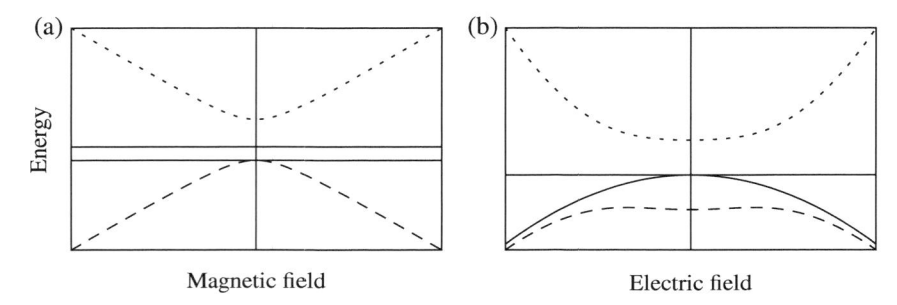

Fig. 4.5 Energy eigenvalues for a $J = 1$ state in orthogonal electric and magnetic fields as a function of (a) magnetic field with a fixed electric field, and (b) electric field with a fixed magnetic field. Solid line represents state $|E_0\rangle$, dotted line $|E_+\rangle$, dashed line $|E_-\rangle$ (Eqs. 4.79–4.81). Plot (a) gives an example of an avoided crossing (see text) at zero magnetic field. The central vertical lines on the plots indicate zero magnetic and electric field, respectively.

Since the Stark shifts are quadratic in the electric field $\mathcal{E}$, the tensor shifts can be described as the contraction of a rank-two tensor α_{ik} with the second rank tensor formed out of the electric field components, $\mathcal{E}_i \mathcal{E}_k$ (see Chapter 19). The *polarizability tensor* α_{ik} must be proportional to the irreducible rank-two tensor formed out of the atomic angular momenta (since there are no other vectors available to describe the atomic system). Therefore, according to Eq. (3.106), we have

$$\alpha_{ik} \propto \frac{1}{2}(F_i F_k + F_k F_i) - \frac{1}{3}\mathbf{F}^2. \tag{4.83}$$

In the presence of a strong magnetic field **B**, the components of **F** not along **B** average out to zero. Thus α_{ik} has only one nonzero component, with matrix element

$$\langle Fm|\alpha_{zz}|Fm\rangle \propto m^2 - \frac{1}{3}F(F+1). \tag{4.84}$$

The second-rank irreducible tensor composed of the electric field is given by

$$\mathcal{E}_{ik} \propto \mathcal{E}_i \mathcal{E}_k - \frac{1}{3}\mathcal{E}^2. \tag{4.85}$$

Since the only nonzero component of the polarizability is α_{zz}, only the zz-component of the electric field tensor $\mathcal{E}_{zz} = \mathcal{E}^2\left(\cos^2\theta - \frac{1}{3}\right)$ is relevant. (Note that this quantity can be constructed by squaring the z-component of $\mathcal{E}$, $\mathcal{E}_z = \mathcal{E}\cos\theta$, and subtracting the average of $\mathcal{E}_z^2$ over all angles. This subtraction can be justified because we expect the tensor shift to vanish upon averaging over all possible directions of the electric field.) Thus the tensor shift is given by

$$\Delta E(M) \propto \frac{3\cos^2\theta - 1}{2}\left[3m^2 - F(F+1)\right], \tag{4.86}$$

where we have included a numerical factor so that the angular factor is equal to the Legendre polynomial $P_2(\theta)$ with the standard normalization. Note that the shifts vanish when the electric field is at the "*magic angle*" (at which $\cos^2\theta = 1/3$) to the magnetic field. These results are valid for both static and oscillating electric fields.

Let us now illustrate this result by an explicit calculation for the case of an $F = 1$ state, using a slightly different approach. The natural quantization axis for this problem is the direction of the magnetic field; transforming the Hamiltonian (4.82) to the corresponding frame using the rotation matrix for $F = 1$ (Sec. 3.4.2)

$$\mathscr{D}(0,\beta,0) = \begin{pmatrix} \frac{1}{2}(1+\cos\beta) & \frac{1}{\sqrt{2}}\sin\beta & \frac{1}{2}(1-\cos\beta) \\ -\frac{1}{\sqrt{2}}\sin\beta & \cos\beta & \frac{1}{\sqrt{2}}\sin\beta \\ \frac{1}{2}(1-\cos\beta) & -\frac{1}{\sqrt{2}}\sin\beta & \frac{1}{2}(1+\cos\beta) \end{pmatrix}, \qquad (4.87)$$

we have

$$H' = \mathscr{D}^{-1}(0,\theta,0) H \mathscr{D}(0,\theta,0)$$

$$= C \begin{pmatrix} \frac{3\cos^2\theta-1}{2} & \frac{3\sin\theta\cos\theta}{\sqrt{2}} & \frac{3(1-\cos^2\theta)}{2} \\ \frac{3\sin\theta\cos\theta}{\sqrt{2}} & -(3\cos^2\theta-1) & -\frac{3\sin\theta\cos\theta}{\sqrt{2}} \\ \frac{3(1-\cos^2\theta)}{2} & -\frac{3\sin\theta\cos\theta}{\sqrt{2}} & \frac{3\cos^2\theta-1}{2} \end{pmatrix}. \qquad (4.88)$$

Because we are considering the case where the electric field shifts are much smaller than those due to the magnetic field, we can neglect the effect of the off-diagonal matrix elements in (4.88) as they do not contribute to energy-level shifts to first order in the electric perturbation ($\propto \mathcal{E}^2$). The electric-field shifts are thus given by the diagonal matrix elements in (4.88), which can be seen to have the form (4.86).

4.5 Atoms in oscillating fields

Atoms in oscillating fields is a broad topic that includes atomic transitions (discussed in Chapter 7). Here, we consider the energy level shifts caused by oscillating electric and magnetic fields, which are in some ways similar to those due to static fields.

We first analyze the shift of atomic energy levels due to an electric field which varies in time, known as the *AC-Stark effect*. The AC-Stark effect is an invaluable tool for *laser trapping and cooling* and is a basic mechanism behind many nonlinear optical effects.

We will also see that the *AC-Zeeman effect*—produced by a weak, oscillating magnetic field transverse to a strong, static leading field—is completely analogous to the AC-Stark effect. As we have seen, there is a fundamental difference between the Zeeman and Stark effects in that magnetic fields lead to first-order energy shifts while electric fields cause shifts only in second order. However, in the presence of a strong leading magnetic field, additional weak transverse fields only lead to second-order energy shifts; this removes the dissimilarity.

4.5.1 AC-Stark effect for a two-level system: the low-frequency case

Consider two opposite-parity states of an atom, a lower state $|a\rangle$ and an upper state $|b\rangle$, separated in energy by $\hbar\omega_0$. For concreteness, we assume that the states are Zeeman sublevels of two different angular-momentum states, with the same value of m. The matrix element of the electric-dipole operator between these states has one nonzero component, d_z (Chapter 7). A sinusoidally varying $\hat{z}$-directed electric field with magnitude $\mathcal{E} = \mathcal{E}_0 \sin\omega t$ is applied to the atom. In this discussion, we neglect the linewidths of the energy levels, i.e., the relaxation rate Γ for the states is much less than all other relevant frequencies in the problem ($|\omega_0 - \omega|, \omega, \omega_0$). The amplitude $\mathcal{E}_0$ of the applied field is small enough that the effect of the electric field may be treated as a weak perturbation, i.e., $d\mathcal{E}_0 \ll \hbar|\omega_0 - \omega|, \hbar\omega, \hbar\omega_0$.

76 *Atoms in external electric and magnetic fields*

Consider first a slow modulation, $\omega \ll \omega_0$. One way to think about this problem is to imagine the atom in an instantaneous static field $\mathcal{E}$. Then the state $|a\rangle$ is shifted down and $|b\rangle$ is shifted up by (see Eq. 4.60)

$$\Delta E \approx \frac{d^2 \mathcal{E}^2}{\hbar \omega_0}. \tag{4.89}$$

Here we represent the oscillating energy levels in terms of a modified level-structure (spectrum) by determining the time-averaged probability of finding the energy level at a particular value of the energy. We treat the atomic energy levels similarly to the description of a frequency-modulated oscillator.

The time dependence of an atomic state is governed by the Schrödinger equation

$$i\hbar \frac{d|\psi(t)\rangle}{dt} = E(t)|\psi(t)\rangle.$$

Neglecting atomic transitions, the only thing that changes in time is the phase of the state. Solving the Schrödinger equation then gives $|\psi(t)\rangle = |\psi(0)\rangle \exp[-i\varphi(t)]$, where the phase $\varphi(t)$ is given by

$$\varphi(t) = \frac{1}{\hbar} \int_0^t E(t') dt'. \tag{4.90}$$

When the field is varying with $\omega \ll \omega_0$, for short times, the electric field is effectively DC. Thus the time-dependent energy-level shifts are simply given by the DC Stark-shift formula (4.89):

$$\Delta E(t) = \pm \frac{d^2 \mathcal{E}_0^2}{\hbar \omega_0} \sin^2 \omega t, \tag{4.91}$$

where we employ a time-dependent field (with a particular choice of phase) in place of a static field and the plus and minus signs refer to the upper and lower states, respectively. The instantaneous energy $E(t)$ of the upper state $|b\rangle$ is thus given by

$$\begin{aligned} E(t) &= E_0 + \Delta E(t) \\ &= E_0 + \hbar \Omega \sin^2 \omega t \\ &= E_0 + \frac{\hbar \Omega}{2} - \frac{\hbar \Omega}{2} \cos 2\omega t, \end{aligned} \tag{4.92}$$

where E_0 is the unperturbed energy and $\hbar \Omega = d^2 \mathcal{E}_0^2 / (\hbar \omega_0)$ is the peak-to-peak amplitude of the modulation of the level energy. The phase $\varphi(t)$ acquired by state $|b\rangle$ in its time evolution is obtained by evaluating the integral in Eq. (4.90):

$$\varphi(t) = \frac{1}{\hbar} \int_0^t E(t') dt' = \left(\frac{E_0}{\hbar} + \frac{\Omega}{2} \right) t - \frac{\Omega}{4\omega} \sin 2\omega t. \tag{4.93}$$

The time-dependent state $|\psi_b(t)\rangle = e^{-i\varphi(t)} |b\rangle$ is then given by

$$|\psi_b(t)\rangle = e^{-i\varphi(t)} |b\rangle = e^{-i(\omega_0 + \Omega/2)t} e^{i(\Omega/4\omega) \sin 2\omega t} |b\rangle. \tag{4.94}$$

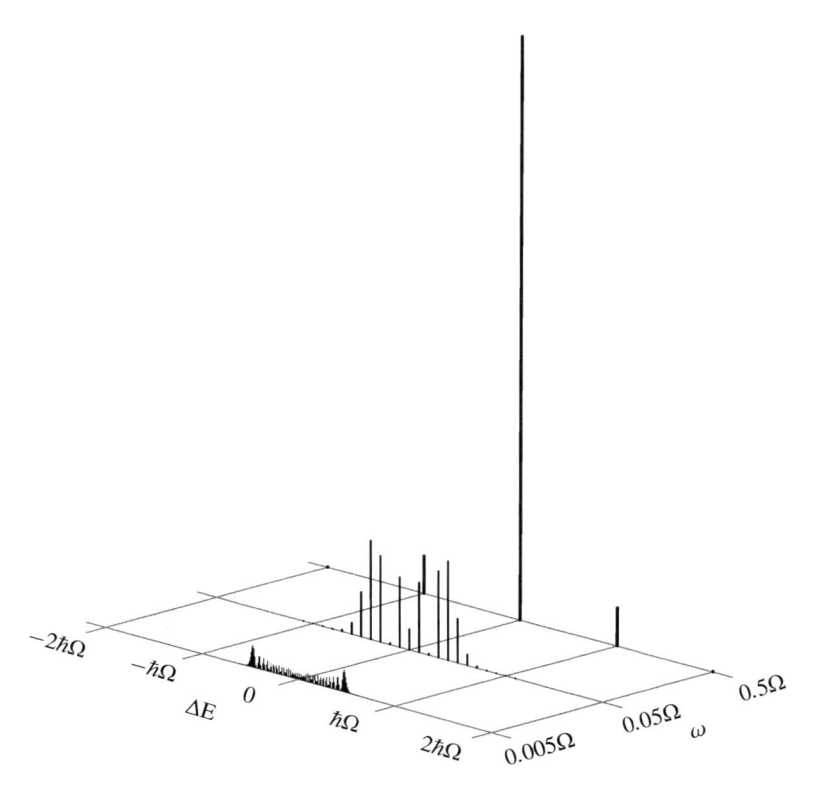

Fig. 4.6 A time-dependent energy level in an electric field oscillating with frequency ω. The level is represented as a spectrum analogous to that of frequency-modulated light. Shown are spectra for the same peak-to-peak amplitude of the level excursion (modulation) $\Delta E_0 = \hbar\Omega$ but different modulation frequencies (the shift oscillates at 2ω) and three different frequencies ω. In all cases, modulation leads to a set of sidebands spaced by $2\hbar\omega$. For example, when $\omega = 0.5\,\Omega$, the first sidebands are at $|\Delta E| = \hbar\Omega$. At low modulation frequencies, the set of many sidebands resembles the classical distribution of positions for an oscillator with sharp peaks at the turning points. At high modulation frequencies, only the sidebands closest to the carrier are pronounced.

We can describe the energy spectrum exhibited by $|\psi_b(t)\rangle$ in terms of *sidebands* by expanding (4.94) in terms of *Bessel functions* $J_n(\alpha)$, where α is the *modulation index*, using the formula (see, for example, Arfken and Weber 2005 or Siegman 1986):

$$e^{i\alpha \sin\theta} = \sum_{n=-\infty}^{\infty} J_n(\alpha) e^{in\theta}. \qquad (4.95)$$

The state is then written as a sum of components,

$$|\psi_b(t)\rangle = \sum_{n=-\infty}^{\infty} J_n\left(\frac{\Omega}{4\omega}\right) e^{-i(\omega_0 + \Omega/2)t} e^{i2n\omega t} |b\rangle, \qquad (4.96)$$

each with amplitude $J_n(\alpha)$, $\alpha = \Omega/(4\omega)$, and frequency $\omega_0 + \Omega/2 - 2n\omega$.

The time dependence in Eq. (4.96) suggests the following interpretation. The energy spectrum is given by a set of sidebands centered around $\omega_0 + \Omega/2$ and spaced by frequency intervals of 2ω.[5] The statistical weight of each energy sideband is given by the square of the Bessel function $J_n(\alpha)$. By statistical weight we mean the following: Suppose we initially put an atom in one of the states, subject to the oscillating electric field, and then measure its energy with, for example, a probe field. The statistical weight gives us the probability that we measure the state to have a particular energy.

The average values of the shifts are thus given by Eq. (4.89) (where we use the average value $\overline{\mathcal{E}^2} = \mathcal{E}_0^2/2$):

$$\overline{\Delta E} \approx \pm \frac{d^2 \mathcal{E}_0^2}{2\hbar\omega_0}. \tag{4.97}$$

The character of the spectrum depends on the modulation index $\alpha = \Omega/(4\omega)$. For $4\omega \ll \Omega$, the modulation index is large. This corresponds to a slow oscillation of the frequency, generating the spectrum shown in the front of Fig. 4.6; the levels are most likely to be found at the maximum or minimum energies, where the rate of change is the slowest. For $4\omega \gg \Omega$, the spectrum is dominated by a single strong line at $\omega_0 + \Omega/2$; the next most prominent sidebands occur at $\pm 2\omega$ about the central frequency (shown at the rear of Fig. 4.6).

So far, we have explicitly assumed an electric field oscillating much slower than the resonance optical frequency. This assumption justified the use of Eq. (4.89) for calculating "instantaneous" energy shifts. We now need to devise a more general approach that will allow us to tackle the cases of the electric field oscillating at a frequency comparable to or even higher than the resonance frequency ω_0. We will do this in Sec. 4.5.3, but first let us discuss the related problem of the AC-Zeeman effect.

4.5.2 AC-Zeeman effect

There is a direct analogy between the AC-Stark shift and the AC-Zeeman shift. Consider a spin-1/2 system (e.g., a paramagnetic atom) subjected to a strong, leading magnetic field $\mathbf{B}_0 = B_0 \hat{\mathbf{z}}$, splitting the Zeeman sublevels by $\hbar\omega_0 = g\mu_B B_0$ (where g is the appropriate Landé factor). A weak, oscillatory transverse field

$$\mathbf{B}_\perp(t) = B_\perp \sin \omega t \, \hat{\mathbf{x}} \tag{4.98}$$

is applied to the atom. Here $\mu_B B_\perp$ is analogous to $d\mathcal{E}_0$ for an electric field.

We will assume that $\mu_B B_\perp \ll |\omega_0 - \omega|$ (weak field limit), and derive the average energy level shifts due to the transverse field by going into a frame rotating at ω in the same sense as the Larmor precession due to B_0 and applying the *rotating-wave approximation*. In order to describe the evolution of angular momentum in such a rotating frame, it may be convenient to introduce an additional, *fictitious field*

$$\mathbf{B}_f = -\frac{\omega_m}{g\mu_B} \tag{4.99}$$

which accounts for the effects of rotation.

---

[5]The factor of 2 appearing here and in the subsequent discussion is related to the fact that the square of the electric field oscillates at twice the frequency of the oscillation of the electric field itself.

The oscillating, transverse magnetic field (4.98) can be written as a sum of two counter-rotating magnetic fields:

$$\mathbf{B}_\perp(t) = \frac{B_\perp}{2}(\sin\omega t\,\hat{\mathbf{x}} + \cos\omega t\,\hat{\mathbf{y}}) + \frac{B_\perp}{2}(\sin\omega t\,\hat{\mathbf{x}} - \cos\omega t\,\hat{\mathbf{y}}). \quad (4.100)$$

The Hamiltonian describing the interaction of $\mathbf{B}_\perp(t)$ with the atom is

$$H_\perp = -\boldsymbol{\mu}\cdot\mathbf{B}_\perp(t), \quad (4.101)$$

and we can write the magnetic moment in terms of the Pauli matrices:

$$\boldsymbol{\mu} = -\frac{g\mu_B}{2}\left[\begin{pmatrix} 0 & 1 \\ 1 & 0 \end{pmatrix}\hat{\mathbf{x}} + \begin{pmatrix} 0 & -i \\ i & 0 \end{pmatrix}\hat{\mathbf{y}} + \begin{pmatrix} 1 & 0 \\ 0 & -1 \end{pmatrix}\hat{\mathbf{z}}\right], \quad (4.102)$$

where we are using the *spinor representation* in which

$$|b\rangle = |+\rangle = \begin{pmatrix} 1 \\ 0 \end{pmatrix}, \quad (4.103a)$$

$$|a\rangle = |-\rangle = \begin{pmatrix} 0 \\ 1 \end{pmatrix}. \quad (4.103b)$$

Employing Eqs. (4.100) and (4.102) in Eq. (4.101), with a bit of algebra we find

$$H_\perp = \frac{g\mu_B B_\perp}{4}\left[\begin{pmatrix} 0 & -ie^{i\omega t} \\ ie^{-i\omega t} & 0 \end{pmatrix} + \begin{pmatrix} 0 & ie^{-i\omega t} \\ -ie^{i\omega t} & 0 \end{pmatrix}\right]. \quad (4.104)$$

Now we go into a frame rotating at ω. This transformation has two effects: (1) it causes one rotating component of the transverse field to lose its time dependence and appear to be a static field while the other component rotates at twice its original frequency, and (2) it affects the perceived evolution of angular momentum—to account for this effect we introduce the fictitious magnetic field from Eq. (4.99). In the rotating frame, the perturbing Hamiltonian looks like

$$H_\perp^{(\mathrm{rot})} = \frac{ig\mu_B B_\perp}{4}\left[\begin{pmatrix} 0 & -1 \\ 1 & 0 \end{pmatrix} + \begin{pmatrix} 0 & e^{-i2\omega t} \\ -e^{i2\omega t} & 0 \end{pmatrix}\right], \quad (4.105)$$

while the Hamiltonian due to the leading field and the fictitious field can be written as

$$H_0^{(\mathrm{rot})} = \begin{pmatrix} \hbar(\omega_0-\omega) & 0 \\ 0 & 0 \end{pmatrix}. \quad (4.106)$$

Our next step is to make the rotating-wave approximation. If ω is sufficiently close to resonance, the fast-oscillating counter-rotating terms will be far off resonance, and can be dropped. This eliminates the second matrix in Eq. (4.105), leaving us with the total Hamiltonian:

$$H_{\mathrm{tot}}^{(\mathrm{rot})} \approx \begin{pmatrix} \hbar(\omega_0-\omega) & -ig\mu_B B_\perp/4 \\ ig\mu_B B_\perp/4 & 0 \end{pmatrix}. \quad (4.107)$$

The factors of 1/2 in the off-diagonal terms of the matrix in Eq. (4.107) are due to the fact that in making the rotating-wave approximation we have "thrown away" half of the transverse field. Solving for the eigenvalues of this matrix gives us the perturbed energies:

80 *Atoms in external electric and magnetic fields*

$$E_\pm = \frac{1}{2}\left(\hbar(\omega_0-\omega) \pm \sqrt{(g\mu_B B_\perp/2)^2 + \hbar^2(\omega_0-\omega)^2}\right). \tag{4.108}$$

Now we make use of the assumption that $\mu_B B_\perp \ll |\omega_0 - \omega|$ to simplify Eq. (4.108):

$$E_+ \approx \hbar(\omega_0-\omega) + \frac{(g\mu_B B_\perp/2)^2}{4\hbar(\omega_0-\omega)}, \tag{4.109a}$$

$$E_- \approx -\frac{(g\mu_B B_\perp/2)^2}{4\hbar(\omega_0-\omega)}. \tag{4.109b}$$

Finally, to go back into the lab frame we eliminate the energy shift due to the fictional field (4.99), and find

$$E_+ \approx \hbar\omega_0 + \frac{(g\mu_B B_\perp/2)^2}{4\hbar(\omega_0-\omega)}, \tag{4.110a}$$

$$E_- \approx -\frac{(g\mu_B B_\perp/2)^2}{4\hbar(\omega_0-\omega)}. \tag{4.110b}$$

4.5.3 The general AC-Stark effect

Having determined the AC-Zeeman shifts under the rotating-wave approximation, we can do the same for the case of the near-resonant AC-Stark effect.

Starting with the two-level Stark Hamiltonian (analogous to Eq. 4.53):

$$H = \begin{pmatrix} 0 & -d\mathcal{E}_0 \sin\omega t \\ -d\mathcal{E}_0 \sin\omega t & \hbar\omega_0 \end{pmatrix}, \tag{4.111}$$

we can transform it by using an analogy with the rotating frame we used for the magnetic case. In our new frame, the positive-frequency component of the oscillating electric field appears static.[6] Assuming that the positive component is near resonance, so that the In this frame, we have (neglecting the far off-resonant negative-frequency components in the off-diagonal matrix elements):

$$H^{(\text{rot})} \approx \begin{pmatrix} 0 & id\mathcal{E}_0/2 \\ id\mathcal{E}_0/2 & \hbar(\omega_0-\omega) \end{pmatrix}. \tag{4.112}$$

This Hamiltonian is completely analogous to that of Eq. (4.107) with the replacement

$$\frac{-ig\mu_B B_\perp}{4} \rightarrow \frac{id\mathcal{E}_0}{2}, \tag{4.113}$$

so we can immediately read off the AC Stark shifts:

$$\Delta E \approx \pm \frac{d^2 \mathcal{E}_0^2}{4(\omega_0-\omega)}, \tag{4.114}$$

where the two signs refer to the upper and the lower state, respectively. The reason we have approximate equality in Eq. (4.114) is that we have neglected the negative-frequency component of the oscillating electric field. This can easily be fixed by adding a second term in Eq. (4.114) with $\omega \to -\omega$.

---

[6]One can attempt to go beyond the formal analogy and assign a physical meaning to the transformation by saying that we are going to a frame "oscillating" together with the electric field.

The energy shifts due to the "counter-rotating" term are known as the *Bloch–Siegert shifts*. Of course, there are similar shifts in the AC-Zeeman case. In fact, the terminology originates from magnetic-resonance experiments.

The result (upon collecting terms) is:

$$\Delta E = \pm \frac{d^2 \mathcal{E}_0^2}{2} \frac{\omega_0}{\omega_0^2 - \omega^2}. \quad (4.115)$$

Note that the sign of the AC-Stark shift changes depending on the detuning. This means that, for example, light fields detuned to the low-frequency side of an atomic resonance push energy levels apart while light fields tuned to the high-frequency side of a transition push them together. Such light shifts are important for laser trapping and cooling.

It is important to note that throughout our entire treatment we have neglected the nonzero width Γ of the energy levels, so the above formulas are inapplicable near resonance ($|\omega_0 - \omega| \lesssim \Gamma$; see Chapter 7 for a discussion of near-resonant interactions).

In the limit $\omega \ll \omega_0$, Eq. (4.115) reduces to Eq. (4.97),

$$\Delta E \approx \pm \frac{d^2 \mathcal{E}_0^2}{2\omega_0}, \quad (4.116)$$

while with $|\omega - \omega_0| \ll \omega_0$ it reduces back to Eq. (4.114). When the electric-field frequency greatly exceeds the transition frequency ($\omega \gg \omega_0$), we have

$$\Delta E \approx \mp \frac{d^2 \mathcal{E}_0^2 \omega_0}{2\omega^2}. \quad (4.117)$$

5
Polarized atoms

Atomic physics experiments are commonly done not with a single atom, but with an ensemble of atoms.[1] Consider an ensemble of atoms in a state with total angular momentum F. If the atoms are all in identical states, i.e., can be characterized by the same wave function, the ensemble is said to be completely *polarized*, or that it is in a *pure state*.[2] The individual atom wave function also describes the ensemble as a whole. On the other hand, it may be the case that the individual wave functions are random, so that there is equal probability to find an atom in any Zeeman sublevel, and there is no fixed phase relationship between any two sublevels. (As we discuss below, this corresponds to *populations* being equal and *coherences* being zero.) In this case the ensemble is *unpolarized*. We will see that this seemingly simple unpolarized state cannot, in fact, be described by a wave function. In order to describe polarized and unpolarized states, we must employ a generalization of the wave function—this generalization is known as the *density matrix*.

5.1 The density matrix

We first consider a simple example that shows that an ensemble cannot always be described by a wave function. The spin part of an atomic wave function in a $S = 1/2$ state can be written

$$|\psi\rangle = c_{\frac{1}{2}}|\tfrac{1}{2}\rangle + c_{-\frac{1}{2}}|\tfrac{1}{2}\rangle = \begin{pmatrix} c_{\frac{1}{2}} \\ c_{-\frac{1}{2}} \end{pmatrix}. \tag{5.1}$$

Here the c_m are expansion coefficients of the wave function over a complete basis (in this case—the Zeeman basis), found as

$$c_m = \langle m|\psi\rangle. \tag{5.2}$$

Since the state can be assumed to be normalized,

$$|c_{\frac{1}{2}}|^2 + |c_{-\frac{1}{2}}|^2 = 1, \tag{5.3}$$

and the overall phase is arbitrary, there are only two free parameters: the relative magnitude and relative phase of $c_{1/2}$ and $c_{-1/2}$. This means that we can write

$$\begin{pmatrix} c_{\frac{1}{2}} \\ c_{-\frac{1}{2}} \end{pmatrix} = \begin{pmatrix} e^{-i\phi/2} \cos\frac{\theta}{2} \\ e^{i\phi/2} \sin\frac{\theta}{2} \end{pmatrix}, \tag{5.4}$$

---

[1] Recently, however, there have been many experiments using one or a small number of particles (see, for example, Dehmelt 1990; Haroche and Raimond 2006, and literature cited therein).

[2] An anisotropy or *polarization* induced in the atomic states is a population difference and/or coherence between Zeeman sublevels (*Zeeman polarization*). More generally, polarization also includes population imbalance and coherences between hyperfine-structure states (*hyperfine polarization*), or indeed any states in an atom.

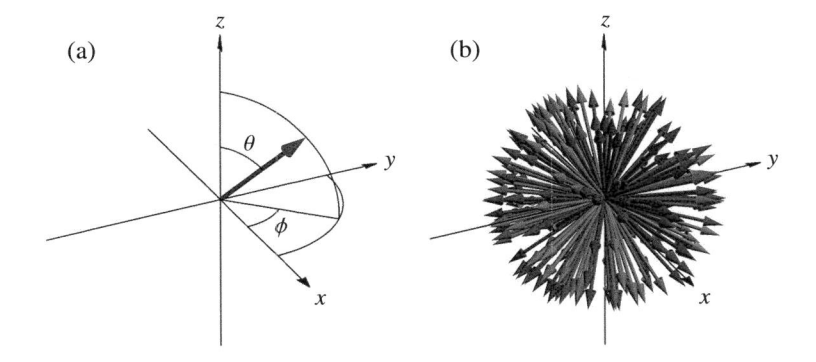

Fig. 5.1 Expectation values of spin for (a) a single atom and (b) an unpolarized ensemble.

where θ parametrizes the relative magnitude and ϕ parametrizes the relative phase. We can evaluate the "expectation vector" for the spin operator, which is a vector whose components are the expectation values of the projections on the Cartesian axes. One can imagine determining such expectation values by performing measurements of each of the components on a subensemble of a large ensemble of particles in the same state. (By performing the measurements of different components on different subensembles, we avoid running afoul of the quantum-mechanical uncertainty rules discussed in Chapter 2.)

We can find the expectation vector of the spin operator by plugging in the Pauli matrices:

$$\langle \mathbf{S} \rangle = \frac{\hbar}{2} \left(\langle \psi | \sigma_x | \psi \rangle, \langle \psi | \sigma_y | \psi \rangle, \langle \psi | \sigma_z | \psi \rangle \right)$$
$$= \frac{\hbar}{2} \left(\langle \psi | \begin{pmatrix} 0 & 1 \\ 1 & 0 \end{pmatrix} | \psi \rangle, \langle \psi | \begin{pmatrix} 0 & -i \\ i & 0 \end{pmatrix} | \psi \rangle, \langle \psi | \begin{pmatrix} 1 & 0 \\ 0 & -1 \end{pmatrix} | \psi \rangle \right). \quad (5.5)$$

Evaluating the x-component as an example, we get:

$$\langle \psi | \sigma_x | \psi \rangle = \begin{pmatrix} e^{i\phi/2} \cos \frac{\theta}{2} & e^{-i\phi/2} \sin \frac{\theta}{2} \end{pmatrix} \begin{pmatrix} 0 & 1 \\ 1 & 0 \end{pmatrix} \begin{pmatrix} e^{-i\phi/2} \cos \frac{\theta}{2} \\ e^{i\phi/2} \sin \frac{\theta}{2} \end{pmatrix}$$
$$= \sin \theta \cos \phi. \quad (5.6)$$

For all three components we have

$$\langle \mathbf{S} \rangle = \frac{\hbar}{2} (\sin \theta \cos \phi, \sin \theta \sin \phi, \cos \theta); \quad (5.7)$$

thus, we find that the expectation vector points in the (θ, ϕ) direction (Fig. 5.1a), and is nonzero for any atomic wave function.

Now consider an *unpolarized ensemble* of N atoms. We might model such an ensemble in the following way. Each atom is described by a wave function; however, the wave function is different for different atoms. We denote the wave function of the i-th atom by $|\psi_i\rangle$. In an unpolarized ensemble, the expectation vector for each atom is equally likely to point in any direction (Fig. 5.1b), so the average spin is zero:

$$\overline{\langle \mathbf{S} \rangle} = \frac{1}{N} \sum_{i=1}^{N} \langle \psi_i | \mathbf{S} | \psi_i \rangle = 0. \quad (5.8)$$

84 Polarized atoms

Thus there are no values of θ and ϕ that correspond to the average state of an unpolarized ensemble. This means that there is no single wave function that describes the ensemble. In principle, we could keep track of the wave function of every atom; however, aside from the computational difficulties involved, we have, in general, no way of measuring the individual atom wave functions. In order to describe the ensemble, then, we would like to find a generalization of the wave function as some sort of average over the atomic wave functions that can represent both polarized and unpolarized states.

What is the information that this representation needs to contain? In an experiment with an atomic ensemble, we perform some measurement, represented by an observable A. The measured value is the expectation value of A averaged over all of the atoms in the ensemble. This can be written

$$\overline{\langle A \rangle} = \frac{1}{N} \sum_{i=1}^{N} \langle \psi_i | A | \psi_i \rangle$$

$$= \sum_m \frac{1}{N} \sum_{i=1}^{N} \langle \psi_i | A | m \rangle \langle m | \psi_i \rangle \quad (5.9)$$

$$= \sum_m \frac{1}{N} \sum_{i=1}^{N} \langle m | \psi_i \rangle \langle \psi_i | A | m \rangle.$$

Here we have inserted the identity operator

$$1 = \sum_m |m\rangle \langle m|, \quad (5.10)$$

where $\{|m\rangle\}$ is any complete set of basis states. Using the definition of the trace of an operator,

$$\operatorname{Tr} \mathcal{O} = \sum_m \langle m | \mathcal{O} | m \rangle, \quad (5.11)$$

we can write (5.9) as

$$\overline{\langle A \rangle} = \operatorname{Tr} \left(\frac{1}{N} \sum_{i=1}^{N} |\psi_i\rangle\langle\psi_i| A \right)$$

$$= \operatorname{Tr}(\rho A), \quad (5.12)$$

where we have defined the new operator

$$\rho = \frac{1}{N} \sum_{i=1}^{N} |\psi_i\rangle\langle\psi_i|, \quad (5.13)$$

called the *density operator*. This operator is independent of A, and represents a kind of average over the atomic wave functions. The density operator evidently contains all of the information of interest about the average state of the ensemble,[3] since the result of any possible physical

---
[3] We have formed here the density operator that describes just the electronic spin states of the atoms. In general, the density operator can describe all possible degrees of freedom of the atomic system.

measurement can be found by multiplying the corresponding operator by ρ and taking the trace.[4]

The matrix elements of the density operator defined by Eq. (5.13) form the *density matrix*, and are given by

$$\rho_{mn} = \langle m|\rho|n \rangle = \frac{1}{N} \sum_{i=1}^{N} \langle m|\psi_i \rangle \langle \psi_i|n \rangle = \frac{1}{N} \sum_{i=1}^{N} c_m^{(i)} c_n^{(i)*}, \qquad (5.14)$$

where we have used Eq. (5.2) to write the result in terms of $c_m^{(i)}$, the m-th expansion coefficient (i.e., $c_{\frac{1}{2}}$ or $c_{-\frac{1}{2}}$ in the spin-1/2 example) for the i-th atom. In fact, we can regard Eq. (5.14) as an alternate definition of the density operator. A given system may require an infinite number of density-matrix elements to describe it completely; however, we will take the common approach of restricting the density matrix to a (finite) number of states of interest for a given problem. The diagonal elements of the density matrix are given by

$$\rho_{mm} = \frac{1}{N} \sum_{i=1}^{N} |c_m^{(i)}|^2, \qquad (5.15)$$

and represent the average probability of finding an atom in the state $|m\rangle$ (the *population* of $|m\rangle$). Off-diagonal density-matrix elements ($m \neq n$) are referred to as *coherences*.

We now show how the density matrix resolves the difficulty encountered above. If the ensemble is in a pure state, the density-matrix elements become simply

$$\rho_{mn} = c_m c_n^*, \qquad (5.16)$$

or, for the spin-1/2 case, using Eq. (5.4),

$$\rho = \begin{pmatrix} \cos^2 \frac{\theta}{2} & \frac{1}{2} e^{-i\phi} \sin \theta \\ \frac{1}{2} e^{i\phi} \sin \theta & \sin^2 \frac{\theta}{2} \end{pmatrix}. \qquad (5.17)$$

The expectation vector of $\mathbf{S}$ is given by $\langle \mathbf{S} \rangle = \text{Tr}(\rho \mathbf{S})$. (From now on we assume that all expectation values refer to the ensemble average.) Again taking the x-component as an example, we have

$$\langle \sigma_x \rangle = \text{Tr}(\rho \sigma_x)$$
$$= \text{Tr} \left[\begin{pmatrix} \cos^2 \frac{\theta}{2} & \frac{1}{2} e^{-i\phi} \sin \theta \\ \frac{1}{2} e^{i\phi} \sin \theta & \sin^2 \frac{\theta}{2} \end{pmatrix} \begin{pmatrix} 0 & 1 \\ 1 & 0 \end{pmatrix} \right] \qquad (5.18)$$
$$= \sin \theta \cos \phi,$$

as before.

For an unpolarized ensemble, on the other hand, it is equally likely to find an atom in the spin-up or spin-down state (equal populations), so the diagonal density-matrix elements are given by $\rho_{\frac{1}{2},\frac{1}{2}} = \rho_{-\frac{1}{2},-\frac{1}{2}} = 1/2$. Meanwhile, the off-diagonal elements for each atom ($c_{\frac{1}{2}}^{(i)} c_{-\frac{1}{2}}^{(i)*}$,

---

[4] Because the trace of a product of matrices is invariant when the matrices are cyclically permuted, $\text{Tr}(\rho A)$ and $\text{Tr}(A\rho)$ are equivalent.

for example) have completely random phases, so that these quantities average to zero over the ensemble. For the unpolarized ensemble, then,

$$\rho = \begin{pmatrix} \frac{1}{2} & 0 \\ 0 & \frac{1}{2} \end{pmatrix}. \tag{5.19}$$

Calculating $\langle \sigma_x \rangle$ in this case gives

$$\begin{aligned} \langle \sigma_x \rangle &= \text{Tr}\,(\rho \sigma_x) \\ &= \text{Tr}\left[\begin{pmatrix} \frac{1}{2} & 0 \\ 0 & \frac{1}{2} \end{pmatrix} \begin{pmatrix} 0 & 1 \\ 1 & 0 \end{pmatrix} \right] \\ &= 0, \end{aligned} \tag{5.20}$$

as required. Thus the density matrix correctly describes both the polarized and unpolarized cases (as well as the general case of partial polarization). We will henceforth use it to describe atomic ensembles.[5]

> We have introduced the concept of the density matrix starting with a model in which each atom in the ensemble is described by a wave function. While this is a convenient way to begin the exploration of the density-matrix apparatus, this model is unnecessarily restrictive. If the system under consideration is a part of a larger system consisting of interacting parts, it is generally impossible to assign wave functions to the individual parts (Landau and Lifshitz 1977, Sec. 14). In this case, even a single atom may need to be described by a density matrix, rather than a wave function. An ensemble-averaged density matrix can be defined for this case by
>
> $$\rho = \sum_k |\psi_k\rangle P_k \langle \psi_k|, \tag{5.21}$$
>
> where P_k is the probability for the state $|\psi_k\rangle$ to exist in the ensemble. Note that the properties of the density matrix of the ensemble do not depend on whether this definition or Eq. (5.13) is used.
>
> A single-atom density matrix written in the basis of position eigenstates $[\rho(\mathbf{x}, \mathbf{x}')]$ is an analog of a single-particle distribution function. A multi-particle density matrix that accounts for correlations between atoms can be written; we neglect such correlations in our treatment.

The density matrix has a number of useful properties. For a normalized density matrix, the total probability of being in any state must be unity. In other words, we have for the trace of the density matrix

$$\text{Tr}\,\rho = \sum_n \frac{1}{N} \sum_{i=1}^N \langle n|\psi_i\rangle\langle\psi_i|n\rangle = \frac{1}{N} \sum_{i=1}^N \sum_n \langle\psi_i|n\rangle\langle n|\psi_i\rangle = \frac{1}{N} \sum_{i=1}^N \langle\psi_i|\psi_i\rangle = 1. \tag{5.22}$$

Also, by construction, the density matrix is *Hermitian*, meaning that it is equal to its conjugate transpose. There is a nice mathematical property that Hermiticity is a necessary and sufficient condition for a matrix to have real eigenvalues. The eigenvalues of the density matrix are the populations in the basis in which the density matrix has no coherences. Such a basis can always be found; in some cases it may simply require a rotation of the quantization axis (see Sec. 5.4). These populations must always be greater than or equal to zero—in other words, the density matrix is *nonnegative definite*.

---

[5]The density matrix was introduced in 1927 independently by L. D. Landau and J. von Neumann. For a comprehensive discussion of density-matrix theory, see the book by Blum (1996).

In general, the matrix element ρ_{mn} depends on the magnitude of ρ_{mm} and ρ_{nn}, as well as the correlation between the two states. We can define the normalized degree of coherence between states $|m\rangle$ and $|n\rangle$ by introducing the quantity

$$\zeta_{mn} = \frac{\rho_{mn}}{\sqrt{\rho_{mm}\rho_{nn}}}. \tag{5.23}$$

The absolute value of the complex parameter ζ_{mn} can be interpreted as the degree of coherence between the states $|m\rangle$ and $|n\rangle$:

$$|\zeta_{mn}| \leq 1. \tag{5.24}$$

This is similar to the approach used in statistics to obtain the normalized correlation matrix from an unnormalized covariance matrix—see, for example, Brandt (1999).

Let us mention that whether density-matrix elements representing coherences are zero or nonzero is not an invariant physical property, but rather a property that generally depends on the choice of the basis. We will illustrate this for specific cases, for example, in Sec. 5.4 and in Chapter 8, where we will further discuss the meaning and importance of coherences.

The density matrix can be used in place of a wave function in any situation, while the converse is true only for ensembles in a pure state. There is a simple test to determine whether a system is in a pure state (Stenholm 2005). A quantum system is in a pure state if and only if it satisfies

$$\rho^2 = \rho. \tag{5.25}$$

To prove this statement, we begin by diagonalizing the normalized density matrix ρ, so that

$$\rho = U^{-1}\rho' U, \tag{5.26}$$

where U performs the change of basis and ρ' is a diagonal matrix. Suppose that ρ represents a pure state. Then ρ' has one nonzero diagonal matrix element, representing the common state of all of the atoms. Since ρ is normalized, this matrix element is equal to one. Then $\rho'^2 = \rho'$, so

$$\rho^2 = U^{-1}\rho' U U^{-1}\rho' U = U^{-1}\rho'^2 U = U^{-1}\rho' U = \rho. \tag{5.27}$$

On the other hand, if ρ does not represent a pure state, ρ' will have more than one nonzero element, each less than one. In this case, $\rho'^2 \neq \rho'$, so

$$\rho^2 = U^{-1}\rho' U U^{-1}\rho' U = U^{-1}\rho'^2 U \neq U^{-1}\rho' U = \rho, \tag{5.28}$$

so $\rho^2 = \rho$ if and only if ρ represents a pure state.

Sometimes the criterion of whether an ensemble is in a pure state is formulated in terms of the *von Neumann entropy*

$$S = -k_B \text{Tr}(\rho \ln \rho), \tag{5.29}$$

where k_B is the Boltzmann constant. Here the logarithm of a matrix can be evaluated by diagonalizing the matrix and then applying the logarithm to the eigenvalues. The entropy of an ensemble in a pure state is zero, as can be shown from Eq. (5.29), while in a mixed state $S > 0$. For a discussion of the von Neumann entropy in a modern context, see, for example, Haroche and Raimond (2006), Secs. 2.4.3 and 4.1.1.

Note that this test works only if the density matrix is normalized, i.e.,

$$\text{Tr}\,\rho = \sum_i \rho_{ii} = 1. \tag{5.30}$$

88 *Polarized atoms*

As an example, we see that an unpolarized ensemble of atoms with $F = 1$, with density matrix

$$\rho = \begin{pmatrix} \frac{1}{3} & 0 & 0 \\ 0 & \frac{1}{3} & 0 \\ 0 & 0 & \frac{1}{3} \end{pmatrix}, \tag{5.31}$$

does not satisfy the pure-state condition

$$\rho^2 = \begin{pmatrix} \frac{1}{9} & 0 & 0 \\ 0 & \frac{1}{9} & 0 \\ 0 & 0 & \frac{1}{9} \end{pmatrix} \neq \rho, \tag{5.32}$$

but a pure "stretched" state

$$\rho = \begin{pmatrix} 1 & 0 & 0 \\ 0 & 0 & 0 \\ 0 & 0 & 0 \end{pmatrix}, \tag{5.33}$$

does satisfy the condition.

5.2 Rotation of density matrices

To see the effect of an active rotation on the density matrix (which is loosely referred to as "rotating the density matrix"), consider the expectation value of an operator A (Eq. 5.12). From Eq. (3.33), we have for the expectation value after a rotation obtained using the rotation operator $\mathscr{D}$:

$$\langle A' \rangle = \text{Tr}\left(\rho A'\right) = \text{Tr}\left(\rho \mathscr{D}^\dagger A \mathscr{D}\right) = \text{Tr}\left(\mathscr{D} \rho \mathscr{D}^\dagger A\right) = \text{Tr}\left(\rho' A\right), \tag{5.34}$$

using the invariance of the trace to cyclic permutation of its arguments, where the rotated density matrix is given by

$$\rho' = \mathscr{D} \rho \mathscr{D}^\dagger. \tag{5.35}$$

This formula can also be obtained by directly rotating the wave functions in the definition (5.13). Note that the formula for rotation of a density matrix differs from that for the rotation of an operator representing an observable, in that the rotation operator is swapped with its Hermitian conjugate.

As a simple example, let us consider the rotation of a density matrix about the z-axis by an angle ϕ. In this case, the rotation matrix $\mathscr{D}$ is given by (Eq. 3.74)

$$[\mathscr{D}_{\hat{z}}(\phi)]_{mm'} = e^{-im\phi} \delta_{mm'}. \tag{5.36}$$

Then the density matrix resulting from the rotation is given by

$$\rho'_{mm'} = e^{-im\phi} \rho_{mm'} e^{im'\phi} = e^{-i(m-m')\phi} \rho_{mm'}, \tag{5.37}$$

so that Zeeman coherences are multiplied by a phase factor and populations are left unchanged.

5.3 Angular-momentum probability surfaces

We now discuss a technique for visualizing the angular-momentum polarization state specified by the density matrix. Consider the density matrix corresponding to the manifold of Zeeman sublevels of a state with total angular momentum F. If we measure the projection of the angular momentum along some axis, the possible outcomes of the measurement, according to quantum mechanics, are the values $-F, -F+1\ldots, F$. The probability of measuring a given value m along the quantization axis is given by $\rho_{mm} = \langle Fm|\rho|Fm \rangle$. We will characterize the polarization state along the quantization axis by recording the probability $\rho_{FF} = \langle FF|\rho|FF \rangle$ of finding the maximum possible projection $m = F$. To find this quantity with respect to some other axis, we rotate the state $|FF\rangle$ using the rotation operator $\mathscr{D}(\phi, \theta, 0)$ (Chapter 3), to obtain

$$|FF(\theta, \phi)\rangle = \mathscr{D}(\phi, \theta, 0)|FF\rangle$$
$$= \sum_{m'} D^{(F)}_{m'F}(\phi, \theta, 0)|Fm'\rangle. \qquad (5.38)$$

Thus the probability of finding the maximum projection of angular momentum along the (θ, ϕ) direction is given by[6]

$$\rho_{FF}(\theta, \phi) = \langle FF(\theta, \phi)|\rho|FF(\theta, \phi)\rangle$$
$$= \sum_{mm'} D^{(F)*}_{mF}(\phi, \theta, 0)\rho_{mm'} D^{(F)}_{m'F}(\phi, \theta, 0). \qquad (5.39)$$

We can plot this probability as a function of the direction of the quantization axis to obtain a visual representation of the density matrix, the *angular-momentum probability surface* (Auzinsh 1997; Rochester and Budker 2001).

As an example, consider atoms in a pure state with $F = 1$, $m = -1$, for which $|\psi\rangle = |1, -1\rangle$. The density matrix for this state is

$$\rho = \begin{pmatrix} 0 & 0 & 0 \\ 0 & 0 & 0 \\ 0 & 0 & 1 \end{pmatrix}. \qquad (5.40)$$

Using the explicit form of the D-matrix given by Eqs. (3.75) and (3.77) (or available from the *Mathematica* package discussed in Appendix F), we obtain the following equation for the distance from the origin for the surface in a direction given by the angles θ and ϕ:

$$\rho_{FF}(\theta, \phi) = \sin^4(\theta/2). \qquad (5.41)$$

This surface is plotted in Fig. 5.2, along with two of its cross-sections. Also shown is the surface for a copy of the density matrix that has been rotated by $\pi/4$ about the vector $\hat{\mathbf{n}} = \hat{\mathbf{x}} + \hat{\mathbf{y}}$. In fact, rotation of the polarization state as a whole is what occurs when a weak magnetic field is applied to the ensemble (Larmor precession). In Fig. 5.3 the probability surface for an unpolarized ensemble is drawn. The volume contained by the surface in Fig. 5.3 is smaller than that contained by the surfaces in Fig. 5.2, even though the normalization of the corresponding density matrices is the same. The volume of the probability surface is generally not the same for different polarization states of the ensemble. For example, if the polarized state shown in Fig. 5.2 gradually relaxes to the unpolarized state in Fig. 5.3, the volume as well as the shape changes in the process of relaxation, even though no atoms are lost from the ensemble.

---

[6]The expression is different from Eq. (5.35) because it corresponds to a passive rather than an active rotation.

90 Polarized atoms

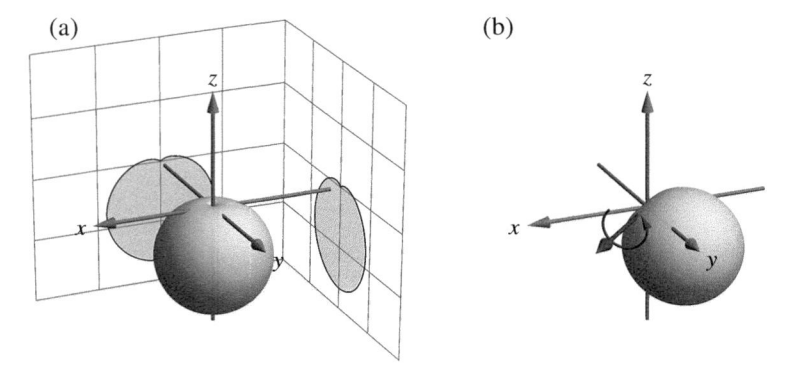

Fig. 5.2 (a) Angular-momentum probability surfaces for the density matrix given in Eq. (5.40) and two of its cross-sections; note that the $\hat{z}$-direction is the only direction along which there is zero probability to measure the maximal component of angular momentum, and the probability is equal to one in the $-\hat{z}$-direction. (b) The same density matrix rotated by $\pi/4$ around the vector $\hat{n} = \hat{x} + \hat{y}$.

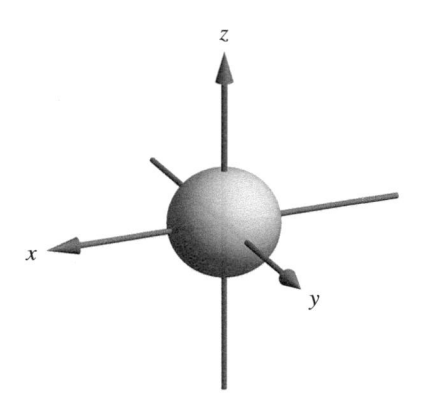

Fig. 5.3 Angular-momentum probability surface for an unpolarized ensemble with $F = 1$. The surface is a sphere with radius 1/3, equal to the population in each Zeeman sublevel.

The immediate utility of the angular-momentum probability surfaces is that they illustrate the rotational symmetry properties of the density matrix. In the examples just given the unpolarized surface is spherically symmetric, because for an unpolarized density matrix there is no preferred direction. For the stretched state $|1, -1\rangle$ the quantization axis is the only preferred direction, while the other two directions are completely equivalent, as reflected by the axial symmetry of the probability surfaces shown in Fig. 5.2. Note that, as shown in Fig. 5.2, a physical rotation changes the symmetry axis, but does not affect the symmetry about that axis.

Why is it useful to know the symmetry of the state? We will see that the symmetry of the polarization state of atoms determines the kind of optical anisotropy they can have. For example, if the distribution is axially symmetric with respect to the direction of propagation of linearly polarized light, the optical properties of the ensemble, such as absorption coefficient or refractive index, will not depend on the direction of the light polarization. We discuss this point in detail in Chapter 10.

5.4 Angular-momentum probability surfaces and the density matrix: equivalence and symmetries

Clearly, the probability surface contains significant information about the polarization of an atomic state, and hence about the corresponding density matrix. By looking at a probability surface, we can tell whether the state is polarized, what kind of polarization it has, what the preferred axes are, etc. One might wonder, how complete is the knowledge of the density matrix that one can attain from the knowledge of the angular-momentum probability surface?

Remarkably, as we show in Sec. 5.7 (see also Alexandrov et al. 2005, Appendix A), the probability surface contains as much information about the atoms as the density matrix. Thus, if the surface is known, the full density matrix can be recovered from it. In mathematical parlance, we can formulate a theorem: *There exists a one-to-one correspondence between the density matrix for the manifold of Zeeman sublevels of a state with total angular momentum F and the angular-momentum probability surface for this state.*

Armed with this result, we can deepen our understanding (while also increasing the utility) of the symmetry properties of the polarization surfaces.

Specifically, the symmetry of the angular-momentum probability surface can reveal, at a glance, important information about specific density-matrix elements. For example, the k-fold symmetry with respect to the quantization axis of the angular-momentum probability surface corresponds to the $|\Delta m| = k$ coherences in the density matrix.

We can formulate the following result (another theorem): *The angular-momentum probability surface has k-fold symmetry about some axis if and only if the density matrix, when written with the quantization axis chosen to lie along this symmetry axis, only has nonzero coherences (i.e., off-diagonal matrix elements) with $\Delta m = kN$, where N is an integer.*

To prove the general result regarding the connection between the coherences and the symmetry of the angular-momentum probability surface, we start by noting that, if the surface is invariant under a certain rotation, then, due to the one-to-one correspondence of the surface and the density matrix, the latter is also invariant under such rotation. Let us consider rotations about the quantization axis. Under a rotation by an angle φ, the $\rho_{mm'}$ element of the density matrix is multiplied by a phase factor $e^{i(m-m')\varphi}$ (Sec. 5.2). If there is k-fold symmetry, under a rotation by an angle $\varphi = 2\pi/k$, the probability surface and thus the density matrix will remain unchanged. Thus

$$e^{i(m-m')2\pi/k}\rho_{mm'} = e^{iN2\pi}\rho_{mm'}, \quad (5.42)$$

where N is an integer, which requires that either $m - m' = kN$ or $\rho_{mm'} = 0$. The proof in the converse direction is similarly straightforward: the density matrix that only has $m - m' = kN$ coherences is invariant with respect to a rotation by $\varphi = 2\pi/k$, and so is the corresponding surface.

Consider a density matrix for $F = 2$:

$$\rho = \begin{pmatrix} \frac{7}{32} & 0 & -\frac{\sqrt{\frac{3}{2}}}{16} & 0 & \frac{7}{32} \\ 0 & \frac{1}{8} & 0 & \frac{1}{8} & 0 \\ -\frac{\sqrt{\frac{3}{2}}}{16} & 0 & \frac{5}{16} & 0 & -\frac{\sqrt{\frac{3}{2}}}{16} \\ 0 & \frac{1}{8} & 0 & \frac{1}{8} & 0 \\ \frac{7}{32} & 0 & -\frac{\sqrt{\frac{3}{2}}}{16} & 0 & \frac{7}{32} \end{pmatrix}. \quad (5.43)$$

In this case all matrix elements happen to be real. A schematic representation of the density matrix and the corresponding angular-momentum probability surface are shown in Fig. 5.4(a). We see that the only nonzero coherences are those with $|\Delta m| = 2$ and 4, and the probability surface is two-fold symmetric with respect to the (vertical) quantization axis.

What if the surface is rotationally symmetric (see, for example, Fig. 5.4b)? Such a surface is k-fold symmetric for arbitrary k. Suppose a coherence $\rho_{mm'}$ is nonzero for some $m \neq m'$. Then we must have $m - m' = kN$ for some integer N and every k. But if we choose $k > |m - m'|$ the equality cannot be satisfied. This means that if the surface is totally symmetric around an axis, the density matrix will have no coherences when written in the basis corresponding to that axis (Fig. 5.4b).

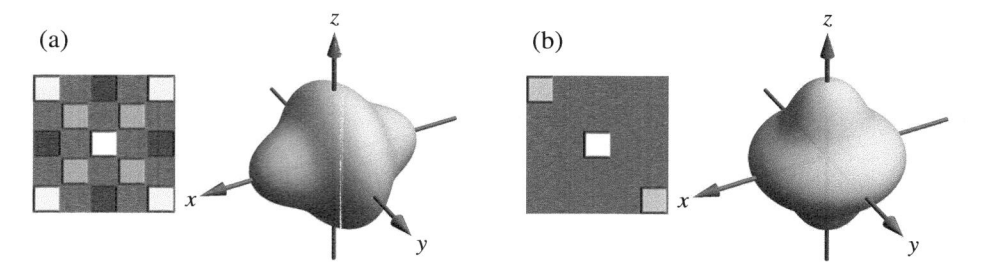

Fig. 5.4 Examples illustrating the relation between the symmetry of the angular-momentum probability surface and the coherences in the density matrix. In these examples, $F = 2$, and the relevant symmetry axis is vertical, which is also chosen to be the quantization axis. (a) The surface is two-fold symmetric; the corresponding density matrix (shown schematically to the left of the surface) has only coherences with $|\Delta m| = 2$ and 4. (b) The surface is rotationally symmetric; the density matrix has no coherences (see text). The magnitudes of the density-matrix elements are indicated by shading: zeros are represented by 50% gray, lighter shades are positive values, and darker shades are negative.

Thus if we have a surface with various axes of symmetry, in order to write the density matrix in the simplest form, we should choose the quantization axis along the axis of highest symmetry. While this axis may be difficult to determine directly from the density matrix, it can be found by simply looking at the angular-momentum probability surface.

5.5 Temporal evolution of the density matrix: the Liouville equations

A state of a system, whether it is described by a wave function or, more generally, by a density matrix, evolves in time. The evolution is governed by the initial conditions, the internal structure of the particles, as well as by external fields such as static electric and magnetic fields, or light fields. In addition, a system under study is generally not completely isolated from the environment, and the interactions with the environment need to be modeled by including phenomenological terms into the evolution equations. These interactions often lead to relaxation and repopulation, for example, radiative decay, collisions, and the departure and arrival of particles from and to the system (for example, atoms flying into and out of the interaction region).

5.5.1 Derivation of the Liouville equation

In an attempt to derive an equation describing the evolution of a density matrix, we first take the time derivative of ρ, using the explicit definition in terms of the average of the individual atomic wave functions

$$i\hbar \frac{d}{dt}\rho = i\hbar \frac{1}{N} \sum_{i=1}^{N} \frac{d}{dt} |\psi_i\rangle\langle\psi_i|$$
$$= \frac{1}{N} \sum_{i=1}^{N} \left[\left(i\hbar \frac{d}{dt}|\psi_i\rangle\right) \langle\psi_i| + |\psi_i\rangle \left(i\hbar \frac{d}{dt}\langle\psi_i|\right) \right].$$
(5.44)

The expression is now in terms of the evolution of the atomic wave functions, and so we can use the time-dependent Schrödinger equation

Temporal evolution of the density matrix: the Liouville equations

$$i\hbar \frac{d}{dt}|\psi\rangle = H|\psi\rangle \qquad (5.45)$$

and its Hermitian conjugate

$$-i\hbar \frac{d}{dt}\langle\psi| = \langle\psi|H. \qquad (5.46)$$

This gives

$$i\hbar \frac{d}{dt}\rho = \frac{1}{N}\sum_{i=1}^{N}(H|\psi_i\rangle\langle\psi_i| - |\psi_i\rangle\langle\psi_i|H)$$
$$= [H, \rho]. \qquad (5.47)$$

This equation is the *Liouville equation* for a density matrix.

5.5.2 Relaxation and repopulation

Our discussion so far contains an implicit but important approximation. We have neglected the interactions between our chosen system and the rest of the world. In fact, including such interactions using the methods described so far would be very difficult, because we would need to work with very large ensembles containing not only the atoms in our system, but every atom in the environment, and describe all the interactions between these atoms. Luckily, in most cases, the interaction with the environment can be described phenomenologically as *relaxation* and *repopulation* of the atoms in the system, without explicitly including atoms in the environment. Examples of processes leading to relaxation include collisions of atoms between themselves and with the walls of the container, as well as radiative decay of excited states. The latter example is rather interesting because the role of the environment is played by the quantum vacuum, a point to which we will return in subsequent discussion.

Suppose that we prepare atoms in an excited state. These atoms will eventually spontaneously decay. Sometimes relaxation of the amplitude of a basis state $|\psi_m\rangle$ at a rate $\Gamma_m/2$ is included in the Schrödinger equation phenomenologically as an additional term on the right-hand side of Eq. (5.45) leading to

$$i\hbar \frac{d}{dt}|\psi_m\rangle = \sum_k H_{mk}|\psi_k\rangle - i\hbar\frac{1}{2}\Gamma_m|\psi_m\rangle. \qquad (5.48)$$

Unfortunately, this remedy helps only in certain cases. It can be done in this form only if the basis-state amplitude relaxes to some other states outside the system that we are analyzing (for a more detailed discussion see, for example, Shore 1990). This assumption can be realistic for some systems, but generally it does not hold. Mathematically, the problem is related to the fact that using Eq. (5.48) effectively means working with a non-Hermitian Hamiltonian, which leads to nonconservation of probabilities and nonorthogonal eigenstates.

It turns out that these difficulties can largely be avoided in the density matrix formalism. If the relaxation of an atomic state $|n\rangle$ is assumed exponential with rate $\Gamma_n/2$, the rate of change of the population of this level due to relaxation is given by $\dot\rho_{nn} = -\Gamma_n\rho_{nn}$. Furthermore, the rate of change of a coherence $\rho_{nn'}$ is given by the average of the decay rates: $\dot\rho_{nn'} = -(\Gamma_n + \Gamma_{n'})\rho_{nn'}/2$. (This is equivalent to the coherence decaying at the sum of the decay rates for the amplitudes in the wave function picture.) These terms can be conveniently included in the Liouville equation using the *relaxation matrix* $\widehat{\Gamma}$ as (Stenholm 2005)

$$i\hbar\frac{d}{dt}\rho = [H,\rho] - i\hbar\frac{1}{2}(\widehat{\Gamma}\rho + \rho\widehat{\Gamma}). \tag{5.49}$$

Here $\widehat{\Gamma}$ is a diagonal matrix with the population decay rate of each state on the diagonal. These rates can include terms due to spontaneous emission and collisions, as well as atoms leaving the experimental volume (*transit relaxation*).

There is also another aspect to each of these processes, *repopulation*, that requires additional terms in the density-matrix equations. For example, as atoms leave the region of interest, other atoms may be entering. Depending on the situation, these "new" atoms may be polarized or unpolarized. For example, if the incoming atoms are unpolarized ground-state atoms with n sublevels in the ground state, the influx term in the Liouville equation is a diagonal matrix, which we will denote by Λ, with the ground-state terms equal to γ/n, where γ is the *transit rate*. This form of the influx term ensures a proper normalization of the density matrix as we can see from the trivial case of no external fields. Then the commutator term in the Liouville equation is zero (i.e., the Hamiltonian commutes with the density matrix), so the Liouville equation reduces to

$$i\hbar\frac{d}{dt}\rho = -i\hbar\frac{1}{2}(\widehat{\Gamma}\rho + \rho\widehat{\Gamma}) + i\hbar\Lambda = 0 \tag{5.50}$$

in the steady state. The upper-state and off-diagonal elements of the density matrix are zero; for the ground-state diagonal elements, the equation is written

$$-i\gamma\rho_{mm} + i\gamma/n = 0, \tag{5.51}$$

with solution $\rho_{mm} = 1/n$ for the n nonzero elements.

Another repopulation term describes atoms that spontaneously decay from the upper states and end up in one of the lower levels described by the density matrix, leading to the transfer of populations and possibly also coherences from the upper to the lower states. We will discuss the specific form of the repopulation matrix for this case in Sec. 12.1. Comprehensive discussion of the inclusion of relaxation in the density-matrix equations is given by, for example, Stenholm (2005); Blum (1996); and Shore (1990).

5.6 Example: alignment-to-orientation conversion

Now that we have introduced the density matrix, the equations describing its evolution, and the polarization probability surfaces, let us consider an example showing how these can be applied in practice. We choose the example in such a way as to provide a bridge to a subsequent discussion of one of the themes of this book—atomic *polarization moments*.

Consider $F = 1$ atoms prepared in the state described by the density matrix (5.40) and shown in Fig. 5.2(a). Suppose an electric field $\mathcal{E}$ is applied along $\hat{\mathbf{x}}$. We will use the Liouville equation to find the evolution of the state and visualize the evolution with the angular-momentum probability surfaces.[7]

---

[7] Note that, in this particular case, a wave function rather than a density-matrix calculation would have been sufficient to describe the system.

Example: alignment-to-orientation conversion

The effective Hamiltonian for the interaction of the electric field with the atoms is given in Eqs. (4.77–4.78) as

$$H = \begin{pmatrix} \frac{1}{4}\alpha_2 \mathcal{E}_x^2 & 0 & -\frac{3}{4}\alpha_2 \mathcal{E}_x^2 \\ 0 & -\frac{1}{2}\alpha_2 \mathcal{E}_x^2 & 0 \\ -\frac{3}{4}\alpha_2 \mathcal{E}_x^2 & 0 & \frac{1}{4}\alpha_2 \mathcal{E}_x^2 \end{pmatrix}, \quad (5.52)$$

where we have included the tensor polarizability α_2 but neglected the scalar polarizability α_0.

Using this Hamiltonian in the Liouville equation (5.47) we obtain the following set of differential equations for the density-matrix elements $\rho_{m,m'}$:

$$\dot{\rho}_{1,1} = i\frac{\omega_S}{2}(\rho_{-1,1} - \rho_{1,-1}), \quad (5.53\text{a})$$

$$\dot{\rho}_{1,0} = i\frac{\omega_S}{2}(\rho_{-1,0} - \rho_{1,0}), \quad (5.53\text{b})$$

$$\dot{\rho}_{1,-1} = i\frac{\omega_S}{2}(\rho_{-1,-1} - \rho_{1,1}), \quad (5.53\text{c})$$

$$\dot{\rho}_{0,1} = i\frac{\omega_S}{2}(\rho_{0,1} - \rho_{0,-1}), \quad (5.53\text{d})$$

$$\dot{\rho}_{0,0} = 0, \quad (5.53\text{e})$$

$$\dot{\rho}_{0,-1} = i\frac{\omega_S}{2}(\rho_{0,-1} - \rho_{0,1}), \quad (5.53\text{f})$$

$$\dot{\rho}_{-1,1} = i\frac{\omega_S}{2}(\rho_{1,1} - \rho_{-1,-1}), \quad (5.53\text{g})$$

$$\dot{\rho}_{-1,0} = i\frac{\omega_S}{2}\alpha_2(\rho_{1,0} - \rho_{-1,0}), \quad (5.53\text{h})$$

$$\dot{\rho}_{-1,-1} = i\frac{\omega_S}{2}(\rho_{1,-1} - \rho_{-1,1}), \quad (5.53\text{i})$$

where

$$\omega_S = \frac{2\pi}{\tau_S} = \frac{3\alpha_2 \mathcal{E}^2}{2\hbar} \quad (5.54)$$

is the Stark splitting, i.e., the energy difference between the eigenstates of the Hamiltonian. (For $F = 1$ there is only one such splitting.) The differential equations can be solved in a straightforward way,[8] yielding for the time-dependent density matrix

$$\rho(t) = \begin{pmatrix} \sin^2\left(\frac{\omega_S t}{2}\right) & 0 & \frac{i}{2}\sin \omega_S t \\ 0 & 0 & 0 \\ -\frac{i}{2}\sin \omega_S t & 0 & \cos^2\left(\frac{\omega_S t}{2}\right) \end{pmatrix}. \quad (5.55)$$

This formula explicitly shows periodic evolution of the density matrix at a frequency ω_S. Since this evolution is induced by Stark splitting, it is known as *Stark beats*, in analogy to the beating that one hears when two closely spaced audio-frequency tones are played simultaneously.

We can visualize the dynamic evolution of the atomic polarization using the probability surface. The radius vector describing the probability surface is (Eq. 5.39; see also Sec. 3.4.2)

$$r(\theta, \phi) = \rho_{FF}(\theta, \phi) = \frac{3}{8} + \frac{1}{8}\cos 2\theta - \frac{1}{2}\cos\theta \cos \omega_S t - \frac{1}{4}\sin 2\phi \sin^2\theta \sin \omega_S t. \quad (5.56)$$

This surface is plotted in Fig. 5.5 for several values of t. The surface exhibits the changing

---

[8] To minimize the amount of effort involved, this can be easily done in a computer algebra system such as *Mathematica*.

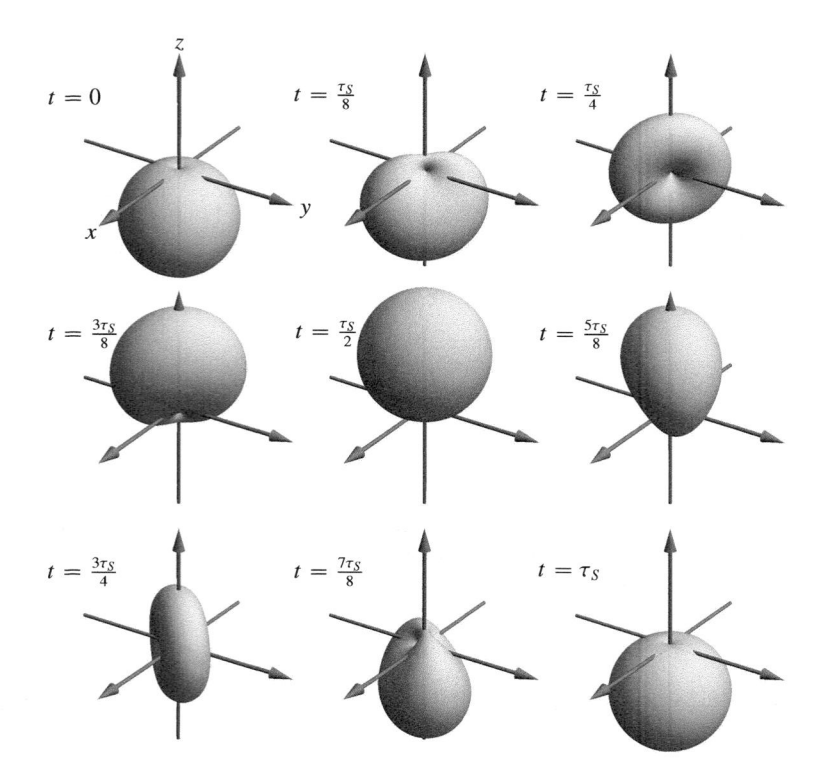

Fig. 5.5 A sequence of probability surfaces representing electric-field-induced evolution of a state with $F=1$. The state is initially ($t=0$) stretched along $-\hat{\mathbf{z}}$ and an electric field is applied along $\hat{\mathbf{y}}$, causing Stark beats with period τ_S. At $t=\tau_S/4$, the atoms are aligned along $\hat{\mathbf{x}}+\hat{\mathbf{y}}$. At $t=\tau_S/2$ the atoms have orientation in the $+\hat{\mathbf{z}}$-direction, at $t=3\tau_S/4$ they are aligned along $\hat{\mathbf{x}}-\hat{\mathbf{y}}$, and, finally, at $t=\tau_S$ the polarization returns to its initial state.

symmetry of the atomic polarization. Initially the ensemble has a preferred direction: the surface is symmetric about the z-axis, but not symmetric with respect to inversion of the z-axis—it "points," or has *orientation*, in the $-\hat{\mathbf{z}}$-direction. At $t=\tau_S/4$ the "doughnut"-shaped surface has a preferred axis (it is symmetric about $\hat{\mathbf{x}}+\hat{\mathbf{y}}$), but no preferred direction (it is the same in both the $\hat{\mathbf{x}}+\hat{\mathbf{y}}$ and $-(\hat{\mathbf{x}}+\hat{\mathbf{y}})$ directions. Polarization with one preferred axis and but no preferred direction is said to be *aligned*. (There are also aligned states which contain orthogonal preferred axes; orientation and alignment are defined more rigorously in terms of *polarization moments* (*PM*), in the next section.) The ensemble continues to oscillate between states that contain orientation and those that contain only alignment. Thus this type of evolution is known as alignment-to-orientation conversion, because it can be responsible for the appearance of orientation in systems in which only alignment was initially created (see Auzinsh and Ferber 1992; Budker *et al.* 2002a, and references therein).

5.7 Multipole moments

In the previous section we discussed orientation and alignment, which are types of atomic polarization having particular rotational symmetries. It can be useful to classify the symmetries

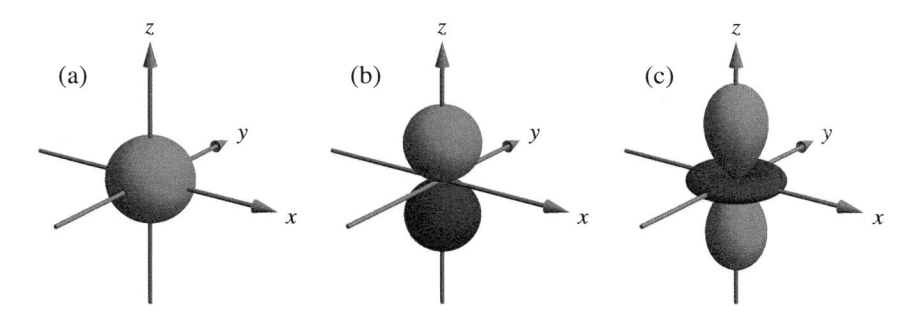

Fig. 5.6 Plots of the spherical harmonics (a) Y_{00}, (b) Y_{10}, (c) Y_{20}. In (b) and (c) the darker shading indicates negative function values.

appearing in a given atomic state; for example, oriented and aligned states interact with polarized light quite differently.

In Chapter 2 we described the spherical harmonics $Y_{lm}(\theta, \varphi)$, which are functions of θ and φ, each with unique symmetry properties determined by l and m. A few of the lowest-rank spherical harmonics are plotted in Fig. 5.6. We see that Y_{10} and Y_{20} have a preferred direction and a preferred axis, respectively, like the oriented and aligned states described in the previous section, while the Y_{00} state is isotropic, like the unpolarized state shown in Fig. 5.3. In fact, the density matrix can be entirely decomposed into components (polarization moments) each of which exhibits symmetry corresponding to a particular spherical harmonic. Geometrically, this could be done by finding the linear combination of Y_{lm}'s that equals the angular-momentum probability surface for a particular state. However, it will often be more useful to us to think of the polarization moments in terms of a decomposition of the density matrix itself. We now describe how to go about this decomposition.

The density matrix ρ was introduced as an operator (Sec. 5.1). When we write the density operator as a density matrix, we are actually performing a decomposition of the density operator over a complete set of basis operators. For an atomic state with angular momentum F we write

$$\rho = \sum_{mm'} \rho_{mm'} |m\rangle\langle m'|, \tag{5.57}$$

where $\rho_{mm'}$ are the density-matrix elements, and $|m\rangle\langle m'|$ are the $(2F+1) \times (2F+1)$ basis operators. For example, the diagonal operators $|m\rangle\langle m|$ are the *projection operators* onto the states $|m\rangle$. As discussed in this chapter, this set of basis operators is of great physical importance, representing the populations of and coherences between the Zeeman sublevels. However, it does not always have the most desirable properties; for example, the same state can be described by a very different set of nonzero density-matrix elements depending on the choice of quantization axis (Fig. 5.4).

What is a more agreeable set of basis operators from the point of view of symmetry? In Chapter 3 we describe the irreducible tensor operators, which are operators that have the rotational symmetries of the spherical harmonics. Because the irreducible tensor operators are orthogonal, we also find that they form a complete basis: since a rank-κ operator has $2\kappa + 1$ components $q = -\kappa, \ldots, \kappa$, we see that the set of operators with rank $\kappa = 0, \ldots, 2F$ have

98 *Polarized atoms*

$$\sum_{\kappa=0}^{2F}(2\kappa+1) = (2F+1)^2 \tag{5.58}$$

independent components, equal to the number of degrees of freedom of an operator on a state with angular momentum F. We therefore define a set of irreducible tensor operators $\mathcal{T}_q^\kappa$, called *polarization operators*, and expand the density matrix over this set:

$$\rho = \sum_{\kappa=0}^{2F}\sum_{q=-\kappa}^{\kappa} \rho^{\kappa q} \mathcal{T}_q^\kappa. \tag{5.59}$$

The *polarization moments* (PM; also known as *state multipoles*) of the density matrix are then defined as the coefficients $\rho^{\kappa q}$ in this expansion. (We will also use the term polarization moment to refer to a term $\rho^{\kappa q}\mathcal{T}_q^\kappa$ of the expansion.) We write ρ^κ to refer to the collection of components $\rho^{\kappa q}$ with $q = -\kappa, \ldots, \kappa$.

The definition of an irreducible tensor, given in Sec. 3.7, is quite restrictive, so that the properties of an irreducible tensor are essentially constrained up to an overall factor. We can therefore define the polarization operators by simply giving a normalization condition

$$\operatorname{Tr} \mathcal{T}_q^\kappa \mathcal{T}_{q'}^{\kappa'\dagger} = \delta_{\kappa\kappa'}\delta_{qq'} \tag{5.60}$$

and a phase convention

$$\mathcal{T}_q^{\kappa\dagger} = (-1)^q \mathcal{T}_{-q}^\kappa = \mathcal{T}^{\kappa q}. \tag{5.61}$$

Using the Wigner–Eckart theorem (Sec. 3.8), we can then find the reduced matrix element of the polarization operators:

$$\begin{aligned}
1 &= \operatorname{Tr} \mathcal{T}_q^\kappa \mathcal{T}_q^{\kappa\dagger} \\
&= \sum_{mm'} \left(\mathcal{T}_q^\kappa\right)_{mm'} \left(\mathcal{T}_q^\kappa\right)_{mm'}^* \\
&= \sum_{mm'} \frac{1}{2F+1} |\langle F \| \mathcal{T}^\kappa \| F \rangle|^2 \langle Fm'\kappa q | Fm \rangle^2 \\
&= \frac{1}{2\kappa+1} |\langle F \| \mathcal{T}^\kappa \| F \rangle|^2,
\end{aligned} \tag{5.62}$$

giving

$$\langle F \| \mathcal{T}^\kappa \| F \rangle = \sqrt{2\kappa+1}, \tag{5.63}$$

where we have used the Clebsch–Gordan sum rule

$$\sum_{m_1 m_3} \langle F_1 m_1 F_2 m_2 | F_3 m_3 \rangle^2 = \frac{2F_3+1}{2F_2+1}, \tag{5.64}$$

and we know that the reduced matrix element is real because $\mathcal{T}_0^{\kappa\dagger} = \mathcal{T}_0^\kappa$ according to Eq. (5.61).

Now that we have the reduced matrix element, we can find the matrix elements of the polarization operators:

$$\langle Fm'|\mathcal{T}_q^\kappa|Fm\rangle = \sqrt{\frac{2\kappa+1}{2F+1}}\langle Fm\kappa q|Fm'\rangle \qquad (5.65)$$
$$= (-1)^{F-m}\langle Fm'F,-m|\kappa q\rangle,$$

where we have used the Clebsch–Gordan identity

$$\langle F_1m_1 F_2m_2|F_3m_3\rangle = (-1)^{F_1-m_1}\sqrt{\frac{2F_3+1}{2F_2+1}}\langle F_3m_3 F_1,-m_1|F_2m_2\rangle. \qquad (5.66)$$

To find the coefficients $\rho^{\kappa q}$ in the polarization-moment expansion, we multiply both sides of Eq. (5.59) by $\mathcal{T}_{q'}^{\kappa'\dagger}$, take the trace, and use the normalization condition:

$$\mathrm{Tr}\left(\rho \mathcal{T}_{q'}^{\kappa'\dagger}\right) = \sum_{\kappa q}\rho^{\kappa q}\,\mathrm{Tr}\left(\mathcal{T}_q^\kappa \mathcal{T}_{q'}^{\kappa'\dagger}\right) = \sum_{\kappa q}\rho^{\kappa q}\delta_{\kappa\kappa'}\delta_{qq'} = \rho^{\kappa'q'}. \qquad (5.67)$$

Using Eq. (5.65) we can find a formula for the polarization moments in terms of the density-matrix elements:

$$\begin{aligned}\rho^{\kappa q} &= \mathrm{Tr}\left(\rho \mathcal{T}_q^{\kappa\dagger}\right) \\ &= \sum_{mm'}\rho_{m'm}\left(\mathcal{T}_q^\kappa\right)_{m'm} \\ &= \sum_{mm'}(-1)^{F-m}\langle Fm'F,-m|\kappa q\rangle\rho_{m'm},\end{aligned} \qquad (5.68)$$

and the inverse formula

$$\begin{aligned}\rho_{m'm} &= \sum_{\kappa q}\rho^{\kappa q}\left(\mathcal{T}_q^\kappa\right)_{m'm} \\ &= \sum_{\kappa q}(-1)^{F-m}\langle Fm'F,-m|\kappa q\rangle\rho^{\kappa q},\end{aligned} \qquad (5.69)$$

which allows reconstruction of the density-matrix elements if the polarization moments are known.

The coefficients $\rho^{\kappa q}$ are contravariant quantities. The corresponding covariant components ρ_q^κ are determined by

$$A = \sum_{\kappa=0}^{2F}\sum_{q=-\kappa}^{\kappa}\rho_q^\kappa \mathcal{T}^{\kappa q}, \qquad (5.70)$$

and can be found as

$$\mathrm{Tr}\left(\rho \mathcal{T}_{q'}^{\kappa'}\right) = \sum_{\kappa q}\rho_q^\kappa\,\mathrm{Tr}\left(\mathcal{T}^{\kappa q}\mathcal{T}_{q'}^{\kappa'}\right) = \sum_{\kappa q}\rho_q^\kappa \delta_{\kappa\kappa'}\delta_{qq'} = \rho_{q'}^{\kappa'}. \qquad (5.71)$$

Explicitly, we have

100   Polarized atoms

$$\rho_q^\kappa = \text{Tr}\left(\rho T_q^\kappa\right)$$
$$= \sum_{mm'} \rho_{m'm} \left(T_q^\kappa\right)_{mm'} \qquad (5.72)$$
$$= \sum_{mm'} (-1)^{F-m'} \langle FmF, -m'|\kappa q\rangle \rho_{m'm},$$

and the inverse formula

$$\rho_{m'm} := \sum_{\kappa q} \rho_q^\kappa \left(T_q^{\kappa\dagger}\right)_{m'm}$$
$$:= \sum_{\kappa q} (-1)^{F-m'} \langle FmF, -m'|\kappa q\rangle \rho_q^\kappa. \qquad (5.73)$$

The polarization moments ρ^κ are given names by analogy with the expansion of a static electric field into multipole moments. In that case 2^κ point charges are required to produce a field configuration consisting of only a rank-κ moment. Thus we obtain the designations for the lowest-rank moments ρ^0—*monopole moment* (equal to the population divided by $\sqrt{2F+1}$), ρ^1—*dipole moment* or *orientation*, ρ^2—*quadrupole moment* or *alignment*, ρ^3—*octupole moment*, ρ^4—*hexadecapole moment*, ρ^5—*triacontadipole moment*, and ρ^6—*hexacontatetrapole moment*. (The utility of these terms diminishes rapidly as κ increases beyond this point.) Polarization moments as high as hexacontatetrapole have been created and detected (see Chapter 18).

It should be noted that there are other definitions of the terms "orientation" and "alignment" in the literature. For example, Zare (1988) identifies alignment with any of the even moments in atomic polarization (quadrupole, hexadecapole, etc.) and orientation with the odd moments (dipole, octupole, etc.). To add to the confusion, nuclear physicists often use the term "polarization" to specifically designate orientation, whereas in our convention, the term is used to describe an ensemble that has any moment with $\kappa > 0$.

We now discuss a few important properties of the polarization moments. Some physical intuition about them can be gained by examining their relation to the Zeeman populations and coherences (Sec. 5.1). This is found from the transformation equations (5.68) and (5.69), and the properties of the Clebsch–Gordan coefficients, in particular the requirement that $m' - m = q$ for the Clebsch–Gordan coefficients in the sums to be nonzero. For example, in Eq. (5.68), setting $q = 0$ implies that only terms with $m = m'$, i.e., the Zeeman-sublevel populations ρ_{mm}, contribute to the sum. Furthermore, setting $m = m'$ in Eq. (5.69) shows that the Zeeman populations can be expressed entirely in terms of the polarization moments with $q = 0$. Thus the polarization moments ρ_0^κ describe polarization along the quantization axis, i.e., *longitudinal polarization* characterizing to the distribution of Zeeman populations. Analogously, a nonzero PM with $q \neq 0$ means that $\rho_{mm'}$ coherences with $m' - m = q$ are nonzero. When there are coherences between the sublevels, $\rho_q^\kappa \neq 0$ for some $q \neq 0$, and the medium is said to have *transverse polarization*..

Table 5.1, presented at the end of this chapter, surveys some of the PMs that may be present in low-F states. Longitudinal (ρ_0^κ) and linear combinations of transverse ($\rho_{\pm q}^\kappa$ for $q > 0$) polarization moments for states with total angular momentum $F = 0, \frac{1}{2}, \ldots, 3$. Particular linear combinations of $\rho_{\pm q}^\kappa$ must be chosen in order for the corresponding density matrix to be Hermitian. (This is a requirement for physically meaningful density matrices, see Sec. 5.1.) For each polarization moment, the density matrix in the mm' basis is displayed. For PMs with $\kappa > 0$, this density matrix is nonphysical, as it has zero trace. To the right is shown the corresponding physical matrix (with unit trace) representing the given PM plus the rank-zero moment. The amplitude of the given PM in this density matrix is chosen to be the largest possible consistent with a physical (i.e., positive definite) density matrix. Also shown for each polarization moment is the AMPS (Sec. 5.3) corresponding to the physical density matrix.

The polarization moments are helpful in understanding the symmetry of the atomic polarization, as each moment has the symmetry of a particular spherical harmonic $Y_{\kappa q}(\theta, \phi)$. Note that the symmetry is an invariant of physical rotations of the system. An example is Larmor precession: the linear Zeeman effect causes polarization moments with the same rank κ and different q to transform into each other, but does not change the rank κ or the symmetry of the angular momentum, as discussed in Sec. 5.3.[9] In the special case of rotations around $\hat{\mathbf{z}}$, no moments are mixed—components are merely altered by a phase factor.

In addition, the multipole expansion is useful in certain situations for reducing the complexity of the density matrix evolution equations, especially for states with large angular momentum. In molecular spectroscopy, one typically deals with states of much larger angular momenta (rotational quantum numbers $\simeq 100$) than for atoms. In this case, the standard Liouville equations of motion form a large coupled system that can be difficult to solve. However, the equations of motion for the multipole expansion coefficients can be much simpler (Auzinsh and Ferber 1991). This idea was introduced by Dyakonov (1964) and Ducloy (1976) and later applied to the analysis of a large variety of nonlinear magneto-optical effects in diatomic molecules (see the book by Auzinsh and Ferber, 1995, and references therein).

To illustrate the simplification afforded by the use of polarization moments in the presence of symmetries in the problem, consider excitation of an $F = 5$ excited state from an unpolarized $F = 4$ initial state with weak x-polarized light. The excited-state density matrix normalized to unit trace is

$$\rho = \begin{pmatrix} \frac{3}{22} & 0 & -\frac{1}{22\sqrt{5}} & 0 & 0 & 0 & 0 & 0 & 0 & 0 & 0 \\ 0 & \frac{6}{55} & 0 & -\frac{\sqrt{3}}{55} & 0 & 0 & 0 & 0 & 0 & 0 & 0 \\ -\frac{1}{22\sqrt{5}} & 0 & \frac{29}{330} & 0 & -\frac{\sqrt{\frac{14}{3}}}{55} & 0 & 0 & 0 & 0 & 0 & 0 \\ 0 & -\frac{\sqrt{3}}{55} & 0 & \frac{4}{55} & 0 & -\frac{\sqrt{\frac{7}{30}}}{11} & 0 & 0 & 0 & 0 & 0 \\ 0 & 0 & -\frac{\sqrt{\frac{14}{3}}}{55} & 0 & \frac{7}{110} & 0 & -\frac{1}{22} & 0 & 0 & 0 & 0 \\ 0 & 0 & 0 & -\frac{\sqrt{\frac{7}{30}}}{11} & 0 & \frac{2}{33} & 0 & -\frac{\sqrt{\frac{7}{30}}}{11} & 0 & 0 & 0 \\ 0 & 0 & 0 & 0 & -\frac{1}{22} & 0 & \frac{7}{110} & 0 & -\frac{\sqrt{\frac{14}{3}}}{55} & 0 & 0 \\ 0 & 0 & 0 & 0 & 0 & -\frac{\sqrt{\frac{7}{30}}}{11} & 0 & \frac{4}{55} & 0 & -\frac{\sqrt{3}}{55} & 0 \\ 0 & 0 & 0 & 0 & 0 & 0 & -\frac{\sqrt{\frac{14}{3}}}{55} & 0 & \frac{29}{330} & 0 & -\frac{1}{22\sqrt{5}} \\ 0 & 0 & 0 & 0 & 0 & 0 & 0 & -\frac{\sqrt{3}}{55} & 0 & \frac{6}{55} & 0 \\ 0 & 0 & 0 & 0 & 0 & 0 & 0 & 0 & -\frac{1}{22\sqrt{5}} & 0 & \frac{3}{22} \end{pmatrix}. \quad (5.74)$$

This matrix can be considerably simplified to

---

[9] The nonlinear Zeeman effect, on the other hand, causes more complicated evolution that does not preserve the polarization symmetry (see Chapter 16).

102 *Polarized atoms*

$$\rho = \begin{pmatrix} 0 & 0 & 0 & 0 & 0 & 0 & 0 & 0 & 0 & 0 & 0 \\ 0 & \frac{3}{55} & 0 & 0 & 0 & 0 & 0 & 0 & 0 & 0 & 0 \\ 0 & 0 & \frac{16}{165} & 0 & 0 & 0 & 0 & 0 & 0 & 0 & 0 \\ 0 & 0 & 0 & \frac{7}{55} & 0 & 0 & 0 & 0 & 0 & 0 & 0 \\ 0 & 0 & 0 & 0 & \frac{8}{55} & 0 & 0 & 0 & 0 & 0 & 0 \\ 0 & 0 & 0 & 0 & 0 & \frac{5}{33} & 0 & 0 & 0 & 0 & 0 \\ 0 & 0 & 0 & 0 & 0 & 0 & \frac{8}{55} & 0 & 0 & 0 & 0 \\ 0 & 0 & 0 & 0 & 0 & 0 & 0 & \frac{7}{55} & 0 & 0 & 0 \\ 0 & 0 & 0 & 0 & 0 & 0 & 0 & 0 & \frac{16}{165} & 0 & 0 \\ 0 & 0 & 0 & 0 & 0 & 0 & 0 & 0 & 0 & \frac{3}{55} & 0 \\ 0 & 0 & 0 & 0 & 0 & 0 & 0 & 0 & 0 & 0 & 0 \end{pmatrix} \quad (5.75)$$

by choosing the quantization axis along the direction of the light polarization. However, in either case, the density matrices in the Zeeman basis disguise the true symmetry of the state: in fact, the excited state is described by only two nonzero polarization moments,

$$\rho_0^0 = \frac{1}{\sqrt{11}}, \quad \rho_0^2 = -\frac{1}{5}\sqrt{\frac{26}{33}}, \quad (5.76)$$

in the case of z-polarized light, and four nonzero polarization moments,

$$\rho_0^0 = \frac{1}{\sqrt{11}}, \quad \rho_0^2 = -\frac{1}{5}\sqrt{\frac{26}{33}}, \quad \rho_{\pm 2}^2 = -\frac{1}{10}\sqrt{\frac{13}{11}}, \quad (5.77)$$

in the case of x-polarized light. In this example, going to the polarization-moment representation does, indeed, greatly reduce the number of parameters that one has to deal with.

We can apply the idea of the polarization moments representing the symmetry of the system to the analysis of relaxation processes. Suppose there is some relaxation mechanism that is on average not associated with any specific directions in space, such as relaxation in atomic collisions. Because the index q describes spatial components, the relaxation rates for different polarization moments can depend on κ but not on q. It turns out, there are certain general inequalities that constrain possible relative values of the relaxation rates for different polarization moments (see, for example, Auzinsh and Ferber 1995, Sec. 5.8).

The angular-momentum probability surface for a density matrix can be written in terms of its polarization moments and the spherical harmonics $Y_{\kappa q}(\theta, \phi)$ (Alexandrov et al. 2005):

$$\rho_{FF}(\theta, \phi) = \sqrt{\frac{4\pi}{2F+1}} \sum_{\kappa=0}^{2F} \sum_{q=-\kappa}^{\kappa} \langle FF\kappa 0|FF\rangle \rho^{\kappa q} Y_{\kappa q}(\theta, \phi). \quad (5.78)$$

Given a probability distribution $\rho_{FF}(\theta, \phi)$, the polarization moments $\rho^{\kappa q}$ and thus the density matrix elements $\rho_{mm'}$ can be recovered using the orthonormality of the spherical harmonics, so all three are complete and equivalent descriptions of the ensemble-averaged polarization. All three descriptions can be useful in calculations, especially in the large-F limit, for which $\rho_{FF}(\theta, \phi)$ corresponds [apart from a normalization factor (Auzinsh and Ferber 1995)] to the classical probability distribution of the angular-momentum direction.

Multipole moments

Another way to visually represent angular-momentum states quite similar to the angular-momentum probability surfaces is to plot a spatial distribution related to the *Wigner function* for angular-momentum states (see, for instance, Dowling *et al.* 1994). The Wigner function is defined with a formula much like Eq. (5.78):

$$W(\theta, \phi) = \sum_{\kappa=0}^{2F} \sum_{q=-\kappa}^{\kappa} \rho^{\kappa q} Y_{\kappa q}(\theta, \phi), \tag{5.79}$$

the essential difference being that the contributions of polarization moments of various ranks are weighted differently. This difference has consequences for the physical interpretation of the surfaces: the angular-momentum probability surfaces represent probability distributions and so are always positive, while the Wigner functions can be negative and thus cannot be interpreted as probability distributions. Wigner functions are sometimes plotted as positive distributions by defining

$$f(\theta, \phi) = 1 + \frac{W(\theta, \phi)}{\hbar\sqrt{F(F+1)}}. \tag{5.80}$$

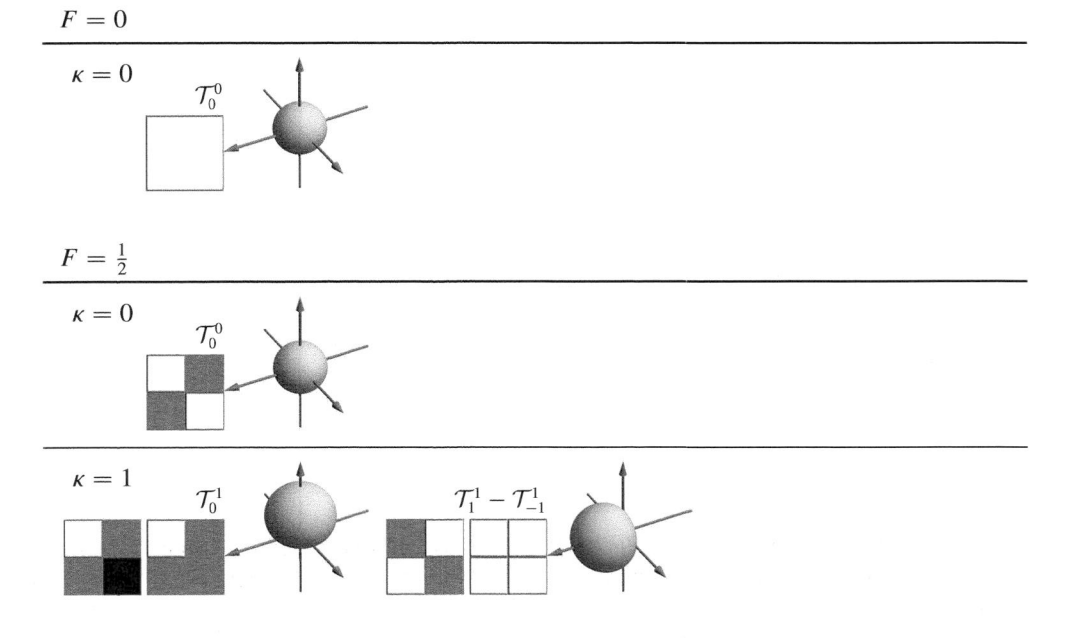

Table 5.1 A survey of polarization moments for states with $F = 0, \frac{1}{2}, \ldots, 3$. Density matrices proportional to various linear combinations of the polarization operators $\mathcal{T}_q^\kappa$ are shown, along with the corresponding AMPS. For $\kappa > 0$, both the nonphysical (with zero trace) and physical (with unit trace) density matrices are shown (see text). For the linear combinations of polarization moments used, the corresponding density matrices are real. The values are indicated by shading: zero is 50% gray, positive values are lighter, and negative values are darker. The AMPS are normalized so that the surfaces representing the $\kappa = 0$ PMs are the same radius for every value of F. [In the standard normalization the surface for an unpolarized density matrix (ρ_0^0) is a sphere of radius $1/(2F+1)$.]

$F = 0$

$\kappa = 0 \qquad \mathcal{T}_0^0$

$F = \frac{1}{2}$

$\kappa = 0 \qquad \mathcal{T}_0^0$

$\kappa = 1 \qquad \mathcal{T}_0^1 \qquad\qquad \mathcal{T}_1^1 - \mathcal{T}_{-1}^1$

104 *Polarized atoms*

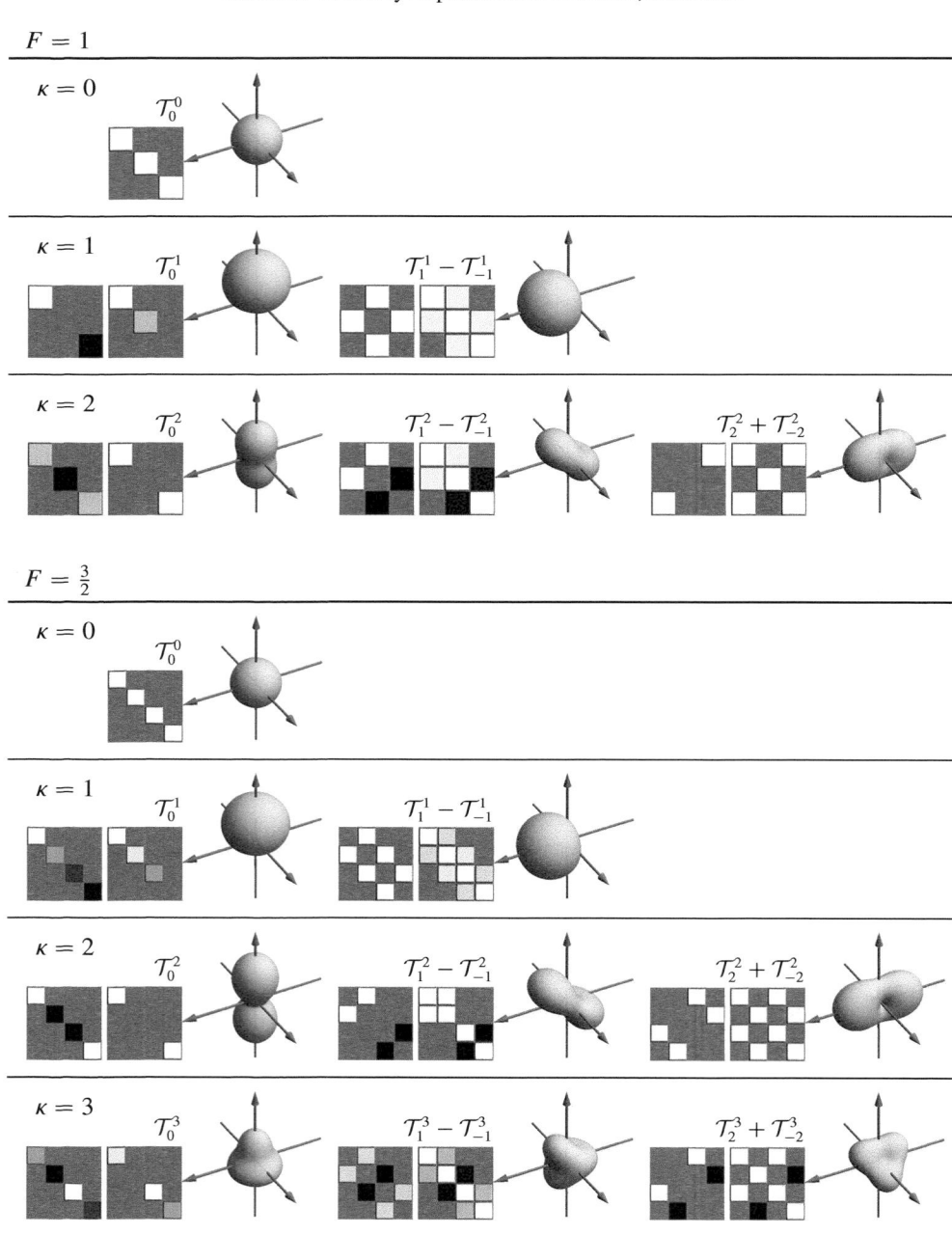

Table 5.1 A survey of polarization moments, continued.

$F = 1$

$\kappa = 0$: $\mathcal{T}_0^0$

$\kappa = 1$: $\mathcal{T}_0^1$; $\mathcal{T}_1^1 - \mathcal{T}_{-1}^1$

$\kappa = 2$: $\mathcal{T}_0^2$; $\mathcal{T}_1^2 - \mathcal{T}_{-1}^2$; $\mathcal{T}_2^2 + \mathcal{T}_{-2}^2$

$F = \frac{3}{2}$

$\kappa = 0$: $\mathcal{T}_0^0$

$\kappa = 1$: $\mathcal{T}_0^1$; $\mathcal{T}_1^1 - \mathcal{T}_{-1}^1$

$\kappa = 2$: $\mathcal{T}_0^2$; $\mathcal{T}_1^2 - \mathcal{T}_{-1}^2$; $\mathcal{T}_2^2 + \mathcal{T}_{-2}^2$

$\kappa = 3$: $\mathcal{T}_0^3$; $\mathcal{T}_1^3 - \mathcal{T}_{-1}^3$; $\mathcal{T}_2^3 + \mathcal{T}_{-2}^3$

Multipole moments 105

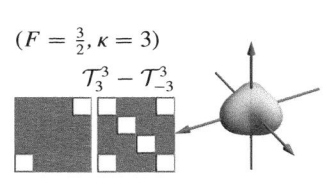

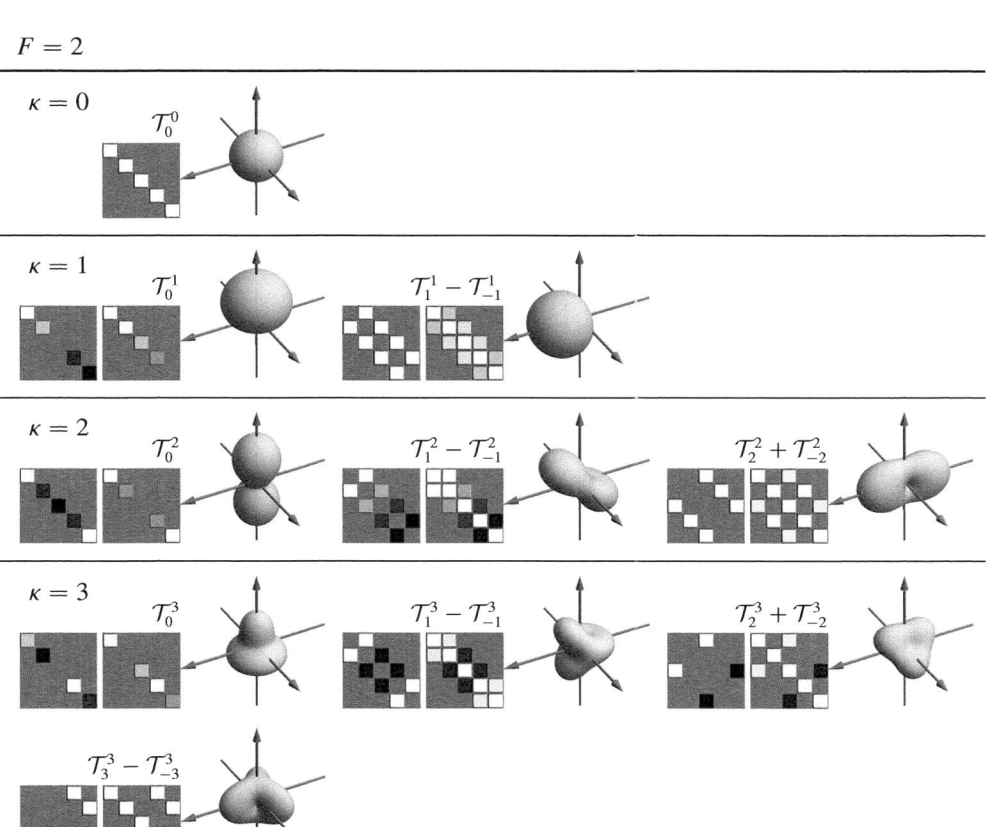

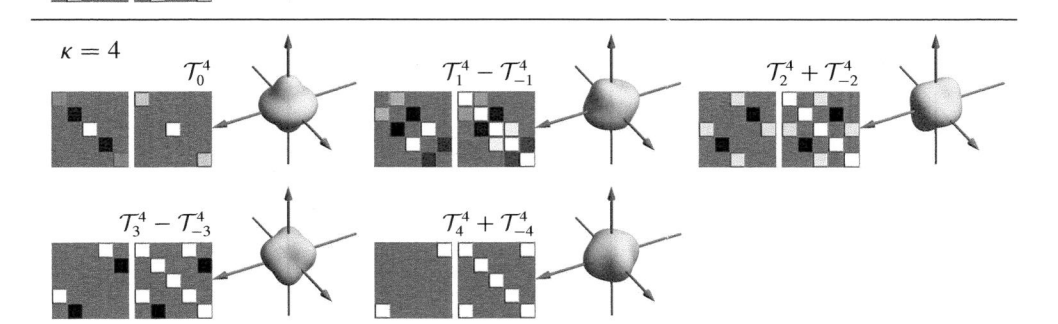

Table 5.1 A survey of polarization moments, continued.

$(F = \frac{3}{2}, \kappa = 3)$

$\mathcal{T}_3^3 - \mathcal{T}_{-3}^3$

$F = 2$

$\kappa = 0$ $\mathcal{T}_0^0$

$\kappa = 1$ $\mathcal{T}_0^1$ $\mathcal{T}_1^1 - \mathcal{T}_{-1}^1$

$\kappa = 2$ $\mathcal{T}_0^2$ $\mathcal{T}_1^2 - \mathcal{T}_{-1}^2$ $\mathcal{T}_2^2 + \mathcal{T}_{-2}^2$

$\kappa = 3$ $\mathcal{T}_0^3$ $\mathcal{T}_1^3 - \mathcal{T}_{-1}^3$ $\mathcal{T}_2^3 + \mathcal{T}_{-2}^3$

$\mathcal{T}_3^3 - \mathcal{T}_{-3}^3$

$\kappa = 4$ $\mathcal{T}_0^4$ $\mathcal{T}_1^4 - \mathcal{T}_{-1}^4$ $\mathcal{T}_2^4 + \mathcal{T}_{-2}^4$

$\mathcal{T}_3^4 - \mathcal{T}_{-3}^4$ $\mathcal{T}_4^4 + \mathcal{T}_{-4}^4$

106 *Polarized atoms*

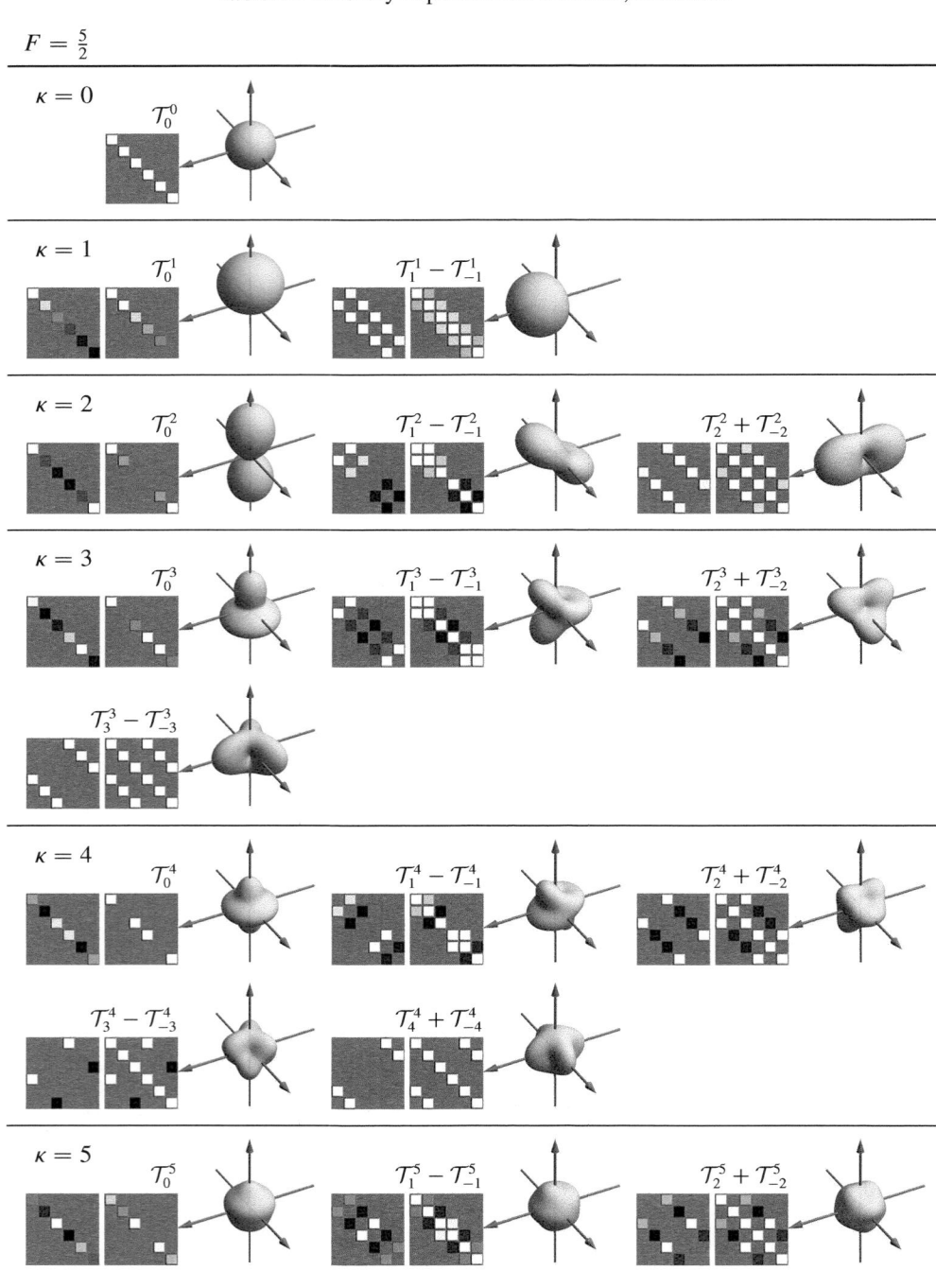

Table 5.1 A survey of polarization moments, continued.

$F = \frac{5}{2}$

$\kappa = 0$ — $\mathcal{T}_0^0$

$\kappa = 1$ — $\mathcal{T}_0^1$, $\mathcal{T}_1^1 - \mathcal{T}_{-1}^1$

$\kappa = 2$ — $\mathcal{T}_0^2$, $\mathcal{T}_1^2 - \mathcal{T}_{-1}^2$, $\mathcal{T}_2^2 + \mathcal{T}_{-2}^2$

$\kappa = 3$ — $\mathcal{T}_0^3$, $\mathcal{T}_1^3 - \mathcal{T}_{-1}^3$, $\mathcal{T}_2^3 + \mathcal{T}_{-2}^3$, $\mathcal{T}_3^3 - \mathcal{T}_{-3}^3$

$\kappa = 4$ — $\mathcal{T}_0^4$, $\mathcal{T}_1^4 - \mathcal{T}_{-1}^4$, $\mathcal{T}_2^4 + \mathcal{T}_{-2}^4$, $\mathcal{T}_3^4 - \mathcal{T}_{-3}^4$, $\mathcal{T}_4^4 + \mathcal{T}_{-4}^4$

$\kappa = 5$ — $\mathcal{T}_0^5$, $\mathcal{T}_1^5 - \mathcal{T}_{-1}^5$, $\mathcal{T}_2^5 + \mathcal{T}_{-2}^5$

Multipole moments 107

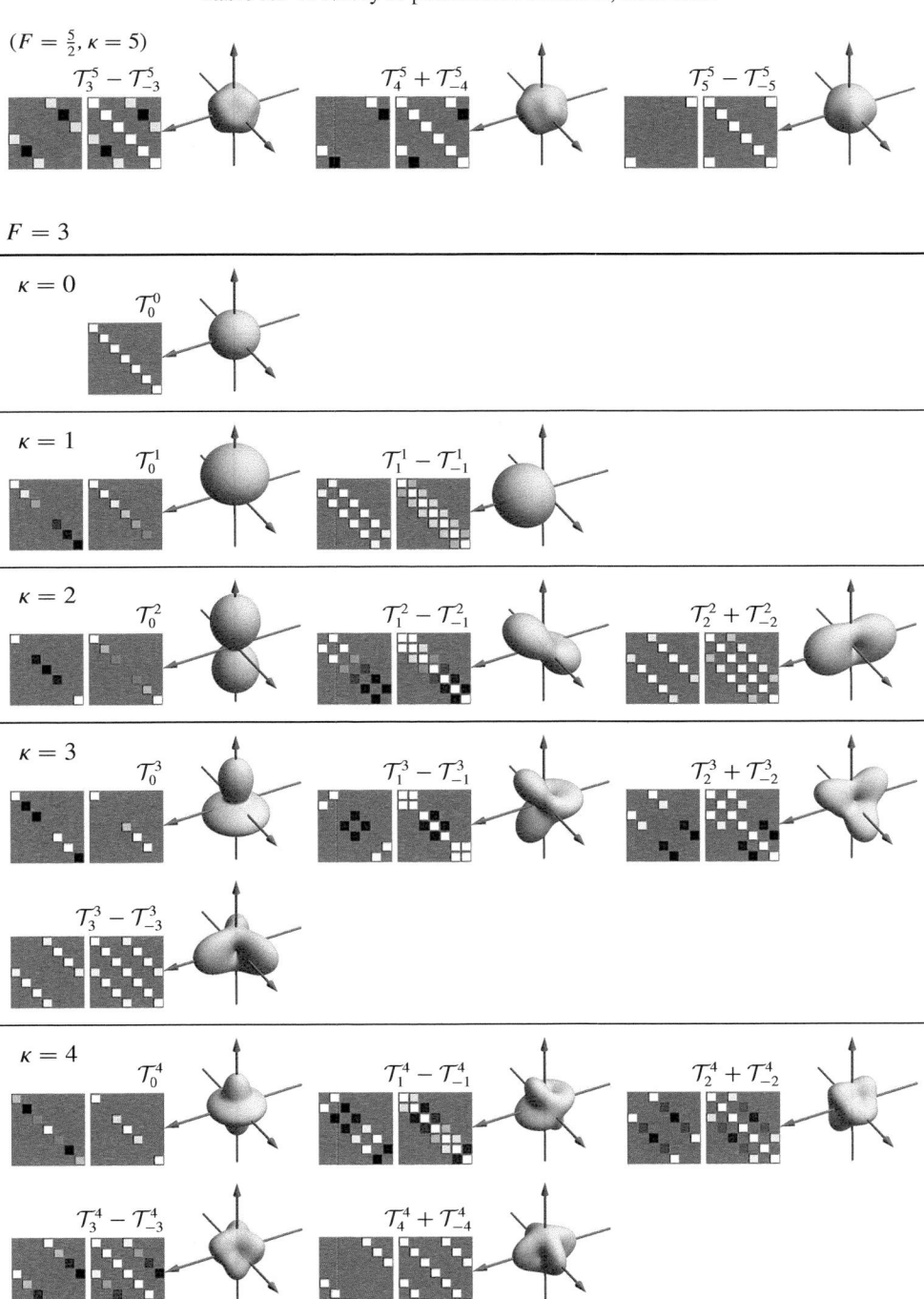

Table 5.1 A survey of polarization moments, continued.

$(F = \frac{5}{2}, \kappa = 5)$

$\mathcal{T}_3^5 - \mathcal{T}_{-3}^5$ $\qquad\qquad \mathcal{T}_4^5 + \mathcal{T}_{-4}^5$ $\qquad\qquad \mathcal{T}_5^5 - \mathcal{T}_{-5}^5$

$F = 3$

$\kappa = 0$ $\qquad \mathcal{T}_0^0$

$\kappa = 1$ $\qquad \mathcal{T}_0^1 \qquad\qquad \mathcal{T}_1^1 - \mathcal{T}_{-1}^1$

$\kappa = 2$ $\qquad \mathcal{T}_0^2 \qquad\qquad \mathcal{T}_1^2 - \mathcal{T}_{-1}^2 \qquad\qquad \mathcal{T}_2^2 + \mathcal{T}_{-2}^2$

$\kappa = 3$ $\qquad \mathcal{T}_0^3 \qquad\qquad \mathcal{T}_1^3 - \mathcal{T}_{-1}^3 \qquad\qquad \mathcal{T}_2^3 + \mathcal{T}_{-2}^3$

$\mathcal{T}_3^3 - \mathcal{T}_{-3}^3$

$\kappa = 4$ $\qquad \mathcal{T}_0^4 \qquad\qquad \mathcal{T}_1^4 - \mathcal{T}_{-1}^4 \qquad\qquad \mathcal{T}_2^4 + \mathcal{T}_{-2}^4$

$\mathcal{T}_3^4 - \mathcal{T}_{-3}^4 \qquad\qquad \mathcal{T}_4^4 + \mathcal{T}_{-4}^4$

108 *Polarized atoms*

Table 5.1 A survey of polarization moments, continued.

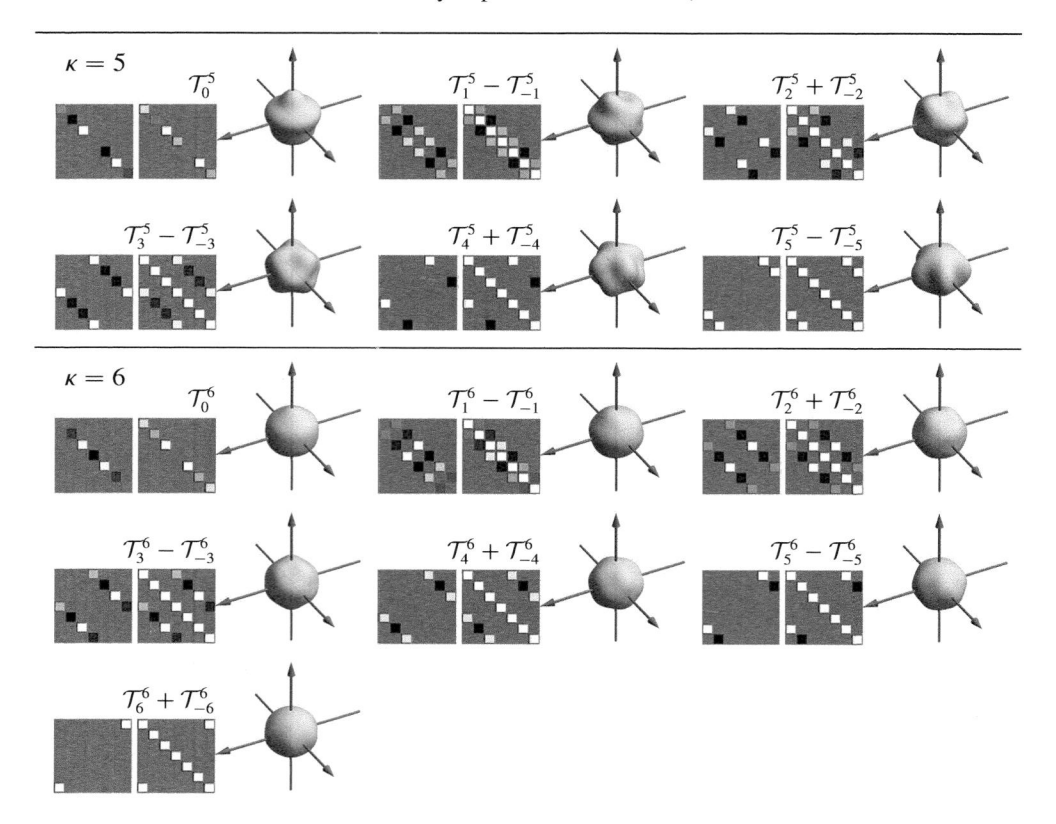

$\kappa = 5$	$\mathcal{T}_0^5$	$\mathcal{T}_1^5 - \mathcal{T}_{-1}^5$	$\mathcal{T}_2^5 + \mathcal{T}_{-2}^5$
	$\mathcal{T}_3^5 - \mathcal{T}_{-3}^5$	$\mathcal{T}_4^5 + \mathcal{T}_{-4}^5$	$\mathcal{T}_5^5 - \mathcal{T}_{-5}^5$
$\kappa = 6$	$\mathcal{T}_0^6$	$\mathcal{T}_1^6 - \mathcal{T}_{-1}^6$	$\mathcal{T}_2^6 + \mathcal{T}_{-2}^6$
	$\mathcal{T}_3^6 - \mathcal{T}_{-3}^6$	$\mathcal{T}_4^6 + \mathcal{T}_{-4}^6$	$\mathcal{T}_5^6 - \mathcal{T}_{-5}^6$
	$\mathcal{T}_6^6 + \mathcal{T}_{-6}^6$		

6
Polarized light

To analyze the interaction of atoms with light we need to take into account not only the angular momenta of the atomic states but, equally importantly, the polarization of the light that interacts with the atoms. In this book (if not explicitly stated otherwise) we will consider light as a plane wave. In other words, we assume that atoms that we excite are far from the source of the light and that the electric and magnetic fields $\mathcal{E}(\mathbf{r},t)$ and $\mathbf{B}(\mathbf{r},t)$ that describe the light wave at any instant are uniform in the plane perpendicular to the direction of the propagation of the light (Fig. 6.1).

In the past, the polarization direction and polarization plane (containing the polarization vector and the propagation direction) were often defined in terms of the magnetic-field vector (Born and Wolf 1999). However, this convention is not often used in the current literature. When the interaction of light with matter is considered, there are compelling reasons to regard $\mathcal{E}(\mathbf{r},t)$ as the vector characterizing the light field. The force $\mathbf{F}$ on an electron is given by the Lorentz law

$$\mathbf{F} = -e\left(\mathcal{E} + \frac{\mathbf{v}}{c} \times \mathbf{B}\right), \qquad (6.1)$$

where e is the magnitude of the electron charge, c is the speed of light, and $\mathbf{v}$ is the velocity of the electron. Hence the electric vector is seen to act even when the electron is at rest, while

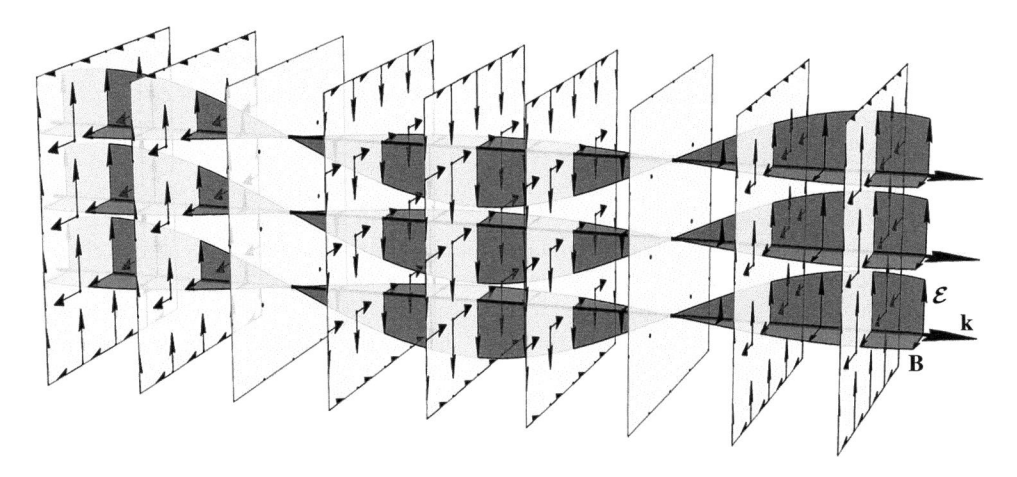

Fig. 6.1 In a plane wave propagating in a transparent homogeneous medium, both electric and magnetic fields are uniform in any plane that is perpendicular to the wave vector. Shown is the case of a running linearly polarized wave; for such a wave the electric and magnetic fields are in phase.

the magnetic vector plays a part only when the particle is in motion; moreover, since v/c is usually rather small, the effect of the magnetic field may often be neglected (Born and Wolf 1999). Thus, we adopt the prevailing terminology of laser spectroscopy and consider an electromagnetic field to be polarized in the direction of its electric-field vector and its polarization plane to be the plane containing the electric vector and the direction of propagation of the light.

6.1 The light polarization ellipse

We first discuss two special cases of light polarization. The electric-field vector of light with *linear* (or *plane*) polarization oscillates in a polarization plane that is fixed in the laboratory reference frame (Fig. 6.1). Linear polarization defines a *light polarization axis* along which the electric field points, but it does not select a preferred direction. Thus, linearly polarized light can be represented by a double-headed arrow indicating the alternating direction of the electric field.

For *circularly polarized* light, the electric-field vector $\mathcal{E}$ has constant magnitude and, at a fixed point in space, rotates around the wave propagation direction. Circularly polarized light can be described as having *chirality*, related to the angular momentum carried by the light beam (Sec. 6.3). Unfortunately, there are two competing conventions for referring to the rotation direction of circularly polarized light. In quantum electrodynamics, the right-hand rule is used: for right-circularly polarized light, if the thumb points in the light propagation direction, the fingers follow the direction of the photon electric field. The traditional terminology in optics to describe the rotation direction, however, is opposite to this convention. In optics, if the electric field in the light beam viewed face-on rotates counterclockwise (as seen in a plane affixed to the observer), the light is said to be *left circularly polarized*. If the electric field rotates in the clockwise direction, this light is said to be *right circularly polarized* (Born and Wolf 1999). In this book we will stick to this latter definition of right- and left-circularly polarized light.

We can also describe polarized light in reference to a fixed coordinate system, independent of the light propagation direction. We will do this by decomposing the light polarization into components along the z-axis and perpendicular to it. In a chosen coordinate system we call linearly z-polarized light π-*polarized light*. Light with polarization that rotates clockwise as viewed from the positive z-direction is called σ^- polarized and counterclockwise rotating polarization is called σ^+. (Thus right-circularly polarized light propagating in the $\hat{\mathbf{z}}$-direction corresponds to σ^- polarization, while $\hat{\mathbf{z}}$ propagating left-circular light corresponds to σ^+.) Arbitrarily polarized light, no matter what its propagation direction, can be decomposed into π, σ^-, and σ^+ polarizations. Note that light can only have a π-polarized component if it does not propagate along the z-axis; pure π-polarized light must propagate perpendicularly to the z-axis, while pure σ^+ and σ^- light must propagate along the z-axis.

Both linearly and circularly polarized light are special cases of *elliptically polarized* light. For general elliptically polarized light, the tip of the electric field vector at a given point in space draws an ellipse (Fig. 6.2). The figure shows several geometrical parameters that can be used to specify the shape of the ellipse and the polarization state of the light. We now discuss the relationships between these parameters, which will be used in our subsequent discussion. The reader may wish to skip the remainder of this section and refer back to it as necessary.

The electric field $\mathcal{E}$ of an arbitrary fully polarized plane wave is conveniently represented as a complex quantity, the real part of which is taken in order to obtain the physical field. Thus we write

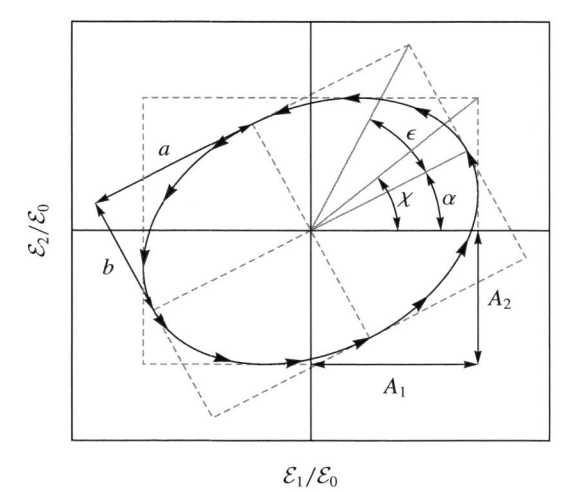

Fig. 6.2 Elliptically polarized light. For a given value of the spatial coordinate **r**, the electric-field vector of the light draws out an ellipse. Various parameters characterizing the ellipse are discussed in the text.

$$\mathcal{E}(\mathbf{r}, t) = \mathcal{E}_0 \operatorname{Re} \hat{\boldsymbol{\varepsilon}} e^{i(\mathbf{k} \cdot \mathbf{r} - \omega t)}, \tag{6.2}$$

where $\mathcal{E}_0$ is the (real) amplitude of the electric field, the light polarization and phase is described by the complex polarization unit vector $\hat{\boldsymbol{\varepsilon}}$, and the time and position dependence is contained in the exponential factor, where **k** is the wave vector and ω is the light frequency.

The polarization vector $\hat{\boldsymbol{\varepsilon}}$ has three independent real parameters—there is a complex coefficient for each of the two directions perpendicular to $\hat{\mathbf{k}}$, and one constraint from the condition $\hat{\boldsymbol{\varepsilon}} \cdot \hat{\boldsymbol{\varepsilon}}^* = 1$. Including the electric-field amplitude, there are four independent parameters specifying the polarization, amplitude, and phase of a light electric field. There are several parametrizations commonly used that express these parameters more explicitly.

First, we can write Eq. (6.2) as

$$\mathcal{E}(\mathbf{r}, t) = \mathcal{E}_0 \operatorname{Re} \left[\left(A_1 \hat{\mathbf{e}}_1 + A_2 e^{i\phi} \hat{\mathbf{e}}_2 \right) e^{i(\mathbf{k} \cdot \mathbf{r} - \omega t + \varphi)} \right], \tag{6.3}$$

where φ is the overall phase, $\hat{\mathbf{e}}_1$ and $\hat{\mathbf{e}}_2$ are two orthogonal unit vectors perpendicular to **k** with $\hat{\mathbf{e}}_2 = \hat{\mathbf{k}} \times \hat{\mathbf{e}}_1$, A_1 and A_2 are the relative positive real amplitudes of two orthogonal components, normalized so that $A_1^2 + A_2^2 = 1$, and ϕ is the relative phase between the components[1] (we take ϕ to be in the range $-\pi \leq \phi \leq \pi$). If ϕ is positive, $\mathcal{E}$ rotates in the counterclockwise direction and the light is left elliptically polarized. If ϕ is negative, $\mathcal{E}$ rotates in the clockwise direction and the light is right elliptically polarized. Together, A_1, A_2, and ϕ specify the light polarization; because of the normalization condition on $A_{1,2}$, there are two independent real parameters between these quantities.

If we define

$$\tan \chi = \frac{A_2}{A_1}, \tag{6.4}$$

with $0 \leq \chi \leq \pi/2$, then $A_1 = \cos \chi$, $A_2 = \sin \chi$, and

$$\mathcal{E}(\mathbf{r}, t) = \mathcal{E}_0 \operatorname{Re} \left[\left(\cos \chi \, \hat{\mathbf{e}}_1 + \sin \chi \, e^{i\phi} \hat{\mathbf{e}}_2 \right) e^{i(\mathbf{k} \cdot \mathbf{r} - \omega t + \varphi)} \right]. \tag{6.5}$$

We will refer to this as the χ–ϕ parametrization.

---

[1] Our notations pertaining to the description of light polarization follow, where possible, those of Huard (1997).

112 Polarized light

We can also parametrize the light electric field using a geometric description of the polarization ellipse. We begin by considering a special case in which the principal axes of the polarization ellipse are along $\hat{\mathbf{e}}_{1,2}$. In this case, the components of the electric field along those axes are $\pm \pi/2$ out of phase. Then we can write the semimajor and semiminor axes as $a = \max(A_1, A_2)$ and $b = \min(A_1, A_2)$, and set $\phi = \pm \pi/2$. From Eq. (6.3) the electric field is given by

$$\mathcal{E}(\mathbf{r}, t) = \mathcal{E}_0 \operatorname{Re}\left[(a\,\hat{\mathbf{e}}_1 \pm ib\,\hat{\mathbf{e}}_2)\, e^{i(\mathbf{k}\cdot\mathbf{r}-\omega t+\varphi)}\right]. \tag{6.6}$$

Similarly to Eq. (6.5), since $a^2 + b^2 = 1$, we can write

$$\mathcal{E}(\mathbf{r}, t) = \mathcal{E}_0 \operatorname{Re}\left[(\cos\epsilon\,\hat{\mathbf{e}}_1 + i \sin\epsilon\,\hat{\mathbf{e}}_2)\, e^{i(\mathbf{k}\cdot\mathbf{r}-\omega t+\varphi)}\right], \tag{6.7}$$

where

$$\epsilon = \pm \arctan b/a \tag{6.8}$$

is the *ellipticity*. This parametrization is useful to describe experiments, for example, Faraday rotation (Chapter 10), in which the change in α as the light traverses an atomic medium is directly related to the particular atomic properties that are being investigated.

In general, the principal axes are not along $\hat{\mathbf{e}}_{1,2}$. We can describe this case by rotating $\mathcal{E}$ by an angle α about $\hat{\mathbf{k}}$. The rotated field is given by

$$\mathcal{R}_\alpha \mathcal{E}(\mathbf{r}, t) = \mathcal{E}_0 \operatorname{Re}\Big\{ [\cos\epsilon\,(\cos\alpha\,\hat{\mathbf{e}}_1 + \sin\alpha\,\hat{\mathbf{e}}_2) \\ + i \sin\epsilon\,(\cos\alpha\,\hat{\mathbf{e}}_2 - \sin\alpha\,\hat{\mathbf{e}}_1)] e^{i(\mathbf{k}\cdot\mathbf{r}-\omega t+\varphi)} \Big\} \tag{6.9a}$$

$$= \mathcal{E}_0 \operatorname{Re}\Big\{ [(\cos\alpha\cos\epsilon - i\sin\alpha\sin\epsilon)\,\hat{\mathbf{e}}_1 \\ + (\sin\alpha\cos\epsilon + i\cos\alpha\sin\epsilon)\,\hat{\mathbf{e}}_2] e^{i(\mathbf{k}\cdot\mathbf{r}-\omega t+\varphi)} \Big\}. \tag{6.9b}$$

We will refer to this as the α–ϵ parametrization, in which we describe the polarization ellipse by the ellipticity ϵ, which is given by the arctangent of the ratio of the minor and major axes of the ellipse, and the angle α between the major axis and the $\hat{\mathbf{e}}_1$-axis. In contrast, the χ–ϕ parametrization describes the polarization ellipse in terms of the ratio of the amplitudes and the relative phase of the electric field oscillations along the $\hat{\mathbf{e}}_1$ and $\hat{\mathbf{e}}_2$ axes.

Lastly, the Stokes parameters are commonly used because of their direct connection with practical light-polarization measurements. In unnormalized form they are defined by

$$P_0 = I = I_1 + I_2, \tag{6.10a}$$
$$P_1 = I_1 - I_2, \tag{6.10b}$$
$$P_2 = I_{+\pi/4} - I_{-\pi/4}, \tag{6.10c}$$
$$P_3 = I_+ - I_-, \tag{6.10d}$$

where I is the total light intensity, $I_{1,2}$ are the intensities of the components along $\hat{\mathbf{e}}_{1,2}$, $I_{\pm\pi/4}$ are the intensities of the linear components at angles $\pm\pi/4$ to $\hat{\mathbf{e}}_1$, and $I_\pm$ are the intensities of the circular components $\sigma^\pm$. We will use the normalized Stokes parameters, which, using Eq. (6.3), are given by

The light polarization ellipse **113**

$$S_1 = \frac{P_1}{P_0} = A_1^2 - A_2^2, \tag{6.11a}$$

$$S_2 = \frac{P_2}{P_0} = \left|\frac{A_1 + A_2 e^{i\phi}}{\sqrt{2}}\right|^2 - \left|\frac{A_1 - A_2 e^{i\phi}}{\sqrt{2}}\right|^2 = 2 A_1 A_2 \cos\phi, \tag{6.11b}$$

$$S_3 = \frac{P_3}{P_0} = \left|\frac{-A_1 + i A_2 e^{i\phi}}{\sqrt{2}}\right|^2 - \left|\frac{A_1 + i A_2 e^{i\phi}}{\sqrt{2}}\right|^2 = 2 A_1 A_2 \sin\phi. \tag{6.11c}$$

From this we can see that $S_1^2 + S_2^2 + S_3^2 = 1$, so $S_{1,2,3}$ represent two independent parameters specifying the light polarization.

To write the light electric field in terms of the normalized Stokes parameters, we invert Eqs. (6.11):

$$A_1 = \sqrt{\frac{1 + S_1}{2}}, \tag{6.12a}$$

$$A_2 = \sqrt{\frac{1 - S_1}{2}}, \tag{6.12b}$$

$$\phi = \arctan\frac{S_3}{S_2}. \tag{6.12c}$$

Substituting into Eq. (6.3), we find

$$\mathcal{E}(\mathbf{r}, t) = \mathcal{E}_0 \operatorname{Re}\left[\left(\sqrt{\frac{1 + S_1}{2}}\,\hat{\mathbf{e}}_1 + \sqrt{\frac{1 - S_1}{2}}\, e^{i \arctan S_3/S_2}\,\hat{\mathbf{e}}_2\right) e^{i(\mathbf{k}\cdot\mathbf{r} - \omega t + \varphi)}\right]. \tag{6.13}$$

It is useful to find formulas to translate between the χ–ϕ, α–ϵ, and Stokes parametrizations. To relate the χ–ϕ and Stokes parametrizations, we equate Eqs. (6.5) and (6.13) [or use Eqs. (6.4) and (6.12)]. This gives

$$\chi = \arctan\sqrt{\frac{1 - S_1}{1 + S_1}} = \frac{1}{2}\arccos S_1, \tag{6.14a}$$

$$\phi = \arctan\frac{S_3}{S_2}. \tag{6.14b}$$

Inverting these formulas gives

$$S_1 = \cos 2\chi, \tag{6.15a}$$

$$S_2 = \cos\phi \sin 2\chi, \tag{6.15b}$$

$$S_3 = \sin\phi \sin 2\chi. \tag{6.15c}$$

To relate the α–ϵ and Stokes parametrizations, we equate Eqs. (6.9) and (6.13). These two formulas have a different overall phase, because of the rotation by the angle α that we applied in Eq. (6.9). Therefore we set the phase in Eq. (6.9) to φ' and allow it to vary when we equate the two formulas. After some trigonometric manipulation, we find

$$S_1 = \cos 2\alpha \cos 2\epsilon, \tag{6.16a}$$

$$S_2 = \sin 2\alpha \cos 2\epsilon, \tag{6.16b}$$

$$S_3 = \sin 2\epsilon, \tag{6.16c}$$

114 *Polarized light*

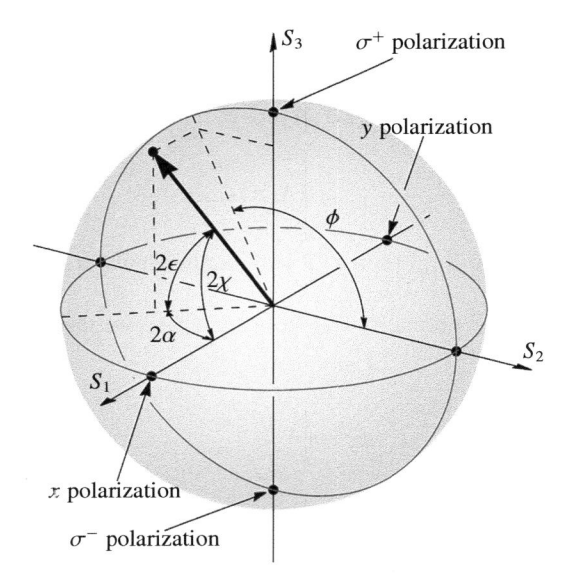

Fig. 6.3 The Poincaré sphere. The polarization state is represented by a point whose coordinates are the normalized Stokes parameters. The figure also illustrates a geometrical interpretation of the various light polarization parameters.

and an overall phase difference of $\varphi' = \varphi + \arctan(\tan\alpha \tan\epsilon)$. Inverting these formulas gives

$$\alpha = \frac{1}{2}\arctan\frac{S_2}{S_1}, \qquad (6.17a)$$

$$\epsilon = \frac{1}{2}\arcsin S_3. \qquad (6.17b)$$

Using Eqs. (6.14–6.17), we can relate the χ–ϕ and α–ϵ parametrizations:

$$\alpha = \frac{1}{2}\arctan(\tan 2\chi \cos\phi), \qquad (6.18a)$$

$$\epsilon = \frac{1}{2}\arcsin(\sin 2\chi \sin\phi), \qquad (6.18b)$$

and

$$\chi = \frac{1}{2}\arccos(\cos 2\alpha \cos 2\epsilon), \qquad (6.19a)$$

$$\phi = \arctan(\csc 2\alpha \tan 2\epsilon). \qquad (6.19b)$$

From Eq. (6.16) we can see that if (S_1, S_2, S_3) are interpreted as the Cartesian coordinates of a point in a three-dimensional space, then $(2\epsilon, 2\alpha)$ are the polar coordinates of that point. The point lies on the surface of a sphere, known as the *Poincaré sphere* (Fig. 6.3). The angles ϕ and 2χ also have a geometrical interpretation on the Poincaré sphere, although it is not as straightforward as the interpretation of 2ϵ and 2α. The Poincaré sphere is also useful for the case of partially polarized light, for which $S_1^2 + S_2^2 + S_3^2 < 1$, and the corresponding point lies inside the sphere (Sec. 6.2).

6.2 Partially polarized light and unpolarized light

If, for a given light beam, we can find a polarization analyzer that transmits the light without loss, then the light is said to be in a pure polarization state. A polarization analyzer is a

Partially polarized light and unpolarized light

device that lets through only light with a certain polarization component (which could be of an arbitrary elliptical polarization). If for a certain light field there is no such analyzer possible, then this light is in a *partially polarized* (or *mixed*) state. If the amount of light that goes through the analyzer is completely independent of the polarization state that it transmits, the light is said to be *unpolarized*.

We can consider unpolarized light to be produced by two independent light sources of equal intensity emitting light with mutually orthogonal linear polarizations (Blum 1996). "Independent" means that the relative phase of the light sources changes in a random way, much more rapidly than the time resolution of the detector used to measure the light intensity transmitted through a polarization analyzer.

> We should keep in mind that even if a plane wave light beam is "unpolarized," the fields are always perpendicular to the propagation direction. Thus, as long as the radiation field is directional, it is anisotropic. Later, we will see that, because of this, "unpolarized" light can transfer angular momentum anisotropy to an ensemble of atoms.
>
> Although truly unpolarized light is difficult to produce experimentally, it is useful as a theoretical construct. One can imagine three equal-intensity light sources that radiate, for example, linearly polarized light with random phases in three orthogonal directions.

By its nature, unpolarized radiation cannot be monochromatic. Consider a model in which the randomness in the light phase is introduced via phase jumps occurring with a uniform probability in time. The frequency of phase jumps of the radiation field is directly related to the spectral width $\Delta \nu$ of this radiation, as discussed below. For example, if the probability of a phase jump (such that the phase after the jump "does not know" what the phase was before the jump) during a time dt is dt/τ, where τ is known as the *correlation time*, then the radiation spectrum has full width at half maximum of $\Delta \nu = 1/(\pi \tau)$ (Townes and Schawlow 1975; Frish 1963).

> Here is a derivation of this result. The Fourier transform of the electric field in the interval between two consecutive phase jumps separated by a time t_s is proportional to
>
> $$\frac{1}{\sqrt{2\pi}} \int_0^{t_s} e^{i(\omega_0 - \omega)t} dt = \frac{1 - e^{-i(\omega - \omega_0)t_s}}{i\sqrt{2\pi}(\omega - \omega_0)}. \tag{6.20}$$
>
> If the probability of a phase jump is constant in time, the probability of having no jump until the time t_s is $e^{-t_s/\tau}$. Integrating over many such jumps, using this probability as a weighting factor, we obtain
>
> $$\frac{1}{i\sqrt{2\pi}} \int_0^\infty e^{-t_s/\tau} \frac{1 - e^{-i(\omega - \omega_0)t_s}}{(\omega - \omega_0)} dt_s/\tau = \frac{1}{\sqrt{2\pi}} \frac{\tau}{1 + i\tau(\omega - \omega_0)}. \tag{6.21}$$
>
> Taking the absolute value squared of this expression gives the intensity spectrum, proportional to
>
> $$\frac{\tau^2}{1 + \tau^2(\omega - \omega_0)^2} = \frac{1}{(\omega - \omega_0)^2 + 1/\tau^2}. \tag{6.22}$$
>
> This spectrum has the functional form of a *Lorentzian lineshape*, and its full width at half maximum is $2/\tau$ [$1/(\pi \tau)$ in ν units].
>
> Note that the Lorentzian spectrum (6.22) also arises from continuous exponential decay of a classical oscillator. In this case, the characteristic time τ is that of the decay of the amplitude of the oscillator. This is a factor of 2 longer than the characteristic decay time of the energy of the oscillator usually referred to as the oscillator's *lifetime*. The Lorentzian spectrum also arises in the case of *phase diffusion* commonly encountered in laser physics, not to mention radiophysics.

116 *Polarized light*

Partially polarized and unpolarized light is discussed further in Sec. 6.5.

6.3 Spin angular momentum of polarized light

Consider a circularly polarized light beam directed at a target. The bound electrons in the target can be thought of classically as oscillators that may be displaced from the equilibrium point in an arbitrary direction. In response to the rotating electric field of the light wave, these are set into circular motion (Blum 1996; Feynman *et al.* 1989, V. 1, Sec. 33-10). This suggests that circularly polarized light carries angular momentum; this idea is confirmed by the theory of quantum electrodynamics.

There are some subtleties involved in the idea of angular momentum carried by a plane wave (Khrapko 2001; Allen and Padgett 2002). In fact, a plane wave with infinite extent cannot carry angular momentum (Heitler 1984). The density of linear momentum associated with an electromagnetic field is

$$\text{Linear momentum density} = \frac{1}{4\pi c} (\mathcal{E} \times \mathbf{B}), \tag{6.23}$$

so the classical angular momentum of the electromagnetic field is given by

$$\text{Angular momentum} = \frac{1}{4\pi c} \int \mathbf{r} \times (\mathcal{E} \times \mathbf{B}) \, d^3 r. \tag{6.24}$$

In a plane electromagnetic wave the electric and magnetic fields are everywhere transverse to the propagation direction, so the momentum is parallel to the propagation direction; thus the resultant angular momentum cannot be in the same direction. This appears to contradict the notion that beams of circularly polarized light carry angular momentum. In reality, even very large, uniform-amplitude beams are effectively apertured (i.e., spatially constrained) by the optical system or the finite size of objects with which the beam interacts. Any aperture introduces a light-intensity gradient, and a detailed analysis of Maxwell's equations shows that aperturing leads to the appearance of field components in the nominal propagation direction. This resolves the contradiction.

The ratio of angular momentum to energy carried by circularly polarized light can be found from the following classical argument (Feynman *et al.* 1989, V. 1, Sec. 33-10), which avoids the conceptual subtlety mentioned just above. Suppose that circularly polarized light shines on an absorbing medium (Fig. 6.4). Absorption of the light puts the electrons into motion, which by symmetry must be circular. Let $\mathbf{r}$ be the displacement of an electron from its equilibrium position. It follows the rotation at frequency ω of the light electric field vector $\mathcal{E}$ with some phase lag φ_0. Now let us look at the work done by the light field on the electron. This work is done on the electron by the component $\mathcal{E}_t$ of the electric field transverse to $\mathbf{r}$ and along the velocity $\mathbf{v}$. The rate at which energy is added to the absorbing medium is

$$\left| \frac{d'W}{dt} \right| = |\mathbf{F} \cdot \mathbf{v}| = e\mathcal{E}_t v = e\mathcal{E}_t \omega r, \tag{6.25}$$

where $\mathbf{F}$ is the force acting on the electron. The rate of transfer of angular momentum to the medium is

$$\left| \frac{dJ}{dt} \right| = |\mathbf{r} \times \mathbf{F}| = e\mathcal{E}_t r. \tag{6.26}$$

Dividing Eq. (6.26) by Eq. (6.25) and integrating, we arrive at

$$|J| = \frac{W}{\omega}. \qquad (6.27)$$

From quantum mechanics, we know that a photon of frequency ω has energy $\hbar\omega$. Thus, after one photon is absorbed, the angular momentum transferred from the field to the electron is $\hbar$.

The fact that the angular momentum per photon cannot exceed $\sim \hbar$ can be obtained from dimensional analysis without considering an absorbing medium. The angular momentum should be formed as a product of a linear momentum and a distance. Suppose we have a light beam of a diameter d and total power P. The diffraction angle is $\theta \sim \lambda/d$, so the energy flux in the transverse direction is $P\lambda/d$. The desired quantity that has the dimensions of linear momentum p is obtained by dividing this by c. Taking the beam diameter as the quantity of the dimensions of distance, we arrive at the scale of angular momentum in the light propagation direction:

$$pd \sim \frac{P}{c}\frac{\lambda}{d}d \sim \hbar\omega\dot{N}\lambda/c \sim \hbar\dot{N}, \qquad (6.28)$$

where $\dot{N}$ is the photon flux in the beam, and $\hbar\omega$ is the energy per photon. This result, within a numerical factor (of 2π), agrees with $\hbar$ per photon.

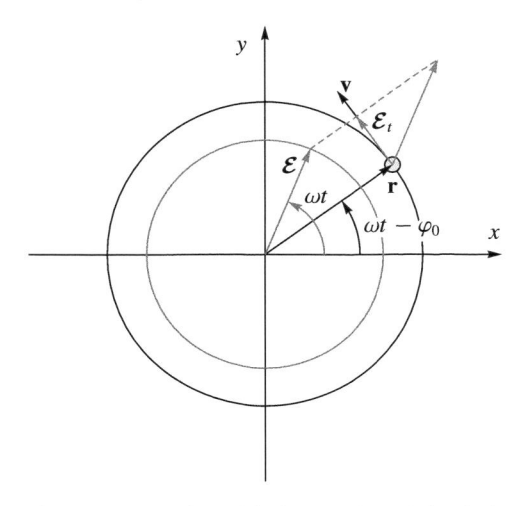

Fig. 6.4 Classical model of absorption of circularly polarized light. The electric field $\mathcal{E}$ of a circularly polarized light wave sets an electron in the medium in motion on a circular trajectory. The electron position is $\mathbf{r}$ and the velocity is $\mathbf{v}$.

6.3.1 Orbital angular momentum of light

The angular momentum carried by the light that we have discussed so far is related to the light polarization. This type of angular momentum, sometimes called the *spin* angular momentum of the light (Padgett *et al.* 2004), was examined in the work of John Poynting at the beginning of the 20th century.

Light can carry another type of angular momentum, associated not with the light polarization, but with the spatial dependence of the light beam. This *orbital* angular momentum of light has attracted interest more recently, at the end of the 20th century, and light beams that carry orbital angular momentum have been produced. In such a beam, the electric and magnetic fields have a phase dependence (Fig. 6.5) that is described by a factor $\exp(-im\varphi)$ (referred to as *azimuthal*). In this case, the "quantum" number m characterizes the orbital angular momentum of the light. This quantum number can have values $0, \pm 1, \pm 2, \ldots$ and the respective angular momentum collinear with the light beam assumes the values $m\hbar$. Because of the azimuthal phase dependence, there is a phase singularity on the axis of the light beam. Consequently, there must be zero light intensity on the axis for $m \neq 0$.

Beams that carry spin angular momentum can be produced by passing linearly polarized light through a *quarter-wave plate*, which transforms it into circularly polarized light. A light

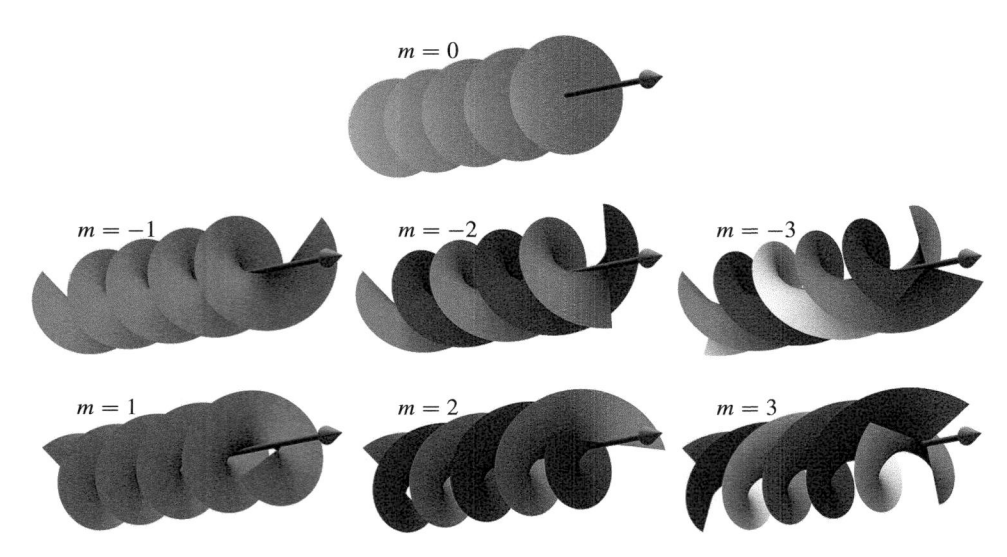

Fig. 6.5 Light beams with orbital angular momentum characterized by the azimuthal dependence of the phase in the form $\exp(-im\varphi)$, where m is the "quantum" number characterizing the projection of the orbital angular momentum on the light propagation direction. For $m \neq 0$, the light intensity on the axis of the beam has to be zero. The plot shows surfaces of constant phase. For $m = 0$, the surfaces are a set of planes separated by the wavelength of the light in the direction of light propagation. For helical ($m \neq 0$) beams the surfaces are $|m|$ helices, rotated around the beam axis by $2\pi/m$ with respect to each other. Note that the pitch (the advance per turn) of an individual branch goes as m.

beam with orbital angular momentum can be produced in a somewhat similar way. We can start with a laser beam that has a specific transverse mode structure—a *Hermite–Gaussian mode* (see, for example, Siegman 1986). Such a beam has a specific intensity distribution in the plane perpendicular to the propagation direction. This distribution has a number of bright spots separated by zero-intensity boundaries. The number of bright spots depends on the specific mode. If this beam passes through two cylindrical lenses that are placed at a certain distance from each other, it is transformed into a beam with nonzero orbital angular momentum. The particular value of orbital momentum that the light-beam acquires depends on the Hermite–Gaussian beam mode of the initial beam (Beijersbergen *et al.* 1993). Although this conversion mechanism is highly efficient, each orbital angular momentum state of the light beam requires a specific initial mode. This requirement limits the application of this conversion method.

Another more universal method to convert Hermite–Gaussian beams into beams with certain orbital angular momenta is based on the use of computer-generated holograms. With this method, it is possible to generate light beams of any orbital angular momentum from the same initial Hermite–Gaussian light beam (Padgett *et al.* 2004).

For more information on orbital angular momentum of light, the reader is referred to the articles by Allen *et al.* (1992) and Friese *et al.* (1996).

6.4 Spherical basis for light polarization

In this section, we discuss an alternate, *spherical-basis* representation of light polarization that is useful for the analysis of the interaction of light with atoms, particularly when employing

Spherical basis for light polarization 119

the Wigner–Eckart theorem (Chapter 3), and more generally the irreducible tensors.
The covariant spherical-basis unit vectors are given in Sec. 3.1.2:

$$\hat{\epsilon}_1 = -\frac{1}{\sqrt{2}}(\hat{x} + i\hat{y}), \tag{6.29a}$$

$$\hat{\epsilon}_0 = \hat{z}, \tag{6.29b}$$

$$\hat{\epsilon}_{-1} = \frac{1}{\sqrt{2}}(\hat{x} - i\hat{y}), \tag{6.29c}$$

where $\hat{x}, \hat{y}, \hat{z}$, are the basis unit vectors in Cartesian coordinates. (Here we use the covariant representation of the spherical-basis vectors; see Sec. 3.1.2.) If we multiply these spherical-basis vectors by the phase factor $\exp(-i\omega t)$ representing the oscillation of the electric field in a light wave, we have for the $\hat{\epsilon}_1$ vector

$$\begin{aligned}\hat{\epsilon}_1 e^{-i\omega t} &= -\frac{1}{\sqrt{2}}\left(\hat{x}e^{-i\omega t} + i\hat{y}e^{-i\omega t}\right) \\ &= -\frac{1}{\sqrt{2}}\left(\hat{x}e^{-i\omega t} + \hat{y}e^{-i(\omega t - \pi/2)}\right).\end{aligned} \tag{6.30}$$

The oscillation along the x-axis is a quarter of a period ahead of that along $\hat{y}$ and so the polarization of light associated with this vector, which propagates in the $\hat{z}$ direction, rotates counterclockwise in the xy plane. This is left-circularly polarized light. Similarly, one can see that the basis vector $\hat{\epsilon}_{-1}$ represents light that propagates along $\hat{z}$ with the electric-field vector rotating in a clockwise direction in the xy plane, i.e., right-circularly polarized light.

Finally, the $\hat{\epsilon}_0$ basis vector represents linear oscillations along the z-axis and can be used to describe linearly z-polarized light. Just as each vector can be represented by its three Cartesian components, it can also be represented by its three spherical components.

As described in Sec. 3.1.2, any vector can be decomposed into components in the spherical basis. For the polarization vector $\hat{\varepsilon}$ of a light wave we have

$$\hat{\varepsilon} = \sum_q \varepsilon^q \hat{\epsilon}_q, \tag{6.31}$$

where the *contravariant spherical components* ε^q of the vector $\hat{\varepsilon}$ are connected to the Cartesian components of this vector according to

$$\varepsilon^1 = -\frac{1}{\sqrt{2}}(\varepsilon_x - i\varepsilon_y), \tag{6.32a}$$

$$\varepsilon^0 = \varepsilon_z, \tag{6.32b}$$

$$\varepsilon^{-1} = \frac{1}{\sqrt{2}}(\varepsilon_x + i\varepsilon_y). \tag{6.32c}$$

Because of the practical importance of this decomposition, we give several explicit examples. Linearly polarized light with the electric-field vector parallel to $\hat{z}$ has only one nonzero spherical-basis component, $\varepsilon^0 = 1$. Similarly, for left-circularly polarized light propagating along $\hat{z}$, the polarization vector has only one nonzero component, namely $\varepsilon^1 = 1$.

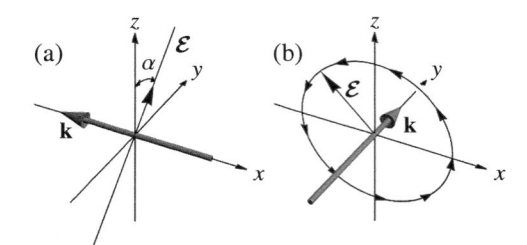

Fig. 6.6 (a) Light with linear polarization at an angle α to the z-axis in the yz plane. (b) Right-circularly polarized light propagating in the positive y-direction.

It is not always convenient to choose the quantization axis along the light polarization or propagation directions. For example, if we are analyzing atom–light interactions in the presence of an external field (e.g., a static electric or magnetic field), it may be convenient to choose $\hat{\mathbf{z}}$ along this field, which is generally different from the light-polarization or the light-propagation direction. A concrete example is depicted in Fig. 6.6(a). Light is linearly polarized with its polarization vector in the yz plane, tilted at an angle α with respect to $\hat{\mathbf{z}}$. As we see from the formulas, the light propagation direction does not actually matter for the decomposition into spherical components (for linearly polarized light, it could be anywhere in the plane perpendicular to the polarization vector); Fig. 6.6(a) shows light propagating along $-\hat{\mathbf{x}}$. The Cartesian components of the polarization vector are $\varepsilon_x = 0$, $\varepsilon_y = \sin\alpha$, and $\varepsilon_z = \cos\alpha$. If we insert these values in Eqs. (6.32), we arrive at the spherical components for the polarization vector:

$$\varepsilon^1 = -\frac{1}{\sqrt{2}}\left(\varepsilon_x - i\varepsilon_y\right) = \frac{i}{\sqrt{2}}\sin\alpha, \tag{6.33a}$$

$$\varepsilon^0 = \varepsilon_z = \cos\alpha, \tag{6.33b}$$

$$\varepsilon^{-1} = \frac{1}{\sqrt{2}}\left(\varepsilon_x + i\varepsilon_y\right) = \frac{i}{\sqrt{2}}\sin\alpha. \tag{6.33c}$$

Let us consider a second example involving circularly polarized light. In this case, the electric-field oscillations along different axes have different phases. Let us analyze right-circularly polarized light that propagates in the positive y-direction (Fig. 6.6b). Recall that this means that if we are looking into the light beam we observe the $\boldsymbol{\mathcal{E}}$ vector rotating clockwise. The rotation is in the xz plane. If we decompose this rotation into oscillations along the z- and x-axes, then the oscillations along the x-axis are a quarter of the oscillation period ahead of the oscillations along $\hat{\mathbf{z}}$. Then $\varepsilon_z = 1/\sqrt{2}$ and $\varepsilon_x = \exp(-i\pi/2)/\sqrt{2} = -i/\sqrt{2}$. If we insert these components in the general expressions for spherical components of the light polarization vector we arrive at

$$\varepsilon^1 = -\frac{1}{\sqrt{2}}\left(\varepsilon_x - i\varepsilon_y\right) = \frac{i}{2}, \tag{6.34a}$$

$$\varepsilon^0 = \varepsilon_z = \frac{1}{\sqrt{2}}, \tag{6.34b}$$

$$\varepsilon^{-1} = \frac{1}{\sqrt{2}}\left(\varepsilon_x + i\varepsilon_y\right) = -\frac{i}{2}. \tag{6.34c}$$

Another method to calculate the components of an arbitrary polarization vector is to use the Wigner D-matrices (Varshalovich *et al.* 1988): we find the polarization vector in a basis in which it is simple, and then rotate the basis to the required direction. The vector component in the new (rotated) coordinate frame ε'^q can be related to the components in the initial frame using the Wigner-D matrices (Sec. 3.4.2)

$$\varepsilon'^q = \sum_{q'} \varepsilon^{q'} D^{(1)*}_{q'q}(\alpha, \beta, \gamma). \tag{6.35}$$

The angles α, β, and γ are the Euler angles.

For example, to find the electric-field vector of right-circularly polarized light propagating along $\hat{\mathbf{y}}$, we first write the vector for light propagating along $\hat{\mathbf{z}}$: $\varepsilon^{-1} = 1, \varepsilon^0 = \varepsilon^1 = 0$. We then perform a rotation with $\alpha = 0, \beta = \gamma = \pi/2$:

$$\varepsilon'^q = D^{(1)*}_{-1q}(0, \pi/2, \pi/2). \tag{6.36}$$

Using the formulas for the Wigner D-matrices (Sec. 3.4.2), we immediately arrive at the values (6.34) for the light-polarization vector components.

6.5 The polarization density matrix

The light polarization vector has three spherical components, which are analogous to the three Zeeman states of a spin-one atom. This is not an accident; although we are using a classical description of the light field, in a fully quantum-mechanical approach the field is described by a spin-one particle: the photon.

As a result, we can describe an arbitrary polarization state of the light in a manner exactly analogous to the description of spin-one atoms, i.e., using the density matrix (Chapter 5). The light polarization matrix, which we denote by Φ, would then take the form

$$\Phi = \overline{\begin{pmatrix} |\varepsilon^1|^2 & \varepsilon^{1*}\varepsilon^0 & \varepsilon^{1*}\varepsilon^{-1} \\ \varepsilon^{0*}\varepsilon^1 & |\varepsilon^0|^2 & \varepsilon^{0*}\varepsilon^{-1} \\ \varepsilon^{-1*}\varepsilon^1 & \varepsilon^{-1*}\varepsilon^0 & |\varepsilon^{-1}|^2 \end{pmatrix}}. \tag{6.37}$$

Here the overbar implies averaging over a time longer than the relevant coherence time between various spectral components of the light beam. Unpolarized light corresponds to the identity matrix, and in general the matrix describes any arbitrary polarization state.

Note that since the *polarization density matrix* corresponds to a $J = 1$ state, it can contain polarization moments of rank up to two, i.e., orientation and alignment (see Sec. 5.7). This property of light is very important for the analysis of light–atom interactions.

The diagonal elements of the density matrix represent the relative intensity of light components with polarization along the spherical-basis vectors, and off-diagonal elements represent the correlation between the oscillation of pairs of spherical-basis components of the electric field vector.

As with the atomic density matrix, there is a simple condition to distinguish between fully polarized and partially polarized light (corresponding to pure and mixed states in the atomic case); namely, light is fully polarized if and only if

$$(\Phi)^2 = \Phi. \tag{6.38}$$

The proof of this is given in Sec. 5.1.

122 *Polarized light*

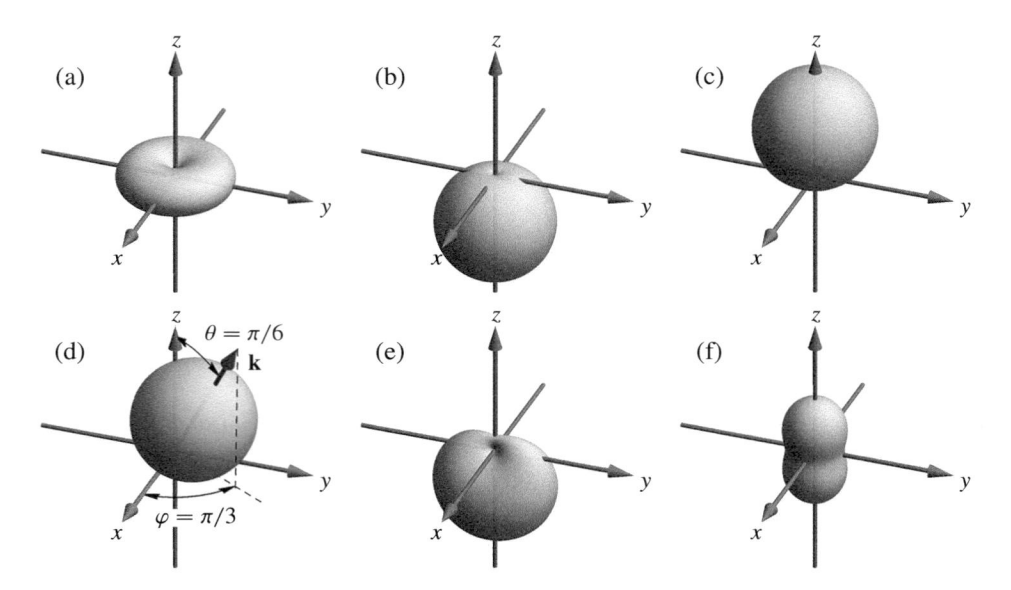

Fig. 6.7 Angular-momentum probability surfaces for (a) light linearly polarized along z, (b) right-circularly polarized light propagating along $\hat{\mathbf{z}}$, (c) left-circularly polarized light propagating along $\hat{\mathbf{z}}$, (d) left-circularly polarized light propagating in the direction $\theta = \pi/6, \varphi = \pi/3$, (e) elliptically polarized light ($\epsilon = \pi/8$) propagating along $\hat{\mathbf{z}}$, and (f) an incoherent superposition of light with two opposite polarizations propagating along $\hat{\mathbf{z}}$. In many important cases, the angular-momentum probability surfaces are highly symmetric. For example, the surfaces (a) and (f) are axially symmetric with respect to $\hat{\mathbf{z}}$, so this is a "preferred" axis in these cases. Note also that, for the surfaces (a) and (f), there is no preferred direction. There are, however, preferred directions for all the other surfaces in figure.

6.6 Angular-momentum probability surfaces for light

Using the polarization density matrix, we can draw the light spin-angular-momentum spatial distribution in the same manner as for atoms (Sec. 5.3). As with the atomic case, we can use these surfaces to illustrate the symmetry of the light polarization state. We draw surfaces defined by $\Phi_{11}(\theta, \phi)$, the probability to find the maximal projection of photon spin in the direction (θ, ϕ).

Let us consider a few examples. For the simple case of linearly z-polarized light we have the following polarization density matrix:

$$\Phi = \begin{pmatrix} 0 & 0 & 0 \\ 0 & 1 & 0 \\ 0 & 0 & 0 \end{pmatrix}. \tag{6.39}$$

The respective probability surface is shown in Fig. 6.7(a). We can see that in this case the spatial distribution of the spin angular momentum does not have a preferred direction in space, but only a preferred axis along z.

If we have right- or left-circular polarized light propagating along $\hat{\mathbf{z}}$ then the respective polarization density matrices are

$$\Phi_{\text{right}} = \begin{pmatrix} 0 & 0 & 0 \\ 0 & 0 & 0 \\ 0 & 0 & 1 \end{pmatrix}, \quad \Phi_{\text{left}} = \begin{pmatrix} 1 & 0 & 0 \\ 0 & 0 & 0 \\ 0 & 0 & 0 \end{pmatrix}, \tag{6.40}$$

and the respective probability surfaces are shown in Fig. 6.7(b) and (c). We see that, in accordance with our discussion at the beginning of the chapter, right-circular polarized light carries angular momentum directed in the opposite direction to the light propagation direction (the z-axis) and left-circular polarized light carries angular momentum directed along the light propagation.

Needless to say, the angular-momentum probability surfaces for light are exactly the same as those we have already seen for angular-momentum-one atoms in Sec. 5.3.

Next, let us consider a slightly more complicated situation—left-circularly polarized light that propagates in the direction characterized by the polar angles $\theta = \pi/6$ and $\varphi = \pi/3$. Determining the polarization density matrix according to Eq. (6.35), we have

$$\Phi_{\text{left}} = \begin{pmatrix} \frac{1}{16}(7 + 4\sqrt{3}) & -\frac{\sqrt{3}+2}{8\sqrt{2}}e^{\frac{i2\pi}{3}} & \frac{1}{16}e^{-\frac{i2\pi}{3}} \\ -\frac{\sqrt{3}+2}{8\sqrt{2}}e^{-\frac{i2\pi}{3}} & \frac{1}{8} & \frac{\sqrt{3}-2}{8\sqrt{2}}e^{\frac{i2\pi}{3}} \\ \frac{1}{16}e^{\frac{i2\pi}{3}} & \frac{\sqrt{3}-2}{8\sqrt{2}}e^{-\frac{i2\pi}{3}} & \frac{1}{16}(7 - 4\sqrt{3}) \end{pmatrix}. \tag{6.41}$$

The corresponding angular-momentum probability surface (Fig. 6.7d) is consistent with the concept of angular momentum carried by the light: a left-circular polarized photon carries angular momentum that is directed along the light beam.

Two additional examples of the probability surfaces are shown in Fig. 6.7(e) and (f), the first of which shows elliptically polarized light, while the second shows an incoherent superposition of two opposite polarizations. Note that, while the latter polarization state is frequently referred to as "unpolarized," it actually has alignment along the light propagation axis.

6.7 Stokes parameters for partially polarized light

In Sec. 6.1 we introduced the Stokes parameters for describing fully polarized light. The Stokes parameters are also useful for describing partially polarized and unpolarized light. In this case the parameters $S_{1,2,3}$ are still defined in terms of normalized intensities, but we no longer have the normalization condition $S_1^2 + S_2^2 + S_3^2 = 1$. Indeed, the quantity

$$p = \sqrt{S_1^2 + S_2^2 + S_3^2} \tag{6.42}$$

is defined to be the *degree of polarization* of the light. From the definition (6.42) we can see that the degree of polarization is the length of the vector describing the light polarization in the Poincaré sphere representation. Thus, while the fully polarized states are represented by points on the surface of the Poincaré sphere, the partially polarized states are represented by points within the volume.

If the light propagates along the z-axis some polarization density matrix elements are always identically zero, because the light field is transverse. Namely, since the z-component is zero, the polarization density matrix has the form

$$\Phi = \begin{pmatrix} \rho_{11} & 0 & \rho_{1-1} \\ 0 & 0 & 0 \\ \rho_{-11} & 0 & \rho_{-1-1} \end{pmatrix}, \tag{6.43}$$

i.e., nonzero elements can only appear in the corners of the matrix. (Note that if we are considering a light beam propagating in a particular direction, there is no component of polarization along the propagation direction. Then a 2×2 matrix can be used to describe the polarization of the beam.) The nonzero matrix elements can be related to the Stokes parameters. The Stokes parameters are defined (Eq. (6.10)) in terms of the light intensities of the linear polarizations I_x, I_y, $I_{+\pi/4}$, $I_{-\pi/4}$ and the circular polarizations I_+ and I_-. From the definition (6.32) of the spherical vector components we have

$$\varepsilon_x = \frac{1}{\sqrt{2}} (\varepsilon_{-1} - \varepsilon_1), \tag{6.44a}$$

$$\varepsilon_y = -\frac{i}{\sqrt{2}} (\varepsilon_{-1} + \varepsilon_1). \tag{6.44b}$$

Thus the intensities I_x and I_y are given by

$$\begin{aligned} I_x &= \mathcal{E}_0^2 \langle \varepsilon_x | \rho | \varepsilon_x \rangle \\ &= \frac{\mathcal{E}_0^2}{2} \begin{pmatrix} -1 & 0 & 1 \end{pmatrix} \begin{pmatrix} \rho_{11} & 0 & \rho_{1-1} \\ 0 & 0 & 0 \\ \rho_{-11} & 0 & \rho_{-1-1} \end{pmatrix} \begin{pmatrix} -1 \\ 0 \\ 1 \end{pmatrix} \\ &= \frac{\mathcal{E}_0^2}{2} (\rho_{11} - \rho_{1-1} - \rho_{-11} + \rho_{-1-1}) \end{aligned} \tag{6.45}$$

and

$$\begin{aligned} I_y &= \mathcal{E}_0^2 \langle \varepsilon_y | \rho | \varepsilon_y \rangle \\ &= \frac{\mathcal{E}_0^2}{2} \begin{pmatrix} 1 & 0 & 1 \end{pmatrix} \begin{pmatrix} \rho_{11} & 0 & \rho_{1-1} \\ 0 & 0 & 0 \\ \rho_{-11} & 0 & \rho_{-1-1} \end{pmatrix} \begin{pmatrix} 1 \\ 0 \\ 1 \end{pmatrix} \\ &= \frac{\mathcal{E}_0^2}{2} (\rho_{11} + \rho_{1-1} + \rho_{-11} + \rho_{-1-1}). \end{aligned} \tag{6.46}$$

Thus the Stokes parameter S_1 is given by

$$S_1 = \frac{1}{\mathcal{E}_0^2} (I_x - I_y) = -(\rho_{-11} + \rho_{1-1}) = -2 \operatorname{Re}(\rho_{-11}). \tag{6.47}$$

Similarly, since

$$\varepsilon_{+\pi/4} = \frac{1}{\sqrt{2}} (\varepsilon_x + \varepsilon_y) \tag{6.48}$$

and

$$\varepsilon_{-\pi/4} = \frac{1}{\sqrt{2}} (-\varepsilon_x + \varepsilon_y), \tag{6.49}$$

we find

$$S_2 = \frac{1}{\mathcal{E}_0^2} (I_{+\pi/4} - I_{-\pi/4}) = -i(\rho_{1-1} - \rho_{-11}) = 2 \operatorname{Im}(\rho_{1-1}). \tag{6.50}$$

Finally,

$$S_3 = \frac{1}{\mathcal{E}_0^2}(I_+ - I_-) = \rho_{11} - \rho_{-1-1}. \tag{6.51}$$

We can invert the above relations, along with the normalization condition

$$\rho_{11} + \rho_{-1-1} = 1, \tag{6.52}$$

to find the polarization density-matrix elements in terms of the Stokes parameters:

$$\Phi = \frac{1}{2}\begin{pmatrix} 1+S_3 & 0 & -S_1 + iS_2 \\ 0 & 0 & 0 \\ -S_1 - iS_2 & 0 & 1-S_3 \end{pmatrix}. \tag{6.53}$$

Finally we remember that in this discussion we have assumed that the quantization (z) axis is along the light propagation direction. It may be convenient to have the quantization axis in another direction, for example, along the direction of some other field in the problem. In this case, the density matrix can be found by performing a rotation using the Wigner D-matrices (Sec. 3.4.2).

7
Atomic transitions

In this chapter, we will consider various selection rules governing atomic transitions. Central to our discussion will be the notions of *transition amplitude* and *transition probability*. In order to introduce these quantities, we will first consider a basic problem in time-dependent perturbation theory, with general application: determining the behavior of a system with two energy eigenstates subject to a harmonic perturbation coupling these states.

7.1 Two-level system under the action of a periodic perturbation

7.1.1 Spin-1/2 magnetic resonance

We begin by considering once more the problem discussed in Sec. 4.5.2 of a spin-1/2 system—such as, for example, a paramagnetic atom—immersed in a strong leading magnetic field $\mathbf{B} = B_0 \hat{\mathbf{z}}$. This system has two energy levels corresponding to the two angular-momentum projections on $\hat{\mathbf{z}}$ separated by an energy interval $\hbar \Omega_L = g \mu_B B_0$. Let us also assume that we have an additional weak field $\mathbf{B}_1$ that lies in the xy plane and rotates around $\hat{\mathbf{z}}$ with frequency ω.

Suppose that at $t = 0$, our system is in one of the energy states, for example, the lower one, which we assume to be the spin-down state. In the spinor representation, this is written as

$$\psi(0) = \begin{pmatrix} 0 \\ 1 \end{pmatrix}. \tag{7.1}$$

The question is: what is the state of the system at a later time?

The already familiar trick to solving this problem is to remove the time dependence of the perturbation by going to the frame with the same z-axis but rotating together with the field $\mathbf{B}_1$. In this frame, the field $\mathbf{B}_1 = B_1 \hat{\mathbf{y}}$ is stationary (we assumed it is directed along the y-axis of the rotating frame), and the effective field along $\hat{\mathbf{z}}$ becomes

$$B_z = B_0 - \frac{\hbar \omega}{g \mu_B}. \tag{7.2}$$

In particular, on resonance, where the field $\mathbf{B}_1$ rotates in the direction of the Larmor precession and at the Larmor frequency Ω_L, the leading field is completely eliminated in the rotating frame, and the entire evolution of the system is simply Larmor precession around $\mathbf{B}_1$. If the rotation frequency ω is different from Ω_L, the evolution in the rotating frame corresponds to Larmor precession of the spin around the vector sum

$$\left\{ B_0 - \frac{\hbar \omega}{g \mu_B} \right\} \hat{\mathbf{z}} + B_1 \hat{\mathbf{y}}. \tag{7.3}$$

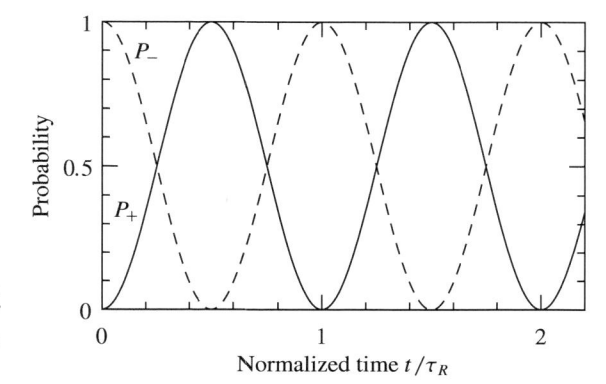

Fig. 7.1 Probabilities of finding the spin in the "up" state (solid line) and the "down" state (dashed line) as a function of time in units of the Rabi period $\tau_R = 2\pi/\Omega_R$.

We can now write down the time evolution of the spin state by recognizing that (in the rotating frame) it is just Larmor precession in the effective field of Eq. (7.3). It is not difficult to evaluate the general solution; however, to keep things as simple as possible, we start with the resonant case, where the spin state just rotates around $\hat{\mathbf{y}}$. At time t, the angle of the rotation is $\beta = \Omega_R t$, where we define the *Rabi frequency* as $\Omega_R = g\mu B_1/\hbar$ as the frequency of Larmor precession in field B_1. The solution is the result of the application of a rotation matrix of Eq. (3.64) to the state of Eq. (7.1):

$$\psi(t) = \begin{pmatrix} \cos\frac{\beta}{2} & -\sin\frac{\beta}{2} \\ \sin\frac{\beta}{2} & \cos\frac{\beta}{2} \end{pmatrix} \begin{pmatrix} 0 \\ 1 \end{pmatrix} = \begin{pmatrix} -\sin\Omega_R t/2 \\ \cos\Omega_R t/2 \end{pmatrix}. \tag{7.4}$$

From this, we see that the probability of finding the spin in the "up" state is

$$P_+ = \sin^2\Omega_R t/2 = \frac{1 - \cos\Omega_R t}{2}, \tag{7.5}$$

while the probability of finding the spin in the initial "down" state is

$$P_- = \cos^2\Omega_R t/2 = \frac{1 + \cos\Omega_R t}{2}, \tag{7.6}$$

i.e., the population oscillates harmonically between the lower and the upper state at the Rabi frequency.

7.1.2 Generalization

Having solved a specific problem of the spin-1/2 magnetic resonance, we have, in fact, solved one of the most general and important problems in all of physics—the evolution of a two-level system under the action of a periodic perturbation. According to the discussion in Sec. 4.5.2, the perturbation Hamiltonian associated with $\mathbf{B}_1$ can be written as

$$H_1 = \begin{pmatrix} 0 & -iVe^{i\omega t} \\ iVe^{-i\omega t} & 0 \end{pmatrix}, \tag{7.7}$$

where $V = g\mu_B B_1/2 = \hbar\Omega_R/2$ characterizes the strength of the coupling. Now, our solution works not just for a spin-1/2 system in a magnetic field, but for any two-level system under a periodic perturbation described by a Hamiltonian of the form (7.7). The most important case for us will be a two-level atom with the two states coupled by an optical field. In the case of an electric-dipole transition, H_1 is the electric-dipole Hamiltonian $H_\mathcal{E} = -\mathbf{d} \cdot \mathcal{E}$ introduced in Sec. 4.3, where $\mathbf{d}$ is the electric-dipole operator, and $\mathcal{E}$ is the electric field. In the rotating-wave approximation (see Sec. 4.5.3), this Hamiltonian can be cast in a form similar to that of Eq. (7.7) with $V = d\mathcal{E}_0/2$, where $\mathcal{E}_0$ is the light-field amplitude, so that the Rabi frequency in this case is $\Omega_R = d\mathcal{E}_0/\hbar$, where d is the electric-dipole matrix element.

Continuing to think about two-level atoms exposed to light, let us look at Rabi oscillations from a somewhat different perspective. Initially, the atoms are in the lower state, and we apply light. At first, i.e., for small t, the probability of finding an atom in the excited state grows quadratically (see Eq. 7.5). What happens here is nothing else but *absorption* of the light quanta (photons), an elementary process in which a photon is removed from the light field, and the atom goes from the lower state to the upper state. Now, as the probability of finding an atom in the upper state increases, it eventually reaches unity. At this point, the atoms can no longer absorb photons. However, upon interaction with radiation, the atoms can *emit* photons, going back to the lower state, and adding an additional photon to the light field in the process. This is *stimulated emission*. At times when the atoms are neither in the upper nor the lower state with unit probability, either stimulated emission or absorption can occur, and which of these processes "wins" depends on the history of the process ("encoded" in the relative phase of the coherence between the upper and the lower state).

> *Spontaneous vs. stimulated transitions*
>
> When an atom in the lower state absorbs radiation and is transferred to the upper state, it is clear that the radiation is the agent which initiates the transition. But what if the atom is initially in the upper state and there is no light shining on it. We know that, eventually, the atom will decay via *spontaneous emission*. The question is: what causes the atom in a stationary state to spontaneously radiate an electromagnetic wave and undergo a transition to the lower final state?
>
> It turns out that the answer to this question is actually beyond nonrelativistic quantum mechanics, and, to be answered properly, requires one to go into the realm of *quantum electrodynamics*, and, more generally, the *quantum field theory*. The gist of the story is that, even in the absence of applied light there is a certain density of electromagnetic-field fluctuations with all possible frequencies and polarizations that exists in what would otherwise be considered the vacuum. This means that if a free atom is in the upper state, there is always a vacuum fluctuation with a frequency and polarization that is needed to connect resonantly the upper and the lower states of the atom, which causes a spontaneous decay to occur. Thus, we can say that, in some sense, spontaneous emission is, in fact, stimulated emission induced by vacuum fluctuations. A thorough introduction to quantum field theory and *quantum vacuum* can be found in the book by Milonni (1994).

It is also instructive to take a closer look at the early times where the upper-state population is very small, and so stimulated emission from the upper state can be neglected. Since the probability of finding atoms in the upper state increases quadratically with time, this means that the amplitude of the upper state increases linearly. This can be thought of as constructive interference. The amplitude of the transition per time interval dt adds to that in the next interval, so the overall amplitude grows linearly with time.

So far, we have just briefly discussed the physics of two-level systems. There is still a lot more to talk about, for example, what happens if there is relaxation; what happens with nonzero detuning (i.e., when the perturbation is not resonant; see also Chapter 4), etc., and, in fact, there are whole monographs written on the subject, e.g., the one by Allen and Eberly (1987); see also the book by Shore (1990). We, however, will postpone these discussions until later; an interested reader may also consult Prob. 3.1 in the book by Budker *et al.* (2008) and the article by Auzinsh (2004).

7.2 Selection rules for electric-dipole transitions

The Rabi frequency of a driven transition discussed in Sec. 7.1 is proportional to the transition matrix element. We will now discuss some general rules, the *selection rules*, that are particularly useful in identifying the cases where the matrix elements are strongly suppressed or even vanish. If all selection rules are satisfied, the transition may be expected to be strong, with the matrix element on the order of a characteristic atomic scale for this quantity (see the detailed discussion below). In this section, we go beyond the two-state approximation we used above, and actually consider a realistic situation with a full account of the Zeeman structure of the levels involved in the transition.

In Chapter 2 we examined the quantum numbers characterizing atomic states, including those related to various atomic angular momenta. We discussed the electron orbital angular momentum **L**, the electron spin angular momentum **S**, and the total electron angular momentum **J**, obtained by coupling the orbital and spin angular momenta. We also considered the nuclear spin angular momentum **I**, and total (nuclear coupled with electronic) angular momentum **F**.

All these angular momenta play important roles in radiative transitions in atoms, both in absorption and emission of light. During these transitions not only energy, but also angular momentum is conserved. This means that, for example, in the process of absorption, if we add the angular momentum carried by the photon to the total angular momentum of the initial atomic state we must arrive at the total angular momentum of the excited (final) atomic state. Similarly, in emission, the photon must carry away the proper amount of angular momentum so that the total angular momentum is conserved.

Let us first analyze this angular-momentum conservation rule qualitatively. As we have seen in Sec. 6.3 and will discuss more thoroughly later in this chapter, a photon carries an angular momentum of $\hbar$ (we initially neglect the photon's orbital angular momentum). If we measure all the angular momenta in units of $\hbar$, this means that the total angular momentum of an atom cannot change by more than one unit in an optical transition. Let us for a moment neglect nuclear spin. The total electronic angular momentum of an atom must obey the following selection rule in an electric-dipole transition between an initial state with angular momentum quantum numbers J_i and a final state with J_f

$$\Delta J = J_f - J_i = 0, \pm 1. \tag{7.8}$$

Moreover, since this selection rule is nothing else but the familiar triangle rule of angular-momentum addition, where the triangle should be formed by the initial and final total electron angular momenta and the angular momentum of the photon, there is an additional selection rule saying that $J_i = 0 \rightarrow J_f = 0$ transitions are not allowed, even though they satisfy Eq. (7.8), because it is impossible to construct a triangle in which two sides are zero and the third is equal to one (corresponding to photon spin).

We remember that the total electronic angular momentum of an atom that is described using Russell–Saunders coupling is formed from two components—the electron orbital angular momentum **L** and the electron spin angular momentum **S**. One may ask: when the total electron angular momentum is changing, which component of it is changing—the spin part, the orbital part, or both? In fact, it is the rearrangement of the orbital motion of an electron that is responsible for absorption and emission of photons. This conclusion is supported by classical electrodynamics which relates absorption or emission of an electromagnetic wave to acceleration of an electric charge. This leads to the selection rule

$$\Delta L = 0, \pm 1. \tag{7.9}$$

The electron spin in an optical transition remains constant:

$$\Delta S = 0. \tag{7.10}$$

Formally, this can be seen by writing the initial wave function as a product of the spatial and the spin wave functions, which can be done when the spin is uncoupled from (or weakly coupled to) the orbital angular momentum. The electric-dipole operator then only affects the spatial part, so the spin wave function is preserved in the course of a transition. Note that when the LS coupling fails, the spatial and the spin wave functions become *entangled*, and this conclusion is no longer true.

Just as for the case of the total angular momentum, $L_i = 0 \to L_f = 0$ transitions are forbidden. Equivalently, we can write

$$J_i + J_f \geq 1 \tag{7.11}$$

and

$$L_i + L_f \geq 1. \tag{7.12}$$

It must be noted that while the selection rule for J is strict (neglecting nuclear spin), the selection rules for total spin and total orbital angular momentum can be violated. The reason for this difference is that (in the absence of nuclear spin) the total electronic angular momentum of an isolated atom is always conserved, whereas the total electronic spin of an atom and its total orbital angular momentum are conserved only approximately. In Chapter 2 we discussed two angular-momentum coupling models in an atom—the Russell–Saunders or LS coupling scheme and the jj coupling scheme. While LS coupling is a reasonable approximation for most atoms, in general, neither of the coupling schemes is exact (this is expressed, for example, as term mixing if the LS scheme is used). Clearly, we can ask for fulfillment of selection rules imposed on angular momenta L and S only to the extent these momenta exist in an atom, or, in other words, to the extent the LS-coupling scheme is fulfilled.

The amplitude for a dipole transition between states ψ_1 and ψ_2 is proportional to

$$\langle \psi_2 | \mathbf{d} | \psi_1 \rangle, \tag{7.13}$$

where $\mathbf{d} = -e\mathbf{r}$ is the dipole-moment operator, e is the electron-charge magnitude and $\mathbf{r}$ is the position vector of the electron with respect to the nucleus. In multi-electron atoms, we need to take the sum over all electrons: $\mathbf{d} = \sum_i \mathbf{d}_i$. Note, however, that such an operator can only connect (i.e., have a nonzero matrix element between) two multi-electron states that differ from each other by the state of only one electron. Thus, transitions between configurations different by more than one electron are forbidden. For example, the following transitions between excited states in the rare-earth element samarium are allowed:

$$(\text{Xe})4f^6 \mathbf{6s6p} \to (\text{Xe})4f^6 \mathbf{6s7s}, \tag{7.14}$$

$$(\text{Xe})4f^6 \mathbf{6s6p} \to (\text{Xe})4f^6 \mathbf{7p6p}, \tag{7.15}$$

while the following is forbidden:

$$(\text{Xe})4f^6 \mathbf{6s6p} \to (\text{Xe})4f^6 \mathbf{7p7s}. \tag{7.16}$$

If the atomic states are mixtures of different configurations, the transition amplitude may be a sum of several terms describing single-electron transitions for the different configurations.

The inner product $\langle\psi_2|\psi_1\rangle$ can be considered a measure of how much of state ψ_1 is contained in state ψ_2. The dipole moment operator acts on the initial state $|\psi_1\rangle$ and creates a new state $\mathbf{d}|\psi_1\rangle$. Equation (7.13) can then be understood as a measure of how similar this new state is to the final state $|\psi_2\rangle$. The greater the similarity, the greater the transition amplitude. These arguments work for both absorption, for which the initial state has lower energy, and for emission, for which the initial state is higher.

Now let us look at the relation between the symmetry of atomic states and the selection rules for dipole transitions. To address this, we need to consider the parity of the wave functions and the transition probability amplitude. The transition amplitude is proportional to the integral

$$\langle\psi_2|\mathbf{d}|\psi_1\rangle = -e\langle\psi_2|\mathbf{r}|\psi_1\rangle = -e\int \psi_2^*(\mathbf{r})\,\mathbf{r}\,\psi_1(\mathbf{r})\,d^3r. \qquad (7.17)$$

The operator $\mathbf{r}$ is odd under spatial inversion ($\mathbf{r} \to -\mathbf{r}$), and the wave functions are either odd or even. For the matrix element to be nonzero, the product $\psi_2^*\mathbf{r}\psi_1$ must be even under spatial inversion because if it is odd, for each contribution to the integral around $\mathbf{r}$, there will be an equal and opposite contribution around $-\mathbf{r}$ that will cancel it. If the functions $\psi_1(\mathbf{r})$ and $\psi_2(\mathbf{r})$ have the same parity, then the product of these functions is an even function and the integrand is odd. The integral then vanishes, meaning that electric-dipole transitions between states of the same parity are strictly forbidden.[1] On the other hand, if the wave functions have opposite parity the integrand is an even function and so the transition amplitude may be nonzero.

The ongoing discussion applies equally well to single-electron atoms as to multi-electron atoms. If the atom contains more than one electron then the parity of the total wave function is the product of the parities of the individual electronic wave functions. If separate electrons have angular momentum quantum numbers $l_1, l_2, l_3, \ldots$ then the parity of the total wave function is $(-1)^{l_1}(-1)^{l_2}(-1)^{l_3}\cdots$. Thus the total wave function is even (odd) if the sum of the angular momentum quantum numbers $l_1 + l_2 + l_3 + \cdots$ is an even (odd) number. Moreover, as we have discussed in Sec. 2.5, even if the individual quantum numbers l_i are not well defined, the overall parity of the multi-electron atomic state is still a good quantum number, and the parity selection rule remains valid.

This means that we can add to the selection rules based on angular momentum conservation a parity selection rule stating that electric-dipole transitions are allowed only between atomic states with opposite parity.

For example, in hydrogen, hydrogenic ions, or other atoms in which the dipole transition is determined by rearrangement of the motion of a single electron that moves in the potential of the nucleus or the nucleus and closed shells of the rest of electrons, transitions with

$$\Delta L = 0 \qquad (7.18)$$

are forbidden. (In this case, l for the single outer electron is the same as L for the whole atom.)

In alkali atoms, we have exactly one electron above the closed shells. In these atoms S-S, P-P, and similar transitions are forbidden following the selection rule defined by Eq. (7.18). (The S-S transitions are also forbidden by the triangle rule.)

---

[1] "Strictly" in this context really means "ignoring atomic parity-violation effects;" see, for example, Prob. 1.13 in the book by Budker *et al.* (2008) and references therein.

132 *Atomic transitions*

In case the atomic nucleus has a nonzero spin I, the selection rules associated with the total angular momentum have to refer to F rather than J:

$$\Delta F = 0, \pm 1, \qquad F + F' \geq 1. \tag{7.19}$$

(In the photon-spin picture, the selection rule for the total angular momentum arises directly from angular-momentum conservation in photon–atom interactions.)

However, the J selection rules still remain valid, to the extent that J is a good quantum number, because optical transitions do not influence the nuclear spin.

> In the treatment of atomic transitions, it is commonly assumed that the transition matrix element is diagonal in the quantum numbers (I, m_I) describing the nucleus. Is this justified?
>
> Indeed, the dipole matrix elements associated with the nucleus are typically much smaller than those associated with atomic electrons, because the nuclear size is typically five orders of magnitude smaller than the atomic size. The suppression of the electric-dipole moment, proportional to the characteristic size of the system, is thus about five orders of magnitude. (Magnetic multipoles discussed below are suppressed because the nuclear magneton is three orders of magnitude smaller than the Bohr magneton. Higher-order multipoles also discussed below containing larger powers of r are suppressed by correspondingly larger factors.) Moreover, optical frequencies correspond to energies in the electron-volt range, which are some five or six orders of magnitude smaller than the characteristic energy scale for nuclear excitation, so that light is very far from the nuclear resonances.
>
> While assuming that the nuclear matrix elements are diagonal in calculating optical transitions is well justified, this does not mean that the nuclear quantum numbers can be ignored. They certainly come in via the hyperfine interactions, not to mention the whole field of nuclear magnetic resonance that we will mention again when we discuss magnetic transitions.
>
> Another way the J selection rules can be violated is through hyperfine-induced mixing of atomic states of different J (but the same F). In recent years, hyperfine-interaction-induced transitions in trapped atoms and ions that are forbidden by the J selection rules have found applications in atomic clocks. These transitions are very weak, and result in exceptionally narrow spectral lines useful for precision time-keeping.

In addition to the selection rules discussed so far having to do with the magnitude of the angular momenta (i.e., quantum numbers L, S, J, F), there are also additional selection rules having to do with the angular-momentum projections characterized by magnetic quantum numbers m (m_L, m_S, m_J, m_F). When an atom in an eigenstate of angular momentum undergoes a transition, both the magnitude of angular momentum and its projection on the quantization axis must be conserved. From Sec. 6.3 we know that the projection of the angular momentum of a photon on the quantization axis is equal to 0 or ± 1, depending on its polarization. The particular value of this projection is determined by the photon polarization (photons that are left and right circularly polarized with respect to the quantization axis have $+1$ and -1 spin projection on that axis, respectively, and photons linearly polarized along the quantization axis have spin projection equal to 0). The immediate conclusion is that for magnetic quantum numbers we have

$$\Delta m_J = 0, \pm 1. \tag{7.20}$$

Moreover, we can also determine which value of Δm_J refers to absorption or emission for specific types of light polarization. Namely, if an atom absorbs left circularly polarized light that propagates in a positive direction along $\hat{\mathbf{z}}$, then the atom's total angular momentum projection increases by 1 upon absorption, whereas right circularly polarized light decreases the angular momentum projection by the same amount, while linearly polarized light leaves this projection unchanged.

In the case of spontaneous emission, exactly the opposite occurs. If an atom radiates a photon along $\hat{\mathbf{z}}$ and the projection of its total angular momentum decreases by 1, the photon must be left circularly polarized. This follows from the symmetry between the absorptive and radiative processes: if a photon of a definite polarization causes a specific atomic transition to occur, then if the atom undergoes the opposite transition it must radiate a photon of the same polarization as it initially absorbed. The same selection rules also govern stimulated-emission transitions. (It may be helpful to think of stimulated emission as absorption "going the other way," or a time-reversed absorption process.)

The preceding qualitative analysis tells us whether an electric dipole transition between two states is allowed or forbidden on the basis of symmetry considerations. However, even among allowed transitions, some occur more easily (are "more allowed") than others. To quantify this, we need a more detailed analysis of the transitions between atomic states. But before we embark on this analysis (Sec. 7.3), let us remark that, based on dimensional analysis, a typical allowed *transition dipole moment* (i.e., the electric-dipole matrix element) in an atom is of the order of $d = ea_0$, where $a_0 = \hbar^2/(m_e e^2)$ is the Bohr radius and m is the electron mass. Some particular values of the transition dipoles will be discussed in Sec. 7.4 (see Table 7.1).

7.3 Probability calculation for electric-dipole transitions

Considering an electric-dipole transition between a pair of particular sublevels, the corresponding transition dipole moment has a particular spatial orientation. The efficiency with which light of a certain polarization can drive this transition, therefore, depends both on the intrinsic characteristics (the dipole moment) of the transition, the intensity of the light, and also on a geometrical factor that describes the relative direction of the electric field of the light and the atomic dipole moment.

7.3.1 Calculation using quantized light

The absorption and emission probabilities are properly calculated using *quantum electrodynamics*. A thorough introduction to the subject can be found, for example, in the book by Cohen-Tannoudji *et al.* (1989). To sketch how this works, we consider here the case of a one-electron atom in the presence of a light field characterized by vector potential $\mathbf{A}(\mathbf{r}, t)$. Neglecting the electron spin (which interacts with the magnetic field of the light) the total Hamiltonian of the atom in the presence of the light field can be written as (see Cohen-Tannoudji *et al.* 1989 or, for example, Prob. 4.51 in the book by Griffiths, 2005)

$$H = \frac{1}{2m_e}\left[\mathbf{p}_c + \frac{e}{c}\mathbf{A}(\mathbf{r},t)\right]^2 - \frac{Ze^2}{r}, \tag{7.21}$$

where $\mathbf{p}_c = -i\hbar\nabla$ is the *canonical momentum* of the electron, m_e is its mass, and Ze is the charge of the nucleus. The plus sign of the vector-potential term is due to our definition of e as the *magnitude* of the electron's charge.

Commonly, the optical field is substantially weaker than the characteristic electric field in an atom.

> In the hydrogen atom, the electric-field strength is of the order of $e/a_0^2 \approx 5 \times 10^9$ V/cm. This is indeed stronger than the light field of most lasers. An important exception, however, is short-pulse lasers. For example, the electric-field strength in a focused nanosecond-laser beam can exceed atomic fields. For modern *femtosecond lasers*, the fields can exceed the atomic fields by several orders of magnitude. In this case, the interaction of electromagnetic radiation with atoms no longer can be described in the

134 *Atomic transitions*

perturbation-theory approximation as we do it here. An analysis of such strong-field interactions can be found, for example, in the book by Hartemann (2002).

If the light field is weaker than the atomic field, we can neglect the term proportional to $\mathbf{A}^2(\mathbf{r}, t)$ [a proof of this can be found, for example, in Prob. 3.3 in the book by Budker *et al.* (2008)] and to consider the cross-terms proportional to $\mathbf{A}(\mathbf{r}, t)$ as a perturbation, H_1. Combining H_1 and the free-atom (unperturbed) Hamiltonian H_0 leads to:

$$H_0 + H_1 = \frac{p_c^2}{2m_e} - \frac{Ze^2}{r} + \frac{e}{2m_e c} \left(\mathbf{p}_c \cdot \mathbf{A} + \mathbf{A} \cdot \mathbf{p}_c \right). \tag{7.22}$$

Under the *Coulomb gauge*, for which $\nabla \cdot \mathbf{A} = 0$, we have

$$\begin{aligned}(\mathbf{A} \cdot \mathbf{p}_c) \psi &= -i\hbar \mathbf{A} \cdot \nabla \psi \\ &= -i\hbar \left[\nabla \cdot (\mathbf{A} \psi) - \psi (\nabla \cdot \mathbf{A}) \right] \\ &= (\mathbf{p}_c \cdot \mathbf{A}) \psi.\end{aligned} \tag{7.23}$$

Thus the perturbation H_1 can be rewritten as

$$H_1 = \frac{e}{m_e c} \mathbf{p}_c \cdot \mathbf{A}. \tag{7.24}$$

The next step in our discussion requires *quantization of the electromagnetic field*. In a sense, we have already used the quantum description of the electromagnetic field in referring to photons, the quanta of the light field. Now, however, the time has come for a more mathematical treatment involving things like the photon *annihilation* and *creation operators*. Since this is described in all standard texts on quantum optics (see also the tutorial discussion in Prob. 3.2 in the book by Budker *et al.*, 2008), we will skip the detailed derivation here, jumping straight to the needed result after briefly recalling the basic physics ideas behind field quantization (see also the book by Migdal, 2000).

Quantization of the electromagnetic field: the basic ideas Let us first take the simple case in which there are no electric charges around. In this case we can set the scalar potential to zero. The energy of the electromagnetic field is the integral over the whole space of the sum of the electric term $\mathcal{E}^2/8\pi$ and the magnetic term $\mathbf{H}^2/8\pi$. We can rewrite the energy in terms of the vector potential by using the basic relations (with zero scalar potential)

$$\mathcal{E} = -\frac{1}{c} \frac{\partial \mathbf{A}}{\partial t}, \tag{7.25}$$

$$\mathbf{H} = \nabla \times \mathbf{A}, \tag{7.26}$$

and by decomposing the vector potential into plane waves:

$$\mathbf{A} = \sum_{\mathbf{k}, \hat{\varepsilon}} \left\{ A_{\mathbf{k}, \hat{\varepsilon}} e^{i \mathbf{k} \cdot \mathbf{r}} + A^*_{\mathbf{k}, \hat{\varepsilon}} e^{-i \mathbf{k} \cdot \mathbf{r}} \right\}. \tag{7.27}$$

Here we label the Fourier components of the vector potential by $\mathbf{k}$ and the unit polarization vector $\hat{\varepsilon}$. There are two independent polarizations for each $\mathbf{k}$. Upon such decomposition, the quantities $A_{\mathbf{k}, \hat{\varepsilon}}$ depend on time, but not on $\mathbf{r}$, and the fields become

$$\mathcal{E} = -\frac{1}{c} \sum_{\mathbf{k}, \hat{\varepsilon}} \left\{ \dot{A}_{\mathbf{k}, \hat{\varepsilon}} e^{i \mathbf{k} \cdot \mathbf{r}} + \dot{A}^*_{\mathbf{k}, \hat{\varepsilon}} e^{-i \mathbf{k} \cdot \mathbf{r}} \right\}, \tag{7.28}$$

$$\mathbf{H} = \sum_{\mathbf{k}, \hat{\varepsilon}} \mathbf{k} \left\{ A_{\mathbf{k}, \hat{\varepsilon}} e^{i \mathbf{k} \cdot \mathbf{r}} + A^*_{\mathbf{k}, \hat{\varepsilon}} e^{-i \mathbf{k} \cdot \mathbf{r}} \right\}. \tag{7.29}$$

From this it follows that the energy of the field per unit volume is

$$\sum_{\mathbf{k},\hat{\varepsilon}} \left\{ \frac{1}{8\pi c^2} |\dot{A}_{\mathbf{k},\hat{\varepsilon}}|^2 + \frac{k^2}{8\pi} |A_{\mathbf{k},\hat{\varepsilon}}|^2 \right\}. \tag{7.30}$$

The key idea is to identify the two terms in this expression with the kinetic- and potential-energy terms of a *harmonic oscillator*. The analogy leads us to identifying the vector potential $A_{\mathbf{k},\hat{\varepsilon}}$ with the "coordinate" of the oscillator, $1/4\pi c^2$ with the "mass" of the oscillator, and $\omega = ck$, not surprisingly, with the frequency of the oscillator. Finally, photon creation, a, and annihilation $a^\dagger$ operators are introduced just as they are for the harmonic oscillator, followed by the introduction of the *photon-number operator* $n = a^\dagger a$, and the discussion of the *zero-point fluctuations* that follow from the fact that the energy of the field in a given mode turns out to be equal to $\hbar\omega(n + 1/2)$.

Considering light in a single mode $(\mathbf{k}, \hat{\varepsilon})$, the vector potential of the quantized electromagnetic field is

$$\mathbf{A} = \sqrt{\frac{2\pi\hbar c^2}{V\omega}} \left[a\hat{\varepsilon}e^{i\mathbf{k}\cdot\mathbf{r}} + a^\dagger \hat{\varepsilon}^* e^{-i\mathbf{k}\cdot\mathbf{r}} \right], \tag{7.31}$$

where V is the volume of the box in which normalization of $\mathbf{A}$ is performed. *Box normalization* is a technique used in quantum mechanics to deal with functions that can have infinite extent. Here ω is the frequency of the field and a, $a^\dagger$ are photon annihilation and creation operators, respectively, and $\hat{\varepsilon}$ is the unit polarization vector.

To obtain the form of the terms that characterize light absorption and emission using first-order perturbation theory, all we need to do is calculate the matrix elements of this perturbation between the initial and final states of the combined system of the atom and the light. We write the combined atom–light wave functions in the form

$$|\xi J m\rangle |n\rangle, \tag{7.32}$$

where $|n\rangle$ are the *photon number states* (ξ stands for the atomic quantum numbers other than J, m).

In fact, it is only under very special circumstances that experimentally we deal with states of light with fixed photon numbers. While it is possible to prepare such states, this is not a trivial matter. Two types of light states that are encountered more frequently are the *chaotic* light produced by incoherent light sources such as stars, light bulbs, etc., and *coherent* light produced by lasers (see the book by Loudon, 2000, for a masterful discussion of these different types of light). However, since all the states of light can be decomposed into number states, the results for the first-order derivations performed with the number states are universally valid. The number states are most convenient for the present discussion.

In absorption, the number of photons in the light field decreases by one, so the initial state $|\xi J m\rangle |n + 1\rangle$ goes to the final state $|\xi' J' m'\rangle |n\rangle$, and we need to calculate

$$\langle n | \langle \xi' J' m' | H_1 | \xi J m \rangle | n + 1 \rangle$$

$$\propto \langle n | \langle \xi' J' m' | a\mathbf{p}_c \cdot \hat{\varepsilon} e^{i\mathbf{k}\cdot\mathbf{r}} + a^\dagger \mathbf{p}_c \cdot \hat{\varepsilon}^* e^{-i\mathbf{k}\cdot\mathbf{r}} | \xi J m \rangle | n + 1 \rangle \tag{7.33}$$

$$= \sqrt{n+1} \langle \xi' J' m' | \mathbf{p}_c \cdot \hat{\varepsilon} e^{i\mathbf{k}\cdot\mathbf{r}} | \xi J m \rangle,$$

where we have used the fact that the creation operator increases the number of photons by one:

$$a^\dagger |n\rangle = \sqrt{n+1} |n+1\rangle, \tag{7.34}$$

and the annihilation operator decreases the number of photons by one:

$$a|n\rangle = \sqrt{n}|n-1\rangle. \tag{7.35}$$

The matrix elements appearing in Eqs. (7.34), and (7.35) can be derived, for example, from the explicit harmonic-oscillator wave functions. In the context of quantized light, these expressions can be understood from the following intuitive argument. When the number of photons n is large, there is not much difference between n and $n+1$. The intensity of the light is proportional to the number of photons, so that the light electric field goes as $\sqrt{n}$. Correspondingly, expressions (7.34), and (7.35) say that the amplitude for emission or absorption of another photon into the mode under consideration is proportional to the field of the photons already present in the mode. This is an *interference effect*: the field of the additional photon either adds to (in the process of emission), or subtracts from (in the process of absorption) the field of the other photons. The resultant intensity goes as the square of the total field. The "1" in $\sqrt{n+1}$ in Eq. (7.34) deserves a special mention. It is responsible for nonzero amplitude for emission into a mode where there are no photons present initially (*vacuum*), and is associated with *spontaneous emission*. Note that there is no such thing as *spontaneous absorption* on account of the absence of "1" in the square root in Eq. (7.35).

If we assume that the wavelength of the radiation is much larger than the dimensions of the atom (the *dipole approximation*; see Sec. 7.6), then $|\mathbf{k} \cdot \mathbf{r}| \ll 1$ and

$$e^{i\mathbf{k} \cdot \mathbf{r}} \approx 1. \tag{7.36}$$

Thus, in Eq. (7.33), we need to evaluate the matrix element of $\langle \xi' J' m' | \mathbf{p}_c \cdot \hat{\boldsymbol{\varepsilon}} | \xi J m \rangle$.

The calculation of the matrix elements of the canonical momentum between the unperturbed atomic states is somewhat subtle. A careful derivation by Cohen-Tannoudji *et al.* (1989) (see also the discussion by Scully and Zubairy, 1997, Chapter 5) gives the result

$$\langle \xi' J' m' | \mathbf{p}_c \cdot \hat{\boldsymbol{\varepsilon}} | \xi J m \rangle = \frac{\omega}{\omega_0} \langle \xi' J' m' | \mathbf{p} \cdot \hat{\boldsymbol{\varepsilon}} | \xi J m \rangle, \tag{7.37}$$

where ω is the frequency of the radiation, and $\omega_0 = (E_{J'} - E_J)/\hbar$. This means that the matrix element of the canonical momentum can be replaced by the matrix element of the mechanical momentum, which is particularly convenient in the perturbative approach. Also, Eq. (7.37) shows that there is actually no difference in the matrix elements for the important case of resonant light.

Returning now to our calculation, we have:

$$\begin{aligned}\langle \xi' J' m' | \mathbf{p}_c \cdot \hat{\boldsymbol{\varepsilon}} | \xi J m \rangle &= \frac{\omega}{\omega_0} \langle \xi' J' m' | \mathbf{p} \cdot \hat{\boldsymbol{\varepsilon}} | \xi J m \rangle \\ &= \frac{\omega}{\omega_0} \frac{im}{\hbar} \langle \xi' J' m' | (H_0 \mathbf{r} - \mathbf{r} H_0) \, \hat{\boldsymbol{\varepsilon}} | \xi J m \rangle \\ &= \omega \frac{im}{e} \langle \xi' J' m' | \mathbf{d} \cdot \hat{\boldsymbol{\varepsilon}} | \xi J m \rangle, \end{aligned} \tag{7.38}$$

where we have made use of the *Heisenberg equation*

$$\mathbf{p} = \frac{im}{\hbar} [H_0, \mathbf{r}]. \tag{7.39}$$

The Heisenberg equation can be understood by assigning quantum-mechanical meaning to the classical expression for the linear momentum

$$\mathbf{p} = m\mathbf{v} = m\frac{d\mathbf{r}}{dt}, \tag{7.40}$$

and by recalling that in quantum mechanics, the time derivative of an operator is given by the commutator of the operator with the Hamiltonian, so that

Probability calculation for electric-dipole transitions

$$\mathbf{p} = m\frac{d\mathbf{r}}{dt} = m\frac{i}{\hbar}[H, \mathbf{r}] = \frac{im}{\hbar}(H\mathbf{r} - \mathbf{r}H). \tag{7.41}$$

Since we are performing a perturbative calculation, we set $H = H_0$.

An important conclusion of our calculation is that the matrix element for absorption is proportional to the matrix element of the electric-dipole operator between the initial and the final states:

$$\langle n|\langle \xi'J'm'|H_1|\xi Jm\rangle|n+1\rangle \propto \langle \xi'J'm'|\mathbf{d} \cdot \hat{\varepsilon}|\xi Jm\rangle. \tag{7.42}$$

A similar calculation for emission yields the relevant matrix element in the form

$$\langle \xi Jm|\mathbf{d} \cdot \hat{\varepsilon}^*|\xi'J'm'\rangle. \tag{7.43}$$

This matrix element differs from that of Eq. (7.42) by the interchange of the initial and final states and by the change from $\hat{\varepsilon}$ to $\hat{\varepsilon}^*$. Note that in the case of spontaneous emission, we can choose to calculate the emission amplitude with a particular polarization of the emitted light, and then write the matrix element in the same way as for stimulated emission.

7.3.2 Spherical-basis expansion and application of the Wigner–Eckart theorem

Let us now return to light absorption. Resolving the unit light-polarization vector $\hat{\varepsilon}$ and the dipole-moment vector $\mathbf{d}$ into their spherical components (see the discussion of the spherical-basis representation for light polarization in Sec. 6.4), one writes the scalar product of these vectors as

$$\mathbf{d} \cdot \hat{\varepsilon} = \sum_{q=-1}^{1} d_q \varepsilon^q. \tag{7.44}$$

Taking the matrix element of $\mathbf{d} \cdot \hat{\varepsilon}$, from Eq. (7.44) we have:

$$\langle \xi'J'm'|\mathbf{d} \cdot \hat{\varepsilon}|\xi Jm\rangle = \sum_q \varepsilon^q \langle \xi'J'm'|d_q|\xi Jm\rangle. \tag{7.45}$$

Now we turn our attention to the matrix element on the right-hand side of Eq. (7.45). Using the Wigner–Eckart theorem (Sec. 3.8), we can separate this matrix element into the angular and dynamic factors:

$$\langle \xi'J'm'|d_q|\xi Jm\rangle = (-1)^{J'-m'} \begin{pmatrix} J' & 1 & J \\ -m' & q & m \end{pmatrix} \langle \xi'J'\|d\|\xi J\rangle. \tag{7.46}$$

The $3j$ symbol, which depends on projections of the angular momenta on the quantization axis, namely on m, m' and q, represents the *angular* (i.e., geometric) part of a transition amplitude, while the reduced matrix element, which makes no reference whatsoever to the angular-momentum projections is sometimes referred to as the *dynamic* part.

For practical calculations it is useful to note that, since the usual phase convention for the spherical harmonics implies that the reduced matrix element of d is real, Eq. (3.120) gives

$$\langle \xi'J'\|d\|\xi J\rangle = (-1)^{J'-J}\langle \xi J\|d\|\xi'J'\rangle. \tag{7.47}$$

As discussed in Sec. 3.5, the rules for calculating $3j$ symbols contain selection rules for adding angular momenta: the triangle rule and the projection rule. In this case the angular

138 *Atomic transitions*

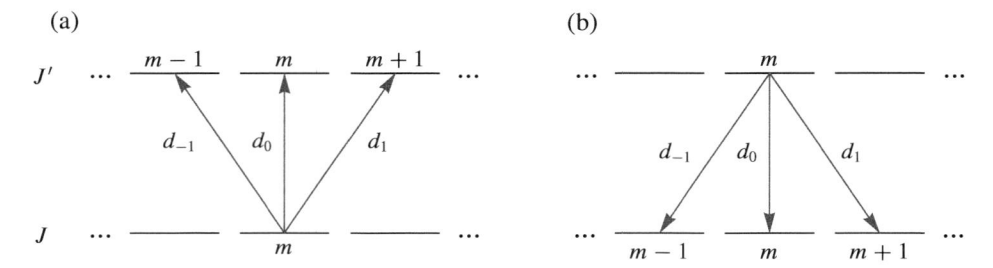

Fig. 7.2 Calculation of sum rules for transition probabilities in absorption (a) and emission (b).

momenta involved are J' which represents the angular momentum of the final state, 1 which is the rank of the dipole operator and represents the angular momentum of the photon (photon spin), and J which represents the angular momentum of the initial state. The triangle inequality then reads

$$|J - 1| \leq J' \leq J + 1, \tag{7.48}$$

which is equivalent to the electric-dipole transition selection rules $\Delta J = 0, \pm 1$ ($J = 0 \not\to J' = 0$) discussed above. Another requirement for a $3j$ symbol to be nonzero is the condition for the angular-momentum projections, which in this case takes the form

$$q + m - m' = 0. \tag{7.49}$$

This rule is equivalent to the $\Delta m = 0, \pm 1$ selection rule discussed above.

It is worth recalling that the photon angular-momentum projection q is related to the polarization of the light (Sec. 6.4). This means that, from the properties of the $3j$ symbols, one can deduce which light polarization is needed to "drive" a particular transition, or whether a particular transition is allowed or forbidden for light of a given polarization. Moreover, the relative strength of different allowed transitions can be calculated.

Using the properties of the $3j$ symbols, one can derive a number of very useful *sum rules*. To illustrate this, let us find the absorption rate for isotropic light. We sum the absorption rates for the three possible dipole transitions originating from one magnetic sublevel of the initial state (Fig. 7.2a), assuming that the intensities of each of the three light-polarization components are the same. Then, the total rate with which the atom leaves the initial state $|\xi J m\rangle$ is proportional to

$$\sum_{q,m'} |\langle \xi' J' m' | d_q | \xi J m \rangle|^2 = \sum_{q,m'} \begin{pmatrix} J' & 1 & J \\ -m' & q & m \end{pmatrix}^2 |\langle \xi' J' \| d \| \xi J \rangle|^2$$
$$= \frac{|\langle \xi' J' \| d \| \xi J \rangle|^2}{2J + 1}, \tag{7.50}$$

where we employed the sum rule for $3j$ symbols of Eq. (3.95). Note that the rate of absorption for unpolarized light is the same for each of the m sublevels.

The assumption that we have equal intensity in all three polarization components of the light is equivalent to the assumption that we are irradiating our atom uniformly from all directions with unpolarized light. The result we have obtained for the absorption rate reflects the spatial symmetry of this problem. Namely, we see that the absorption rate is the same for

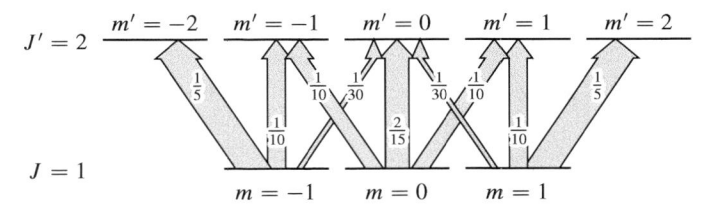

Fig. 7.3 Transition strengths in units of $|\langle\xi'J'\|d\|\xi J\rangle|^2$ for $J = 1 \to J' = 2$ absorption.

all magnetic sublevels of the initial state. As we have discussed in Chapter 6, unpolarized light is not easy to obtain experimentally. One method is to use an optical sphere—an ideally diffusely reflecting cavity of spherical shape. Atoms are placed inside the sphere, and incoherent light—for example, from a gas-discharge lamp—is injected into the sphere.

We now turn to the rates of spontaneous or induced emission from an upper state J' to a lower-lying state J. For spontaneous emission, according to Eq. (7.43), we need to calculate the matrix elements of $\mathbf{d} \cdot \hat{\boldsymbol{\varepsilon}}^*$:

$$\mathbf{d} \cdot \hat{\boldsymbol{\varepsilon}}^* = \sum_q d_q (\varepsilon^*)^q. \tag{7.51}$$

In the case of spontaneous emission, all three possible spontaneous transition channels $\Delta m = 0, \pm 1$ (Fig. 7.2b) are uncorrelated. This is because vacuum fluctuations in different polarizations are uncorrelated, and can also be argued from isotropy of space and thermodynamic considerations. In order to calculate the spontaneous transition rate, we, therefore, need to add transition probabilities (not amplitudes!) for different polarizations, meaning that the rate is proportional to

$$\sum_{q,m} |\langle\xi J m|d_q|\xi' J' m'\rangle|^2 = \sum_{q,m} \begin{pmatrix} J & 1 & J' \\ -m & q & m' \end{pmatrix}^2 |\langle\xi J\|d\|\xi' J'\rangle|^2$$

$$= \frac{|\langle\xi J\|d\|\xi' J'\rangle|^2}{2J'+1}, \tag{7.52}$$

where the sum of Eq. (3.95) has once again been applied in the last step.

This result does not depend on whether this is a $\Delta J = -1, 0$ or 1 transition, and is also independent of the initial magnetic quantum number m'. The latter result is consistent with the principle of isotropy of space that dictates that all sublevels of a given state decay with the same rate, meaning that spontaneous decay cannot change the symmetry of the magnetic-sublevel population distribution.

Now let us investigate the relative magnitudes of absorption rates for some specific values of J and J'. We will calculate the squares of the dipole-transition matrix elements between $|\xi J m\rangle$ and $|\xi' J' m'\rangle$ states, called *transition strengths*.. These can be found from Eq. (7.46) in terms of the reduced dipole matrix element by evaluating the $3j$ symbol for particular values of the quantum number.

Let us start with a $J = 1 \to J' = 2$ transition. The numerical values of the transition strengths in units of $|\langle\xi'J'\|d\|\xi J\rangle|^2$ are presented in Fig. 7.3. The transitions with $m' = m - 1$ are excited by σ^--polarized light, the $m' = m$ transitions are excited by π-polarized light, and the $m' = m + 1$ transitions are excited by σ^+-polarized light.

140 *Atomic transitions*

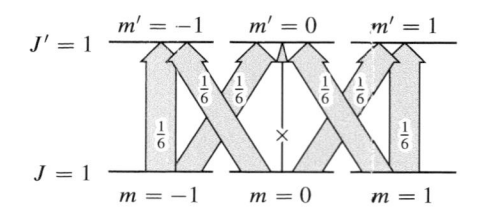

$J' = 1$ $\quad\underline{\qquad m'=-1 \qquad\qquad m'=0 \qquad\qquad m'=1\qquad}$

$J = 1$ $\quad\underline{\qquad m=-1 \qquad\qquad m=0 \qquad\qquad m=1 \qquad}$

Fig. 7.4 Transition strengths in units of $|\langle \xi' J' \| d \| \xi J \rangle|^2$ for $J = 1 \rightarrow J' = 1$ absorption.

We can repeat the same calculations for the transition $J = 1 \rightarrow J' = 1$. The transition strengths for this case are presented in Fig. 7.4. Note that even though the transition between the magnetic sublevels $m' = m = 0$ is allowed by the projection selection rule $\Delta m = 0, \pm 1$, the probability for this transition is zero. This is generally true for $J' = J$ transitions—a qualitative explanation for this can be summarized as follows.

In elementary-particle physics, particles with intrinsic angular momentum equal to one are called *vector particles*. Their internal state can be completely characterized by a *polarization vector*. The relation between the direction of the polarization vector and the quantum number q is discussed for the case of the photon in Sec. 6.4 (see also Budker *et al.* 2008, Prob. 9.5). In this case, we have an $m' = m$ transition, which is driven by π-polarized light. This polarization corresponds to $q = 0$, i.e., the polarization vector of the light is $\hat{\varepsilon}_{\text{photon}} = \hat{\varepsilon}_0 = \hat{z}$. The atom has $J = 1$ so it is also a spin-one particle. Here, the projection quantum number m takes the place of q. Since $m = 0$, the atom is described by the same polarization vector $\hat{\varepsilon}_{\text{atom}} = \hat{z}$.

When we combine two spin-one particles into a composite system, we need to build the wave function of the combined system out of the wave functions of the constituents. The composite wave function has to be linear in the wave functions of the constituents.

Combining two angular-momentum-one particles, we can get total angular momentum zero (scalar), one (vector), or two (tensor). In this case, we need to combine a spin-one initial atomic state with a photon to obtain the final spin-one state, which is also a vector. Now, given polarization vectors $\hat{\varepsilon}_{\text{atom}}$ and $\hat{\varepsilon}_{\text{photon}}$, there is only one way to form a vector that is linear in both of them:

$$\hat{\varepsilon}_{\text{resultant}} \propto \hat{\varepsilon}_{\text{atom}} \times \hat{\varepsilon}_{\text{photon}}. \tag{7.53}$$

But $\hat{\varepsilon}_{\text{atom}}$ and $\hat{\varepsilon}_{\text{photon}}$ are parallel to each other, so $\hat{\varepsilon}_{\text{resultant}}$ is identically zero. The amplitude of the process, in which two spin-one particles combine into a spin-one atomic final state with polarization vector $\hat{\varepsilon}_{\text{final}}$, should be a scalar linear in all three polarization vectors. This is obtained by taking the dot product of $\hat{\varepsilon}_{\text{resultant}}$ with $\hat{\varepsilon}_{\text{final}}$:

$$A \propto \left(\hat{\varepsilon}_{\text{atom}} \times \hat{\varepsilon}_{\text{photon}} \right) \cdot \hat{\varepsilon}_{\text{final}}, \tag{7.54}$$

which vanishes for all three vectors along the z-axis. The selection rule discussed in this problem is a particular case of a more general selection rule

$$\begin{pmatrix} J & 1 & J' \\ 0 & 0 & 0 \end{pmatrix} = 0, \tag{7.55}$$

which means that $m = 0 \rightarrow m' = 0$ dipole transitions are forbidden for any $J = J'$ transition (see Budker *et al.* 2008, Prob. 9.5) for a more detailed discussion].

Finally, let us calculate the transition strengths for the $J = 2 \rightarrow J' = 1$ transition. The results for all three polarization types are summarized in Fig. 7.5. We can now take sums of the transition strengths in each of the three figures discussed above to explicitly verify that the sum rules of Eqs. (7.50) and (7.52) are fulfilled.

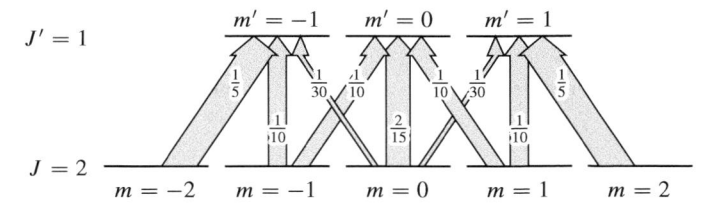

Fig. 7.5 Transition strengths in units of $|\langle \xi' J' \| d \| \xi J \rangle|^2$ for $J = 2 \to J' = 1$ absorption.

7.4 Line strength

The reduced matrix element $\langle \xi' J' \| d \| \xi J \rangle$ appearing in several of the above equations characterizes the dynamic part of the amplitude of an atomic transition. In other words, it characterizes how easy it is to induce the transition (i.e., to induce oscillations of the atomic transition dipole), apart from such details as the mutual spatial orientation of the transition dipole and the light electric field.

The numerical values of the reduced matrix elements differ for different atoms and different transitions in one atom. In general, these matrix elements depend on the specific wave functions of the states that are connected by the dipole transition. As it turns out, in some cases, we can go still further and single out a part of a reduced matrix element associated with angular momenta, and calculate this part exactly. As we have discussed in Sec. 7.2, in the case of the Russell–Saunders (LS) coupling, L changes in an electric-dipole transition, but S does not. By means of angular-momentum algebra (Varshalovich *et al.* 1988; Sobelman 1992, for example), this can be expressed as

$$\langle \xi' L' S J' \| d \| \xi L S J \rangle = (-1)^{L'+S+J+1} \sqrt{(2J+1)(2J'+1)} \begin{Bmatrix} L' & J' & S \\ J & L & 1 \end{Bmatrix} \langle \xi' L' \| d \| \xi L \rangle, \tag{7.56}$$

where the curly brackets denote the $6j$ symbol introduced in Sec. 3.6. The factor $\langle \xi' L' \| d \| \xi L \rangle$ is again a reduced matrix element, but now it is taken between the states of total orbital angular momentum—the part having to do with electron spin is completely described by the prefactor in Eq. (7.56). As discussed in Sec. 3.6, the $6j$ symbols obey a number of selection rules. From these, selection rules for dipole transitions between fine-structure levels can be derived, including $\Delta L = 0, \pm 1$ ($L = 0 \nrightarrow L' = 0$) and $\Delta J = 0, \pm 1$ ($J = 0 \nrightarrow J' = 0$). These triangle conditions assume LS coupling in both the initial and the final state of the transition.

As an example showing the utility of such a separation of the system into a part that interacts with the light and a part that does not, we discuss optical transitions between fine-structure levels of alkali atoms.

Alkali atoms have a $^2S_{1/2}$ ground state and a fine-structure doublet, $^2P_{1/2}$ and $^2P_{3/2}$, in the first excited state. Let us compare the optical transitions $^2S_{1/2} \to {}^2P_{1/2}$ (traditionally called $D1$ transitions) and $^2S_{1/2} \to {}^2P_{3/2}$ ($D2$ transitions). Which of these transitions is stronger and by how much? Let us compare the absolute-value-squared reduced matrix elements $S(JJ') = |\langle \xi' J' \| d \| \xi J \rangle|^2$ (a.k.a. *line strengths*) for these two transitions. For the $^2S_{1/2} \to {}^2P_{1/2}$ transition,

142 Atomic transitions

$$\left|\langle \xi', J' = \tfrac{1}{2} \|d\| \xi, J = \tfrac{1}{2}\rangle\right|^2$$
$$= \left|\langle \xi'(L' = 1, S = \tfrac{1}{2})J' = \tfrac{1}{2} \|d\| \xi(L = 0, S = \tfrac{1}{2})J = \tfrac{1}{2}\rangle\right|^2 \qquad (7.57)$$
$$= (-1)^{2(1+\tfrac{1}{2}+\tfrac{1}{2}+1)} 4 \begin{Bmatrix} 1 & \tfrac{1}{2} & \tfrac{1}{2} \\ \tfrac{1}{2} & 0 & 1 \end{Bmatrix}^2 \left|\langle \xi', L' = 1 \|d\| \xi, L = 0\rangle\right|^2,$$

and for the $^2S_{1/2} \to {}^2P_{3/2}$ transition, we have:

$$\left|\langle \xi', J' = \tfrac{3}{2} \|d\| \xi, J = \tfrac{1}{2}\rangle\right|^2 = \left|\langle \xi'(L' = 1, S = \tfrac{1}{2})J' = \tfrac{3}{2} \|d\| \xi(L = 0, S = \tfrac{1}{2})J = \tfrac{1}{2}\rangle\right|^2$$
$$= (-1)^{2(1+\tfrac{1}{2}+\tfrac{1}{2}+1)} 8 \begin{Bmatrix} 1 & \tfrac{3}{2} & \tfrac{1}{2} \\ \tfrac{1}{2} & 0 & 1 \end{Bmatrix}^2 \left|\langle \xi', L' = 1 \|d\| \xi, L = 0\rangle\right|^2. \qquad (7.58)$$

Because both final states $^2P_{1/2}$ and $^2P_{3/2}$ originate from the same orbital angular momentum state, both transitions depend on the same reduced matrix element $\langle \xi' L' = 1 \|d\| \xi L = 0\rangle$. This allows us to evaluate the relative transition strength of these transitions. The numerical values of the $6j$ symbols are

$$\begin{Bmatrix} 1 & \tfrac{1}{2} & \tfrac{1}{2} \\ \tfrac{1}{2} & 0 & 1 \end{Bmatrix} = \frac{1}{\sqrt{6}} \qquad (7.59)$$

and

$$\begin{Bmatrix} 1 & \tfrac{3}{2} & \tfrac{1}{2} \\ \tfrac{1}{2} & 0 & 1 \end{Bmatrix} = -\frac{1}{\sqrt{6}}. \qquad (7.60)$$

Thus we arrive at

$$\frac{\left|\langle \xi', J' = \tfrac{3}{2} \|d\| \xi, J = \tfrac{1}{2}\rangle\right|^2}{\left|\langle \xi', J' = \tfrac{1}{2} \|d\| \xi, J = \tfrac{1}{2}\rangle\right|^2} = 2, \qquad (7.61)$$

which means that for the alkali atoms the $D2$ line is twice as strong as the $D1$ line.

This result is obtained in the LS-coupling approximation, so it is interesting to see how good this approximation actually is. The experimental value of a dipole-transition matrix element can be obtained from the measured lifetime τ of an atomic level if the *branching ratio*, or fraction of the spontaneous decays of the upper state that go via this transition (also *channel* or *branch*), is known. The simplest situation is when there is only one decay channel of the upper state, such as our present case of transitions between the first excited state and the ground state. For such transitions, the reduced matrix element is related to the lifetime according to (Sobelman 1992; Loudon 2000; Budker *et al.* 2008):

$$\left|\langle \xi' J' \|d\| \xi J\rangle\right|^2 \frac{1}{2J' + 1} \frac{4\omega_0^3}{3\hbar c^3} = \frac{1}{\tau}, \qquad (7.62)$$

where ω_0 is the transition frequency. The reduced matrix elements $\langle \xi' J' \|d\| \xi J\rangle$ calculated in this way for the $D1$ and $D2$ transitions in several alkali atoms are collected in Table 7.1. We see that, while the observed ratios are close to 2, as predicted for LS coupling, there is a deviation that grows with the atomic weight. This is consistent with the LS coupling scheme being a better approximation for light atoms, but still being fairly good even for the heavier alkalis.

Line strength

Table 7.1 Experimental values and ratios of matrix elements for the $D1$ and $D2$ transitions in alkali atoms. The matrix elements are listed in Appendix C.

Alkali atom	Atomic number	$\left(\frac{|\langle 3/2\|d\|1/2\rangle|}{|\langle 1/2\|d\|1/2\rangle|}\right)^2$
Na	11	1.999(4)
K	19	1.999(7)
Rb	37	1.996(4)
Cs	55	1.985(7)
Fr	87	1.902(12)

As discussed in Sec. 2.4, the fine-structure levels in atoms are further split by nuclear-spin-dependent interactions into hyperfine states of total angular momentum F. We now use, once again, the fact that, in an electric-dipole transition, only the electron state can be affected, but not the nuclear spin. This means that we can again factor the matrix element into light-interacting and "spectator" parts. Following the approach we used above for the calculation of the reduced matrix elements $\langle \xi' L' S' J' \| d \| \xi L S J \rangle$, we write

$$\langle \xi' J' I F' \| d \| \xi J I F \rangle = (-1)^{J'+I+F+1} \sqrt{(2F+1)(2F'+1)} \begin{Bmatrix} J' & F' & I \\ F & J & 1 \end{Bmatrix} \langle \xi' J' \| d \| \xi J \rangle. \quad (7.63)$$

As above, the $6j$ symbol supplies the selection rules for transitions involving hyperfine-structure levels.

The quantity $S(FF') = |\langle \xi' J' I F' \| d \| \xi J I F \rangle|^2$ is the line strength for transitions involving the hyperfine-structure levels. Consider as an example the $D1$ and $D2$ transitions of alkali atoms with $I = 3/2$: ^{7}Li, ^{23}Na, ^{39}K, ^{41}K, and ^{87}Rb. Hyperfine interactions split the fine-structure states of these atoms so that from the $^2S_{1/2}$ state there arise two hyperfine levels with $F = 1$ and 2, while from the excited fine-structure states $^2P_{1/2}$ and $^2P_{3/2}$ there arise hyperfine levels $F' = 1, 2$ and $F' = 0, 1, 2, 3$, respectively. Applying Eq. (7.63), we obtain the line strengths for these transitions shown in Fig. 7.6.

The line strengths in Fig. 7.6 exhibit a pair of sum rules that is satisfied in general for the transitions between hyperfine multiplets, namely

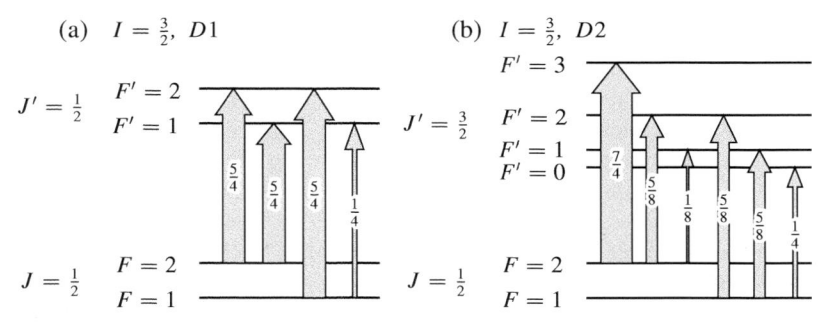

Fig. 7.6 Line strengths in units of $|\langle \xi' J' \| d \| \xi J \rangle|^2$ for transitions involving hyperfine levels for an atom with nuclear spin $I = 3/2$. (a) A $J = 1/2 \rightarrow J' = 1/2$ transition corresponding to the atomic $D1$ line and (b) a $J = 1/2 \rightarrow J' = 3/2$ transition corresponding to the $D2$ line.

144 Atomic transitions

$$\sum_{F} S(FF') = \frac{2F'+1}{2J'+1} \left|\langle \xi' J' \| d \| \xi J \rangle\right|^2, \tag{7.64a}$$

$$\sum_{F'} S(FF') = \frac{2F+1}{2J+1} \left|\langle \xi' J' \| d \| \xi J \rangle\right|^2. \tag{7.64b}$$

Here it is important to keep in mind that relative line strengths—such as presented in Fig. 7.6—allow one to compare different transitions in a particular manifold with each other, but do not allow direct comparison between different manifolds. This is because of the overall factor $\left|\langle \xi' J' \| d \| \xi J \rangle\right|^2$ in the absolute line strength, which varies with J and J'. As demonstrated above, the $D2$ line (Fig. 7.6b) is two times stronger than the $D1$ line (Fig. 7.6a). In order to compare transitions between hyperfine levels belonging to different pairs of fine-structure states of a particular $L \to L'$ transition, one must resolve the reduced matrix elements $\langle \xi' J' \| d \| \xi J \rangle$ further. The recipe for this is given by Eq. (7.56). Collecting the matrix elements that characterize the fine and hyperfine line strengths in a single expression, we arrive at

$$\langle \xi' L' S J' I F' \| d \| \xi L S J I F \rangle = (-1)^{L'+S+J+1} \sqrt{(2J+1)(2J'+1)} \begin{Bmatrix} L' & J' & S \\ J & L & 1 \end{Bmatrix}$$

$$\times (-1)^{J'+I+F+1} \sqrt{(2F+1)(2F'+1)} \begin{Bmatrix} J' & F' & I \\ F & J & 1 \end{Bmatrix} \tag{7.65}$$

$$\times \langle \xi L' \| d \| \xi L \rangle.$$

7.5 Higher-multipole radiative transitions

So far we have considered electric-dipole transitions, which are generally the strongest optical transitions in an atom. Usually, if the electric-dipole transition selection rules discussed so far in this chapter are not fulfilled for a certain transition, it is said that the transition is *forbidden*. However, even if electric-dipole transitions are not allowed, other kinds of transitions may occur. Just as electric-dipole radiation is associated with a time-varying electric-dipole moment, other time-varying electric and/or magnetic moments may also produce radiation. In fact, a whole *multipole expansion* can be built for electromagnetic radiation, which we will discuss in Sec. 7.6. In this expansion, electric-dipole transitions are referred to as E1. The two most frequently encountered transition types other than E1 are magnetic-dipole (M1) and electric-quadrupole (E2), each associated with the corresponding time-varying electromagnetic moment.

The probabilities of optical M1 and E2 transitions are typically around 10^5 times smaller than those of allowed E1 transitions, as can readily be found from an order-of-magnitude analysis. We have seen that the amplitude of an E1 transition is proportional to the electric-dipole moment, with a typical value on the order of $d = ea_0 = \hbar^2/(m_e e)$ (Sec. 7.2). Similarly, an M1 transition, governed by the Hamiltonian $H_B = -\boldsymbol{\mu} \cdot \mathbf{B} = \mu_B (\mathbf{L} + 2\mathbf{S}) \cdot \mathbf{B}/\hbar$, has a typical transition amplitude similarly proportional to the M1 matrix element on the order of a Bohr magneton $\mu_B = e\hbar/(2m_e c)$ (Sec. 2.2.3). To go from the matrix elements to the transition amplitudes, one needs to multiply the matrix elements by the corresponding electric- or magnetic-field strengths of the light. In Gaussian units, these strengths are conveniently equal for a traveling light wave.[2] Consequently, the ratio of the transition amplitudes is on the order of

---

[2]This, in fact, is a major advantage of Gaussian units as compared to SI units.

$$\frac{d}{\mu_B} = \frac{2\hbar c}{e^2} = \frac{2}{\alpha}, \tag{7.66}$$

where $\alpha \approx 1/137$ is the fine-structure constant. Because the transition strength is given by the square of the transition amplitude, we have a ratio of E1 to M1 transition strength on the order of $(2 \times 137)^2 \approx 10^5$. We discuss the ratios of the matrix elements for various multipoles in more detail in Sec. 7.6.

Magnetic-dipole (M1) transitions

Even though M1 transitions are typically much weaker than E1, they are important for states between which electric-dipole transitions are not allowed. For example, consider a transition between two states that is allowed by the E1 angular-momentum selection rules, but is forbidden because the states have the same parity. We will see that this is exactly the case where the transition may go as M1.

Magnetic dipole transitions have the same $\Delta J = 0, \pm 1$ angular momentum selection rule as E1 transitions, but the M1 parity selection rule is different. An electric dipole is created by two separated charges of opposite sign. Inverting the coordinate axes (a parity transformation) is equivalent to reversing the relative positions of the charges, i.e., flipping the sign of the dipole. In other words, the electric-dipole moment is odd under parity reversal (P-odd); this is the origin of the parity selection rule discussed in Sec. 7.2. A P-odd vector is known as a *proper vector* or *polar vector*. On the other hand, a magnetic dipole is proportional to the angular momentum of a charged particle. Classically, we can think of a current loop, for example, a charged particle moving in a circle. The classical angular momentum of the particle is $\mathbf{r} \times \mathbf{p}$; if the coordinate axes are inverted, both $\mathbf{r}$ and $\mathbf{p}$ flip sign, so that the sign of the angular momentum does not change. Thus an angular momentum (and therefore a magnetic dipole) is a P-even vector, known as a *pseudovector* (or *axial vector*). This means that our analysis that dealt with parity selection rules for electric-dipole transitions would give exactly the opposite results for magnetic-dipole transitions. In other words, M1 transitions are allowed between states with the same parity, and, neglecting parity violation, are forbidden between opposite-parity states.

An additional selection rule for M1 transitions is that they can only connect states with the same principal quantum number n. The M1 Hamiltonian is composed of angular-momentum operators, so it only operates on the angular parts of the atomic wave function, and does not affect the radial part (in contrast to the E1 Hamiltonian that contains $\mathbf{r}$). Because the radial wave functions for different values of n are orthogonal, the M1 matrix elements are zero except between states with the same n.

An important class of magnetic-dipole transitions are the transitions between hyperfine substates of the same electronic state. Probably the most well-known magnetic-dipole transition of this kind is the 21-cm-wavelength transition between the hyperfine levels of the ground electronic state of hydrogen. The 21-cm line ($\nu = 1420.40058$ MHz) was first observed in radio astronomy by Ewen and Purcell (1951a;b) using a large horn-type antenna. Later, based on the observation of this magnetic-dipole transition, the first maps of neutral hydrogen in the Galaxy were made, which revealed, for the first time, the spiral structure of the Milky Way. Magnetic-dipole transitions between ground-state hyperfine levels are also used in atomic clocks. An M1 transition between the ground-state hyperfine levels in Cs presently provides the *primary frequency standard* defining the length of the second as the unit of time (see, for example, Audoin and Guinot 2001).[3]

Another important type of M1 transitions are radio-frequency transitions between magnetic sublevels of a given state that have been split by the Zeeman effect (Sec. 4.1). Such transitions are forbidden as E1 since they connect levels of the same parity. In the case of nuclear levels, such transitions are essential for *nuclear magnetic resonance* (NMR) and *magnetic-resonance imaging* (MRI) (see, for example, the book by Slichter, 1996). In the case of ground-state atomic levels, transitions between Zeeman sublevels are employed in *optical-pumping magnetometers*.

> Historically, transitions between Zeeman sublevels were called *Hertzian resonances*. Similarly, Zeeman coherences used to be called *Hertzian coherences*. French physicist Alfred Kastler received the Nobel prize in physics in 1966 *for the discovery and development of optical methods for studying Hertzian resonances in atoms* (http://nobelprize.org/nobel_prizes/physics/laureates/1966/).

Electric-quadruple (E2) transitions

The next type of transition we consider is the *electric quadrupole* (E2), associated with a time-varying quadrupole moment. The quadrupole moment of a system of N particles is a second-rank tensor defined as (cf. Sec. 3.7):

$$Q_{ij} = \sum_{n=1}^{N} q^{(n)} \left(3 r_i^{(n)} r_j^{(n)} - r^2 \delta_{ij} \right), \tag{7.67}$$

where $q^{(n)}$ and $\mathbf{r}^{(n)}$ are the charge and position, respectively, of the n-th particle. Because the tensor rank of the quadrupole operator is higher than that of the rank-one (vector) dipole operator, the selection rules for E2 transitions are different from those for E1. Namely, these transitions obey the following angular-momentum selection rules

$$\Delta J = 0, \pm 1, \pm 2, \quad J + J' \geq 2, \tag{7.68a}$$

$$\Delta m_J = 0, \pm 1, \pm 2. \tag{7.68b}$$

In addition, since the electric quadrupole moment is P-even (as with the magnetic dipole), transitions are allowed only between same-parity states. Finally, to the extent that the Russell–Saunders (LS) coupling scheme is valid, the additional selection rules

$$\Delta L = 0, \pm 1, \pm 2, \quad L + L' \geq 2, \tag{7.69a}$$

$$\Delta S = 0, \tag{7.69b}$$

must be obeyed.

Note that for certain pairs of levels, both E2 and M1 selection rules may be simultaneously satisfied. In this case, transitions between such state are of *mixed* M1/E2 type.

This analysis can be continued for higher-rank transitions. It can be shown that electric octupole and magnetic quadrupole transitions can take place only between states of opposite parity. Electric octupole transitions obey

$$\Delta J = 0, \pm 1, \pm 2, \pm 3, \quad J + J' \geq 3, \tag{7.70}$$

and magnetic quadrupole transitions obey

$$\Delta J = 0, \pm 1, \pm 2, \quad J + J' \geq 2. \tag{7.71}$$

Corresponding selection rules for the Russell–Saunders coupling scheme also apply.

---

[3] Other kinds of electric-dipole-forbidden transitions in the optical and UV range are emerging as new, higher accuracy, standards that are likely to replace the Cs standard in the future (see, for example, Margolis 2009).

Higher-multipole radiative transitions

Let us discuss a practical problem involving an E2 transition and use it as an occasion to compare and contrast the E1 and E2 cases.

(a) Consider an electric-dipole (E1) transition between states of total angular momenta $F = 1/2$ and $F' = 3/2$. We wish to draw and/or list all possible $m \rightarrow m'$ transitions, indicate their relative strengths, and explain which type of plane-wave light (polarization, propagation direction) is needed to drive each of the transitions.

(b) We then wish to do the same for an electric-quadrupole (E2) transition.

Let us address these questions in order.

(a) The Zeeman components of the upper and the lower state are shown in Fig. 7.7(a) along with the E1-allowed transitions. The strength of each of the transitions indicated on the figure is just the square of the appropriate $3j$ symbol (see Sec. 7.3.2):

$$\begin{pmatrix} 3/2 & 1 & 1/2 \\ -m' & q = m' - m & m \end{pmatrix}^2. \tag{7.72}$$

Each of the indicated transitions is driven by light corresponding to the particular spherical component of the polarization vector ε^q (see Sec. 6.4). The direction of light propagation matters only to the extent that the light polarization must be transverse to the propagation direction.

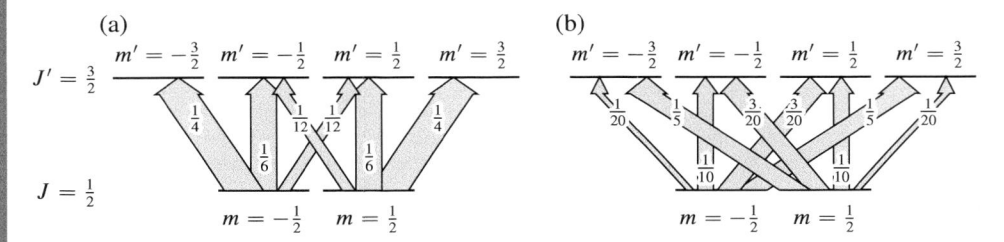

Fig. 7.7 Relative strengths of (a) E1 and (b) E2 $F = 1/2 \rightarrow F' = 3/2$ transitions.

(b) Proceeding in a similar fashion for the E2 case, we arrive at the transitions shown in Fig. 7.7(b). In this case, the $|\Delta m| = 2$ transitions are also allowed, and the strengths of the $m \rightarrow m'$ transitions are given by

$$\begin{pmatrix} 3/2 & 2 & 1/2 \\ -m' & q = m' - m & m \end{pmatrix}^2, \tag{7.73}$$

where the only difference from the E1 case is that the rank in the second column is 2 instead of 1.

In order to figure out what light polarization and direction we need to drive a particular transition, we recall that the E2-interaction Hamiltonian is proportional to the scalar product $\sum_{i,j} Q_{ij} \nabla_j \mathcal{E}_i$. Here, we have rank-two operators $\nabla \mathcal{E}$ and Q. For a plane monochromatic wave, the electric-field gradient reduces to

$$\nabla \mathcal{E} = i\mathbf{k}\mathcal{E}, \tag{7.74}$$

where the notation $\mathbf{k}\mathcal{E}$ (sometimes referred to as a *dyadic*) represents a tensor and is not to be confused with the scalar product.

The treatment that results in the transition strengths shown in Fig. 7.7(b) assumes that we work with spherical components. Therefore, in analogy with the E1 case, a Δm transition is "driven" by the $\kappa = 2$, $q = m' - m$ contravariant component of the tensor (7.74). The expression for this component

148 *Atomic transitions*

can be obtained in the following way. We start by writing the light-polarization vector and the wave vector in the spherical basis:

$$k_{\pm 1} = \mp \frac{1}{\sqrt{2}} (k_x \pm i k_y), \quad k_0 = k_z, \tag{7.75a}$$

$$\varepsilon_{\pm 1} = \mp \frac{1}{\sqrt{2}} (\varepsilon_x \pm i \varepsilon_y), \quad \varepsilon_0 = \varepsilon_z. \tag{7.75b}$$

The next step is to expand this direct product in irreducible tensors (Eq. 3.113). Using the definition (3.112) of the tensor product, and taking into account that $\mathbf{k} \cdot \hat{\varepsilon} = 0$, we find

$$(\mathbf{k}\hat{\varepsilon})_0^2 = \sqrt{\frac{3}{2}} k_z \varepsilon_z, \tag{7.76a}$$

$$(\mathbf{k}\hat{\varepsilon})_{\pm 1}^2 = \mp \frac{1}{2} \left[k_z \varepsilon_x + k_x \varepsilon_z \pm i \left(k_z \varepsilon_y + k_y \varepsilon_z \right) \right], \tag{7.76b}$$

$$(\mathbf{k}\hat{\varepsilon})_{\pm 2}^2 = \frac{1}{2} \left[k_x \varepsilon_x - k_y \varepsilon_y \pm i \left(k_x \varepsilon_y + k_y \varepsilon_x \right) \right]. \tag{7.76c}$$

These expressions can also be obtained directly from Eqs. (3.114).

From Eqs. (7.76), we see that, for example, in order to drive the $|\Delta m| = 2$ transition, we need to have light propagating perpendicularly to the quantization axis with the electric field also perpendicular to the quantization axis. Since the light electric field must be perpendicular to the propagation direction, we see that $|\Delta m| = 2$ transitions are driven when the light propagation direction, the light polarization and the quantization axis are all mutually perpendicular. We also see that it is impossible to choose such a combination of the light propagation direction and polarization as to solely drive, say, the $\Delta m = 2$ transition without driving other transitions.

An interesting observation is that E2 transitions with $\Delta m = 0$ cannot be driven by plane-wave light propagating either along or perpendicular to the quantization axis. This is because the $\Delta m = 0$ transition amplitude is proportional to $k_z \varepsilon_z$, which vanishes in both these cases.

In a standing wave, the situation is somewhat different because Eq. (7.74) no longer holds. In a standing wave, the transition rates depend on the spatial location (this is true both for E1 and E2 transitions). For example, for the E2 case, if the atom is located at the node of a standing wave the average electric field is always zero, but the electric-field gradient, which can drive an E2 transition, is at a maximum.

7.6 Multipole expansion

In the E1 approximation (see Sec. 7.3), we neglect the variation of the light field over the size of the atom and set $\exp(i\mathbf{k} \cdot \mathbf{r}) = 1$. This approximation is good for typical atomic transitions in valence shells because

$$kr \sim \frac{\omega}{c} a_0 \sim \frac{Ry}{\hbar c} a_0 = \frac{m_e e^4}{2\hbar^3 c} \frac{\hbar^2}{m_e e^2} = \frac{e^2}{2\hbar c} = \frac{\alpha}{2} \ll 1, \tag{7.77}$$

where we used the Rydberg as an estimate of the characteristic energy of an atomic transition and the Bohr radius for atomic size.

Another approximation that we made in our discussion of E1 transitions (see Sec. 7.3) is that we neglected the coupling of electromagnetic fields to the intrinsic magnetic moment of the particles. Including such coupling leads to a generalization of the interaction Coulomb-gauge Hamiltonian of Eq. (7.24):

$$H_1 = \frac{e}{m_e c} \mathbf{p}_c \cdot \mathbf{A} - \boldsymbol{\mu} \cdot \mathbf{B}, \tag{7.78}$$

where $\mathbf{B} = \nabla \times \mathbf{A}$ is the magnetic field of the electromagnetic wave, and μ is the intrinsic magnetic moment of the particle.

It is the corrections to these approximations that are responsible for the higher-multipole transitions discussed in the previous section. A straightforward way to analyze the higher-multipole transitions is to expand the exponential factor in the expression for the vector potential of an electromagnetic wave according to

$$e^{i\mathbf{k}\cdot\mathbf{r}} = 1 + i\,(\mathbf{k}\cdot\mathbf{r}) - \frac{(\mathbf{k}\cdot\mathbf{r})^2}{2} + \ldots \tag{7.79}$$

Taking only the leading term in this expansion, which does not contain $\mathbf{k}$, corresponds to neglecting the spatial dependence of the vector potential of the light, while the $\mathbf{k}$-containing terms account for such dependence. The multipolarity of a transition is defined according to the lowest-order term in Eq. (7.79) associated with this particular transition type. It is customary to subdivide the multipoles into *electric* and *magnetic* multipoles, Eκ and Mκ, respectively. The leading term in Eq. (7.79) associated with an Eκ multipole contains $(\mathbf{k}\cdot\mathbf{r})^\kappa$. This has a clear physical meaning that can be understood as follows. As we have discussed in the context of E1 transitions, a transition amplitude has the units of energy and is proportional to the matrix element of $\mathbf{d}\cdot\boldsymbol{\mathcal{E}}$ in the E1 case (this is derived from the expansion of the first term in Eq. 7.78). Taking into account the spatial variation of the electric field allows us to construct other quantities of the same dimension, for example, $Q_{ij}\nabla_j\mathcal{E}_i$, where Q_{ij} are the components of the quadrupole-moment operator, and $\nabla_j\mathcal{E}_i$ characterizes spatial gradients of the electric field of the electromagnetic wave. From the meaning of the expansion of Eq. (7.79), we can, therefore, identify E2 (*electric-quadrupole*) transitions as those arising, to the leading order, from the second term in Eq. (7.79). We can continue this reasoning and associate the subsequent terms with E3 (*electric octupole*), E4, etc. There is a little bit of complication arising in this analysis because, in addition to these electric multipoles, the terms in the expansion of Eq. (7.79) also lead to magnetic multipoles. Magnetic-multipole transitions arise from the interaction of the magnetic field of the wave with the magnetic-dipole, quadrupole, octupole, etc. moments of the system. A set of magnetic- and electric-multipole amplitudes also arises from the second term in Eq. (7.78).

A way to sort this out can be derived from a somewhat complementary view of the multipole transitions to what we have just presented. Here, instead of the semiclassical picture in which the atom is treated quantum mechanically while the electromagnetic field is treated classically, we view the transition as a result of an interaction between an atom and a photon. For example, in absorption, we have an atom (or a molecule, or a nucleus) in the lower atomic state interacting with a light field with some number of photons, which produces an upper-state atom and a light field with one photon fewer. A photon has intrinsic spin $s = 1$, and orbital angular momentum $l = 0, 1, 2, \ldots$ with respect to the atom. The spin and the orbital angular momentum add to produce a total angular momentum $j = l, l+1$, or $l-1$ according to the usual rules of angular-momentum addition. The definition of multipole transition can then be given in the following way:

1. The multipolarity is determined by the total angular momentum of the photon: dipole for $j = 1$, quadrupole for $j = 2$, octupole for $j = 3$, etc.
2. Whether the transition is of electric or magnetic type is determined by whether $j = l \pm 1$ (electric), or $j = l$ (magnetic).

Table 7.2 Summary of the photon quantum numbers and selection rules for various transition types (DeMille *et al.* 2000).

Multipole	E1	M1	E2	M2	E3
power of kr	0,2,...	1,3,...	1,3,...	2,4,...	2,4,...
photon l	0,2	1	1,3	2	2,4
photon j	1	1	2	2	3
atom ΔJ	$\pm 1, 0$	$\pm 1, 0$	$\pm 2, \pm 1, 0$	$\pm 2, \pm 1, 0$	$\pm 3, \pm 2, \pm 1, 0$
	$0 \leftrightarrow 0$	$0 \leftrightarrow 0$	$0 \leftrightarrow 0, 1$	$0 \leftrightarrow 0, 1$	$0 \leftrightarrow 0, 1, 2; 1 \leftrightarrow 1$
photon parity $(-1)^{l+1}$	-1	1	1	-1	-1
atom parity change	yes	no	no	yes	yes

Such a picture is very convenient for formulating selection rules for various types of transitions, and we will do this shortly. However, let us first discuss the connection between the particle-interaction picture and the semiclassical picture based on Eq. (7.79). It turns out that all we need to do to establish this connection is expand the plane wave (which can be thought of as the photon wave function)[4] into partial waves according to *Rayleigh's formula*, to be found in standard textbooks on quantum mechanics and/or electrodynamics:

$$e^{i\mathbf{k}\cdot\mathbf{r}} = \sum_{l=0}^{\infty} i^l (2l+1) j_l(kr) P_l(\cos\theta). \quad (7.80)$$

Here $j_l(kr)$ are the *spherical Bessel functions*, $P_l(\cos\theta)$ are the *Legendre polynomials*, and θ is the angle between **k** and **r**. According to the properties of spherical Bessel functions (see, for example, mathworld.wolfram.com or Arfken and Weber 2005), if $j_l(kr)$ is expanded into powers of kr, the nonzero terms in the expansion are those containing $(kr)^l$, $(kr)^{l+2}$, $(kr)^{l+4}$, etc. For an E1 transition, $l = 0$ or 2, and it first appears in the term independent of kr. For an electric quadrupole, we have $j = 2$, $l = 1$ or 3, and this transition first appears in the term containing kr to the first power. For an M1 transition, $j = 1$, $l = 1$, and, once again, it appears in the term linear in kr. Thus we see a consistency between the photon picture and our earlier discussion.

Let us now turn to the selection rules for various multipoles. The rule associated with the angular-momenta are straightforward: the total angular-momenta of the atom in the initial state, the atom in the final state, and the photon should satisfy the triangle rule (see Sec. 7.2). In addition, taking into account the fact that a photon has intrinsic parity of (-1),[5] and the parity of $P_l(\cos\theta)$ is $(-1)^l$, we can formulate a general rule that parity should change for odd-rank electric-multipole transitions and even-rank magnetic-multipole transitions, and should not change for even-rank electric-multipole transitions and odd-rank magnetic-multipole transitions.

A summary of the results we have discussed so far as applied to the lowest-multipolarity electric and magnetic transitions is given in Table 7.2.

Next, we briefly discuss relative intensities of multipole transitions. A detailed discussion and formulas can be found, for example, in Sobelman (1992). While the transition intensity

---

[4]The question of how to assign a wave function to a photon, and even whether a photon can be described by a wave function, has been a subject of heated debates; see the summary and references in the book by Scully and Zubairy (1997).

[5]This follows from the fact that the vector potential is a polar vector.

depends on the parameters of the specific transition (via the matrix element, wavelength, etc.), the general trend can be captured from the qualitative considerations presented above in the following way.

Consider an electron or a nucleon of charge e and mass m localized in a system (atom, nucleus, ...) with characteristic dimensions R. We can then crudely estimate the allowed Eκ and Mκ matrix elements as

$$\langle \ldots |E\kappa| \ldots \rangle \sim eR^\kappa, \tag{7.81a}$$

$$\langle \ldots |M\kappa| \ldots \rangle \sim \frac{e\hbar}{2mc} R^{\kappa-1}. \tag{7.81b}$$

Generalizing the result of Eq. (7.62), we can estimate the spontaneous-emission rates at a transition frequency ω due to these multipoles as a product of the modulus-squared of the corresponding matrix element, an appropriate power of k (corresponding to the amplitude having the dimension of energy). The result is an estimate for the spontaneous emission rates (completely ignoring any additional numerical coefficients):

$$\Gamma(E\kappa) \sim e^2 R^{2\kappa} \frac{k^{2\kappa+1}}{\hbar}, \tag{7.82a}$$

$$\Gamma(M\kappa) \sim \left(\frac{e\hbar}{2mc}\right)^2 R^{2\kappa-2} \frac{k^{2\kappa+1}}{\hbar}. \tag{7.82b}$$

For atomic valence-electron transitions, $R \sim a_0$, and we find for the typical ratio of the multipole-transition rates to that of an allowed E1 transition of a comparable frequency:

$$\frac{\Gamma(E\kappa)}{\Gamma(E1)} \sim (ka_0)^{2\kappa-2} \sim \left(\frac{\alpha}{2}\right)^{2\kappa-2}, \tag{7.83a}$$

$$\frac{\Gamma(M\kappa)}{\Gamma(E1)} \sim \left(\frac{\hbar}{2a_0 mc}\right)^2 (ka_0)^{2\kappa-2} \sim \left(\frac{\alpha}{2}\right)^{2\kappa}. \tag{7.83b}$$

It is interesting to note that in light nuclei, a typical energy for a γ-ray transition is on the order of an MeV, corresponding to

$$k = \frac{\omega}{c} = \frac{\hbar\omega}{\hbar c} \sim \frac{1 \text{ MeV}}{197 \text{ MeV fm}}, \tag{7.84}$$

while the nuclear size R is on the order of a few femtometers (1 fm = 10^{-13} cm). Thus, $kr \ll 1$, so that, as with atoms, the higher the multipole, the weaker the transitions, in general. However, there are many cases in nuclear physics where high-multipolarity γ-transitions are important. This typically occurs when E1 transitions are forbidden by some selection rules. Examples of this include high-angular-momentum excited states known as *nuclear isomers*, whose E1 decays to low-angular-momentum lower states are forbidden by the angular-momentum selection rules. There are also selection rules specific to nuclear physics that suppress certain E1 transitions; these selection rules are related to *isospin symmetry*, which describes the fact that the *strong nuclear interaction* is largely the same for protons and neutrons. A detailed explanation of this effect, examples, and a general discussion of multipole radiation in nuclei can be found in a rather old but still wonderful nuclear-physics textbook by DeBenedetti (1964).

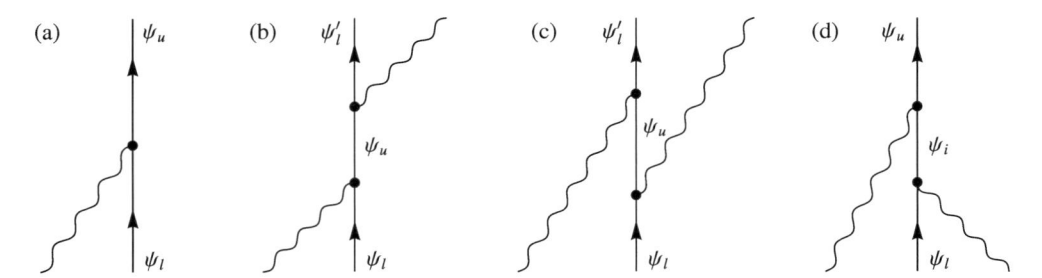

Fig. 7.8 Feynman diagrams illustrating (a) absorption, (b,c) scattering, and (d) a two-photon transition.

7.7 Two-photon and multi-photon transitions

We will now briefly discuss another broad class of atomic transitions that are associated with a *nonlinear response* of the medium to a time-varying electromagnetic field, and in particular, to an optical field. For such transitions, the transition probability, to lowest order in the light intensity, goes as the second or higher power of the intensity. This is in contrast to the single-photon transitions discussed above for which the transition probability goes as the first power of light intensity.

A convenient way to see the differences between single- and multi-photon transitions is to consider the *Feynman diagrams* (Fig. 7.8) that were devised to illustrate quantum-electrodynamical (QED) and, more generally, any kind of *field-theory* processes. In fact, such diagrams arguably revolutionized 20th-century physics by providing a visual representation of field-theory processes that is at once intuitive and mathematically rigorous. A Feynman diagram is useful as a guide to understand a process, but it can also be read as a formula that allows the calculation of a contribution to the transition amplitude.

> See, for example, Appendix H of the book by Budker *et al.* (2008) for an introduction to Feynman diagrams relevant to light–atom interactions. The book also contains a number of problems illustrating the use of the diagrams and the associated rules for calculating the amplitudes of electromagnetic processes.

A Feynman diagram is essentially a graph of the state of a system as a function of time. There are different conventions for drawing Feynman diagrams—here we present the diagrams as they are most frequently used in atomic physics. Time is assumed to increase from bottom to top. The evolution of atomic states is represented by vertical solid lines, and photons are represented by wavy lines propagating at an angle to the atomic state line (the *trunk*) of the diagram. The intersection point between a photon line and the trunk is called a *vertex*. A photon line ending at a vertex represents absorption, and one beginning at a vertex represents emission.

Figure 7.8(a) shows the elementary process of absorption. In the beginning, we have an atom in an initial lower state ψ_l and a photon. In the end, the photon has been absorbed and is no longer present; the atom is in the final upper state ψ_u. In Fig. 7.8(b), one of the possible diagrams for photon scattering is shown. Here the first part of the process is the same as in Fig. 7.8(a), but there is another vertex in which the atom goes from the intermediate upper state to a final lower state ψ'_l with emission of another photon.

There is another process by which the state ψ_u can make a transition to ψ'_l, diagrammed in Fig. 7.8(c). Here the final photon is emitted before the initial photon is absorbed. This process may be counter-intuitive, because the final photon is emitted as the atom makes a transition to a higher-energy state, thus seemingly grossly violating the conservation of energy.

If this seems disturbing to the reader, here is an attempt at an explanation. The time–energy uncertainty condition says that if we are dealing with a short interval of time Δt (here the interval between the emission of the final photon and the absorption of the initial photon), there is a corresponding energy uncertainty $\Delta E = \hbar/\Delta t$. Thus, the intermediate or *virtual* states may take on the "wrong" energy, as long as they exist for only an appropriately short time. This is why the process shown in Fig. 7.8(c) is possible.

Now, to the two-photon processes. Two-photon absorption is illustrated in Fig. 7.8(d). In the beginning, we have an atom in the initial lower state ψ_l and two photons. Then, one of the photons is absorbed, and the atom goes to an intermediate state ψ_i. Then, the second photon is absorbed, and the atom is promoted to the final upper state ψ_u. The total amplitude is found by summing the amplitudes corresponding to the diagrams for all possible intermediate states and the order of the photon absorption.

Depending on the multipolarity of each of the single-photon-absorption subprocesses, the two-photon process can be classified as E1–E1, E1–M1, etc. The selection rules for two-photon absorption can be derived from the corresponding one-photon rules. For example, for an E1–E1 transition, the parity of the initial and the final state should be the same, and the total angular momentum should change by no more than 2 (the maximum angular momentum of two photons in the E1–E1 case is 2).

> There are certain selection rules that are specific to two-photon transitions and have no analogs in the one-photon case. One example concerns two-photon transitions between states of total angular-momentum $F = 0$ and $F'' = 1$. It is fairly straightforward to show that the two photons must have different polarizations for such a transition to be allowed. However, a more subtle requirement is that the two photons must have different energies—the transition is strictly forbidden in the case of *degenerate* (equal-energy) photons. This selection rule results from the fact that photons are bosons, and so a state consisting of two degenerate photons must be symmetric with respect to their exchange. It turns out that in a two-photon system it is impossible to construct a spin-one state with this property (see, for example, the paper by DeMille *et al.*, 2000, for a proof). The selection rule is closely related to the so-called *Landau–Yang* theorem, well known to elementary-particle physicists, that states that a particle with intrinsic spin one cannot decay into two photons. In fact, both the Landau–Yang theorem and the two-photon selection rule are employed in searches for possible small violations of the usual bosonic properties of photons (DeMille *et al.* 2000).

An estimate of two-photon transition probabilities is discussed, for example, in Problem 3.15 of the book by Budker *et al.* (2008), which also gives further references on the subject. A practical approach and concrete examples of how to analyze two-photon transitions with the help of the quantum density matrix can be found in a paper by Auzinsh (2004).

7.8 Visualization of atomic transitions

In the quantum-mechanical description of atomic transitions that we have been discussing, an electron in a given state emits radiation when it makes a transition to a lower-energy state. How can we visualize this process? Classically, a charged particle radiates when it oscillates, for example, when it moves while constrained by a spring. Linear oscillation of the particle produces linearly polarized radiation, while circular or elliptical motion in two dimensions produces circularly or elliptically polarized light. The frequency of the emitted radiation is just the oscillation frequency of the particle.[6] Can we form a similar picture in quantum mechanics?

---

[6] Some physicists assert that all of atomic physics and the physics of light–atom interactions can be understood from the electron-on-a-spring picture. We do not believe this to be quite true—in some instances one needs to employ two electrons on a spring!

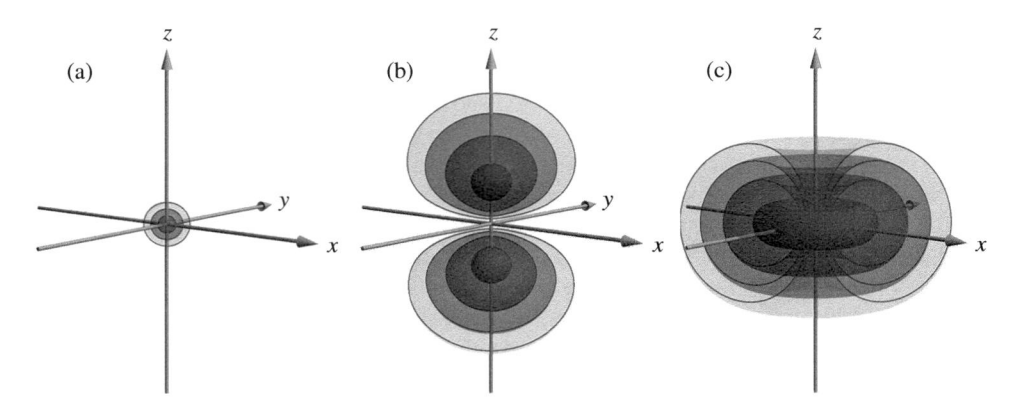

Fig. 7.9 Contour plots of the electron density in pure (a) $|1S\rangle$, (b) $|2P, m = 0\rangle$ and (c) $|2P, m = \pm 1\rangle$ states of hydrogen. Surfaces with constant values of $|\psi_{100}|^2$, $|\psi_{210}|^2$, and $|\psi_{211}|^2$, respectively, are plotted; the opacity of each surface is proportional to the electron density on the surface. The electron density in each of these plots is symmetric with respect to the nucleus. Thus the average position of the positive and negative charges in the atom coincide, and there is no net electric-dipole moment.

The electron on a spring forms a time-varying electric-dipole moment. If we look for such a dipole moment in an atom, we see that an atom in a given energy eigenstate does not have a dipole moment, and thus can hardly be associated with an oscillating or rotating dipole. In Fig. 7.9 the electron density of hydrogen in the 1S and 2P states is shown (see Chapter 2). We see that the electron density in these states is distributed in a symmetric way around the nucleus, so, indeed, there is no electric-dipole moment in these states. Consequently, an atom in an energy eigenstate does not radiate.

We can, however, create a time-dependent electric-dipole moment in an atom by forming a *superposition* of energy eigenstates, and, as we saw in Sec. 7.1, this is just what happens during an atomic transition. Consider spontaneous decay on the $|2P, m' = 0\rangle \to |1S\rangle$ transition in a hydrogen. During the transition, the atom is in a coherent superposition of these two states:

$$\psi = a|1S\rangle + e^{-i(E_2 - E_1)t/\hbar} a'|2P, m' = 0\rangle, \tag{7.85}$$

where $|1S\rangle$ and $|2P, m' = 0\rangle$ represent the wave functions ψ_{100} and ψ_{210} discussed in Chapter 2. The factor $e^{-i(E_2 - E_1)t/\hbar}$ describes the relative phase evolution of the energy eigenstates ($E_{1,2}$ are the energies of the $n = 1, 2$ states). We know that $\omega = (E_2 - E_1)/\hbar$ is the angular frequency of the light emitted during such a transition, so taking the example of the classical picture above, we should look for an electric dipole oscillating at the optical frequency ω. The coefficients a and a' are the amplitudes of each eigenstate in the total wave function. As usual, we demand the wave function to be normalized: $|a|^2 + |a'|^2 = 1$. In the course of the transition, the amplitude a' will decrease (the atom will leave the $|2P, m' = 0\rangle$ state) and the amplitude a will increase (the atom will arrive in the $|1S\rangle$ state). The period of optical oscillation $T = 2\pi/\omega$ is typically of the order of 10^{-15} s and the lifetime of an excited state (giving the time scale of evolution of a and a') is rarely shorter then 10^{-8} s. Thus, if we follow the evolution of the wave function during an optical period, we can safely ignore the time-dependence of a and a'.

Choosing an intermediate point in the transition when both amplitudes are equal:

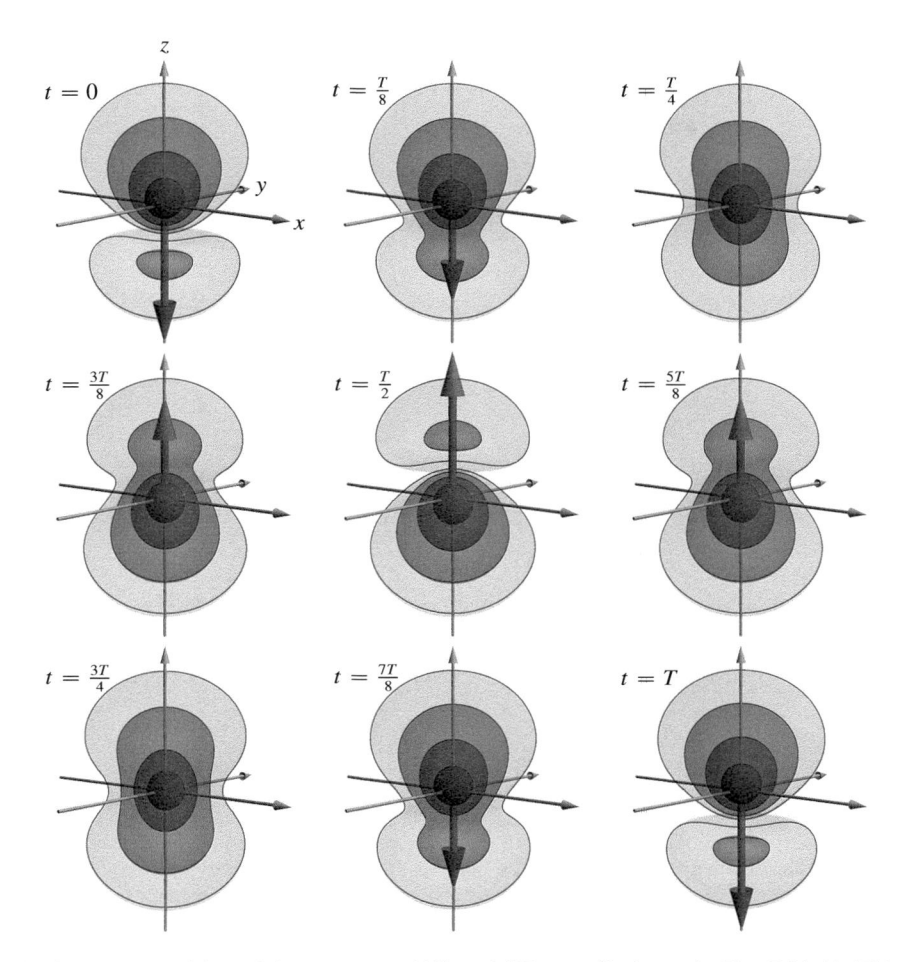

Fig. 7.10 A superposition of the two states $|1S\rangle$ and $|2P, m = 0\rangle$ shown in Fig. 7.9(a,b). This superposition corresponds to a displacement of the electron along the z-axis. The electron density and the corresponding electric-dipole moment (shown with an arrow representing the instantaneous direction and magnitude of the electric-dipole moment) oscillates along the z-axis with a frequency corresponding to the energy separation between the 1S and 2P states. Linearly polarized light is radiated in this case. In this figure, one period of such oscillation is shown.

$$a = a' = \frac{1}{\sqrt{2}}, \tag{7.86}$$

we plot in Fig. 7.10 the electron density corresponding to the wave function (7.85) during one optical period. The arrow on this figure depicts the magnitude and direction of the instantaneous electric-dipole moment

$$\langle \mathbf{d} \rangle = -e \int \mathbf{r} \, |\psi|^2 \, d^3r. \tag{7.87}$$

We see that the temporal evolution of the electron density closely resembles a linear oscillator that oscillates along the z-axis. This picture matches our understanding that atoms undergoing

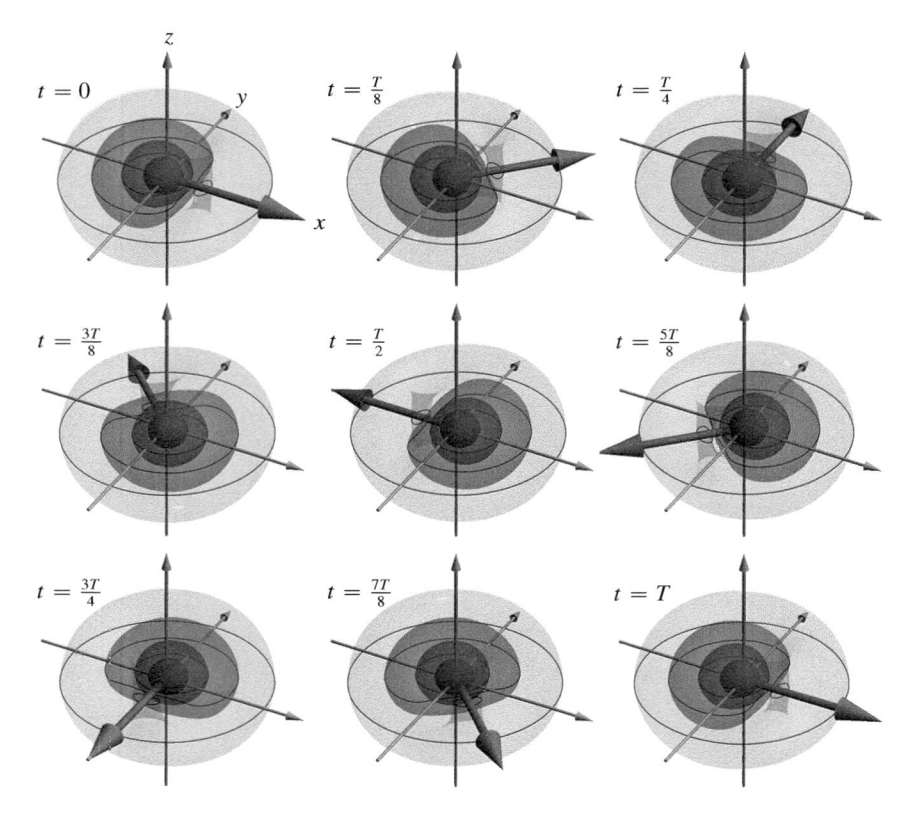

Fig. 7.11 A superposition of the two states $|1S\rangle$ and $|2P, m = 1\rangle$ shown in Fig. 7.9(a,c) corresponding to an electron displacement to one side of the nucleus. The electron density and the corresponding electric-dipole moment (shown by an arrow) rotate around the nucleus at a frequency corresponding to the energy separation (transition frequency). Circularly polarized light is radiated in this case.

an electric-dipole transition corresponding to $\Delta m = 0$ radiate light with linear polarization along $\hat{\mathbf{z}}$.

Now consider the $|2P, m' = 1\rangle \rightarrow |1S\rangle$ transition. The electron density in the superposition state

$$\psi = \frac{1}{\sqrt{2}}|1S\rangle + e^{-i(E_2-E_1)t/\hbar}\frac{1}{\sqrt{2}}|2P, m' = 1\rangle \tag{7.88}$$

is depicted in Fig. 7.11. We see that, in this case, the electron density is shifted to one side of the nucleus. This creates an electric-dipole moment in the opposite direction (the dipole moment points from negative to positive charge), which rotates around the nucleus. The rotation frequency corresponds to the energy separation between the atomic eigenstates and thus corresponds to the frequency of the optical transition. The rotation is in the counterclockwise direction if observed from the positive direction of the z-axis. This corresponds to left-circularly polarized light radiated in this direction. Once again, this picture matches our understanding of the corresponding emission process: an atom that undergoes a $|2P, m' = 1\rangle \rightarrow |1S\rangle$ transition ($\Delta m = m' - m = 1$) radiates a left-circularly polarized photon as observed from the positive direction of the z-axis. If we were to consider the $|2P, m' = -1\rangle \rightarrow |1S\rangle$ transition, we would have exactly the same picture with the exception of the fact that the electron-density rotates in the opposite direction, corresponding to radiation of right-circularly polarized light.

At this point, we need to briefly discuss an important issue. If an atom is initially prepared in the 2P state, in order for it to undergo a transition to the 1S state, some agent is needed to admix the 1S state to the initially prepared 2P state. Otherwise, as we have just discussed, there is no electric-dipole moment, and no transition. It turns out that, in spontaneous decay, vacuum fluctuations act as such an agent. In fact, this initial "push" is the thing that is hardest to find a classical analogy for, and it is frequently said that spontaneous emission from an atomic eigenstate is a truly quantum effect. As the ongoing discussion shows, stimulated emission is a "more classical" phenomenon than spontaneous emission.

For the case of absorption, one would consider a situation in which initially the atom is in a lower state—large magnitude of a in the case of Eq. (7.85)—which in the course of a transition diminishes, and the magnitude of the amplitude a' increases. As a result, energy from the radiation mode of a specific polarization is absorbed by the atom and converted into its internal energy. There is clearly an asymmetry between absorption and emission processes (confirmed by quantum-electrodynamical analysis) in that there is no "spontaneous absorption" that would be prohibited, for example, by energy conservation, and in that vacuum fluctuations do not play a direct role in the process.

A similar analysis can be performed for two levels that are connected by a magnetic-dipole transition. For a $|2P, m' = 0\rangle \to |2P, m' = 1\rangle$ transition, the E1 transition amplitude is zero, but the M1 transition is allowed. During an intermediate stage of this transition, the wave function takes the form

$$\psi = \frac{1}{\sqrt{2}}|2P, m' = 0\rangle + e^{-i\Delta E t/\hbar}\frac{1}{\sqrt{2}}|2P, m' = 1\rangle, \tag{7.89}$$

where ΔE is the energy splitting between the two magnetic sublevels, induced, for example, by a $\hat{\mathbf{z}}$-directed magnetic field. The oscillation frequency $\Delta E/\hbar$ in this case is the Larmor frequency. The electron density corresponding to this wave function is depicted in Fig. 7.12. As the figure indicates, the electron density is symmetric with respect to the nucleus at the origin. Thus the coherent superposition of the two states does not possess an electric-dipole moment. This is the reason E1 transitions between these states are forbidden. On the other hand, we can calculate the magnetic-dipole moment $\boldsymbol{\mu}$ in this coherent superposition of states according to

$$\boldsymbol{\mu} = \frac{1}{2c}\int \mathbf{r} \times \mathbf{j}(\mathbf{r})d^3r, \tag{7.90}$$

where $\mathbf{j}(\mathbf{r})$ is the electric-current density, which, as described in standard electrodynamics and quantum-mechanics texts, can be found from

$$\mathbf{j}(\mathbf{r}) = -\frac{e\hbar}{2mi}\left[\psi^*(\boldsymbol{\nabla}\psi) - (\boldsymbol{\nabla}\psi^*)\psi\right]. \tag{7.91}$$

We see that the system has a nonzero magnetic-dipole moment whose instantaneous direction is tilted with respect to the z-axis and rotates around this axis at the Larmor frequency (Fig. 7.12). This rotation causes the emission of circularly polarized radiation associated with the M1 transition.

158 *Atomic transitions*

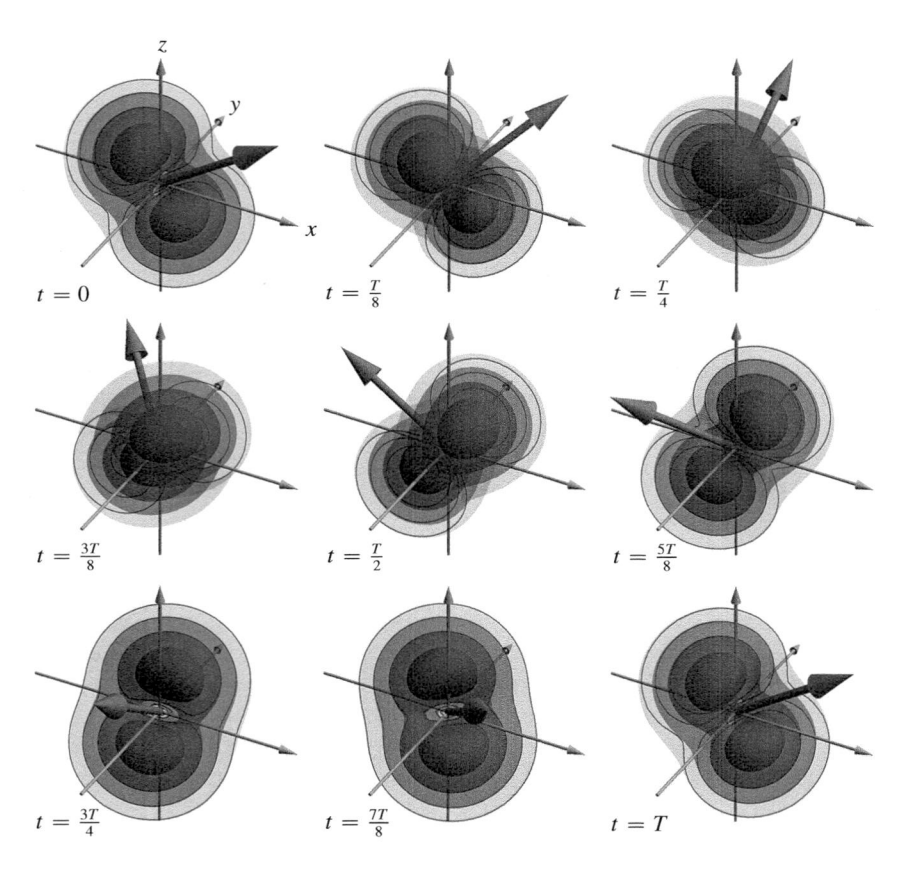

Fig. 7.12 Electron density for a coherent superposition of the $|2P, m = 0\rangle$ and $|2P, m = 1\rangle$ states. The time dependence is due to an energy splitting between these states, which can result, for example, from the application of a magnetic field along $\hat{\mathbf{z}}$. One period of Larmor precession is shown. The arrow indicates the instantaneous direction and magnitude of the magnetic-dipole moment.

8
Coherence in atomic systems

As discussed in Chapter 5, there can be a degree of coherence between any two quantum states in an ensemble, represented by an off-diagonal element of the density matrix. Pure states, which can be represented using wave functions, are in a sense inherently coherent in that coherence is exhibited whenever they are written as a superposition of more than one basis state. The coherence may be between states that are widely separated in energy, as in the case of the upper and lower state of an optical transition, or between degenerate states, for example, between Zeeman sublevels of a given state (*Zeeman coherence*). In this chapter, we will discuss various phenomena associated with Zeeman coherence created by light–atom interactions.

We will see that in some cases, in particular, in our initial examples, the use of coherence to describe the physical process can be avoided by a judicious choice of the basis. However, these examples can be instructive from the point of view of understanding the effect of coherences, which becomes a necessity in general. In this chapter, we consider examples involving pure states, for which we do not need to use the density-matrix formalism. We will, however, turn to the density matrices in later chapters.

8.1 Dark and bright states

We now examine the effects that near-resonant light–atom interactions have on the atoms, and how these effects, in turn, influence further light–atom interactions.

We start with a two-level atom that has a $J = 0$ lower state and a $J = 1$ upper state (Fig. 8.1). We will assume that the light is of low intensity.

Consider the case of weak linearly polarized light with electric field $\mathcal{E} = \mathcal{E}_0 \operatorname{Re} \hat{\varepsilon} e^{-i\omega t}$, where $\hat{\varepsilon} = \hat{z}$. Light of this polarization drives $\Delta m = 0$ transitions, so only the $m' = 0$ magnetic sublevel is excited. To see this using the wave-function formalism, we calculate the amplitudes of the excited state sublevels. For weak light, these are proportional to a sum over the transition dipole moments (Chapter 7) originating from the ground state, given by

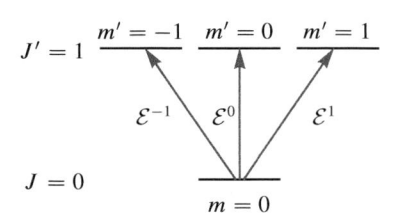

Fig. 8.1 Excitation of a $J = 0 \rightarrow J' = 1$ transition.

$$a_{m'} \propto \sum_m \langle \xi' J' m' | \hat{\varepsilon} \cdot \mathbf{d} | \xi J m \rangle. \tag{8.1}$$

In our particular case for the $J = 0 \rightarrow J' = 1$ transition and z-polarized light, we have

$$a_0 \propto \langle \xi' 10 | d_z | \xi 00 \rangle, \tag{8.2}$$

160 Coherence in atomic systems

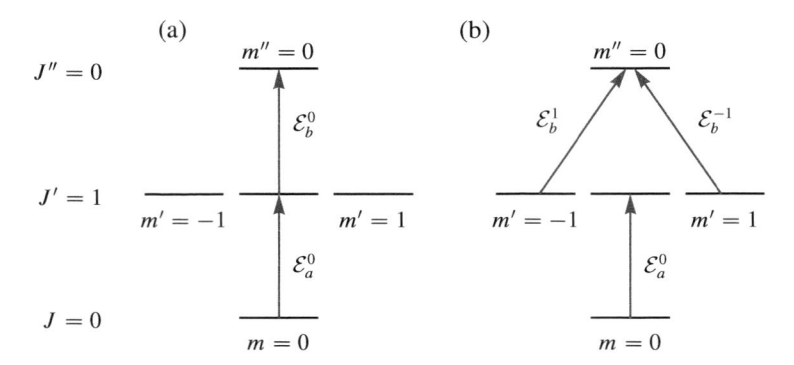

Fig. 8.2 (a) Two-step excitation on a $J = 0 \to J' = 1 \to J'' = 0$ transition using two z-polarized light fields. (b) If the second light field has polarization orthogonal to z, two-step excitation is not possible.

which according to the Wigner–Eckart theorem is equal to

$$a_0 \propto \langle \xi' 1' 0' | d_z | \xi 0 0 \rangle$$
$$= - \begin{pmatrix} 1 & 1 & 0 \\ 0 & 0 & 0 \end{pmatrix} \langle J' \| d \| J \rangle \qquad (8.3)$$
$$= \frac{1}{\sqrt{3}} \langle J' \| d \| J \rangle,$$

where we use the fact that $d_z = d_0$ in spherical coordinates. Similarly, we find that $a_{-1} = a_1 = 0$. After absorption of the z-polarized light, the upper state of the atom can be characterized by the wave function

$$\psi_{J'} = a_0 |\xi' 1 0\rangle. \qquad (8.4)$$

In this case, the excited state consists of just one Zeeman sublevel. Note that the norm of this wave function is much less than unity since most atoms remain in the initial state.

Consider now a transition from the state (8.4) to a higher-lying $J'' = 0$ state driven by a second light field (Fig. 8.2). From the figure it is clear that the absorption of this probe field depends on its polarization. If the probe field is z-polarized (Fig. 8.2a), there will be absorption. However, if the probe field is x- or y-polarized (Fig. 8.2b), there will be no absorption, simply because the $m' = \pm 1$ sublevels are "empty."

What happens if, instead of z-polarized light driving the $J = 0 \to J' = 1$ transition, we have, say, x-polarized light? Because this change is equivalent to just choosing a new quantization axis, it can have no physical consequences (beyond requiring a corresponding relabeling of any other fields in the problem). However, we will now see that, instead of just one sublevel, x-polarized light excites a *coherent superposition* of two Zeeman sublevels. We can see this by performing a quantum-mechanical rotation, or directly, as we do below. While considering the light to be x-polarized causes a bit more complication in the description of the light field, there are numerous situations in which such a description will be convenient; for example, in the analysis of certain magneto-optical effects, we may choose the quantization axis along the magnetic field rather than the light field.

We denote the electric field of the $\hat{x}$ polarized light (referred to as the *pump* light) by $\mathcal{E}_a$. The polarization vector $\hat{\varepsilon}_a$ of the pump light can be expanded in the spherical basis as (Sec. 6.4)

Dark and bright states **161**

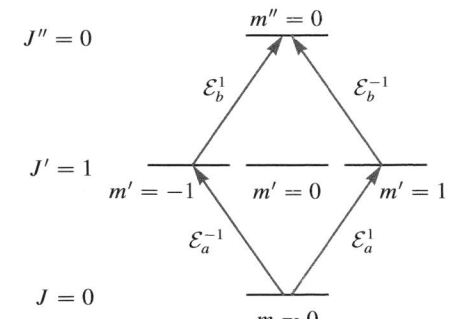

Fig. 8.3 Two-step excitation using light polarized orthogonally to z. Two pathways are possible: $m = 0 \to m' = -1 \to m'' = 0$ and $m = 0 \to m' = 1 \to m'' = 0$.

$$\hat{\varepsilon}_a = \hat{x} = -\frac{1}{\sqrt{2}}\hat{\epsilon}_1 + \frac{1}{\sqrt{2}}\hat{\epsilon}_{-1}, \tag{8.5}$$

so that the spherical-basis components of the polarization vector are given by

$$\varepsilon_a^1 = -\frac{1}{\sqrt{2}}, \quad \varepsilon_a^0 = 0, \quad \varepsilon_a^{-1} = \frac{1}{\sqrt{2}}. \tag{8.6}$$

In this basis, the light is seen to be in a coherent superposition of left- and right-circular polarizations. The two components have the same magnitude—there is a relative minus sign between them that can be interpreted as a phase difference of π. We will examine the significance of this phase shortly. The two components drive different atomic transitions: if an atom absorbs a $\hat{\epsilon}_1$ photon, it undergoes a $\Delta m = m' - m = 1$ transition; if it absorbs a $\hat{\epsilon}_{-1}$ photon it undergoes an $\Delta m = -1$ transition. What is the excited state produced by this light?

After absorbing x-polarized light, each atom is itself in a coherent superposition of atomic states whose general form is:

$$|\psi_{J'}\rangle = \sum_{m'} a_{m'} |\xi' J' m'\rangle. \tag{8.7}$$

For our particular light polarization and atomic transition we have

$$a_{m'} \propto \varepsilon_a^{m'} \langle \xi' 1 m' | d_{q=m'} | \xi 0 0 \rangle, \tag{8.8}$$

which leads to an atomic state

$$\begin{aligned}|\psi_{J'}\rangle &\propto \left[-\frac{1}{\sqrt{2}} \begin{pmatrix} 1 & 1 & 0 \\ 1 & -1 & 0 \end{pmatrix} |\xi', 1, -1\rangle + \frac{1}{\sqrt{2}} \begin{pmatrix} 1 & 1 & 0 \\ -1 & 1 & 0 \end{pmatrix} |\xi' 1 1\rangle \right] \langle J' \| d \| J \rangle \\ &= \frac{1}{\sqrt{6}} \left(|\xi' 1, -1\rangle - |\xi' 1 1\rangle \right) \langle J' \| d \| J \rangle. \end{aligned} \tag{8.9}$$

We see that we have created a coherent superposition of atomic states $|\xi', 1, -1\rangle$ and $|\xi' 1 1\rangle$. The minus sign in Eq. (8.9) comes from the minus sign between the light-field components, and can also be interpreted as a phase difference: here the components of the wave function are π out of phase.

Now let us see why this phase relation is important. Suppose that we have a second light field $\mathcal{E}_b$ (the *probe* field) resonant with a transition between the intermediate state J' and a higher-lying state $J'' = 0$ (Fig. 8.3). We will assume that $\mathcal{E}_b$ is arbitrarily weak so that it

162   *Coherence in atomic systems*

does not affect the intermediate atomic state, and consider the cases in which it has linear polarization along x or y. We want to know how efficiently the probe field excites the upper state J'' from the initially prepared intermediate state J' of Eq. (8.9). To calculate this, we need to evaluate the following matrix element:

$$a_{m''} \propto \langle \xi'' J'' m'' | \hat{\varepsilon}_b \cdot \mathbf{d} | \psi_{J'} \rangle, \tag{8.10}$$

where $|\psi_{J'}\rangle$ is given by Eq. (8.9).

Let us do this first for an x-polarized probe. We require the spherical components of the light polarization vector, which are given by Eq. (8.6). Evaluating the dipole-operator matrix elements as above, we obtain

$$a_0'' \propto \sum_{m'q} \varepsilon_b^q \langle \xi'' 0 0 | d_q | \psi_{J'} \rangle$$

$$\propto \frac{1}{\sqrt{6}} \left(\varepsilon_b^1 \langle \xi'' 0 0 | d_1 | \xi' 1, -1 \rangle \right.$$
$$\left. - \varepsilon_b^{-1} \langle \xi'' 0 0 | d_{-1} | \xi' 1 1 \rangle \right) \langle J' \| d \| J \rangle \tag{8.11}$$

$$= \frac{1}{\sqrt{6}} \left(-\frac{1}{\sqrt{2}} \frac{1}{\sqrt{3}} - \frac{1}{\sqrt{2}} \frac{1}{\sqrt{3}} \right) \langle J' \| d \| J \rangle \langle J'' \| d \| J' \rangle$$

$$= -\frac{1}{3} \langle J' \| d \| J \rangle \langle J'' \| d \| J' \rangle.$$

Let us postpone the analysis of this result and first calculate the same amplitude for y-polarized light. This light has the following polarization components:

$$\varepsilon_b^1 = \frac{i}{\sqrt{2}}, \quad \varepsilon_b^0 = 0, \quad \varepsilon_b^{-1} = \frac{i}{\sqrt{2}}. \tag{8.12}$$

The ε_b^1 and ε_b^{-1} components have the same sign, meaning that they are in phase (the factors of i represent an overall phase that can be ignored).

Performing the calculation as above, we have

$$a_0'' \propto \sum_{m'q} \varepsilon_b^q \langle \xi'' 0 0 | d_q | \psi_{J'} \rangle$$

$$\propto \frac{1}{\sqrt{6}} \left(\varepsilon_b^1 \langle \xi'' 0 0 | d_1 | \xi' 1 - 1 \rangle \right.$$
$$\left. - \varepsilon_b^{-1} \langle \xi'' 0 0 | d_{-1} | \xi' 1 1 \rangle \right) \langle J' \| d \| J \rangle \tag{8.13}$$

$$= \frac{1}{\sqrt{6}} \left(\frac{1}{\sqrt{2}} \frac{1}{\sqrt{3}} - \frac{1}{\sqrt{2}} \frac{1}{\sqrt{3}} \right) \langle J' \| d \| J \rangle \langle J'' \| d \| J' \rangle$$

$$= 0.$$

From Eqs. (8.11) and (8.13) we see that an atom prepared by the pump field in the level $J' = 1$ is in a *bright* (absorbing) state for the x-polarized probe field, while the same atom in the same state is in a *dark* (nonabsorbing) state for the y-polarized probe. This is a typical quantum-interference effect. The pump field puts the atomic states $|\xi' 1, -1\rangle$ and $|\xi' 1 1\rangle$ in a

Dark and bright states

coherent superposition. The relative phase between these two states dictates that for an x-polarized probe field, the two possible quantum paths from the initial $J = 0$ to the final $J'' = 0$ states (one via the $m' = -1$, and another via the $m' = 1$ intermediate state) give contributions to the transition amplitude that are in phase, i.e., the two paths interfere constructively and reinforce each other. On the other hand, for a y-polarized probe, the two quantum paths interfere destructively and cancel each other in the total transition amplitude.

The particular superposition state that the pump field prepares in the state J' depends on the polarization of the light field and the transition dipole moments (dipole transition matrix elements) for the transitions involved. When the second (probe) light field is applied, the absorption due to atoms in the intermediate state J' depends on the superposition of Zeeman sublevels that has been prepared, as well as the light polarization and relevant transition matrix elements.

When the probe field in our example is x-polarized, it can be seen from Eq. (8.11) that the fact that atomic states $|\xi'1,-1\rangle$ and $|\xi'11\rangle$ are prepared in an out-of-phase coherent superposition is compensated by the fact that the light fields in two possible transition pathways ($m' = -1 \to m'' = 0$ and $m' = 1 \to m'' = 0$, see Fig. 8.3), multiplied by the transition matrix elements for these pathways, are also out of phase.

On the contrary, for a y-polarized probe field, the light-field components multiplied by the respective matrix elements for the two pathways are in phase, and due to the fact that the intermediate atomic states $|\xi'1,-1\rangle$ and $|\xi'11\rangle$ are out of phase, the interference between the two pathways is destructive. There is zero probability for a y-polarized probe field to excite the $|\xi''00\rangle$ state, despite the fact that there are atoms excited to the intermediate state. This is because the intermediate-state atoms are in a dark state for y-polarized light.

It is instructive to draw the electron probability density $|\psi_{J'}|^2$ for the intermediate state. For simplicity we will consider the 1S, 2P, and 3S states of the spinless hydrogen atom. From Fig. 8.4 we see that the x-polarized pump light, exciting atoms on the 1S $\to$ 2P transition, has prepared an "electron cloud" in the 2P state that is aligned along the x-axis. In general, optical pumping with linearly polarized light will produce a symmetry axis along the polarization axis, although the shape of the electron distribution depends on the particular transition involved. Now we apply the probe field, resonant with the 2P $\to$ 3S transition. How can we understand the dependence of absorption of this field on its polarization?

Absorption is associated with the excitation of an oscillating dipole. Recalling our discussion in Sec. 7.8 of spontaneous emission of linearly polarized light on a $J' = 1 \to J = 0$ transition (Fig. 7.10), we can say by analogy that the direction of the oscillation of the dipole must be along the same axis along which the 2P electron density is stretched.

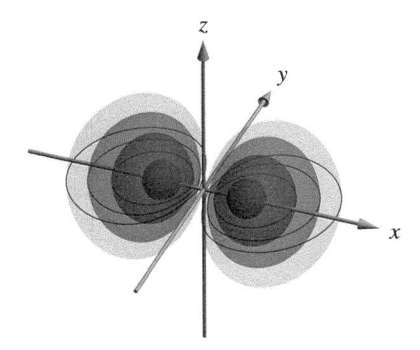

Fig. 8.4 Contour plot of the electron probability density $|\psi_{J'}|^2$ for the intermediate state prepared by the x-polarized pump light. For simplicity we assume the state to be the 2P state of the spinless hydrogen atom. The probability density is constant over each surface. The lines represent contours in the xy plane.

Light that is polarized along that axis can drive this oscillation; light with orthogonal polarization cannot. This gives a visual illustration of why a probe field with the same polarization as the pump field (x) is absorbed, while orthogonally polarized light (y or z) is not.

By changing the polarization of the first light field, we can vary the coherent superposition of $|\xi'J'm'\rangle$ states that are prepared, and by changing the polarization of the probe light we can change the state of the atoms that are excited from the prepared intermediate state. This presents many possibilities for manipulating atoms. A coherent superposition of atomic states is the basis for many interesting and useful effects in atomic and molecular physics (and also effects that have drawn popular attention): *electromagnetically induced transparency*(EIT; see reviews by Harris 1997; Matsko *et al.* 2001), *stimulated Raman adiabatic passage* (*STIRAP*, Bergmann *et al.* 1998), *coherent control* of chemical reactions (Gordon and Rice 1997; Warren *et al.* 1993), *"slow"* and *"superluminal"* light (Milonni 2005), *coherent population trapping* (*CPT*, Arimondo 1996), and *lasing without inversion* (Kocharovskaya 1992), are just some of these effects.

8.2 Quantum beats

If we were to examine the time dependence of the wave functions discussed in the previous section, we would see that they are constant in time. This is because each one can be written as a superposition of energy eigenstates all having the same energy. The time evolution of an eigenstate of energy E is found by multiplying the wave function by factor $e^{-iEt/\hbar}$—if all of the states making up a superposition state have the same energy, this factor is identical for each of them and can be disregarded as a meaningless overall phase factor.

However, the situation changes if the component states have different energies. Suppose that in the example of the previous section, an external DC magnetic field **B** is applied along **z**, causing the magnetic sublevels to experience a Zeeman energy shift. Now each component in a coherent superposition of magnetic sublevels has a different phase factor $e^{-iE_{m'}t/\hbar}$, so that the total wave function is

$$|\psi_{J'}\rangle = \sum_{m'} a_{m'}|\xi'J'm'\rangle e^{-iE_{m'}t/\hbar}. \tag{8.14}$$

Since the energies $E_{m'}$ are different, the time dependence cannot be factored out of Eq. (8.14) and the state will evolve in time.

Let us consider a particular example. A straightforward way to observe the time evolution is to prepare all of the atoms at the same time, using a pulse of pump light that is short compared to the evolution time scale, discussed below. It is also important that the pulse has a wide enough spectral range so that it can excite all of the magnetic sublevels, even though they do not have exactly the same energy. Assuming that the pump light field $\mathcal{E}_a$ is linearly polarized along x and that the pulse occurs at time $t = 0$, we can find the intermediate-state wave function as in Sec. 8.1:

$$|\psi_{J'}\rangle = a_{-1}|\xi'1,-1\rangle e^{-i\omega_{-1}t} + a_1|\xi'11\rangle e^{-i\omega_1 t}$$
$$\propto \frac{1}{\sqrt{6}}\left(|\xi'1,-1\rangle e^{-i\omega_{-1}t} - |\xi'11\rangle e^{-i\omega_1 t}\right)\langle J'\|d\|J\rangle. \tag{8.15}$$

Here $\omega_{m'} = m'\Omega_L$ is the linear Zeeman shift (in frequency units) of each magnetic sublevel, where $\Omega_L = g_{J'}\mu_B B/\hbar$ is the Larmor frequency, μ_B is the Bohr magneton, and $g_{J'}$ is the Landé factor.

Now, as in Sec. 8.1, we apply a probe field $\mathcal{E}_b$, continuous rather than pulsed, and examine the amplitude of the upper-state wave function $|\xi''0''0''\rangle$. We will assume that this probe light is also spectrally broad enough to satisfy the same requirement that we imposed upon the pump light. If the probe field is also x-polarized, then the amplitude a_0'' can be calculated as

$$\begin{aligned}
a_0''(t) &\propto \sum_q \varepsilon_b^q \langle \xi''0''0''|d_q|\psi_{J'}\rangle \\
&= \frac{1}{\sqrt{6}} \left(\varepsilon_b^1 \langle \xi''00|d_1|\xi'1,-1\rangle e^{-i\omega_{-1}t} \right. \\
&\quad \left. - \varepsilon_b^{-1} \langle \xi''00|d_{-1}|\xi'11\rangle e^{-i\omega_1 t} \right) \langle J'\|d\|J\rangle \\
&= -\frac{1}{6} \left(e^{-i\omega_{-1}t} + e^{-i\omega_1 t} \right) \langle J''\|d\|J'\rangle \langle J'\|d\|J\rangle.
\end{aligned} \qquad (8.16)$$

We can take the absolute value squared of this amplitude to find the probability of exciting the upper state to be

$$|a_0''(t)|^2 \propto \frac{1}{18} [1 + \cos(\omega_1 - \omega_{-1})t] \, |\langle J'\|d\|J\rangle|^2 \, |\langle J''\|d\|J'\rangle|^2. \qquad (8.17)$$

The excitation probability is harmonically modulated at a frequency $\omega_1 - \omega_{-1} = 2\Omega_L$ corresponding to the splitting between the two magnetic sublevels involved in the two possible excitation paths. We are observing the difference frequency between two oscillatory processes, analogous to the beat frequency heard when two guitar strings that are slightly out of tune are played simultaneously. In the current example the beats are between two components of a wave function, leading to the terminology *quantum beats*.

Figure 8.5 shows the time evolution of the electron density $|\psi_{J'}|^2$ corresponding to the wave function (8.15). For simplicity we again consider this to be the 2P state of spinless hydrogen. The atom precesses around the magnetic field direction ($\hat{z}$) at the Larmor frequency Ω_L, with a period $T = 2\pi/\Omega_L$. We recall from the previous section that the atomic state must be aligned non-orthogonally to the polarization axis of the probe field in order for the upper state to be excited. If the probe light is x-polarized, it is clear from the figure that this condition is initially satisfied, so that the excitation probability at $t = 0$ is high. As time progresses, the Larmor precession causes the alignment axis to rotate toward the y-axis, converting the bright state into a dark state and causing the excitation probability to drop. As the precession continues, the state becomes a bright state again, and the excitation probability oscillates as described by Eq. (8.17).

Performing the same excitation-probability calculation for the case in which the probe light is y-polarized, we find

$$|a_0''(t)|^2 \propto \frac{1}{18} (1 - \cos 2\Omega_L t) \, |\langle J'\|d\|J\rangle|^2 \, |\langle J''\|d\|J'\rangle|^2. \qquad (8.18)$$

Looking again at Fig. 8.5, we see that the initial state produced by the pump field is a dark state for y-polarized light, and so the upper-state excitation probability at $t = 0$ is zero. Due to the Larmor precession, the dark state is converted to a bright state and back again, and the excitation probability oscillates as before. The only difference between this case and the previous one is the phase of the quantum beats, as can be seen by rotating the plots in Fig. 8.5 by $\pi/2$ about the z-axis.

Quantum beats were discovered at the beginning of the 1960s independently by E. B. Aleksandrov (1964) in the Soviet Union and by J. N. Dodd *et al.* (1964) and coworkers in the

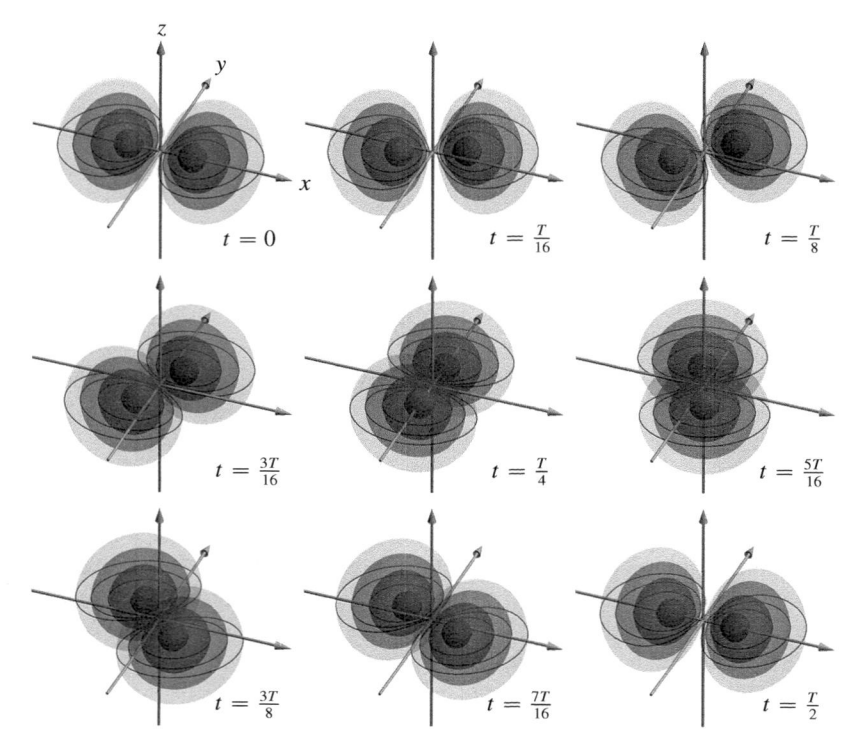

Fig. 8.5 The time evolution of the electron density of the intermediate state $\psi_{J'}$ in a magnetic field after pulsed excitation. The 2P state of the spinless hydrogen atom is drawn. A half-period of the wave function time evolution (Larmor precession) is shown.

United Kingdom. In subsequent years quantum beats proved to be a powerful tool in atomic and molecular spectroscopy (see, for example, the book by Aleksandrov *et al.*, 1993).

8.3 The Hanle effect

Let us now introduce relaxation into the example discussed in the previous section. Suppose it is observed that, in the absence of any light fields, an atom in the intermediate state $\psi_{J'}$ decays to a lower state with an average lifetime of τ. This lifetime places a time limit on the coherent evolution of the state.

First, assume that τ is much shorter than the quantum beat period T discussed in the previous section, or equivalently, that the relaxation rate $\Gamma = 1/\tau$ is much greater than the Larmor frequency Ω_L. Then the intermediate state will decay away before any other evolution can occur, and quantum beats will not be observed. This is equivalent to the case of no external field discussed in Sec. 8.1, since the effect of the magnetic field has been effectively negated. The wave function of the intermediate state does not change from its initial form; the electron density for this case is plotted in Fig. 8.6(a) (equivalent to Fig. 8.4).

Now consider the opposite case in which Γ is much smaller than Ω_L. In the limit of $\Gamma \to 0$ we have no relaxation, i.e., the situation of Sec. 8.2. In that section we assumed pulsed excitation of the intermediate state. Why did we do this? In the absence of relaxation, atoms undergoing quantum beat evolution, such as the Larmor precession shown in Fig. 8.5, will

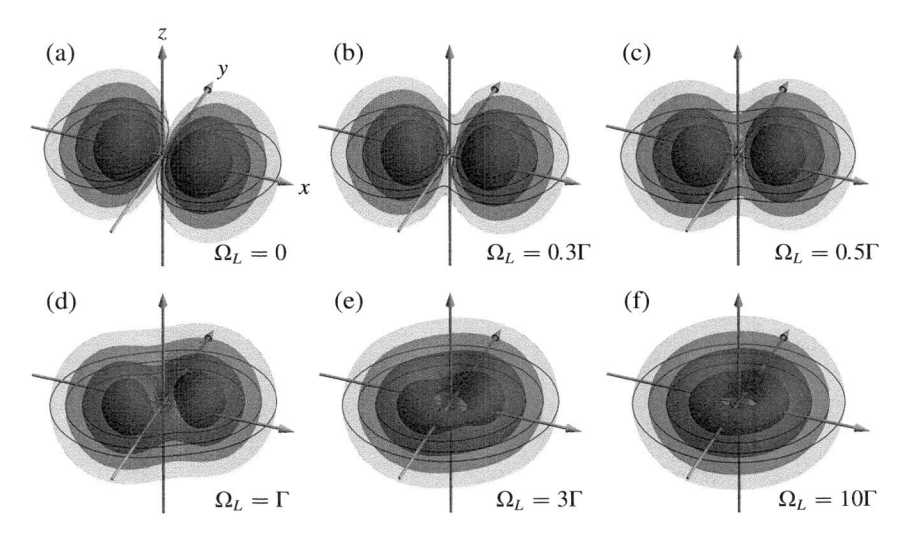

Fig. 8.6 Average electron density plots for an ensemble of spinless hydrogen atoms in the 2P state, subject to a $\hat{z}$-directed magnetic field. The atoms are continuously excited by x-polarized light from the 1S state. In parts (a)–(f) various values of Ω_L/Γ are assumed.

continue to evolve forever. Suppose that the intermediate state is excited continuously. At some instant, the wave function shown in the first frame of Fig. 8.5 is excited, and begins to precess. As it precesses, and becomes the wave function shown in the second frame, additional atoms are prepared by the light in the first-frame wave function. Now the wave functions in the first and second frames are both present simultaneously (as well as wave functions intermediate between the two), and the electron density of the ensemble is found by summing the contributions from each of these. As this process continues, all of the stages of the evolution become present in equal measure, and the average electron density is symmetric about the z-axis, as shown in Fig. 8.6(f) ($\Omega_L = 10\Gamma$ was assumed for this plot). The coherence that produced the distinction between the x- and y-axes does not exist in the average state of the ensemble. Without the coherence, the quantum beats cannot be observed, and the upper-state excitation probability is the same for x- and y-polarized light.

In the intermediate case in which Ω_L and Γ are comparable, on the other hand, we can observe the effect of the coherence, even with continuous excitation. We can see this by looking at Figs. 8.5 and 8.6. As the Larmor frequency is increased from zero, the quantum beats happen faster, so that the atoms can precess farther before they decay. Thus the average alignment axis of the ensemble moves away from the x-axis, as seen in Fig. 8.6(b),(c). As we have discussed in the previous sections, this will cause the excitation probability of the upper state to change. When the Larmor frequency is increased past Γ, the coherences begin to average out, as we have just described. This can be seen in Fig. 8.6(d)–(f). At this point, the dependence of the excitation probability on the Larmor frequency begins to diminish.

To analyze this quantitatively, we write the time-dependent excitation probability of the upper state due to atoms excited into the intermediate state at time t_0. Considering both the pump and probe light fields to be x-polarized, we can find this amplitude by shifting the time variable in Eq. (8.17) and multiplying by a factor $e^{-\Gamma(t-t_0)}$ describing the exponential decay of the intermediate state. This gives

$$\left|a_0''(t-t_0)\right|^2 \propto \frac{1}{18}\left[1+\cos 2\Omega_L(t-t_0)\right]e^{-\Gamma(t-t_0)}\left|\langle J''\|d\|J'\rangle\right|^2\left|\langle J'\|d\|J\rangle\right|^2. \quad (8.19)$$

If the atoms are excited continuously, the total upper-state excitation probability at time t is the sum of the contributions from the atoms excited to the intermediate state at every time t_0 prior to t. This is found by performing an integral of Eq. (8.19) over t_0:

$$\overline{|a_0''|^2} \propto \int_{-\infty}^{t} \left|a_0''(t-t_0)\right|^2 dt_0$$

$$\propto \frac{1}{18}\left(1+\frac{\Gamma^2}{\Gamma^2+4\Omega_L^2}\right)\left|\langle J''\|d\|J'\rangle\right|^2\left|\langle J'\|d\|J\rangle\right|^2. \quad (8.20)$$

Note that this result is independent of t, meaning that quantum beats will not be directly observed in the signal.

The excitation probability contains a characteristic factor

$$1+\frac{\Gamma^2}{\Gamma^2+4\Omega_L^2} \quad (8.21)$$

that depends only on the ratio Ω_L/Γ. We see that the excitation probability drops with increasing Ω_L, as indicated by the qualitative analysis above. From this dependence, a measurement of the excitation probability as a function of a DC magnetic field can be used to determine the lifetime $\tau=1/\Gamma$ of the intermediate atomic state.

The same analysis for a y-polarized probe field leads to

$$\overline{|a_0''|^2} \propto \int_{-\infty}^{t} \left|a_0''(t-t_0)\right|^2 dt_0$$

$$\propto \frac{1}{18}\left(1-\frac{\Gamma^2}{\Gamma^2+4\Omega_L^2}\right)\left|\langle J''\|d\|J'\rangle\right|^2\left|\langle J'\|d\|J\rangle\right|^2. \quad (8.22)$$

Here the excitation probability is initially zero and then increases as the Larmor frequency is increased from zero.

The effect described in this section was discovered by the German physicist Wilhelm Hanle as early as 1924 (Hanle 1924) and bears his name.

9
Optical pumping

Atomic polarization, discussed in Chapter 5, can be created by applying polarized light (Chapter 6) to an atomic transition (Chapter 7). We have seen examples of this in Chapter 8, in which the atoms were prepared in various polarized states by the application of an initial polarized light field. The creation of atomic ground-state polarization by this method is known as *optical pumping*. In the simplest mechanism for optical pumping, polarized light is absorbed by atoms in one or more bright states but does not interact with those in the remaining dark states (Chapter 8), resulting in polarization due to the excess of atoms in the dark states.

The importance of the technique of optical pumping partly lies in the fact that the relaxation rates of ground-state polarization can be significantly slower than excited-state relaxation rates. This means that effects due to ground-state polarization created by optical pumping can be used to perform much more sensitive measurements than could be done otherwise, as we will see in Chapter 10.

> In the late 1940s and early 50s, Alfred Kastler pointed out that polarized resonant light can create strong polarization of atoms in their ground state. Kastler later reviewed the history of the early research on the interaction of polarized light with atoms in his Nobel-prize lecture (Kastler 1967). A more recent review of optical pumping is given by Happer (1972).

Optical pumping is an example of what is known more broadly as a *nonlinear light–atom interaction*, defined as a light-induced process that changes the optical properties of an atomic medium. In this chapter we will first discuss nonlinear interactions and contrast them with linear processes, and then focus on optical pumping in various systems.

9.1 Linear and nonlinear processes; saturation parameters

In a *linear optical process*, the optical properties of the medium through which light is propagating do not depend on the light field. For example, suppose that the amount of light transmitted through a medium is proportional to the input light intensity. This means that the *absorption coefficient* describing the fractional attenuation of the beam by a thin layer of the medium must be independent of the light intensity; the process is therefore linear. The optical properties of a medium can also be observed indirectly: if fluorescence is observed to be proportional to the light intensity, this also indicates that the absorption coefficient is independent of the light intensity.

Linear optical processes are the norm when the light is off-resonant and/or weak (i.e., when the light intensity is below a characteristic *saturation intensity*, to be discussed shortly). As the light intensity is increased, a linear process will, as a rule, become *nonlinear*, with the optical properties of the medium acquiring dependence on the light field.

170 *Optical pumping*

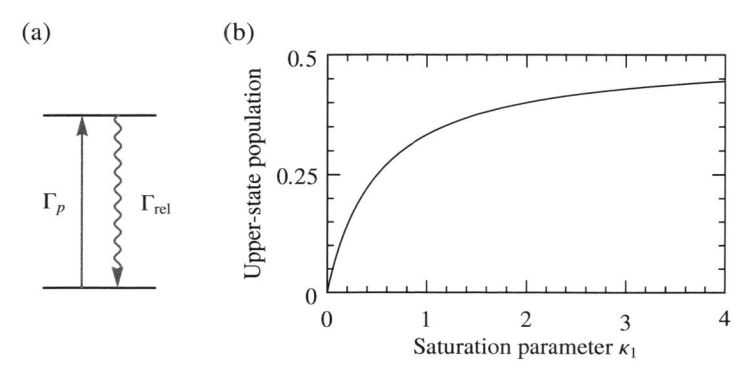

Fig. 9.1 (a) A two-level system subject to resonant light. Atoms are excited to the upper state at a rate $\Gamma_p = \Omega_R^2/\Gamma$, where Ω_R is the Rabi frequency and Γ is the natural width of the transition, and they relax via spontaneous decay at a rate $\Gamma_{rel} = \Gamma$. (b) Fractional upper-state population as a function of the saturation parameter κ_1.

> Nonlinear optical processes are usually associated with high light intensities. However, as we will see below, the saturation intensity depends on the specific system. Furthermore, there are experimental arrangements in which any observed signal indicates a nonlinear effect, so that these effects can show up even with weak illumination. Indeed, conventional spectral lamps were used in early (pre-laser-era) optical pumping experiments with resonant vapors to demonstrate nonlinear effects.

The simplest example of this phenomenon is *saturated absorption.*. Consider a two-level atom subject to resonant light (Fig. 9.1a). If the light power is low, atoms are excited to the upper state only rarely, and they return to the ground state relatively quickly via spontaneous emission. In this case the number of atoms in the ground state is largely unaffected by the light. Since the absorption coefficient is proportional to the ground-state population, we see that it is independent of the light intensity, i.e., the process is linear. If the light intensity is increased, on the other hand, the excitation rate can become comparable to the spontaneous decay rate, with the atoms spending an appreciable fraction of the time in the excited state. The resulting depletion of the ground-state population reduces the absorption coefficient. This means that the absorption coefficient is dependent on the light intensity, indicating that the absorption process in this regime is nonlinear.

According to the above discussion, there is a gradual transition from a linear to a nonlinear process that occurs as the ratio of the excitation rate to the spontaneous decay rate passes through unity. This ratio is an example of a *saturation parameter*, commonly denoted by κ (not to be confused with tensor rank), which takes the general form

$$\kappa = \frac{\Gamma_p}{\Gamma_{rel}} = \frac{\text{excitation rate}}{\text{relaxation rate}}. \tag{9.1}$$

We will denote the saturation parameter for the population of the upper state in this case by κ_1. As an indicator of the transition between the linear and nonlinear regimes, we can study the fractional population of the upper state, which can be related to the observed absorption and fluorescence. When κ_1 is small, an atom that is excited decays on average in the lifetime $\tau_{rel} = 1/\Gamma_{rel}$ and then is excited again after a total time $\tau_p = 1/\Gamma_p$ has elapsed. Thus the fraction of the time that an atom spends in the excited state is $\tau_{rel}/\tau_p = \kappa_1$. For an ensemble, κ_1 is the fraction of the total population in the upper state. This is the linear regime. When κ_1 is

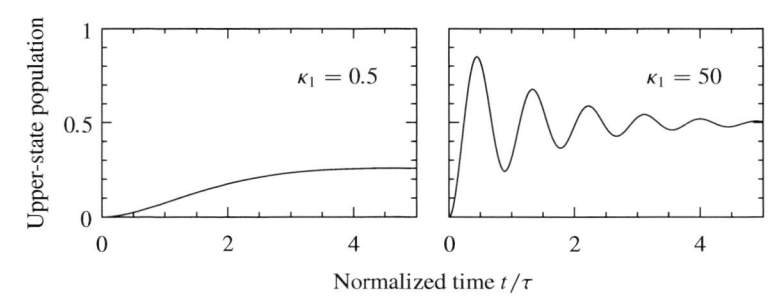

Fig. 9.2 Fractional upper-state population as a function of time in units of the upper-state lifetime $\tau = 1/\Gamma$ for two values of the saturation parameter κ_1.

larger than unity, an atom that has been excited is more likely to interact with the light field again and undergo stimulated emission than it is to spontaneously decay (the stimulated emission rate is the same as the excitation rate). In the limit that the relaxation can be disregarded, an atom spends equal amounts of time in the upper and ground states, and so the fractional upper-state population is 1/2. This is the nonlinear regime. The upper-state population *saturates*, because increasing the light intensity does not further increase the population inversion.[1] In fact, the fractional upper-state population in this example can be shown to be $\kappa_1/(1 + 2\kappa_1)$, which exhibits the saturation behavior that we have described (Fig. 9.1b). Another way to look at this is that, when relaxation is negligible compared to the Rabi frequency, the atoms undergo the Rabi oscillations discussed in Sec. 7.1, with average upper-state occupation probability of 1/2. Note, however, that if there is any relaxation at all in the system, the ensemble-average oscillations will eventually damp out and stabilize at their average value. In other words, for $\kappa_1 \gg 1$ the system is an underdamped oscillator; for $\kappa_1 \ll 1$ the system is an overdamped oscillator (Fig. 9.2).

In order to determine the saturation parameter in our example, we need to find the excitation rate, taking into account the finite width Γ of the upper state. The exact calculation is fairly complex, but there are various methods for obtaining an estimate. One derivation can be obtained from *Fermi's golden rule*, which says that the transition probability per unit time is given, up to a numerical factor, by the square of the transition matrix element (the strength of the transition), times the number of final states per unit energy in the energy range of interest (the number of ways to make the transition), divided by $\hbar$. In this case, there is only one final state, but because of the finite width of the upper state, the state is "smeared out" over an energy range $\hbar\Gamma$. Thus we use $1/\Gamma$ for the "density of states" in frequency units. As discussed in Sec. 7.1, the transition matrix element in frequency units is the Rabi frequency $\Omega_R = d\mathcal{E}_0/\hbar$, where d is the relevant dipole matrix element and $\mathcal{E}_0$ is the optical electric-field amplitude. Using these factors, we have $\Gamma_p = \Omega_R^2/\Gamma$ as an estimate for the transition rate.

> Another estimate can be obtained from the discussion in Sec. 7.1, which neglects relaxation. Without relaxation it is not clear what the transition rate is, because the atoms oscillate between the ground and excited state in a sinusoidal fashion. However, we found in Sec. 7.1 that at short times, the upper-state occupation probability goes as $(\Omega_R t)^2$. In the case of a relaxation rate $\Gamma \gg \Omega_R$, we are interested in the average transition rate over a time $\tau = 1/\Gamma$, after which the atom decays. If we ignore the other effects of relaxation and use the estimate $(\Omega_R \tau)^2$ for the transition probability at time τ, we can divide by the time to obtain Ω_R^2/Γ for the transition rate.

Using the estimate of the excitation rate, we have for the saturation parameter

---
[1] A standard discussion of optical nonlinearity and saturation for a two-level atom is given by Allen and Eberly (1987).

172 Optical pumping

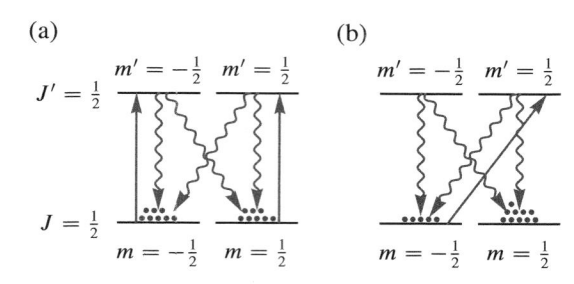

Fig. 9.3 Optical pumping on a $J = 1/2 \to J' = 1/2$ transition with (a) linearly and (b) circularly polarized light.

$$\kappa_1 = \frac{\Gamma_p}{\Gamma} = \frac{\Omega_R^2}{\Gamma^2}. \tag{9.2}$$

This result can be used, for example, to find the intensity of light required to saturate the transition: setting $\kappa_1 = 1$ we have $\Omega_R^2 = \Gamma^2$, or a saturation intensity of $I_s \sim c\mathcal{E}_0^2 = c(\hbar\Gamma/d)^2$, neglecting numerical factors.

> For a closed transition, d can also be written in terms of Γ, using a relation that takes the general form (cf. Eq. 7.62)
>
> $$\Gamma = \frac{4\omega^3}{3\hbar c^3} \frac{1}{2J'+1} |\langle J\|d\|J'\rangle|^2 \tag{9.3}$$
>
> for a $J \to J'$ transition with resonance frequency ω. Dropping numerical factors, we can use this formula to obtain the estimate
>
> $$I_s \approx \frac{\hbar\Gamma\omega^3}{c^2} \tag{9.4}$$
>
> for the saturation intensity.

Depending on the situation of interest, the saturation parameter can take different forms. One system that is described by the saturation parameter κ_1 just discussed is a $J = 1/2 \to J' = 1/2$ transition subject to linearly polarized light (Fig. 9.3a). Suppose that instead of linearly polarized light, circularly polarized light is used (Fig. 9.3b). Atoms will be excited from one ground-state sublevel (the bright state, Sec. 8.1), and not from the other sublevel (the dark state). Atoms will tend to be trapped in the dark state in the process of such optical pumping. If there is no external mechanism to transfer atoms from the dark state back to the bright state, any amount of resonant light will eventually transfer all of the atoms to the dark state, leaving no atoms in the bright state or the upper state; the concept of an equilibrium saturation parameter is meaningless in this limiting case. In a real system, however, there is generally some process, such as collisions, that redistributes the atoms between the ground-state sublevels. Apparently the saturation parameter in this case must depend on the redistribution process. Suppose that the ground-state atoms are redistributed at a rate γ, which we assume to be much slower than Γ. What is the saturation parameter in this case? The excitation rate is the same as above. The upper-state atoms still decay at a rate Γ; however, they may be transferred back into the bright state, ready to be excited again, only at the slower rate γ. Thus γ is the limiting relaxation rate for the system and should be used in writing the saturation parameter. (Even though half of the excited atoms decay directly back to the bright state, it is the slower relaxation rate that determines the behavior of the system at longer time scales.) Thus the saturation parameter is given in this case by

$$\kappa_2 = \frac{\Omega_R^2}{\Gamma\gamma}. \tag{9.5}$$

Because γ can be much slower than Γ, a process governed by κ_2 can become important at a much lower light intensity than one governed by κ_1. It turns out that κ_1 still plays a role in our second example—it can be used to describe the ratio of the population of the exited state to that of the bright sublevel of the ground state. The saturation parameters have additional uses: in this system κ_1 appears in the description of light shifts of the transition, while κ_2 can be used to characterize power broadening.

The ground-state population imbalance created by optical pumping corresponds to atomic polarization (Chapter 5). We can also determine a saturation parameter for the generation of this polarization. It is created as atoms are excited to the upper state at the rate Ω_R^2/Γ; the relaxation rate for the polarization is the ground-state relaxation rate γ. Thus the ground-state polarization is described by the saturation parameter κ_2. Any process depending on the creation of atomic polarization by the light is by definition nonlinear, because, as described in Sec. 8.1, atomic polarization affects the optical properties of the medium, making it, for example, anisotropic. The relatively long lifetime of ground-state polarization (it persists for a time $1/\gamma$ even after the light is turned off) can make it useful for sensitive measurements of energy-level splittings, as we will see in Chapter 10.

Nonlinear effects can also be obtained using multiple light beams—generally one or more pump beams to prepare the medium, and a probe beam to measure its properties. For example, atomic polarization produced by a pump beam could undergo quantum beats (Sec. 8.2) in an external field, and then be probed by a second beam. Processes requiring multiple beams are by their nature nonlinear, because for the effects of the pump beam to be observed in the probe, it must modify the properties of the medium. In some cases, for example in the observation of the Hanle effect via fluorescence described in Sec. 8.3, the pump beam must alter the atomic medium (excite atoms to the intermediate state) for any signal to be observed at all. There is, however, a linear form of the Hanle effect that can be observed in a different experimental configuration in which there is only one excitation light field and fluorescence is observed from the state excited by this light.

9.2 Optical pumping on closed transitions

We now consider various examples of optical pumping on a $J \rightarrow J'$ transition. When no other external fields but the light are present, the problem can be greatly simplified by a judicious choice of the basis. In the properly chosen basis, all of the coherences between sublevels of either the upper or the lower state are identically zero and can be eliminated from consideration. In this basis all that optical pumping does is to redistribute the sublevel populations.

The reason for this is the following. Assuming that we are dealing with E1 transitions, the light polarization vector is the only vector in the problem that breaks spatial isotropy. In the case of circularly polarized light, the polarization vector rapidly rotates around the axis along the propagation direction of the light. Unless we are interested in very fast processes with time scales comparable to the optical period, this direction is the one preferred direction in the problem. In the case of linear polarization, we do not even have a preferred spatial direction, just a preferred axis along which the electric-field vector of the light oscillates (linearly polarized light is aligned, but not oriented; see Sec. 6.6).

For each of these cases, the choice of a convenient basis is obvious: the Zeeman basis with quantization axis along (or opposite to) the light-propagation direction for circularly polarized light, and along the light-polarization axis for linear polarization.

174 *Optical pumping*

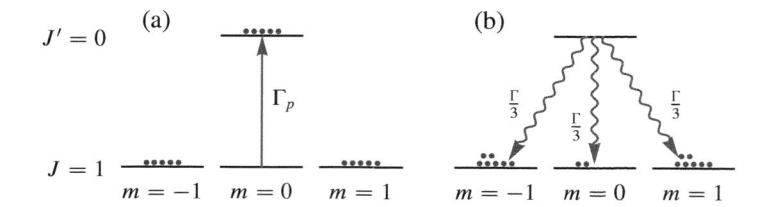

Fig. 9.4 Optical pumping on a closed $J = 1 \rightarrow J' = 0$ transition with z-polarized light.

> The situation is slightly more complicated in the case of elliptically polarized light. Here, it would appear that there are, in fact, two independent preferred axes in the problem: the light propagation direction, and the major axis of the ellipse. As a result, it is impossible to choose a quantization axis for which there are no coherences between Zeeman sublevels. However, there is still a basis consisting of *superpositions* of Zeeman sublevels in which all coherences are zero. Although such a basis can be found for a system with any combination of fields by diagonalizing the density matrix, the basis in general depends on the field strengths. When only a single light field is present the basis is independent of the field strength. This new basis is not necessarily very convenient, because the basis states are not themselves Zeeman sublevels, so that the conventional angular-momentum algebra is not directly applicable. However, there are situation where such states can be useful, for example, for understanding the physics associated with the *elliptical dark states* (Milner and Prior 1998) and *self-rotation* of the polarization ellipse in a resonant atomic medium (Rochester *et al.* 2001).

If, in addition to the light field, some other field—for example, a DC magnetic field—is applied, the best choice for the basis may not be as clear. If the field is not collinear with the symmetry axis dictated by the light field, then there are two different directions in space that break the rotational symmetry of the problem. In this case, there is in general no choice of the quantization axis that allows elimination of coherences from the problem. Commonly, the quantization axis is directed along the external DC field. We will encounter this case in later chapters. In this chapter we consider only a light field, and usually choose the basis in order to eliminate the Zeeman coherences.

We first consider atoms with a $J = 1$ ground state and a $J' = 0$ upper state (Fig. 9.4). The atoms are irradiated with weak, polarized resonant light. Here "weak" means that the rate of photon absorption by an atom, Γ_p (discussed in Sec. 9.1), is slower than the excited-state decay rate Γ, i.e., the saturation parameter κ_1 described in Sec. 9.1 is smaller than one. We also assume that the relaxation rate of ground-state polarization is much smaller than Γ_p and can be ignored. We first consider the case of a *closed transition*, in which all atoms that are excited to the upper state return to the initial state via spontaneous decay.

We could calculate the decay rates from the excited state to each of the three ground-state magnetic sublevels using the method described in Chapter 7. However, it can be seen from only symmetry arguments that the three rates $\Gamma_{m' \rightarrow m}$ are the same and equal to one third of the total decay rate of the excited state: $\Gamma_{m' \rightarrow m} = \Gamma/3$.

Consider excitation with resonant light linearly polarized along the z-axis. This light has only the $\mathcal{E}^0$ component in the spherical basis and so drives the $m = 0 \rightarrow m' = 0$ transition. After an atom is excited from the $m = 0$ ground-state sublevel to the upper state, it has an equal probability to return to any one of the ground-state magnetic sublevels. If the atom ends up in the $m = \pm 1$ sublevels it can no longer interact with the light. If, on the contrary, the atom decays back to the magnetic sublevel $m = 0$ it will again absorb a photon and will be transferred to the excited state. The average time the atom spends in the $m = 0$ ground-state sublevel is determined by the absorption rate, which depends the light intensity. After enough absorption–decay cycles, all of the atoms will be optically pumped to the $m = \pm 1$ ground-state

Fig. 9.5 Optical pumping on a closed $J = 1 \to J' = 0$ transition with (a) z-polarized and (b) x-polarized light. The dashed arrow represents coherence between the two Zeeman sublevels.

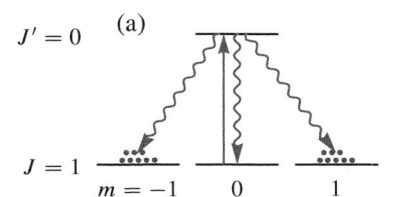

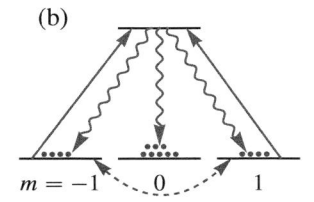

magnetic sublevels and the $m = 0$ sublevel will be empty (Fig. 9.5a). By symmetry, we can see that the $m = \pm 1$ sublevels will be equally populated and, furthermore, cannot have any coherence between them. The ground-state density matrix must therefore be given by

$$\rho = \begin{pmatrix} \tfrac{1}{2} & 0 & 0 \\ 0 & 0 & 0 \\ 0 & 0 & \tfrac{1}{2} \end{pmatrix}. \tag{9.6}$$

The ground-state angular-momentum probability distribution that corresponds to this density-matrix can be found as described in Sec. 5.3 and is shown in Fig. 9.6. As we see from the figure, the angular momentum of this ensemble of atoms is predominantly directed along the z-axis, consistent with the fact that the atoms have zero probability to be in the $m = 0$ state.

Atoms that have been optically pumped into the state described by the density matrix (9.6) will no longer absorb z-polarized light. In the terminology discussed in Sec. 8.1, optical pumping has transferred atoms from the $m = 0$ bright state to the $m = \pm 1$ dark states. In this particular example atoms end up in an incoherent mixture of the $m = \pm 1$ sublevels, as evidenced by the absence of off-diagonal elements in the density matrix.

It is instructive (Shore 1995) to repeat the above analysis in a coordinate frame in which the quantization axis is not along the light polarization direction. It goes without saying that none of the actual physics depends on the choice of the coordinate frame; however, the description of the same physics in different frames can be quite different. Specifically, the ongoing examples will illustrate that—as we have seen in Sec. 8.1—the same atomic polarization can be described either as a difference in populations between Zeeman sublevels (as discussed just above) or Zeeman coherences (as shown just below) depending on the choice of the basis.

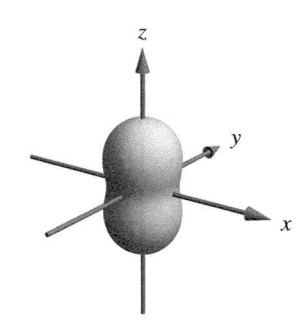

Fig. 9.6 The AMPS produced by the optical-pumping process shown in Fig. 9.4.

> This ambiguity has led to some debate and confusion related to the phenomenon of *lasing without inversion* (see the review by Kocharovskaya 1992), since in many cases what appears as lasing without inversion in one basis looks like lasing with inversion in another.

Consider a basis in which the light field is polarized along x rather than z. We can obtain the results in the new basis from those in the old by performing a rotation that takes $\hat{z}$ into $\hat{x}$. This rotation can be accomplished using the Euler angles $\alpha = 0$, $\beta = \pi/2$, and $\gamma = 0$ (Sec. 3.1.3). Applying the transformation of Eq. (5.35) to the density matrix (9.6), we find the density matrix

176 Optical pumping

$$\rho = \begin{pmatrix} \frac{1}{4} & 0 & \frac{1}{4} \\ 0 & \frac{1}{2} & 0 \\ \frac{1}{4} & 0 & \frac{1}{4} \end{pmatrix}. \tag{9.7}$$

Insofar as the physical situation has not changed, this density matrix still represents an ensemble in which all atoms are in dark states. One of the dark states can be identified as the $m = 0$ sublevel: there are no coherences associated with it in the density matrix, so it must itself be either a bright or a dark state—its nonzero population indicates that it is a dark state. The other dark state, however, is now represented by a coherent superposition of the $m = \pm 1$ sublevels; thus both the corresponding populations and coherences in the density matrix (9.7) are nonzero. In fact, what we have here is the maximum possible coherence between the $m = \pm 1$ sublevels given their population.[2] The coherent superposition of the $m = \pm 1$ sublevels with the opposite phase is the bright state. We see that even though the light drives both the $m = \pm 1 \rightarrow m' = 0$ transitions it does not affect all of the atoms in the $m = \pm 1$ ground-state sublevels. The atoms in the bright superposition of the two sublevels are excited, leaving the atoms in the dark superposition behind (Fig. 9.5).

For completeness, let us also analyze optical pumping with y-polarized light. In this case, the Euler angles that put the light-polarization axis along y are $\alpha = 0, \beta = \pi/2$, and $\gamma = \pi/2$. The corresponding density-matrix transformation leads to the density matrix

$$\rho = \begin{pmatrix} \frac{1}{4} & 0 & -\frac{1}{4} \\ 0 & \frac{1}{2} & 0 \\ -\frac{1}{4} & 0 & \frac{1}{4} \end{pmatrix}. \tag{9.8}$$

This matrix is similar to that of Eq. (9.7) with the exception of the signs of the coherences. These signs correspond to the phases between the $m = \pm 1$ sublevels, which we saw in Chapter 8 determine the direction of polarization in the xy plane.

While the descriptions of optical pumping in these alternate bases are entirely equivalent to the one in which the light polarization is along z, they can be somewhat less intuitive because of the need to consider coherences. It is a general rule that if the applied field has an axis of symmetry that coincides with the quantization axis, then no coherences will be produced in the Zeeman basis by optical pumping (assuming that no other fields are present). For linearly polarized light the symmetry axis is the polarization axis, while for circularly polarized light it is the propagation axis. (In the general case of elliptical polarization, the field does not have a symmetry axis.) For the remainder of this chapter we will choose the light-field symmetry axis along the quantization axis in order to give the clearest description of optical pumping.

The above analysis can be generalized from excitation on a $J = 1 \rightarrow J' = 0$ transition to any $J \rightarrow J' = J - 1$ transition. Figure 9.7(a) shows the excitation of such a transition with z-polarized light. The dark states in this situation are the ground-state sublevels with $m = \pm J$, and they will eventually trap all of the atoms. The ground-state density matrix will then be given by

---

[2]From the definition of the density matrix (Eq. 5.14) it is clear that the magnitude of the coherence between two states cannot exceed the square root of the product of their populations (see also Eq. 5.23).

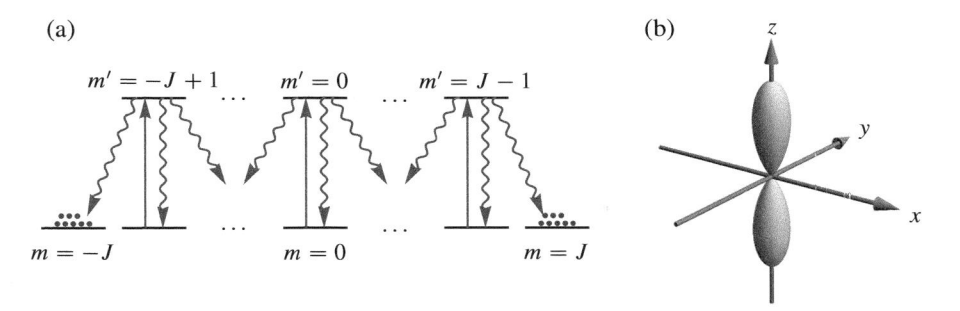

Fig. 9.7 (a) Optical pumping with z-polarized light on a $J \to J' = J - 1$ transition with arbitrary J. (b) Angular-momentum probability distribution corresponding to the density matrix of Eq. (9.9) for $J = 10$.

$$\rho = \begin{pmatrix} \frac{1}{2} & 0 & & & 0 \\ 0 & \ddots & & & \\ & & 0 & \ddots & \\ 0 & & & \ddots & 0 \\ & & & 0 & \frac{1}{2} \end{pmatrix}. \tag{9.9}$$

All of the elements of this matrix are zero except for those that characterize the population of the Zeeman sublevels $|m = \pm J\rangle$. The angular-momentum probability distribution corresponding to this density matrix is shown for the case of $J = 10$ in Fig. 9.7(b). An angular-momentum quantum number this large does not occur in the ground state of atoms, but it does occur in atomic excited states and in molecules.

It is interesting to note that as the magnitude of the angular momentum increases, the angular-momentum distribution begins to take on classical characteristics. In the classical limit, there is no uncertainty in the angular-momentum direction: in an ensemble with equal amounts of $|m = \pm J\rangle$, each particle would have an angular-momentum vector strictly along $\pm \hat{z}$, and the angular-momentum distribution would consist of delta functions in the $\pm \hat{z}$-directions. Comparison of the distributions for $J = 1$ (Fig. 9.6) and $J = 10$ (Fig. 9.7b) shows that for higher J the angular-momentum distribution is concentrated in a smaller solid angle around $\pm \hat{z}$, approaching the classical limit. One way to think about this is to consider that the number of possible angular-momentum projections on the quantization axis for a given J is $2J + 1$. The more projections that are possible, the finer the resolution with which an angle relative to the quantization axis can be specified (here we can think of the vector model of angular momentum introduced in Sec. 2.2.4). This means that as J increases the uncertainty with which the angular-momentum direction can be described becomes smaller. This can also be seen by expressing the angular-momentum probability distribution as a sum of spherical harmonics corresponding to the multipole moments present in the density matrix (Sec. 5.7). Because the highest-rank polarization moment possible in a density matrix is $\kappa = 2J$, the angular-momentum distributions for density matrices with higher J can have more spherical harmonics with higher degrees of symmetry. This allows better approximation of the infinitely sharp features of a classical distribution. As another example of this, the angular-momentum

178 *Optical pumping*

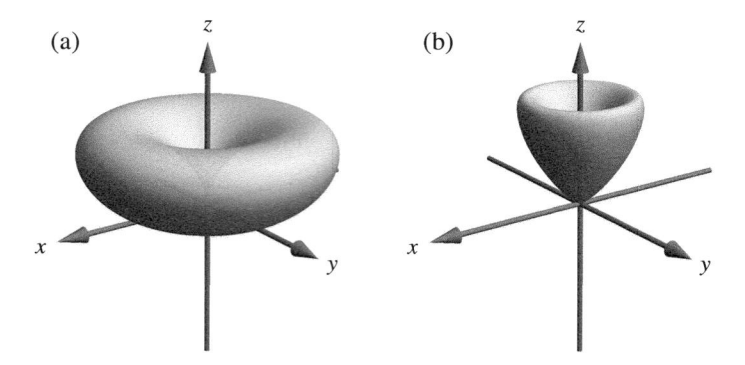

Fig. 9.8 AMPS for the state $|J, m = J - 1\rangle$ with (a) $J = 2$ and (b) $J = 10$. The sharper angular features in part (b) begin to resemble the cones appearing in the vector model of angular momentum (Sec. 2.2.4).

probability distribution of the state $|J, m = J - 1\rangle$ is plotted in Fig. 9.8 for $J = 2$ and $J = 10$. For higher J the plot begins to resemble the cones appearing in the vector model of angular momentum, which is itself a semiclassical model.

Now let us discuss optical pumping with circularly polarized light—for definiteness we will consider right circularly polarized light propagating along $\hat{z}$. This light, according to the discussion in Chapter 6, has only one nonzero spherical component, $\mathcal{E}^{-1}$, which induces $m \to m' = m - 1$ transitions. This situation is diagrammed for a $J = 1 \to J' = 0$ transition in Fig. 9.9(a).

Because atoms excited to the upper state have equal probability to decay to any of the ground-state magnetic sublevels, eventually all of the atoms will be pumped out of the $m = 1$ sublevel and equally distributed between the $m = -1$ and $m = 0$ sublevels. The density matrix for this state is

$$\rho = \begin{pmatrix} \frac{1}{2} & 0 & 0 \\ 0 & \frac{1}{2} & 0 \\ 0 & 0 & 0 \end{pmatrix}. \tag{9.10}$$

As with the case of linear polarization, this density matrix represents an incoherent superposition of two dark states, which in this basis appear as $|m = -1\rangle$ and $|m = 0\rangle$. The corresponding angular-momentum probability surface is depicted in Fig. 9.9(b). The fact that the surface is elongated toward $-\hat{z}$ shows that the angular momentum of the light, which is directed opposite to the propagation direction for right circularly polarized light (Sec. 6.6), has been transferred to the atoms.

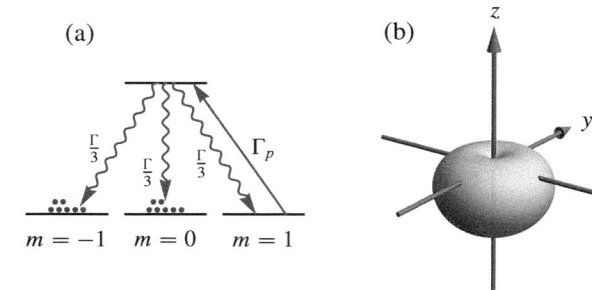

Fig. 9.9 (a) Optical pumping on a closed $J = 1 \to J' = 0$ transition with right circularly polarized light. (b) The AMPS for atoms obtained in this process.

Optical pumping on closed transitions

As with the case of linearly polarized light, these conclusions can be generalized to any $J \to J-1$ transition. For any J, all atoms will be pumped to the $m=-J$ and $m=-J+1$ sublevels. As J increases, the angular-momentum probability surfaces become more and more stretched along $-\hat{z}$ (Fig. 9.10).

We now examine transitions between states of the same angular momentum. We first consider a $J=1 \to J'=1$ transition (Fig. 9.11a). If we excite this transition with z-polarized light, the selection rule for $J \to J'=J$ transitions (Sec. 7.3.2) indicates that only the $m=-1 \to m'=-1$ and $m=1 \to m'=1$ transitions occur, while the $m=0 \to m'=0$ transition does not. As a result, as shown in Fig. 9.11(a), all atoms are optically pumped to the $m=0$ sublevel. The corresponding density matrix is

$$\rho = \begin{pmatrix} 0 & 0 & 0 \\ 0 & 1 & 0 \\ 0 & 0 & 0 \end{pmatrix}. \quad (9.11)$$

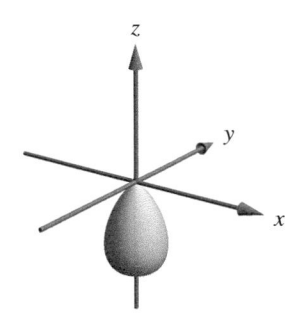

Fig. 9.10 AMPS for atoms optically pumped with right circularly polarized light on a $J=10 \to J'=9$ transition.

In this case there is only one dark state in the ground state as opposed to two for the $J \to J'=J-1$ transitions.

The angular-momentum probability surface for an ensemble of atoms that have the density matrix of Eq. (9.11) is depicted in Fig. 9.11(b). The surface has a pancake shape, with a high probability of measuring the maximum projection of the angular momentum occurring only for directions close to the xy plane. This angular-momentum distribution also reflects the fact that, according to the density matrix of Eq. (9.11), all atoms in this ensemble have zero angular-momentum projection on the quantization axis.

Generalizing this result to a $J \to J'=J$ transition with arbitrary J, we see that the population is driven by optical pumping into the $m=0$ ground-state magnetic sublevel. The ensemble of atoms is then described by a $(2J+1) \times (2J+1)$ density matrix with only one nonzero element, ρ_{00}, which corresponds to the population of the magnetic sublevel $m=0$. The angular-momentum probability distribution for this density matrix is shown in Fig. 9.12 for the case $J=10$. Again, for large J the distribution begins to approach the classical limit. For a classical particle, specifying $m=0$ means that the angular momentum has strictly zero projection on the quantization axis; i.e., the angular-momentum direction must lie in the xy plane. We see that for $J=10$ the angular-momentum probability surface is a thin pancake that

Fig. 9.11 (a) Optical pumping of a closed $J=1 \to J'=1$ transition with z-polarized light. (b) AMPS for the ground state density matrix produced in this process.

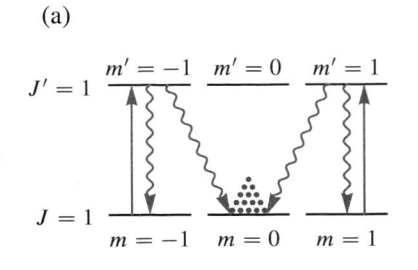

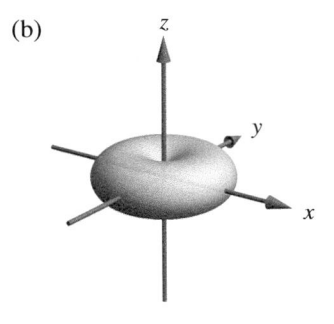

180 *Optical pumping*

is confined to a volume very close to the plane. (Incidentally, the equation for this surface is $\rho_{JJ}(\theta,\varphi) \propto \sin^{20}\theta$; for arbitrary J, it is $\rho_{JJ}(\theta,\varphi) \propto \sin^{2J}\theta$.)

Finally, we consider transitions for which the angular momentum of the upper state is larger than that of the ground state. It turns out that optical pumping on $J \to J' = J+1$ transitions is quite different in some ways from what has been discussed so far.

Excitation with z-polarized light on a $J = 1 \to J' = 2$ transition is diagrammed in Fig. 9.13(a). The relative transition rates for absorption and branching ratios for spontaneous decay, proportional to the squares of the corresponding $3j$ symbols, are shown in Fig. 9.13(a). We see that for this transition, in contrast to the $J \to J-1$ and $J \to J$ cases, the light interacts with all of the ground-state magnetic sublevels, and there are no dark substates of the ground state. This means that there are no states that can trap atoms during optical pumping. Nevertheless, optical pumping still causes redistribution of the population between the ground-state sublevels. This is due to the fact that the excitation and spontaneous decay rates are different for different pairs of ground- and excited-state sublevels.

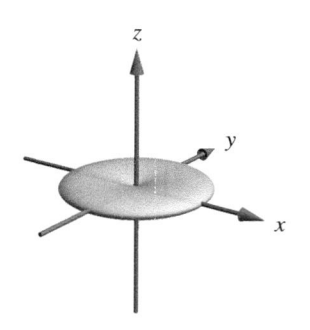

Fig. 9.12 AMPS for atoms with ground-state angular momentum $J = 10$ optically pumped into the $m = 0$ magnetic sublevel.

Because atoms are not simply trapped in dark states, it will be a little more work for us to find the ground-state density matrix compared to the cases discussed so far. For the most general solution we would find the steady-state density matrix by solving the Liouville equation for the evolution of the density matrix, as will be done in Chapter 10. However, in the case of z-polarized light and under our assumption of weak excitation ($\Gamma_p \ll \Gamma$, i.e., $\kappa_1 \ll 1$) the evolution equations can be reduced to a much simpler form, known as *rate equations*. As discussed above, with z-polarized light no ground-state coherences are generated, so that only the sublevel populations need be considered. Furthermore, the assumption $\Gamma_p \ll \Gamma$ will allow us to exclude the excited-state magnetic sublevels from explicit consideration, as we

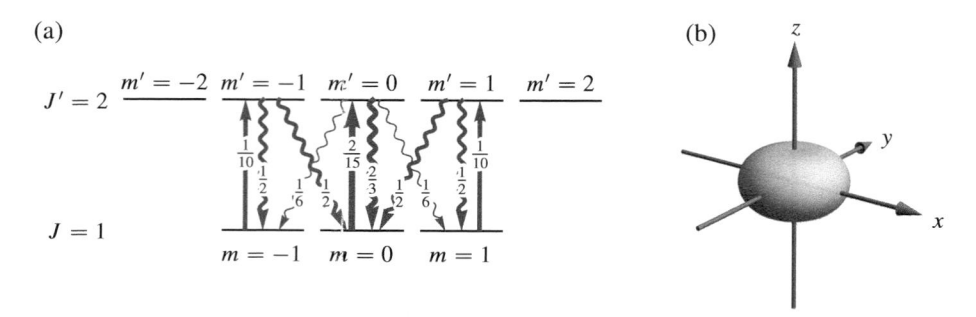

Fig. 9.13 (a) Optical pumping on a closed $J = 1 \to J' = 2$ transition with z-polarized light. The arrows indicating absorption are labeled with the corresponding transition strengths $|\langle \xi' J'm'|d|\xi Jm\rangle|^2$ in units of $|\langle \xi' J'\|d\|\xi J\rangle|^2$. Branching ratios for decay transitions are also indicated, found by normalizing $|\langle \xi Jm|d|\xi' J'm'\rangle|^2$ so that the sum of the branching ratios from a particular excited-state sublevel is unity. (b) The corresponding angular-momentum probability surface in the case of optical pumping with weak z-polarized light.

Fig. 9.14 The effective transition scheme for optical pumping of the $J = 1 \to J' = 2$ transition with z-polarized light in the case of $\tilde{\Gamma}_p \ll \Gamma$. The rate of each excitation-decay path connecting two ground state sublevels (or one with itself) is given in units of $\tilde{\Gamma}_p$.

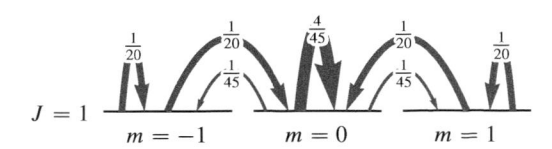

demonstrate below. In our analysis, the excited-state magnetic sublevels will merely serve as nodes where population that is taken from the ground-state is branched and immediately returned back, according to the branching ratios, to certain ground-state magnetic sublevels.

For example, consider the ground-state magnetic sublevel $|m = 0\rangle$. As discussed in Sec. 9.1, population is removed from it at a rate $\Gamma_p = \Omega_R^2/\Gamma = (d\mathcal{E}_0)^2/\Gamma$, where Ω_R is the Rabi frequency. Writing the matrix element of d in terms of the reduced matrix element $\|d\|$ using the Wigner–Eckart theorem, and evaluating the $3j$ symbol, we have $\Gamma_p = (2/15)\|d\|^2\mathcal{E}_0^2/\Gamma$. We will define a nominal pumping rate in terms of the reduced matrix element as $\tilde{\Gamma}_p = \|d\|^2\mathcal{E}_0^2/\Gamma$; the rate of pumping out of the $m = 0$ sublevel is then $2\tilde{\Gamma}_p/15$ (Fig. 9.13a). The excited atoms subsequently decay back to the ground state. Evaluating the appropriate matrix elements of d, it is found that the spontaneous-decay rates of the excited state magnetic sublevel $|m' = 0\rangle$ are such that $1/6$ of this removed population is transferred to the $m = -1$ ground-state sublevel, $1/6$ to the $m = 1$ sublevel, and $4/6$ to the initial sublevel $|m = 0\rangle$. A similar analysis for all three ground-state magnetic sublevels produces the numbers shown in Fig. 9.13(a).

Under the assumption $\tilde{\Gamma}_p \ll \Gamma$, we can assume that each atom decays immediately after being excited. We can then eliminate the excited state from Fig. 9.13(a) by considering each possible excitation-decay path connecting a ground-state sublevel with itself or another sublevel; the rate for each of these processes is the excitation rate times the branching ratio of the decay. When all of the possible transition paths are listed (Fig. 9.14) the excited state need no longer be considered.

This list of transition rates can be converted into a system of rate equations by noting that the number of atoms transferred per unit time is the transition rate times the number of atoms being pumped, and the rate of change of the population of a particular sublevel is equal to the rate that atoms are transferred into it minus the rate that atoms are transferred out. From Fig. 9.14, we find the system of equations

$$\dot{n}_{-1} = -\frac{1}{10}\tilde{\Gamma}_p n_{-1} + \frac{1}{45}\tilde{\Gamma}_p n_0 + \frac{1}{20}\tilde{\Gamma}_p n_{-1}, \qquad (9.12a)$$

$$\dot{n}_0 = -\frac{2}{15}\tilde{\Gamma}_p n_0 + \frac{1}{20}\tilde{\Gamma}_p n_{-1} + \frac{1}{20}\tilde{\Gamma}_p n_1 + \frac{4}{45}\tilde{\Gamma}_p n_0, \qquad (9.12b)$$

$$\dot{n}_1 = -\frac{1}{10}\tilde{\Gamma}_p n_1 + \frac{1}{45}\tilde{\Gamma}_p n_0 + \frac{1}{20}\tilde{\Gamma}_p n_1. \qquad (9.12c)$$

The first and the last equations in this system are identical, with the substitution $n_{-1} \leftrightarrow n_1$. This is to be expected from the symmetry of the level and excitation scheme. The symmetry allows us to reduce the number of independent equations by writing one equation for n_0 and another for $n_{\pm 1}$, where $n_{\pm 1} = n_1 = n_{-1}$.

To solve these equations for the steady state, we set the time derivatives to zero and supplement the system with an equation requiring the total population to be conserved. Normalizing the total population to one, we have

182 Optical pumping

$$-\frac{1}{10}n_{\pm 1} + \frac{1}{45}n_0 + \frac{1}{20}n_{\pm 1} = 0, \tag{9.13a}$$

$$-\frac{2}{15}n_0 + \frac{1}{10}n_{\pm 1} + \frac{4}{45}n_0 = 0, \tag{9.13b}$$

$$2n_{\pm 1} + n_0 = 1. \tag{9.13c}$$

The first two equations are linearly dependent. Solving this system gives $n_0 = 9/17$ and $n_{\pm 1} = 4/17$.

This is a rather interesting result. Due to optical pumping, the largest population (more than half of the total) is accumulated in the $m = 0$ magnetic sublevel. This happens to be the magnetic sublevel that has the largest absorption rate. Thus, after optical pumping the atoms will absorb z-polarized light more strongly than they did when the light was first applied. This is precisely the opposite of what happens for $J \to J - 1$ and $J \to J$ transitions, in which the atoms are pumped into dark states that do not absorb light. This distinctive behavior of $J \to J + 1$ transitions has observable consequences in many contexts in atomic physics including nonlinear magneto-optical rotation (Chapter 11).

We can write the density matrix representing the ground state as

$$\rho = \begin{pmatrix} \frac{4}{17} & 0 & 0 \\ 0 & \frac{9}{17} & 0 \\ 0 & 0 & \frac{4}{17} \end{pmatrix}; \tag{9.14}$$

the corresponding angular-momentum probability surface is shown in Fig. 9.13(b). This surface can be compared with that for the excitation of a $J = 1 \to J' = 0$ transition with light of the same polarization (Fig. 9.6). In contrast to Fig. 9.6, in Fig. 9.13(b) we see an increased probability to find the maximum projection of the angular momentum in the xy plane.

The absorption coefficient for weak light with a certain polarization is proportional to a sum over the ground-state sublevels of the population times the absorption rate for each sublevel. Using this we can calculate the relative absorption coefficient for atoms before and after optical pumping. Prior to the application of light, the atoms are in thermal equilibrium and the ground-state populations are all equal to $1/3$. The sum is given by

$$\frac{1}{3}\frac{1}{10}\widetilde{\Gamma}_p + \frac{1}{3}\frac{2}{15}\widetilde{\Gamma}_p + \frac{1}{3}\frac{1}{10}\widetilde{\Gamma}_p = \frac{1}{9}\widetilde{\Gamma}_p. \tag{9.15}$$

After optical pumping, it is

$$\frac{4}{17}\frac{1}{10}\widetilde{\Gamma}_p + \frac{9}{17}\frac{2}{15}\widetilde{\Gamma}_p + \frac{4}{17}\frac{1}{10}\widetilde{\Gamma}_p = \frac{2}{17}\widetilde{\Gamma}_p. \tag{9.16}$$

This means that the absorption coefficient has increased by a factor $18/17$ ($\approx 6\%$). The effect can be larger ($\sim 40\%$) for larger values of the angular momentum (Papoyan et al. 2002).

For optical pumping on a $J \to J' = J + 1$ transition with circularly polarized light, a similar phenomenon occurs. In this case it is clear that the atoms are driven to either the $m = J$ or $m = -J$ sublevels, depending on whether right or left circularly polarized light is used. Again, the population ends up in the magnetic sublevel with the strongest absorption of the excitation light. The result is an increase of absorption over that due to an unpolarized ground state by a factor (Kazantsev et al. 1984; Papoyan et al. 2002)

$$\frac{3(2J+1)}{2J+3}. \tag{9.17}$$

This is a factor of 3 in the large-J limit, but it is relatively large even for $J = 1$ (a factor of 1.8).

Given what we have seen in this section, it seems plausible that, even in the general case of arbitrary light polarization, absorption decreases as a result of optical pumping on $J \to J-1$ and $J \to J$ transitions (due to pumping into dark states) but increases for $J \to J+1$ (due to migration of population to states with the strongest absorption). This has been shown in the seminal work of Kazantsev *et al.* (1984) and can be verified with arguments similar to the ones above. As mentioned above, these properties are important, for example, for understanding nonlinear magneto-optical effects, as we further discuss in Chapter 10.

9.3 Optical pumping on open transitions

In atomic systems with only two levels, atoms excited to the upper state must decay back to a sublevel of the initial state, as in the closed systems discussed in the previous section. If there are multiple levels, however, it may be that excited-state atoms decay to some other level than the one that they were excited from. For example, the $D1$ and $D2$ lines in the alkali atoms are $J = 1/2 \to J' = 1/2$ and $J = 1/2 \to J' = 3/2$ transitions, respectively, that can be treated as two-level closed systems as long as their hyperfine structure is unresolved. If the Doppler width, light spectral width, and other forms of broadening are small enough that the hyperfine structure is resolved, on the other hand, atoms excited on one hyperfine transition, say $F = 2 \to F' = 2$, may decay to a different hyperfine ground state, for example $F' = 2 \to F = 1$. This is often a worry in experiments using laser excitation because atoms can be transferred out of the hyperfine ground state being observed into the other hyperfine state. If there is no light pumping atoms back out of the other hyperfine ground state, atoms that decay there are "lost" from the experiment.

The limiting case in which none of the excited atoms decay back to the initial state is known as an *open transition*. This case does not generally occur for fully allowed transitions involving atomic ground states, but could be seen in situations involving weakly allowed transitions and/or metastable states. In any event, examining open transitions is useful for the understanding of optical pumping.

> The model of an open transition describes, with good accuracy, transitions in a diatomic molecule from a ground-electronic-state vibrational–rotational level to an excited-state level. Because there are typically a large number of allowed spontaneous transitions to the many rotational–vibrational levels in the ground electronic state of the molecule, the probability for the molecule to spontaneously decay back to the initial level is very small and can often be neglected (see, for example, Auzinsh and Ferber 1995).

Let us analyze a $J = 1 \to J' = 0$ transition subject to linearly polarized light (Fig. 9.4a) again, this time assuming that it is open. We again assume weak excitation and initially neglect relaxation in the ground state.

As before, absorption of z-polarized light results in depletion of the $m = 0$ magnetic sublevel. In contrast to the case of a closed transition, however, the $m = 0$ population is not transferred to the $m = \pm 1$ sublevels, but is removed from the system. The resulting ground-state density matrix is

184   *Optical pumping*

$$\rho = \begin{pmatrix} \frac{1}{3} & 0 & 0 \\ 0 & 0 & 0 \\ 0 & 0 & \frac{1}{3} \end{pmatrix}, \tag{9.18}$$

which is the same as the density matrix of Eq. (9.6), up to overall normalization. The corresponding angular-momentum probability surface is thus the same (up to normalization) as that shown in Fig. 9.6.

This similarity between open and closed transitions is also found for a $J = 1 \to J' = 1$ transition, and, in fact, for all $J \to J - 1$ and $J \to J$ transitions.

However, there are substantial differences between open and closed $J \to J + 1$ transitions. Since there are no dark states in the ground state for any choice of the excitation-light polarization, in the absence of ground-state relaxation optical pumping will completely remove all atoms from the ground state.

In practice, there is always some relaxation in the ground state, although it can be very slow compared to other rates in the problem. Consider a $J \to J + 1$ transition optically pumped with a nominal pumping rate $\widetilde{\Gamma}_p = \|d\|^2 \mathcal{E}_0^2 / \Gamma$. Upon excitation, atoms quickly spontaneously decay to some other level, where they remain. A source of ground-state relaxation may be as follows: Suppose that the atoms are contained in an evacuated vapor cell (in which we can neglect collisions between the atoms), and that the light beam is spatially limited so that its diameter is much smaller than the size of the cell. Due to thermal motion, the atoms enter and leave the light beam (interaction region). As a simple model, this transit through the light beam can be described with a single effective relaxation rate γ (the *transit rate*).[3] We assume that the atoms coming into the interaction region from outside are unpolarized. Normalizing the density of atoms to one, the concentration of atoms in a given sublevel outside the interaction region is $1/w$, where $w = 2J + 1$ is the number of ground-state sublevels. This means that the rate that atoms enter the interaction region into a particular sublevel is γ/w. The rate equation for the population of a magnetic sublevel within the interaction region is then

$$\dot{n}_m = -\Gamma_p^{(m,q)} n_m - \gamma n_m + \frac{\gamma}{w}, \tag{9.19}$$

for which the steady-state solution is

$$n_m = \frac{\gamma}{\Gamma_p^{(m,q)} + \gamma} \frac{1}{w}. \tag{9.20}$$

Here

$$\Gamma_p^{(m,q)} = \widetilde{\Gamma}_p \begin{pmatrix} J' & 1 & J \\ -m' & q & m \end{pmatrix}^2 \tag{9.21}$$

is the pumping rate for the sublevel $|m\rangle$ for light with a polarization corresponding to a spherical component q, where $m' = m + q$ is the magnetic quantum number of the excited state. Note that the quantity $\widetilde{\Gamma}_p/\gamma$ is nominally the saturation parameter κ_2 discussed in Sec. 9.1. Denoting this quantity by $\widetilde{\kappa}_2$, and specializing to z-polarized light ($q = 0$), $J = 1$, and $J' = 2$, we have for the steady-state populations of the ground-state magnetic sublevels:

$$n_{-1} = \frac{1}{3 + \frac{3}{10}\widetilde{\kappa}_2}, \qquad n_0 = \frac{1}{3 + \frac{2}{5}\widetilde{\kappa}_2}, \qquad n_1 = \frac{1}{3 + \frac{3}{10}\widetilde{\kappa}_2}. \tag{9.22}$$

---
[3] In fact, in many cases this model may be accurate enough for quantitative simulation of actual experiments.

In contrast to the closed-transition case, where for the same excitation polarization we had the largest population in the $m = 0$ magnetic sublevel, now this sublevel has the smallest population. This is precisely because this sublevel has the highest excitation rate due to the light. The dependence of the result on a saturation parameter reflects the fact that the introduction of ground-state relaxation was necessary to obtain this result.

The density matrix that corresponds to the population distribution given by Eq. (9.22) for $\tilde{\kappa}_2 = 30$ is

$$\rho = \begin{pmatrix} \frac{1}{12} & 0 & 0 \\ 0 & \frac{1}{15} & 0 \\ 0 & 0 & \frac{1}{12} \end{pmatrix}. \quad (9.23)$$

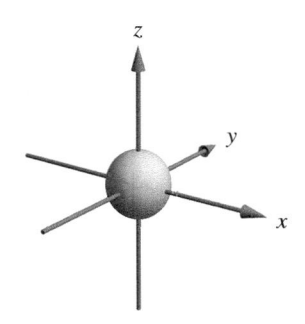

Fig. 9.15 Ground-state AMPS for optical pumping of an open $J = 1 \rightarrow J' = 2$ transition with z-polarized light.

Figure 9.15 depicts the corresponding angular-momentum probability surface, which can be contrasted with that for the closed-transition case (Fig. 9.13b).

> The nonequilibrium polarization of the ground state in the case of an open transition results from the differences in the excitation rates for different magnetic sublevels. Optical polarization of atoms using such a process is known as *depopulation pumping*. As we have seen for closed transitions, however, the transfer of the excited-state polarization back to the ground state, a.k.a. *repopulation pumping*, plays an equally important role in that case.

10
Light–atom interaction observed in transmitted light

In this chapter, we discuss experiments in which polarized light is transmitted through an atomic medium. The light causes the atoms in the medium to become polarized and this polarization in turn affects the propagation of the light. The experimental signals consist of the changes in the properties of the light as measured after its exit from the medium.

We first find the propagation equation (Sec. 10.1) for the optical field in the medium. This equation can be used to relate the atomic density matrix to the changes in the light parameters. Then, using the Liouville equation, we find the steady-state atomic density matrix for a specific system (Sec. 10.2). We analyze the resulting predicted experimental signals in various regimes.

As we will see, light is directly affected by coherences between the lower and upper states of an atomic dipole transition. The lowest-order (linear) optical effects that are observed depend only on these coherences. Higher-order (nonlinear) effects result from the creation of other types of atomic polarization. In Sec. 10.3, we use the perturbative approach to solving the Liouville equation to systematically categorize the effects of various orders.

10.1 Effect of atoms on transmitted light

The changes in the electric field of a plane light wave as it traverses an atomic medium can be described in terms of the parametrization schemes given in Sec. 6.1. We will concentrate on the α–ϵ parametrization, in which the electric field $\mathcal{E}$ of a plane wave of frequency ω and wave vector **k** is written as (Eq. 6.9)

$$\mathcal{E}(\mathbf{r},t) = \text{Re}\left\{ \mathcal{E}_0 e^{i(\mathbf{k}\cdot\mathbf{r} - \omega t + \varphi)} [(\cos\alpha\cos\epsilon - i\sin\alpha\sin\epsilon)\,\hat{\mathbf{e}}_1 \\ + (\sin\alpha\cos\epsilon + i\cos\alpha\sin\epsilon)\,\hat{\mathbf{e}}_2] \right\}, \quad (10.1)$$

where $\hat{\mathbf{e}}_1$ and $\hat{\mathbf{e}}_2 = \hat{\mathbf{k}} \times \hat{\mathbf{e}}_1$ are two orthogonal unit vectors perpendicular to **k**, $\mathcal{E}_0$ is the electric-field amplitude, φ is an overall phase, α is the polarization angle (*azimuth*) with respect to the $\hat{\mathbf{e}}_1$ axis, and ϵ is the ellipticity (equal up to a sign to the arctangent of the ratio of the minor to the major axis of the polarization ellipse).

As light propagates through a medium, it undergoes changes in total electric field characterized by the change in the electric field amplitude $\Delta\mathcal{E}_0$ (related to the *absorption* or *gain* in the medium), the *phase shift* $\Delta\varphi$, the *polarization rotation* $\Delta\alpha$, and the *change in ellipticity* $\Delta\epsilon$. These changes are described by the *wave equation* in a medium, which governs the propagation of the optical field. Because the wave equation depends on the dipole polarization of the

Effect of atoms on transmitted light **187**

medium, which in turn depends on the atomic density matrix, we can use the wave equation to relate the changes in the optical parameters to the density matrix.

The wave equation follows directly from the Maxwell equations for electromagnetic fields in a medium:

$$\nabla \cdot \mathbf{D} = 4\pi \varrho, \tag{10.2a}$$

$$\nabla \times \boldsymbol{\mathcal{E}} = -\frac{1}{c}\frac{\partial \mathbf{B}}{\partial t}, \tag{10.2b}$$

$$\nabla \cdot \mathbf{B} = 0, \tag{10.2c}$$

$$\nabla \times \mathbf{H} = \frac{4\pi}{c}\mathbf{j} + \frac{1}{c}\frac{\partial \mathbf{D}}{\partial t}. \tag{10.2d}$$

We are interested in cases in which the free electric charge density ϱ and the free current density $\mathbf{j}$ are zero, and we will assume that the medium is nonmagnetic, which implies that $\mathbf{H}$ is equal to the magnetic field $\mathbf{B}$. The *electric displacement* $\mathbf{D}$ is given by

$$\mathbf{D} = \boldsymbol{\mathcal{E}} + 4\pi \mathbf{P}, \tag{10.3}$$

where $\mathbf{P}$ is the polarization of the medium, i.e., the dipole moment per unit volume. The dipole moment can be found from the density matrix as the expectation value of the dipole operator (Sec. 5.1): $\mathbf{P} = n\,\mathrm{Tr}\,\rho\mathbf{d}$, where n is the atomic density. With these assumptions we have

$$\nabla \cdot \boldsymbol{\mathcal{E}} + 4\pi \nabla \cdot \mathbf{P} = 0, \tag{10.4a}$$

$$\nabla \times \boldsymbol{\mathcal{E}} = -\frac{1}{c}\frac{\partial \mathbf{B}}{\partial t}, \tag{10.4b}$$

$$\nabla \cdot \mathbf{B} = 0, \tag{10.4c}$$

$$\nabla \times \mathbf{B} = \frac{1}{c}\frac{\partial \boldsymbol{\mathcal{E}}}{\partial t} + \frac{4\pi}{c}\frac{\partial \mathbf{P}}{\partial t}. \tag{10.4d}$$

We then eliminate $\mathbf{B}$ by taking the curl of both sides of Eq. (10.4b) and substituting for $\nabla \times \mathbf{B}$ with Eq. (10.4). This gives

$$\nabla \times \nabla \times \boldsymbol{\mathcal{E}} = -\frac{1}{c^2}\frac{\partial^2 \boldsymbol{\mathcal{E}}}{\partial t^2} - \frac{4\pi}{c^2}\frac{\partial^2 \mathbf{P}}{\partial t^2}. \tag{10.5}$$

We now use the vector identity

$$\nabla \times \nabla \times \boldsymbol{\mathcal{E}} = \nabla(\nabla \cdot \boldsymbol{\mathcal{E}}) - \nabla^2 \boldsymbol{\mathcal{E}}. \tag{10.6}$$

In general, the divergence of $\boldsymbol{\mathcal{E}}$ is not zero for nonlinear media, but under the assumption that $\boldsymbol{\mathcal{E}}$ is a transverse plane wave, the field never points along the direction in which it varies, so $\nabla \cdot \boldsymbol{\mathcal{E}} = 0$. With this we arrive at the wave equation:

$$\nabla^2 \boldsymbol{\mathcal{E}} - \frac{1}{c^2}\frac{\partial^2 \boldsymbol{\mathcal{E}}}{\partial t^2} = \frac{4\pi}{c^2}\frac{\partial^2 \mathbf{P}}{\partial t^2}. \tag{10.7}$$

The medium polarization $\mathbf{P}$ is induced by the electric field of the light, so it oscillates at the light frequency. (We will show directly that that this is true below.) As with the light field, we can pull out the spatial and temporal dependence of the polarization and write the amplitude, directional, and phase information in terms of four real parameters:

$$\mathbf{P} = \text{Re}\left\{e^{i(\mathbf{k}\cdot\mathbf{r}-\omega t+\varphi)}\left[(P_1 - iP_2)\hat{\mathbf{e}}_1 + (P_3 - iP_4)\hat{\mathbf{e}}_2\right]\right\}, \tag{10.8}$$

where the P_i are the *in phase* and *quadrature components* of the polarization (not to be confused with the Stokes parameters designated with the same notation). Here we have chosen to reference the overall phase to that of the light in order to simplify the subsequent algebra; φ is not an independent parameter in this expression.

For a plane wave, the light-field parameters only vary along the propagation direction $\hat{\mathbf{k}}$. Using this fact and taking into account the time dependence of $\mathcal{E}$ and $\mathbf{P}$ given by Eqs. (10.1) and (10.8), the wave equation reduces to

$$\frac{\partial^2 \mathcal{E}}{\partial \ell^2} + k^2 \mathcal{E} = -4\pi k^2 \mathbf{P}, \tag{10.9}$$

where ℓ is the distance along the light propagation direction (the optical path), and we have used $k = \omega/c$.

We now substitute the parametrized expressions for $\mathcal{E}$ and $\mathbf{P}$ into Eq. (10.9). Taking the second derivative of the electric field with respect to ℓ results in many terms containing derivatives of the light-field parameters. There are second-order terms containing factors such as $d^2\alpha/d\ell^2$ or $(d\phi/d\ell)(d\alpha/d\ell)$, and first-order terms containing factors such as $k(d\alpha/d\ell)$. The first-order terms must each contain a factor of k in order for the units of the first-order and second-order terms to be the same. Comparing the general form of these terms, we see that if the first derivatives such as $d\alpha/d\ell$ are much smaller than k, i.e., if the fractional change of the light-field parameters is small over a distance equal to the wavelength of the light, the second-order terms can be neglected. Under this approximation, we can solve the wave equation to find expressions for the change of the light-field parameters per unit distance:

$$\frac{1}{\mathcal{E}_0}\frac{d\mathcal{E}_0}{d\ell} = \frac{2\pi\omega}{\mathcal{E}_0 c}\left[\sin\alpha\left(-P_1 \sin\epsilon + P_4 \cos\epsilon\right) + \cos\alpha\left(P_2 \cos\epsilon + P_3 \sin\epsilon\right)\right], \tag{10.10a}$$

$$\frac{d\varphi}{d\ell} = \frac{2\pi\omega}{\mathcal{E}_0 c}\sec 2\epsilon\left[\cos\alpha\left(P_1 \cos\epsilon + P_4 \sin\epsilon\right) + \sin\alpha\left(-P_2 \sin\epsilon + P_3 \cos\epsilon\right)\right], \tag{10.10b}$$

$$\frac{d\alpha}{d\ell} = \frac{2\pi\omega}{\mathcal{E}_0 c}\sec 2\epsilon\left[\cos\alpha\left(P_1 \sin\epsilon + P_4 \cos\epsilon\right) - \sin\alpha\left(P_2 \cos\epsilon - P_3 \sin\epsilon\right)\right], \tag{10.10c}$$

$$\frac{d\epsilon}{d\ell} = -\frac{2\pi\omega}{\mathcal{E}_0 c}\left[\sin\alpha\left(P_1 \cos\epsilon + P_4 \sin\epsilon\right) + \cos\alpha\left(P_2 \sin\epsilon - P_3 \cos\epsilon\right)\right]. \tag{10.10d}$$

As an example, let us consider light linearly polarized along $\hat{\mathbf{x}}$, propagating along $\hat{\mathbf{z}}$. Choosing $\hat{\mathbf{e}}_1 = \hat{\mathbf{x}}$, $\hat{\mathbf{e}}_2 = \hat{\mathbf{y}}$, the initial values of α and ϵ are zero, and we have

$$\frac{1}{\mathcal{E}_0}\frac{d\mathcal{E}_0}{dz} = \frac{2\pi\omega}{\mathcal{E}_0 c}P_2, \tag{10.11a}$$

$$\frac{d\varphi}{dz} = \frac{2\pi\omega}{\mathcal{E}_0 c}P_1, \tag{10.11b}$$

$$\frac{d\alpha}{dz} = \frac{2\pi\omega}{\mathcal{E}_0 c}P_4, \tag{10.11c}$$

$$\frac{d\epsilon}{dz} = \frac{2\pi\omega}{\mathcal{E}_0 c}P_3. \tag{10.11d}$$

We will calculate the polarization components P_i explicitly later in this chapter. Here we have assumed that the parameters $\mathcal{E}_0$, φ, α, ϵ do not change appreciably over the length of the medium (i.e., the medium is *optically thin*), so that they can be approximated by their initial values in the right-hand sides of Eqs. (10.10). In the *optically thick* case, we must solve the differential equations for these parameters as the light propagates through the medium. Since the density matrix itself depends on the light field, some complications in the analysis may arise, especially when atoms can travel between regions with different light parameters. These issues will be discussed in Sec. 11.6.

From Eqs. (10.11), it is particularly apparent that, as one would expect, the polarization components along the electric field are responsible for absorption and phase shift, while the perpendicular components cause changes in polarization. However, it is less intuitive that the out-of-phase components of polarization P_2 and P_4 are responsible for absorption and rotation, which involve the change in amplitude of components of the in-phase field.

This can be made a little clearer by considering the case of a *linear medium*, for which the induced complex polarization is proportional to the complex electric field:

$$\widetilde{\mathbf{P}} = \overleftrightarrow{\chi} \cdot \widetilde{\mathcal{E}}, \qquad (10.12)$$

where the *linear susceptibility tensor* $\overleftrightarrow{\chi}$ is the complex proportionality constant. Here $\widetilde{\mathcal{E}}$ and $\widetilde{\mathbf{P}}$ are defined as in Eqs. (10.1) and (10.8), respectively, but without dropping the imaginary parts of the expressions. The real part of $\overleftrightarrow{\chi}$ describes the in-phase response of the medium, and the imaginary part describes polarization produced out of phase with the light field. For our case of a light field propagating along $\hat{\mathbf{z}}$ and linearly polarized along $\hat{\mathbf{x}}$, the χ_{xx} component (related to P_1 and P_2) describes the polarization induced in the direction of the electric field, and the χ_{yx} component (related to P_3 and P_4) describes polarization induced transverse to the electric field. It is clear from symmetry that if $\chi_{yx} = 0$, the light field will remain strictly polarized along $\hat{\mathbf{x}}$ and no optical rotation or change in ellipticity will occur. Taking this case for simplicity, the wave equation (10.9) for the x-component of the light field becomes

$$\frac{\partial^2 \widetilde{\mathcal{E}}_x}{\partial z^2} + k^2(1 + 4\pi\chi_{xx})\widetilde{\mathcal{E}}_x = 0. \qquad (10.13)$$

If χ_{xx} is real, the result of the polarization is to change the effective value of k, i.e., to change the wavelength of light in the medium. This is equivalent to imposing a phase shift that is proportional to the distance traveled in the medium. This effect of the in-phase response of the medium is analogous to changing the spring constant k_s of a simple harmonic oscillator described by

$$m\ddot{x} + k_s x = 0, \qquad (10.14)$$

where m is the mass of the particle acted on by the spring and x is the position. Here, changing the spring constant changes the frequency of oscillation and does not induce any damping. Similarly, changing the in-phase polarization response of the medium affects the "springiness" of the medium, leading to the effective change in the wave number k.

Now consider the case in which χ_{xx} is imaginary, so that the induced polarization is out of phase with the electric field. Because the first space derivative of $\mathcal{E}_x$, assuming slow variation of the light parameters, is given by $\partial \widetilde{\mathcal{E}}_x/\partial z = ik\widetilde{\mathcal{E}}_x$, we can rewrite the polarization term of Eq. (10.13) to obtain

$$\frac{\partial^2 \tilde{\mathcal{E}}_x}{\partial z^2} + 4\pi \, \mathrm{Im} \, (\chi_{xx}) \, k \frac{\partial \tilde{\mathcal{E}}_x}{\partial z} + k^2 \tilde{\mathcal{E}}_x = 0. \tag{10.15}$$

In this form we can see that this term is analogous to a retarding force $-b\dot{x}$ that opposes the motion of a damped harmonic oscillator:

$$m\ddot{x} - b\dot{x} + k_s x = 0. \tag{10.16}$$

This force continuously removes kinetic energy from the harmonic oscillator (or continuously adds it if the sign of b is negative). Likewise, the out-of-phase polarization always either opposes or enhances the rate of change of the electric field with respect to distance, causing either absorption or gain, depending on the sign of $\mathrm{Im}\,(\chi_{xx})$.

As noted above, the effect of a continuously changing phase is to change the wave number in the medium. Thus the phase shift can alternatively be represented by an index of refraction n:

$$\mathcal{E}_0(z) e^{i[kz - \omega t + \varphi(z)]} = \mathcal{E}_0(z) e^{i(nkz - \omega t)}, \tag{10.17}$$

where we have taken z as the propagation axis. For a thin medium, $\varphi(z) = (d\varphi/dz)z$. Therefore $nkz = kz + (d\varphi/dz)z$, so

$$n = 1 + \frac{d\varphi}{k\,dz}. \tag{10.18}$$

If a complex index of refraction $\tilde{n}$ is used, it can account for both the phase shift and the attenuation:

$$\mathcal{E}_0 e^{i(\tilde{n}kz - \omega t)} = \mathcal{E}_0 e^{-\mathrm{Im}(\tilde{n})kz} \, e^{i[\mathrm{Re}(\tilde{n})kz - \omega t]}. \tag{10.19}$$

Taking the space derivative shows that we can define the imaginary part of $\tilde{n}$ by

$$\mathrm{Im}\,\tilde{n} = -\frac{1}{\mathcal{E}_0} \frac{d\mathcal{E}_0}{k\,dz}. \tag{10.20}$$

The real and imaginary parts of the complex index of refraction can be used along with the changes in the polarization angle and ellipticity as the four quantities measuring changes in the optical field. Another approach is to define the complex index of refraction for two complementary light polarizations (for example, left- and right-circular). These two complex quantities then characterize the modifications to the light field. For example, as will be discussed in Sec. 10.2.5, linear optical rotation can be interpreted as a phase shift between left- and right-circularly polarized light, i.e., the difference between the real part of the index of refraction for the two polarizations.

We will primarily discuss optical signals in terms of $\mathcal{E}_0$, φ, α, and ϵ, and refer to the index of refraction picture where appropriate.

10.2 Magneto-optical effects with linearly polarized light

We now explicitly calculate the effect of an atomic medium on transmitted light in a simple case—that of linearly polarized light resonant with a $F = 1 \to F' = 0$ transition (Fig. 10.1). We choose the quantization axis to be along the light propagation direction $\hat{z}$ and the light polarization axis to be along $\hat{x}$; we also assume that there is a static magnetic field applied along $\hat{z}$.

Our aim is to model an ensemble in which the atoms have a range of velocities, for example, thermally distributed atoms in a vapor cell (Fig. 10.2). The effect of atomic motion along the light propagation direction is to Doppler-shift the light frequency in the reference frame of the atom; this leads to Doppler broadening of the observed signals. To approach this problem, we will initially neglect the motion along the z-axis, as if the atoms have been laser-cooled along this axis. The signals for this case can then be integrated over the Doppler shifts to find the result for the thermal ensemble. (Additional complications arise if the vapor cell is coated or contains buffer gas—this will be described in Sec. 11.5.)

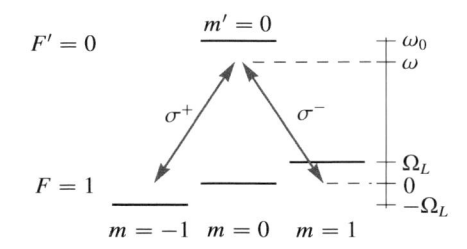

Fig. 10.1 A $F = 1 \rightarrow F' = 0$ transition of frequency ω_0. The lower sublevels are split by an energy corresponding to the Larmor frequency Ω_L. The arrows indicate the interaction with light of frequency ω polarized perpendicular to the quantization axis. The upper state spontaneously decays at rate Γ.

We begin by finding the density-matrix evolution equation for the atoms in the laser beam. We assume that the atoms that leave the laser beam have their polarization destroyed before entering the beam again; thus only the atoms in the laser beam need be considered (this is the usual case for an uncoated, buffer-gas-free vapor cell). For simplicity, we will assume that the light-intensity profile is uniform, and will model the atoms' transit through the beam by assuming a uniform relaxation rate γ equal to the inverse of the average transit time (see Sec. 5.5.2).

Using the Hamiltonians for the magnetic-field–atom and light–atom interaction (under the rotating-wave approximation, discussed in Sec. 10.2.2) we write the evolution equations for the density matrix describing both the lower and upper states. (This is in contrast to our discussion in earlier chapters, where we have mostly considered density matrices for only one state.) We then solve the equations in the steady-state condition and use the results of Sec. 10.1 to find the effect of the atoms on the transmitted light.

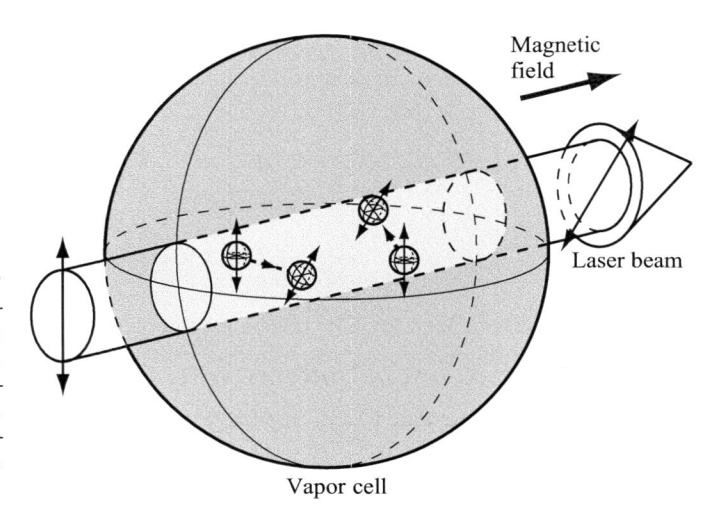

Fig. 10.2 Atoms in a vapor cell subject to linearly polarized light and a static magnetic field. The atoms fly through the light beam at a rate γ. The interaction between the atoms and the light alters the light amplitude and polarization at the output.

10.2.1 The Hamiltonian

We use the basis states $|\xi F m\rangle$, where ξ represents additional quantum numbers, denoted by

$$|\xi F, 1\rangle = \begin{pmatrix} 1 \\ 0 \\ 0 \\ 0 \end{pmatrix}, \quad |\xi F, 0\rangle = \begin{pmatrix} 0 \\ 1 \\ 0 \\ 0 \end{pmatrix}, \quad |\xi F, -1\rangle = \begin{pmatrix} 0 \\ 0 \\ 1 \\ 0 \end{pmatrix}, \quad |\xi' F', 0\rangle = \begin{pmatrix} 0 \\ 0 \\ 0 \\ 1 \end{pmatrix}. \quad (10.21)$$

In indices of density-matrix elements, we will refer to these states as simply m for the lower-state sublevels and m' for the upper-state sublevels. Thus the density matrix takes the form

$$\rho = \begin{pmatrix} \rho_{1,1} & \rho_{1,0} & \rho_{1,-1} & \rho_{1,0'} \\ \rho_{0,1} & \rho_{0,0} & \rho_{0,-1} & \rho_{0,0'} \\ \rho_{-1,1} & \rho_{-1,0} & \rho_{-1,-1} & \rho_{-1,0'} \\ \rho_{0',1} & \rho_{0',0} & \rho_{0',-1} & \rho_{0',0'} \end{pmatrix}. \quad (10.22)$$

The total Hamiltonian H is the sum of the unperturbed Hamiltonian H_0, the light–atom-interaction Hamiltonian H_l, and the magnetic-field–atom-interaction Hamiltonian H_B. Taking the energy of the lower state to be zero, the unperturbed Hamiltonian H_0 is given by

$$H_0 = \begin{pmatrix} 0 & 0 & 0 & 0 \\ 0 & 0 & 0 & 0 \\ 0 & 0 & 0 & 0 \\ 0 & 0 & 0 & \hbar\omega_0 \end{pmatrix}, \quad (10.23)$$

where ω_0 is the transition frequency.

An x-polarized optical electric field $\mathcal{E}$ is written

$$\mathcal{E} = \mathcal{E}_0 \cos \omega t \, \hat{\mathbf{x}}, \quad (10.24)$$

where $\mathcal{E}_0$ is the electric field amplitude and ω is the light frequency. We assume that the atomic medium is optically thin, so that we can neglect the change in light polarization and intensity inside the medium when calculating the state of the medium. The light–atom interaction Hamiltonian is given by (Sec. 4.3)

$$\begin{aligned} H_l &= -\mathcal{E} \cdot \mathbf{d} \\ &= -\mathcal{E}_0 \cos(\omega t)\, d_x \\ &= -\frac{1}{\sqrt{2}} \mathcal{E}_0 \cos(\omega t)\, (d_{-1} - d_{+1}), \end{aligned} \quad (10.25)$$

where $\mathbf{d}$ is the dipole operator. The matrix elements of d_{+1} and d_{-1} for this transition can be written using the Wigner–Eckart theorem (Sec. 3.8) as

$$\langle \xi_1 F_1 m_1 | d_{\pm 1} | \xi_2 F_2 m_2 \rangle = (-1)^{F_1 - m_1} \langle \xi_1 F_1 \| d \| \xi_2 F_2 \rangle \begin{pmatrix} F_1 & 1 & F_2 \\ -m_1 & \pm 1 & m_2 \end{pmatrix}. \quad (10.26)$$

Reduced matrix elements with different ordering of states are related by (Eq. 3.120):

$$\langle \xi_1 F_1 \| T^\kappa \| \xi_2 F_2 \rangle = (-1)^{F_1 - F_2} \langle \xi_2 F_2 \| T^\kappa \| \xi_1 F_1 \rangle^*, \quad (10.27)$$

and since the reduced dipole matrix element is real,

$$\langle \xi F \| d \| \xi' F' \rangle = -\langle \xi' F' \| d \| \xi F \rangle. \tag{10.28}$$

Using Eqs. (10.26) and (10.28), we have

$$d_{-1} = \frac{\langle \xi 1 \| d \| \xi' 0' \rangle}{\sqrt{3}} \begin{pmatrix} 0 & 0 & 0 & 0 \\ 0 & 0 & 0 & 0 \\ 0 & 0 & 0 & 1 \\ -1 & 0 & 0 & 0 \end{pmatrix}, \tag{10.29a}$$

$$d_{1} = \frac{\langle \xi 1 \| d \| \xi' 0' \rangle}{\sqrt{3}} \begin{pmatrix} 0 & 0 & 0 & 1 \\ 0 & 0 & 0 & 0 \\ 0 & 0 & 0 & 0 \\ 0 & 0 & -1 & 0 \end{pmatrix}. \tag{10.29b}$$

Thus H_l is given in matrix form by

$$H_l = \frac{\hbar \Omega_R \cos \omega t}{\sqrt{2}} \begin{pmatrix} 0 & 0 & 0 & 1 \\ 0 & 0 & 0 & 0 \\ 0 & 0 & 0 & -1 \\ 1 & 0 & -1 & 0 \end{pmatrix}, \tag{10.30}$$

where $\Omega_R = \langle \xi F \| d \| \xi' F' \rangle \mathcal{E}_0 / (\sqrt{3}\hbar)$ is the optical Rabi frequency.

The magnetic field interaction Hamiltonian H_B for a $\hat{z}$-directed magnetic field **B** is given by

$$\begin{aligned} H_B &= -\boldsymbol{\mu} \cdot \mathbf{B} \\ &= g\mu_0 \, \mathbf{F} \cdot \mathbf{B} \\ &= g\mu_0 F_z B \\ &= \hbar \Omega_L \begin{pmatrix} 1 & 0 & 0 & 0 \\ 0 & 0 & 0 & 0 \\ 0 & 0 & -1 & 0 \\ 0 & 0 & 0 & 0 \end{pmatrix}, \end{aligned} \tag{10.31}$$

where $\Omega_L = g\mu_0 B/\hbar$ is the Larmor frequency. Thus the total Hamiltonian is given by

$$\begin{aligned} H &= H_0 + H_l + H_B \\ &= \hbar \begin{pmatrix} \Omega_L & 0 & 0 & \frac{1}{\sqrt{2}} \Omega_R \cos \omega t \\ 0 & 0 & 0 & 0 \\ 0 & 0 & -\Omega_L & -\frac{1}{\sqrt{2}} \Omega_R \cos \omega t \\ \frac{1}{\sqrt{2}} \Omega_R \cos \omega t & 0 & -\frac{1}{\sqrt{2}} \Omega_R \cos \omega t & \omega_0 \end{pmatrix}. \end{aligned} \tag{10.32}$$

10.2.2 Rotating-wave approximation

We have now come up against the problem that the Hamiltonian has time dependence at the optical frequency. This oscillation is a vital part of the system—as we have seen, the medium polarization that induces changes in the light field must oscillate at the light frequency. However,

194   Light–atom interaction observed in transmitted light

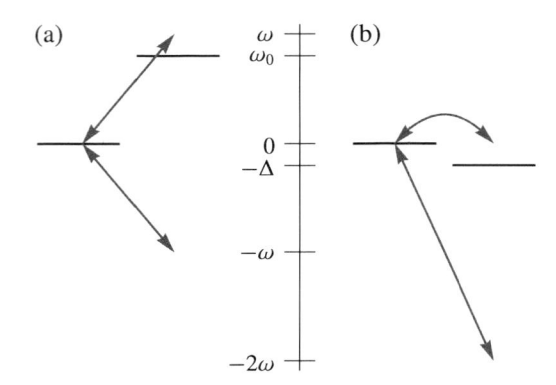

Fig. 10.3 A two-level system subject to an oscillating external field. An energy scale in frequency units is shown. (a) In the laboratory frame, the atomic resonance frequency is ω_0 and the oscillation frequency is ω. One component of the oscillating field is detuned by $\Delta = \omega - \omega_0$ from the atomic resonance, while the other component is tuned 2ω below the first. (b) In the rotating frame, all oscillation and atomic resonance frequencies have been shifted by ω. The near-resonant component of the field is now at zero frequency, and the formerly upper state now lies Δ *below* the lower state, so that the relationship between the oscillation frequency and the resonance frequency remains intact. If the counter-rotating component at frequency -2ω is neglected, the problem is reduced to that of a static field coupling two close-lying levels. Note that as the frequency of the oscillating field is changed, the rotating frame must also be adjusted in order for the co-rotating component to remain static. The change in detuning is then reflected in the position of the "upper" state.

we would like to remove it from direct consideration if possible. Intuitively, it seems that it must be possible to avoid considering this time dependence, because its time scale is much shorter than any other time scale in the problem, and short enough so that any measurement will be an average over many cycles of the optical oscillation. Indeed, in Secs. 4.5.2 and 4.5.3 we found that this could be done for a two-state system with a Hamiltonian of the form

$$H = \hbar \begin{pmatrix} 0 & \Omega \sin \omega t \\ \Omega \sin \omega t & \omega_0 \end{pmatrix} = \hbar \begin{pmatrix} 0 & \frac{i\Omega}{2}\left(e^{-i\omega t} - e^{i\omega t}\right) \\ \frac{i\Omega}{2}\left(e^{-i\omega t} - e^{i\omega t}\right) & \omega_0 \end{pmatrix}. \qquad (10.33)$$

The oscillating field with frequency ω and coupling strength $\hbar\Omega$ induces transitions between the two states with frequency splitting ω_0. The field can be written as the sum of two complex components with frequencies of opposite sign; in the case of a magnetic field coupling two Zeeman sublevels, one of these components corresponds to a magnetic field rotating with the Larmor precession and the other to a field rotating in the opposite direction. This Hamiltonian could also represent the coupling of two states by an electric field, in which case the physical interpretation of a rotation would not apply; however, the terminology of a "rotating wave" is still used by analogy. In any case, if $\omega \approx \omega_0$, one of the components (the "co-rotating") is near-resonant with the transition, while the second ("counter-rotating") is detuned from the first by 2ω (Fig. 10.3a). When the transition frequency is much greater than the linewidth of the transition (nearly always the case for optical transitions), the counter-rotating component can be considered far off-resonant and can be neglected.

In Sec. 4.5 we saw that the co-rotating component of the oscillating field can be rendered static by a suitable transformation into a new frame. For the magnetic-field case under consideration, this frame rotates around the z-axis along with the co-rotating component. According to

Eq. (3.74) this transformation is produced by a rotation matrix of the form $D_{m'm} = e^{im\omega t}\delta_{m'm}$; we can multiply this matrix by an overall phase to obtain the transformation matrix

$$U = \begin{pmatrix} 1 & 0 \\ 0 & e^{-i\omega t} \end{pmatrix}. \tag{10.34}$$

As discussed in Sec. 3.2, this matrix, like any matrix that transforms one basis to another, is unitary, meaning that it satisfies $U^\dagger U = 1$. Since we are rotating the frame, this transformation is intended to act on the basis, so that its effect on a state vector is given by

$$|\widetilde{\psi}\rangle = U^\dagger|\psi\rangle. \tag{10.35}$$

What is the Hamiltonian in the rotating frame? Under a change of basis, an operator typically transforms as $\widetilde{\mathcal{O}} = U^\dagger \mathcal{O} U$. However, it is convenient to consider the Hamiltonian to transform slightly differently. The rotating frame, obtained using the time-dependent U, is not inertial and so the evolution equation in this frame must be modified. This can be accounted for by defining a modified *effective Hamiltonian*

$$\widetilde{H}_{\text{eff}} = U^\dagger H U - i\hbar U^\dagger \frac{\partial U}{\partial t} \tag{10.36}$$

in the rotating frame. The first term on the right-hand side is the standard operator transformation, while the second term, independent of the Hamiltonian, corrects for the effect of the noninertial frame. It is analogous to a fictitious force in classical mechanics used to account for the modifications that Newton's laws undergo in an accelerating frame.

To derive the form (10.36) for the effective Hamiltonian, we start with the time-dependent Schrödinger equation in the laboratory frame:

$$H|\psi\rangle = i\hbar \frac{\partial}{\partial t}|\psi\rangle. \tag{10.37}$$

Multiplying both sides by $U^\dagger$ and inserting the identity operator $UU^\dagger$ in front of the state ket on each side, we obtain

$$U^\dagger H U U^\dagger |\psi\rangle = i\hbar U^\dagger \frac{\partial}{\partial t}\left(UU^\dagger|\psi\rangle\right). \tag{10.38}$$

Using Eq. (10.35), we find

$$\begin{aligned} U^\dagger H U |\widetilde{\psi}\rangle &= i\hbar U^\dagger \frac{\partial}{\partial t}\left(U|\widetilde{\psi}\rangle\right) \\ &= i\hbar U^\dagger \left[\left(\frac{\partial U}{\partial t}\right)|\widetilde{\psi}\rangle + U\frac{\partial}{\partial t}|\widetilde{\psi}\rangle\right]. \end{aligned} \tag{10.39}$$

This can be rearranged to form

$$\left(U^\dagger H U - i\hbar U^\dagger \frac{\partial U}{\partial t}\right)|\widetilde{\psi}\rangle = i\hbar \frac{\partial}{\partial t}|\widetilde{\psi}\rangle, \tag{10.40}$$

and comparison with Eq. (10.37) shows that the evolution in the rotating frame can be described by a Schrödinger equation with an effective Hamiltonian given by Eq. (10.36). From now on we will refer to this effective Hamiltonian simply as the rotating-frame Hamiltonian and denote it by $\tilde{H}$.

For the two-level system described by the Hamiltonian (10.33), the rotating-frame Hamiltonian is found to be

$$\tilde{H} = \hbar \begin{pmatrix} 0 & \frac{i\Omega}{2}\left(1 - e^{-2i\omega t}\right) \\ -\frac{i\Omega}{2}\left(1 - e^{2i\omega t}\right) & -\Delta \end{pmatrix}, \qquad (10.41)$$

where $\Delta = \omega - \omega_0$ is the detuning from resonance. Insofar as Δ is much smaller than ω, we now have two nearly degenerate levels coupled by a static field and a far-off-resonant oscillating field (Fig. 10.3b). (In the magnetic-field case, from the point of view of the frame moving with the co-rotating component of the field, the counter-rotating component rotates at twice the oscillation frequency.) If ω is much greater than the linewidth of the transition, the effect of the oscillating field is negligible and can be ignored (the rotating-wave approximation), resulting in

$$\tilde{H} \simeq \hbar \begin{pmatrix} 0 & \frac{i\Omega}{2} \\ -\frac{i\Omega}{2} & -\Delta \end{pmatrix}. \qquad (10.42)$$

The Hamiltonian is now static; the factors of $1/2$ in the coupling terms have resulted from the fact that we have discarded half of the external field.

In the cases that we will consider, the unitary matrix U can always be written in the form

$$U = e^{-iAt}, \qquad (10.43)$$

where A is a Hermitian matrix. For example, for the transformation matrix (10.34), we have

$$A = \begin{pmatrix} 0 & 0 \\ 0 & \omega \end{pmatrix}. \qquad (10.44)$$

Using Eq. (10.43) in the definition of the rotating-frame Hamiltonian, we have

$$\tilde{H} = U^\dagger H U - \hbar A. \qquad (10.45)$$

This shows that the effect of the correction term in the rotating-wave Hamiltonian is to shift the atomic resonance frequencies by the same amount as the oscillating-field frequencies are shifted. We can also formulate a rule of thumb for choosing the transformation to go into the rotating frame: choose the diagonal elements of A so that when it is subtracted from the laboratory-frame Hamiltonian the atomic resonance frequencies are replaced by the oscillating field detunings. In the optical-field case, A will generally resemble the unperturbed Hamiltonian H_0 divided by $\hbar$, with optical frequencies in place of the atomic transition frequencies.

Following this procedure for the system described in Sec. 10.2.1, we choose the matrix

$$A = \begin{pmatrix} 0 & 0 & 0 & 0 \\ 0 & 0 & 0 & 0 \\ 0 & 0 & 0 & 0 \\ 0 & 0 & 0 & \omega \end{pmatrix}, \qquad (10.46)$$

resulting in the static rotating-frame Hamiltonian

$$\widetilde{H} \simeq \hbar \begin{pmatrix} \Omega_L & 0 & 0 & \frac{1}{2\sqrt{2}}\Omega_R \\ 0 & 0 & 0 & 0 \\ 0 & 0 & -\Omega_L & -\frac{1}{2\sqrt{2}}\Omega_R \\ \frac{1}{2\sqrt{2}}\Omega_R & 0 & -\frac{1}{2\sqrt{2}}\Omega_R & -\Delta \end{pmatrix}. \qquad (10.47)$$

Transforming the density matrix with the matrix U, we can find that the density matrix in the laboratory frame is given in terms of the rotating-frame density-matrix elements by

$$\rho = U\tilde{\rho}U^\dagger = \begin{pmatrix} \tilde{\rho}_{1,1} & \tilde{\rho}_{1,0} & \tilde{\rho}_{1,-1} & e^{i\omega t}\tilde{\rho}_{1,0'} \\ \tilde{\rho}_{0,1} & \tilde{\rho}_{0,0} & \tilde{\rho}_{0,-1} & e^{i\omega t}\tilde{\rho}_{0,0'} \\ \tilde{\rho}_{-1,1} & \tilde{\rho}_{-1,0} & \tilde{\rho}_{-1,-1} & e^{i\omega t}\tilde{\rho}_{-1,0'} \\ e^{-i\omega t}\tilde{\rho}_{0',1} & e^{-i\omega t}\tilde{\rho}_{0',0} & e^{-i\omega t}\tilde{\rho}_{0',-1} & \tilde{\rho}_{0',0'} \end{pmatrix}. \qquad (10.48)$$

This formula will be useful when we need to interpret the results of the calculation of the rotating-frame density matrix. We see that the optical coherences in the laboratory and rotating frames differ by a phase factor oscillating at the optical frequency, while all other density-matrix elements are the same in the two frames.

10.2.3 Relaxation and repopulation

As discussed in Sec. 5.5.2, the effect of relaxation on the density-matrix evolution is not accounted for by the Hamiltonian and must be added "by hand." In the most easily described case, atoms in each basis state $|n\rangle$ relax at particular rates Γ_n, and there are no additional dephasing effects. Then the effect of relaxation can be written

$$\left.\frac{d\rho}{dt}\right|_{\text{relax}} = -\frac{1}{2}(\widehat{\Gamma}\rho + \rho\widehat{\Gamma}), \qquad (10.49)$$

where the relaxation matrix $\widehat{\Gamma}$ is a diagonal matrix with the diagonal elements given by $\widehat{\Gamma}_{nn} = \Gamma_n$.

For the $F = 1 \to F' = 0$ transition under consideration, each sublevel undergoes relaxation at a rate γ due to the exit of atoms from the light beam. In addition, the upper state undergoes spontaneous decay at a rate Γ. The relaxation matrix is then given by

$$\widehat{\Gamma} = \begin{pmatrix} \gamma & 0 & 0 & 0 \\ 0 & \gamma & 0 & 0 \\ 0 & 0 & \gamma & 0 \\ 0 & 0 & 0 & \gamma + \Gamma \end{pmatrix}. \qquad (10.50)$$

Note that this relaxation matrix is the same in the laboratory and rotating frames.

If the number of atoms in the system is conserved, there must be repopulation processes corresponding to the relaxation processes in order to replenish the atoms. These processes can be described by a repopulation matrix Λ, which may depend on ρ:

$$\left.\frac{d\rho}{dt}\right|_{\text{repop}} = \Lambda. \qquad (10.51)$$

As atoms fly out of the beam, there are other atoms, assumed to be in the ground state but otherwise unpolarized, flying into the beam at the same rate. Since the atomic density is normalized to unity, atoms arrive in each of the three ground-state sublevels at a rate $\gamma/3$. Note that this is the simplest possible model for transit relaxation and repopulation. In this model, if the light is turned off and the system is left to relax via this mechanism, the ground-state population would return to its equilibrium state exponentially. This is not always the case experimentally. For a more in-depth discussion of transit relaxation, see, for example, the book by Auzinsh and Ferber (1995).

In addition, the atoms that spontaneously decay from the upper state also repopulate the ground state. In general, the transition rate between various pairs of upper- and lower-state sublevels can be different and coherences as well as population can be transferred by spontaneous decay (Sec. 12.1). In this case, however, there is only one upper-state sublevel, so there are no coherences in the upper state and the spontaneous transition rates to all three lower-state sublevels are the same (i.e., the radiation is isotropic) and equal to $\Gamma \rho_{0'0'}/3$. Thus the repopulation matrix takes the form

$$\Lambda = (\gamma + \Gamma \rho_{0'0'}) \begin{pmatrix} \frac{1}{3} & 0 & 0 & 0 \\ 0 & \frac{1}{3} & 0 & 0 \\ 0 & 0 & \frac{1}{3} & 0 \\ 0 & 0 & 0 & 0 \end{pmatrix}. \tag{10.52}$$

This repopulation matrix is also the same in the laboratory and rotating frames.

10.2.4 Solution of the Liouville equation

The complete Liouville equation for the rotating-frame density matrix is written

$$i\hbar \frac{d}{dt}\tilde{\rho} = \left[\tilde{H}, \tilde{\rho}\right] - i\hbar \frac{1}{2}(\widehat{\Gamma\tilde{\rho}} + \widehat{\tilde{\rho}\Gamma}) + i\hbar \Lambda. \tag{10.53}$$

Using the explicit matrices (10.47), (10.50), and (10.52), we find the evolution equations for the density-matrix elements in the rotating frame:

$$\dot{\tilde{\rho}}_{-1,-1} = -\gamma \tilde{\rho}_{-1,-1} + \frac{1}{3}(\gamma + \Gamma \tilde{\rho}_{0',0'}) + \frac{1}{\sqrt{2}}\Omega_R \, \text{Im}\, \tilde{\rho}_{-1,0'}, \tag{10.54a}$$

$$\dot{\tilde{\rho}}_{0,0} = -\gamma \tilde{\rho}_{0,0} + \frac{1}{3}(\gamma + \Gamma \tilde{\rho}_{0',0'}), \tag{10.54b}$$

$$\dot{\tilde{\rho}}_{1,1} = -\gamma \tilde{\rho}_{1,1} + \frac{1}{3}(\gamma + \Gamma \tilde{\rho}_{0',0'}) - \frac{1}{\sqrt{2}}\Omega_R \, \text{Im}\, \tilde{\rho}_{1,0'}, \tag{10.54c}$$

$$\dot{\tilde{\rho}}_{1,0} = \dot{\tilde{\rho}}^*_{0,1} = -(\gamma + i\Omega_L)\tilde{\rho}_{1,0} - \frac{i}{2\sqrt{2}}\Omega_R \tilde{\rho}_{0',0}, \tag{10.54d}$$

$$\dot{\tilde{\rho}}_{0,-1} = \dot{\tilde{\rho}}^*_{-1,0} = -(\gamma + i\Omega_L)\tilde{\rho}_{0,-1} - \frac{i}{2\sqrt{2}}\Omega_R \tilde{\rho}_{0,0'}, \tag{10.54e}$$

$$\dot{\tilde{\rho}}_{1,-1} = \dot{\tilde{\rho}}^*_{-1,1} = -(\gamma + 2i\Omega_L)\tilde{\rho}_{1,-1} - \frac{i}{2\sqrt{2}}\Omega_R (\tilde{\rho}_{1,0'} + \tilde{\rho}_{0',-1}), \tag{10.54f}$$

$$\dot{\tilde{\rho}}_{1,0'} = \dot{\tilde{\rho}}^*_{0',1} = -\left(\gamma + \frac{1}{2}\Gamma + i(\Omega_L + \Delta)\right)\tilde{\rho}_{1,0'}$$
$$- \frac{i}{2\sqrt{2}}\Omega_R (\tilde{\rho}_{0',0'} + \tilde{\rho}_{1,-1} - \tilde{\rho}_{1,1}), \tag{10.54g}$$

$$\dot{\tilde{\rho}}_{-1,0'} = \dot{\tilde{\rho}}^*_{0',-1} = -\left(\gamma + \frac{1}{2}\Gamma + i(\Delta - \Omega_L)\right)\tilde{\rho}_{-1,0'}$$
$$- \frac{i}{2\sqrt{2}}\Omega_R(-\tilde{\rho}_{0',0'} + \tilde{\rho}_{-1,-1} - \tilde{\rho}_{-1,1}),$$
(10.54h)

$$\dot{\tilde{\rho}}_{0,0'} = \dot{\tilde{\rho}}^*_{0',0} = -\left(\gamma + \frac{1}{2}\Gamma + i\Delta\right)\tilde{\rho}_{0,0'} - \frac{i}{2\sqrt{2}}\Omega_R(\tilde{\rho}_{0,-1} - \tilde{\rho}_{0,1}),$$
(10.54i)

$$\dot{\tilde{\rho}}_{0',0'} = -(\gamma + \Gamma)\tilde{\rho}_{0',0'} - \frac{1}{\sqrt{2}}\Omega_R(\operatorname{Im}\tilde{\rho}_{-1,0'} - \operatorname{Im}\tilde{\rho}_{1,0'}).$$
(10.54j)

Each term in these equations has a physical meaning that derives from our prior discussion of the Hamiltonian and the relaxation and repopulation matrices. For example, Eqs. (10.54a) and (10.54c) describe the rate of change of the population of the $m = -1$ and $m = 1$ sublevels of the ground state, respectively. The first term in each of these equations describes relaxation at a rate γ due to transit of the atoms out the beam, the second term describes repopulation due to transit and spontaneous decay, and the last term accounts for the interaction of the light field (proportional to the Rabi frequency Ω_R) and the electric-dipole moment that depends on the coherences between the $m = -1$ or $m = 1$ and the excited $m' = 0$ state. Because the density matrix is Hermitian, we have $\operatorname{Im}\rho_{m,m'} = (\rho_{mm'} - \rho^*_{mm'})/(2i) = (\rho_{mm'} - \rho_{m'm})/(2i)$. Note that Eqs. (10.54a) and (10.54c) are real, consistent with the mathematical fact that the diagonal elements of a Hermitian matrix are real and the physical fact that these elements represent populations.

To find the steady-state solution of Eqs. (10.54), we set the time derivatives on the left-hand side to zero, and solve the resulting system of linear equations. This would be a somewhat arduous task by hand, but it can be accomplished using a computer algebra system such as *Mathematica*. For experiments using allowed optical transitions, the transit rate γ is usually much slower than the upper-state spontaneous decay rate Γ. Using this as a simplifying assumption, the steady-state solution is given by

$$\tilde{\rho}_{-1,-1} = \frac{1}{D}\Big\{8\gamma^2\Gamma^2\left[-4\Delta\kappa_2\Omega_L + 3\kappa_2^2\Omega_L^2 + \Delta^2(\kappa_2 + 2)(\kappa_2 + 4)\right]$$
$$+ \gamma^2\Gamma^4(\kappa_2 + 2)^3 + 128\gamma^2\Delta^4 + 8\gamma\Gamma^3\Delta\kappa_2^2\Omega_L$$
$$+ 16\Omega_L^2\left[4(\Delta - \Omega_L)^2 + \Gamma^2\right]\left[8(\Omega_L + \Delta)^2 + \Gamma^2(\kappa_2 + 2)\right]\Big\},$$
(10.55a)

$$\tilde{\rho}_{0,0} = \frac{2}{D}\Big\{4\Omega_L^2\left[16\Gamma^2(\kappa_2 + 2)(\Omega_L^2 + \Delta^2) + 64(\Delta^2 - \Omega_L^2)^2 + \Gamma^4(\kappa_2 + 2)^2\right]$$
$$+ \gamma^2\left[\Gamma^2(\kappa_2 + 1) + 4\Delta^2\right]\left[\Gamma^2(\kappa_2 + 2)^2 + 16\Delta^2\right]\Big\},$$
(10.55b)

$$\tilde{\rho}_{1,1} = \frac{1}{D}\Big\{8\gamma^2\Gamma^2\left[4\Delta\kappa_2\Omega_L + 3\kappa_2^2\Omega_L^2 + \Delta^2(\kappa_2 + 2)(\kappa_2 + 4)\right]$$
$$+ \gamma^2\Gamma^4(\kappa_2 + 2)^3 + 128\gamma^2\Delta^4 - 8\gamma\Gamma^3\Delta\kappa_2^2\Omega_L$$
$$+ 16\Omega_L^2\left[4(\Omega_L + \Delta)^2 + \Gamma^2\right]\left[8(\Delta - \Omega_L)^2 + \Gamma^2(\kappa_2 + 2)\right]\Big\},$$
(10.55c)

$$\tilde{\rho}_{1,0} = \tilde{\rho}^*_{0,1} = 0,$$ (10.55d)
$$\tilde{\rho}_{0,-1} = \tilde{\rho}^*_{-1,0} = 0,$$ (10.55e)

$$\tilde{\rho}_{1,-1} = \tilde{\rho}_{-1,1}^* = \frac{\gamma\Gamma\kappa_2}{D}\left\{\gamma\left[\Gamma^3(\kappa_2+2)^2 + 16\Gamma\Delta^2\right] - 4\Omega_L\left[2\Gamma^2(\kappa_2+2)\Omega_L\right.\right.$$
$$\left.\left. + 8i\Gamma\left(\Omega_L^2+\Delta^2\right) + 16\Omega_L\left(\Omega_L^2-\Delta^2\right) + i\Gamma^3(\kappa_2+2)\right]\right\}, \quad (10.55\text{f})$$

$$\tilde{\rho}_{1,0'} = \tilde{\rho}_{0',1}^* = \frac{\sqrt{2\gamma\Gamma}\kappa_2}{D}\left\{8\Omega_L^2\left[2(\Omega_L+\Delta) + i\Gamma\right]\left[8(\Delta-\Omega_L)^2 + \Gamma^2(\kappa_2+2)\right]\right.$$
$$\left. - 2\gamma\Gamma^2\kappa_2\Omega_L[\Gamma(\kappa_2+2) - 4i\Delta]\right. \quad (10.55\text{g})$$
$$\left. + \gamma^2(2\Delta+i\Gamma)\left[-8i\Gamma\Omega_L + 16\Delta(\Delta-\Omega_L) + \Gamma^2(\kappa_2+2)^2\right]\right\},$$

$$\tilde{\rho}_{-1,0'} = \tilde{\rho}_{0',-1}^* = \frac{\sqrt{2\gamma\Gamma}\kappa_2}{D}\left\{-2\gamma\Gamma^2\kappa_2\Omega_L[\Gamma(\kappa_2+2) - 4i\Delta]\right.$$
$$\left. - i\gamma^2(\Gamma-2i\Delta)\left[8i\Gamma\Omega_L + 16\Delta(\Omega_L+\Delta) + \Gamma^2(\kappa_2+2)^2\right]\right. \quad (10.55\text{h})$$
$$\left. + 8\Omega_L^2(2\Omega_L - i\Gamma - 2\Delta)\left[8(\Omega_L+\Delta)^2 + \Gamma^2(\kappa_2+2)\right]\right\},$$

$$\tilde{\rho}_{0,0'} = \tilde{\rho}_{0',0}^* = 0, \quad (10.55\text{i})$$

$$\tilde{\rho}_{0',0'} = \frac{2\gamma\Gamma\kappa_2}{D}\left\{8\Omega_L^2\left[8(\Omega_L^2+\Delta^2) + \Gamma^2(\kappa_2+2)\right] + \gamma^2\left[\Gamma^2(\kappa_2+2)^2 + 16\Delta^2\right]\right\}, \quad (10.55\text{j})$$

where the common denominator D is given by

$$D = 8\Omega_L^2\left[32\Gamma^2(\kappa_2+3)(\Omega_L^2+\Delta^2) + 192(\Delta^2-\Omega_L^2)^2 + \Gamma^4(\kappa_2+2)(\kappa_2+6)\right]$$
$$+ 2\gamma^2\left[\Gamma^2(\kappa_2+2)^2 + 16\Delta^2\right]\left[\Gamma^2(2\kappa_2+3) + 12\Delta^2\right]. \quad (10.56)$$

Here $\kappa_2 = \Omega_R^2/(\Gamma\gamma)$ is the optical-pumping saturation parameter. These solutions are complicated, but they describe the behavior of the system in complete generality with the exception of the assumption $\gamma \ll \Gamma$. We will analyze these results in various ways below. It is interesting to note that the coherences involving the $m=0$ ground-state sublevel vanish identically. Indeed, this is to be expected because this sublevel is not coupled to the upper state by light, nor is it coupled to the $m=\pm 1$ ground-state sublevels by the $\hat{\mathbf{z}}$-directed magnetic field. In fact, we could have excluded this sublevel from our consideration completely, and taken its effect into account by assuming that the upper state spontaneously decays into "unobserved states" with a branching fraction of $1/3$.

10.2.5 Observable signals

The changes in the light parameters induced by propagation through a medium found in Sec. 10.1 are written in terms of the medium polarization in the laboratory frame, rather than the rotating frame. We can write one density matrix in terms of the other using Eq. (10.48). Calculating the expectation value of the optical polarization of the medium $\mathbf{P} = n\,\text{Tr}\rho\mathbf{d}$ (where n is the atomic density), we find

$$\mathbf{P} = n\,\text{Re}\left(\sqrt{\frac{2}{3}}\langle\xi 1\|d\|\xi'0'\rangle e^{-i\omega t}\left[(\tilde{\rho}_{0',-1} - \tilde{\rho}_{0',1})\hat{\mathbf{x}} + i(\tilde{\rho}_{0',-1} + \tilde{\rho}_{0',1})\hat{\mathbf{y}} + \sqrt{2}\tilde{\rho}_{0',0}\hat{\mathbf{z}}\right]\right).$$
$$(10.57)$$

Because the light is propagating along $\hat{\mathbf{z}}$, the component of induced polarization in that direction must be zero. Consulting Eq. (10.55i) we see that this is indeed the case. We can thus write the medium polarization in terms of the parameters $P_{1,2,3,4}$ of Eq. (10.8) with $\hat{\mathbf{e}}_1 = \hat{\mathbf{x}}$ and $\hat{\mathbf{e}}_2 = \hat{\mathbf{y}}$. Comparing Eqs. (10.8) and (10.57), we find

$$P_1 = \sqrt{\frac{2}{3}} \langle \xi 1 \| d \| \xi' 0' \rangle n \, \mathrm{Re}\, (\tilde{\rho}_{-1,0'} - \tilde{\rho}_{1,0'}), \tag{10.58a}$$

$$P_2 = \sqrt{\frac{2}{3}} \langle \xi 1 \| d \| \xi' 0' \rangle n \, \mathrm{Im}\, (\tilde{\rho}_{-1,0'} - \tilde{\rho}_{1,0'}), \tag{10.58b}$$

$$P_3 = \sqrt{\frac{2}{3}} \langle \xi 1 \| d \| \xi' 0' \rangle n \, \mathrm{Im}\, (\tilde{\rho}_{-1,0'} + \tilde{\rho}_{1,0'}), \tag{10.58c}$$

$$P_4 = -\sqrt{\frac{2}{3}} \langle \xi 1 \| d \| \xi' 0' \rangle n \, \mathrm{Re}\, (\tilde{\rho}_{-1,0'} + \tilde{\rho}_{1,0'}). \tag{10.58d}$$

This is all we need in order to calculate the effect of the atomic medium on the light. The observables are found in terms of the rotating-frame density matrix by substituting Eq. (10.58) into Eqs. (10.11):

$$\frac{1}{\mathcal{E}_0} \frac{d\mathcal{E}_0}{d\ell} = \frac{n\Gamma\lambda^2 \, \mathrm{Im}\, (\tilde{\rho}_{-1,0'} - \tilde{\rho}_{1,0'})}{4\sqrt{2}\pi\Omega_R}, \tag{10.59a}$$

$$\frac{d\varphi}{d\ell} = \frac{n\Gamma\lambda^2 \, \mathrm{Re}\, (\tilde{\rho}_{-1,0'} - \tilde{\rho}_{1,0'})}{4\sqrt{2}\pi\Omega_R}, \tag{10.59b}$$

$$\frac{d\alpha}{d\ell} = -\frac{n\Gamma\lambda^2 \, \mathrm{Re}\, (\tilde{\rho}_{-1,0'} + \tilde{\rho}_{1,0'})}{4\sqrt{2}\pi\Omega_R}, \tag{10.59c}$$

$$\frac{d\epsilon}{d\ell} = \frac{n\Gamma\lambda^2 \, \mathrm{Im}\, (\tilde{\rho}_{-1,0'} + \tilde{\rho}_{1,0'})}{4\sqrt{2}\pi\Omega_R}, \tag{10.59d}$$

where we have used Eq. (7.62) with $\Gamma = 1/\tau$.

Substituting the solution (10.55) for the density-matrix elements into Eq. (10.59), we find

$$\frac{\ell_0}{\mathcal{E}_0} \frac{d\mathcal{E}_0}{d\ell} = -\frac{3\Gamma^2}{D} \left\{ 8\Omega_L^2 \left[8\Omega_L^2 + \Gamma^2(\kappa_2 + 2) + 8\Delta^2 \right] + \gamma^2 \left[\Gamma^2(\kappa_2 + 2)^2 + 16\Delta^2 \right] \right\}, \tag{10.60a}$$

$$\ell_0 \frac{d\varphi}{d\ell} = -\frac{6\Gamma\Delta}{D} \left\{ 8\Omega_L^2 \left[-8\Omega_L^2 + \Gamma^2(\kappa_2 + 2) + 8\Delta^2 \right] + \gamma^2 \left[\Gamma^2(\kappa_2 + 2)^2 + 16\Delta^2 \right] \right\}, \tag{10.60b}$$

$$\ell_0 \frac{d\alpha}{d\ell} = -\frac{6\Gamma\Omega_L}{D} \{ 8\Omega_L^2 \left[8\Omega_L^2 + \Gamma^2(\kappa_2 + 2) - 8\Delta^2 \right]$$
$$+ \gamma \left[4\gamma \left(\Gamma^2 - 4\Delta^2 \right) - \Gamma^3 \kappa_2(\kappa_2 + 2) \right] \}, \tag{10.60c}$$

$$\ell_0 \frac{d\epsilon}{d\ell} = \frac{24\Gamma^2 \Delta \Omega_L}{D} \left[\gamma(\Gamma\kappa_2 - 4\gamma) - 16\Omega_L^2 \right], \tag{10.60d}$$

where D is given by Eq. (10.56). We have written the results in terms of the *unsaturated absorption length* on resonance

$$\ell_0 = -\left(\frac{1}{\mathcal{I}} \frac{d\mathcal{I}}{d\ell} \right)^{-1} = -\left(\frac{2}{\mathcal{E}_0} \frac{d\mathcal{E}_0}{d\ell} \right)^{-1} = \frac{6\pi}{\lambda^2 n}, \tag{10.61}$$

found by setting κ_2, Ω_L, and Δ to zero.

The resonant absorption length can be written in terms of the *photon-absorption cross-section* σ according to

$$\ell_0 = \frac{1}{n\sigma}. \tag{10.62}$$

Equation (10.61) then tells us that

$$\sigma = \frac{1}{3}\frac{\lambda^2}{2\pi}, \tag{10.63}$$

i.e., that the absorption cross-section on resonance does not depend on anything except the wavelength of the transition and the ratio of the degeneracies of the initial and final state that determines the numerical coefficient, equal to 1/3 in this case. See, for example, the book by Budker *et al.* (2008, Prob. 3.5) for an alternative derivation and additional discussion.

10.2.6 Linear effects

Note that so far we have made no assumptions about the strength of the light field, i.e., the value of the saturation parameter κ_2. In order to make the distinction between the linear and nonlinear processes discussed in Sec. 9.1, it is helpful to take the limit of vanishingly small light power, $\kappa_2 \to 0$. In this case, the nonlinear effects (which are first order in κ_2 for small κ_2) can be neglected, and the linear effects (which are independent of κ_2) remain.

To find the signals in the linear case, we can set $\kappa_2 = 0$ to obtain

$$\frac{\ell_0}{\mathcal{E}_0}\frac{d\mathcal{E}_0}{d\ell} = -\frac{1}{2}\frac{\Gamma^2\left(4\Delta^2 + 4\Omega_L^2 + \Gamma^2\right)}{16\left(\Delta^2 - \Omega_L^2\right)^2 + 8\Gamma^2\left(\Delta^2 + \Omega_L^2\right) + \Gamma^4}, \tag{10.64a}$$

$$\ell_0\frac{d\varphi}{d\ell} = -\frac{\Gamma\Delta\left(4\Delta^2 - 4\Omega_L^2 + \Gamma^2\right)}{16\left(\Delta^2 - \Omega_L^2\right)^2 + 8\Gamma^2\left(\Delta^2 + \Omega_L^2\right) + \Gamma^4}, \tag{10.64b}$$

$$\ell_0\frac{d\alpha}{d\ell} = -\frac{\Gamma\Omega_L\left(4\Omega_L^2 - 4\Delta^2 + \Gamma^2\right)}{16\left(\Delta^2 - \Omega_L^2\right)^2 + 8\Gamma^2\left(\Delta^2 + \Omega_L^2\right) + \Gamma^4}, \tag{10.64c}$$

$$\ell_0\frac{d\epsilon}{d\ell} = -\frac{4\Gamma^2\Delta\Omega_L}{16\left(\Delta^2 - \Omega_L^2\right)^2 + 8\Gamma^2\left(\Delta^2 + \Omega_L^2\right) + \Gamma^4}. \tag{10.64d}$$

These solutions apply as long as $\kappa_2 \ll 1$; note that they are also independent of γ (under the assumption $\gamma \ll \Gamma$).

In experiments of the type under examination here, measurements are commonly carried out by varying either the magnetic field strength or the light frequency while holding all other parameters fixed. For the Doppler-free linear effects there is considerable symmetry between these two approaches, as is indicated by Eqs. (10.64) and illustrated by Fig. 10.4. In both cases there is resonant behavior with a characteristic width given by the natural width Γ. In the left-hand column of Fig. 10.4, the Larmor frequency (proportional to the magnetic field strength) is fixed at $\Omega_L = 0.2\Gamma$ and the light detuning Δ is varied. In the right-hand column, the light frequency detuning from resonance is fixed at $\Delta = 0.2\Gamma$ and Ω_L is varied. The normalized optical signals are plotted. We see that the attenuation coefficient $d\mathcal{E}_0/d\ell$ (top row) and the change in ellipticity $d\epsilon/d\ell$ (bottom row) are unchanged under the exchange of Δ and Ω_L, while the phase shift $d\varphi/d\ell$ (second row) and optical rotation $d\alpha/d\ell$ (third row) are transformed into each other.

The symmetry can be understood by considering the resonant behavior of the two transitions driven by the σ^+ and σ^- components of the light, respectively (Fig. 10.1). Each of these

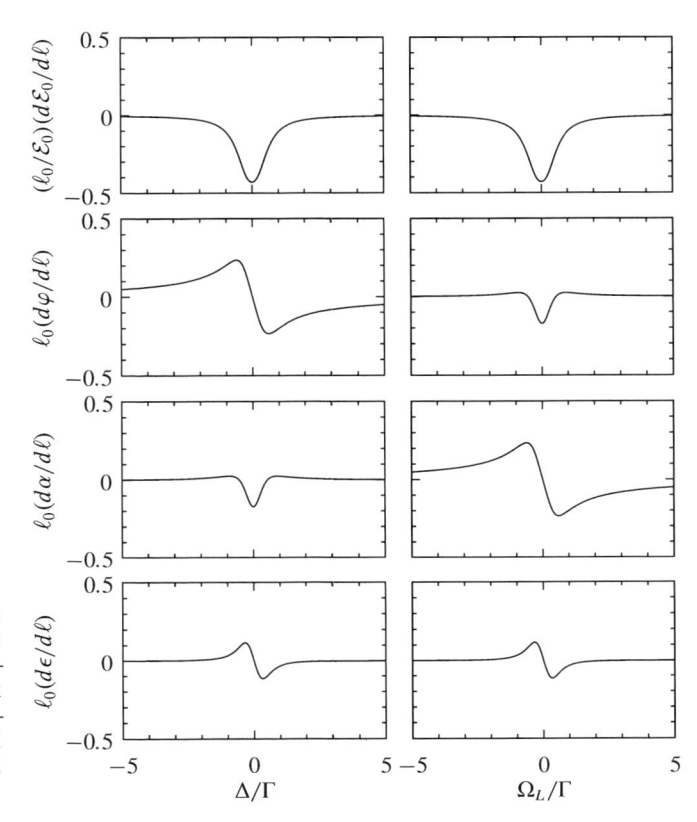

Fig. 10.4 Signals observed in transmitted light resonant with a $F = 1 \rightarrow F' = 0$ atomic transition in the linear case (weak light power; $\kappa_2 = 0$). The parameters are $\Omega_L = 0.2\Gamma$ (left-hand plots), and $\Delta = 0.2\Gamma$ (right-hand plots).

transitions induces absorption with a Lorentzian lineshape and a phase shift with a dispersive lineshape, as can be seen by setting $\Omega_L = 0$ in Eqs. (10.64):

$$\frac{\ell_0}{\mathcal{E}_0}\frac{d\mathcal{E}_0}{d\ell} = -\frac{1}{2}\frac{\Gamma^2}{\Gamma^2 + 4\Delta^2}, \tag{10.65a}$$

$$\ell_0\frac{d\varphi}{d\ell} = -\frac{\Gamma\Delta}{\Gamma^2 + 4\Delta^2}. \tag{10.65b}$$

(This dependence can in fact be obtained using a classical model of a damped harmonic oscillator.) These two lineshapes are close to those shown in the top two plots of the left-hand column of Fig. 10.4 (the difference arises because the magnetic field is not zero in the figure). Some important features of these lineshapes are: they each have a characteristic width Γ; the absorptive curve is symmetric, while the dispersive curve is antisymmetric; the absorptive curve is maximum on resonance, while the dispersive curve is zero on resonance; at large detunings the absorptive curve falls off as Δ^{-2}, while the dispersive curve falls off as Δ^{-1}.

The signals in Fig. 10.4 are each proportional to either the sum or difference of the absorption or phase shift signals from the two transitions. As an example, we consider the optical-rotation signal. The optical rotation in the linear case is generated by *circular birefringence*—a phase shift between the two circular components of the light field upon traversal

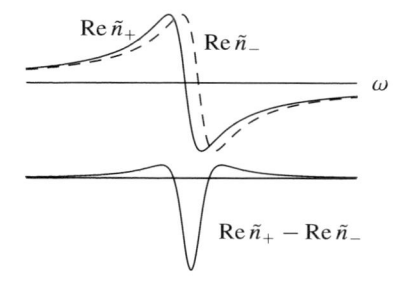

Fig. 10.5 (Top) Real parts of the complex refractive indices as a function of light frequency for σ^+ and σ^- light. The curves are shifted with respect to one another due to the magnetic field. (Bottom) Linear optical rotation is proportional to the difference between Re $\tilde{n}_+$ and Re $\tilde{n}_-$.

of the medium.[1] When two circular components are combined to create linearly polarized light, it is the phase relationship between the two components that determines the direction of linear polarization. If the relative phase changes, the polarization angle rotates. Therefore, the optical-rotation signal is given by the difference between the dispersive phase-shift lineshapes for the σ^+ and σ^- transitions. According to the discussion in Sec. 10.1, this quantity can also be written in terms of the difference between the real parts of the complex indices of refraction for left- and right-circularly polarized light:

$$\frac{d\alpha}{d\ell} = k \left(\text{Re}\, \tilde{n}_+ - \text{Re}\, \tilde{n}_- \right), \tag{10.66}$$

where $\tilde{n}_\pm$ are the refractive indices for the corresponding circular polarizations.[2]

> The general effect of polarization rotation upon light propagation in a medium in the presence of a longitudinal magnetic field was discovered by Michael Faraday and bears his name. However, it was the work of Italian physicists Macaluso and Corbino (1898) that uncovered the resonant character of the magneto-optical effect (as a function of light detuning from resonance) for an atomic vapor. The same authors also connected the phenomenon to the Zeeman effect (Macaluso and Corbino 1899).

Changes in the Larmor frequency and in the light detuning can both be interpreted in terms of level shifts: the relative distance between the σ^+ and σ^- resonances is given by twice the Larmor frequency, whereas changing the light detuning effectively shifts the two resonances together. If the magnetic field is zero the two resonances overlap exactly and the difference between them is zero; there is no optical rotation in this case. If the magnetic field is nonzero, as in Fig. 10.5, the difference between the two curves gives the characteristic spectral shape of linear optical rotation. Note that although the optical-rotation lineshape is symmetric, as is an absorptive Lorentzian, the fact that it arises from the individual dispersive lineshapes means that the total area under the curve is zero. This is important when considering mechanisms such as Doppler broadening, which tends to "wash out" effects with this property. In fact, this is the case for all of the linear signals except for absorption (Fig. 10.4), for similar reasons.

Now we consider the magnetic-field dependence of optical rotation. From the point of view of the individual σ^+ and σ^- resonances, changing the magnetic field shifts the transition resonance frequency, producing the same effect as detuning the light. Thus each phase-shift resonance has a dispersive lineshape in a magnetic field, and the optical rotation signal is the difference between them, as before. However, the magnetic field shifts the resonances in opposite directions to each other, meaning that one of the dispersive curves must be flipped

---

[1] We will see that optical rotation can also be induced by *linear dichroism*—a difference in absorption for two orthogonal linear polarizations.

[2] See also the detailed tutorial discussion of this given by Budker *et al.* (2008, Prob. 4.1).

about the vertical axis. Since the dispersive curves are antisymmetric, this has the effect of changing its sign. In other words, we can find the magnetic-field dependence of the optical rotation signal by adding, rather than subtracting, the dispersive curves. This results in the dispersive shape shown in Fig. 10.4.

Note that the spectral dependence of the total phase shift $d\varphi/d\ell$ is the sum of the phase shifts due to the two resonances. Thus the detuning dependence of the phase shift should have the same shape as the magnetic-field dependence of optical rotation (Fig. 10.4), and vice versa, as can be shown with similar reasoning to that given above.

The total absorption signal depends on the sum of the σ^+ and σ^- absorptive resonances, while the change-of-ellipticity signal depends on their difference, i.e., the *circular dichroism* of the medium. (Linear polarization contains equal amounts of right- and left-circular polarization; if more of one is absorbed than the other, the polarization will no longer be linear.) The above line of reasoning can be continued to show that the absorption and change-of-ellipticity signals transform into themselves upon interchange of Δ and Ω_L.

The high degree of symmetry between the light-detuning and magnetic-field dependencies does not hold over to the nonlinear forms of the optical signals. While the linear effects are understood in terms of level shifts, the nonlinear effects that we will be concerned with are more properly understood in terms of the evolution of atomic polarization. For low light power, this evolution is the Larmor precession due to the magnetic field. This means that the system has the potential to be much more sensitive to changes in the Larmor frequency than changes in the light frequency. We take up this case in the following section.

10.2.7 Nonlinear effects

We now turn our attention to higher light powers, for which nonlinear effects become prominent. Plotting Eqs. (10.60) with a nonzero saturation parameter ($\kappa_2 = 5$) yields some interesting developments (Fig. 10.6). In the right column, the magnetic resonances of width Γ due to the linear effect are still visible, but superimposed on them are much narrower resonances with widths on the order of the transit width $\gamma = 0.05\Gamma$. Evidently, these narrow resonances constitute the nonlinear signals. From Fig. 10.6 we see that, for a given light intensity, the magnetic field can be used to switch the system between the linear and nonlinear regimes. Keeping this in mind, in the left column a small magnetic field ($\Omega_L = \gamma$) is chosen, so that the optical spectra of the nonlinear effects can be examined. We see that no sharp features appear—the symmetry between the detuning and the magnetic-field dependencies is broken in the nonlinear case. (This can also be seen in Fig. 10.7, which shows 3D plots of the same signals as a function of both Δ and Ω_L.) The left column of Fig. 10.6 also shows that the absorption and dispersion spectra are largely the same as in the linear case. However, some differences arise in the spectra of magneto-optical rotation and induced ellipticity. The spectra have the opposite sign as for the linear effects; this is to be expected from comparing the signs of the linear and nonlinear magnetic-field resonances in the right column. In addition, the shapes of the resonances are somewhat different—in particular, the area under the optical-rotation curve is no longer zero. In the following, we discuss the mechanisms that result in these features.

Because the characteristic width of the resonances in the optical spectrum is Γ while the nonlinear magnetic-field resonances are of width γ, we define the dimensionless parameters $u = 2\Omega_L/\gamma$ and $v = 2\Delta/\Gamma$, which can be considered to have magnitudes on the order of unity or smaller. Let us concentrate on the optical-rotation signal. Using the assumption that $|u|$ and $|v|$ are not much greater than unity, and our previous assumption $\gamma \ll \Gamma$, we find from Eq.

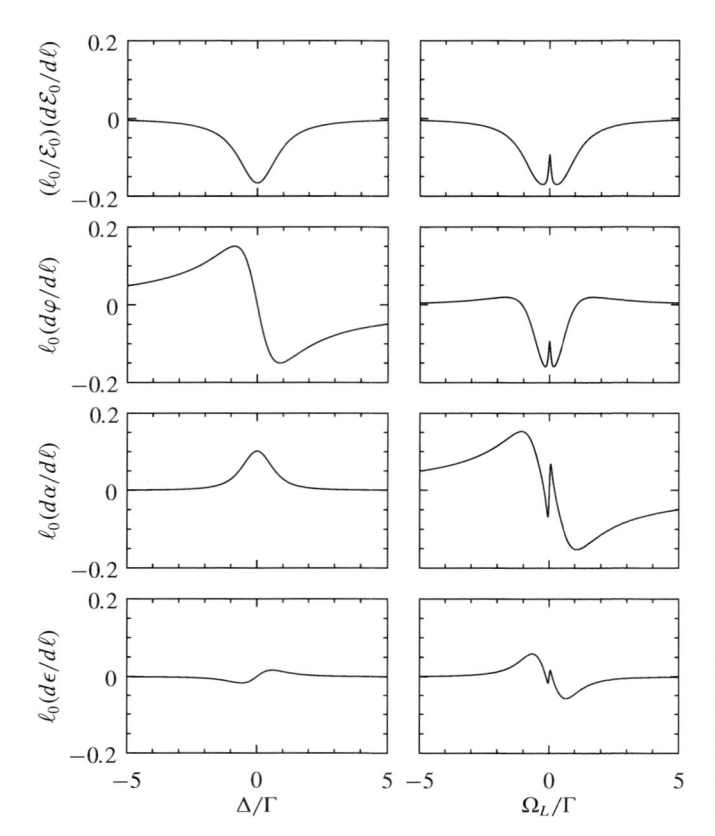

Fig. 10.6 Signals observed in transmitted light resonant with a $F = 1 \to F' = 0$ atomic transition. Parameters are $\Delta = 0.5\,\Gamma$ (left-hand plots), $\Omega_L = 0.05\,\Gamma$ (right-hand plots), $\gamma = 0.05\,\Gamma$, and $\kappa_2 = 5$.

(10.60c):

$$\ell_0 \frac{d\alpha}{d\ell} = \frac{1}{8} \frac{u\kappa_2(\kappa_2 + 2)}{u^2 \left(v^2 + 1 + \frac{\kappa_2}{2}\right) \left(v^2 + 1 + \frac{\kappa_2}{6}\right) + \left(v^2 + 1 + \frac{2\kappa_2}{3}\right) \left[v^2 + (1 + \frac{\kappa_2}{2})^2\right]}. \quad (10.67)$$

Note that this signal is zero when κ_2 is zero, i.e., the linear effect (independent of κ_2) studied in the previous section is missing. This is because the assumption that $|\Omega_L|$ is on the order of γ means that $|\Omega_L| \ll \Gamma$, which—as can be seen in Fig. 10.6—means that the linear effect is very small compared to the nonlinear effect. (Of course, if κ_2 were set to zero, this assumption would not be valid, and the linear term would need to be included.)

The dependence of Eq. (10.67) on u shows that the *nonlinear magneto-optical rotation (NMOR)* signal is a dispersive Lorentzian in the magnetic field for any value of the saturation parameter or detuning. The dependence on v and κ_2 is, however, more complicated. To gain a basic understanding, we can expand the nonlinear signal to lowest order in the saturation parameter:

$$\ell_0 \frac{d\alpha}{d\ell} = \frac{u\kappa_2}{4\left(u^2 + 1\right)\left(v^2 + 1\right)^2} + O\left(\kappa_2^2\right). \quad (10.68)$$

The signal is first order in the saturation parameter (given that the linear effect has been neglected). To lowest order, the optical spectrum of the signal is the square of a Lorentzian

Magneto-optical effects with linearly polarized light 207

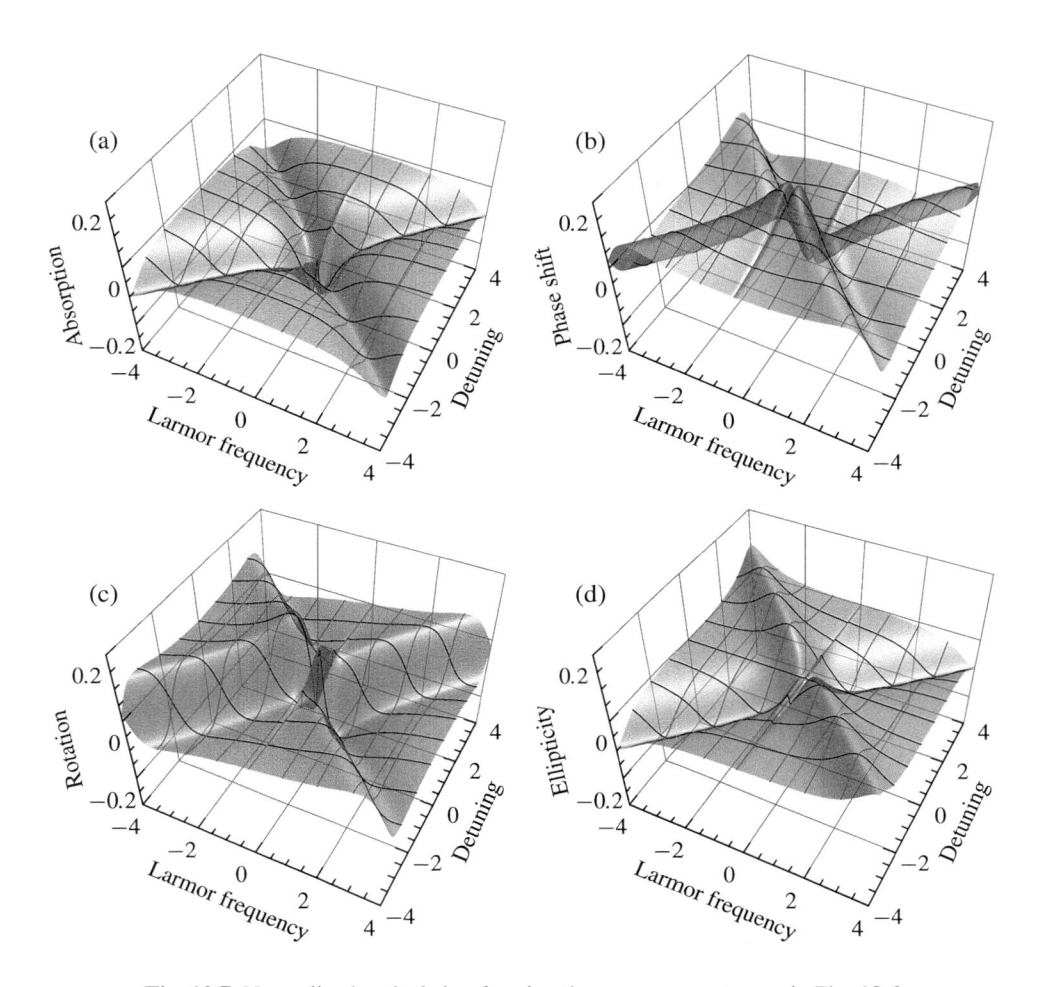

Fig. 10.7 Normalized optical signals using the same parameters as in Fig. 10.6.

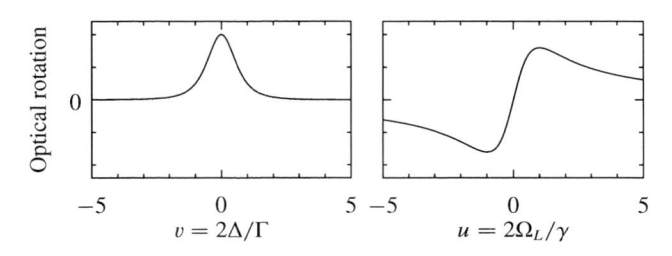

Fig. 10.8 Lowest-order nonlinear magneto-optical rotation in arbitrary units (Eq. 10.68). Left-hand plot is the optical spectrum (square of an absorptive Lorentzian) with $u = 0.5$; right-hand plot is the magnetic-field dependence (dispersive Lorentzian) with $v = 0.5$.

(Fig. 10.8). As we will see, this is because two consecutive optical processes are required to produce the nonlinear signal, each of which have a Lorentzian spectrum. As noted above, this spectrum is qualitatively different from that for linear optical rotation.

> In the geometry considered here, with the magnetic field collinear with the propagation direction of linearly polarized light, NMOR is also known as the *nonlinear Faraday effect* or *nonlinear Faraday rotation*. There were several independent discoveries and rediscoveries of this effect, beginning with precursor observations in the 1970s (and even earlier). The story of these observations and the eventual understanding of them is related in a review paper by Budker *et al.* (2002a). The observation of the nested features in the magnetic-field dependence of the signal as in Fig. 10.6 played a crucial part in these developments. Nonlinear Faraday rotation experiments can nowadays be found in undergraduate physics laboratories; a description of the one at Berkeley is given by Budker *et al.* (1999).

To understand the nonlinear magneto-optical rotation, it is convenient to think of it as occurring in three stages. First, atoms are optically pumped by the linearly polarized light into an aligned state, as discussed in Sec. 9.2. This state is dark (nonabsorbing) with respect to the x-polarized pump light. However, as in the discussion in Chapter 8, this same state is bright with respect to y-polarized light. In fact, the atoms will absorb y-polarized light even more strongly after optical pumping, because when atoms are pumped out of the bright state and into the dark state by x-polarized light, from the point of view of y-polarized light they are pumped from the dark into the bright state. This difference in absorption for light of orthogonal linear polarizations is called *linear dichroism*. The medium thus resembles a dichroic polarizer, e.g., a Polaroid film, absorbing one linear polarization and transmitting the other. In the second stage of the process, the atomic alignment precesses in the magnetic field, rotating the axis of dichroism. Finally, in the third stage of the process, the light polarization is rotated by interaction with the dichroic atomic medium, which tends to rotate light toward its transmitting axis (which is no longer along the initial light polarization axis). The third, "probing," step does not require high light intensity, and can be performed either by a weak probe beam or by the same pump light present in the first step, as in our present example. This *rotating-polarizer model* was introduced by Kanorsky *et al.* (1993).

In the physical system that we are examining, the three stages described above happen simultaneously and continuously, rather than sequentially. Nevertheless, the rotating-polarizer model presents an accurate physical picture of the mechanism for the lowest-order NMOR. How does it account for the dispersive shape of the magnetic resonance? The optically pumped atoms relax after an average time $1/\gamma$. As the magnetic field is increased, the atoms precess farther in this time, increasing the angle of the polarizer and thus the optical-rotation angle. This means that for small fields, the NMOR signal is proportional to the magnetic field. When the magnetic field becomes high enough that the atoms can precess on the order of a full revolution before relaxing, on the other hand, the atomic polarization begins to average out, reducing the signal. The shape of the resonance is antisymmetric simply because when the magnetic field is reversed, the atoms precess in the opposite direction. The spectral lineshape is also described by the rotating-polarizer model: both the pumping and probing stages are absorptive processes, so the lineshape is the product of the two absorptive Lorentzians.

> In our discussion above, we associate nonlinear magneto-optical rotation with linear dichroism of the medium. Yet, in Eq. (10.59c) we see that the rotation is proportional to the real part of the sum of optical coherences corresponding to the two circular components. This form is suggestive of the difference in refractive indices for the two circular components (as for linear magneto-optical rotation). So do we have linear dichroism or circular birefringence in the nonlinear case? (This question, not too long ago, was a subject of debate among researchers.)

Mathematically, the description of the system can be carried out in any complete basis, so on one hand both approaches are correct. On the other hand, in terms of the physical picture there is a clear distinction. The intuitive definition of the complex index of refraction is based on the idea that, when light of a particular polarization propagates through the medium, the medium induces changes in the amplitude and phase of the light. This concept only makes sense, however, if the polarization of the light in this mode does not change upon traversal of the medium, i.e., if this particular polarization is a *polarization eigenstate* of the medium. Otherwise, the index of refraction for this polarization mode will depend on the polarization of the total light field. This means that, in order to describe a process in a physically meaningful way, we must confine ourselves to considering the indices of refraction of the polarization eigenstates. The polarization eigenstates in the case of low-power NMOR are two orthogonal linear polarizations, and so the appropriate description is in terms of linear dichroism.

Despite the success of the rotating-polarizer model for low-power NMOR, it turns out that it fails at higher light powers. For $\kappa_2 \gtrsim 1$ the effect of the AC-Stark shifts due to the light (Sec. 4.5.3) in combination with the magnetic field can induce *alignment-to-orientation conversion (AOC)*, previously discussed in Sec. 5.6. Atomic orientation does not result in linear dichroism. However, it does produce circular birefringence, which, as we saw in Sec. 10.2.6, can also induce optical rotation. This process will be discussed in more detail in Sec. 11.3. We note here that in certain situations rotation due to AOC can have very different characteristics than low-power rotation. In particular, for $F \to F' = F + 1$ transitions the sign of rotation due to AOC is opposite to that predicted by the rotating-polarizer model. For the $F = 1 \to F' = 0$ transition under consideration, however, the effect of AOC is not striking; the main effect of high light power is broadening of the spectral and magnetic resonances due to effective relaxation induced by optical pumping. We now analyze this effect, known as *power broadening*, for the $F = 1 \to F' = 0$ system.

When the optical pumping rate $\Gamma_p = \Omega_R^2/\Gamma$ is small, transit relaxation is the dominant ground-state relaxation mechanism, and the additional relaxation induced by the light itself can be ignored. However, when the optical pumping rate is comparable to or greater than the transit rate γ, i.e., when $\Gamma_p/\gamma = \kappa_2 \gtrsim 1$, polarized atoms may have their polarization destroyed by an additional optical pumping cycle before they have a chance to relax due to transit relaxation. This will only occur if the polarization of the atom is different than that produced by the light; otherwise optical pumping will have no apparent effect.

We can estimate the relaxation rate due to optical pumping as the pumping rate $\Gamma_p = \gamma \kappa_2$. If this relaxation were isotropic, as is transit relaxation, i.e., if it caused atoms to relax at the same rate no matter their polarization, we could add this relaxation rate to the transit rate to obtain the total relaxation rate $\gamma(1 + \kappa_2)$. Because the width of the magnetic-field resonance in NMOR is given by the ground-state relaxation rate, this would then be the estimate of the power broadened width.

In fact, optical-pumping-induced relaxation is not isotropic, as noted above. In our case we can assume that the x-polarized light relaxes atomic polarization that is along the y-axis, but not that along x. This means that power broadening has a somewhat weaker dependence on κ_2. A simple rate-equation model that keeps track of atomic polarization along x and y, relaxation due to transit and the light, and the magnetic-field induced precession gives the power-broadened width as $\gamma\sqrt{1 + \kappa_2}$. This is precisely the result obtained from the full density-matrix calculation for the case of an open system and with detuning set to zero. (An open system is slightly simpler than the closed system we have been considering, because there is only one optical pumping mechanism, depopulation pumping, rather than both depopulation and repopulation pumping.)

For the system considered in this section, we can obtain the power-broadened width of the NMOR resonance from formula (10.67), which is valid for arbitrary saturation parameter. First note that, according to this formula, the magnetic-field dependence of the optical rotation is a dispersive Lorentzian for any value of the saturation parameter, i.e., it takes the form

$$\frac{abu}{u^2+b^2}, \qquad (10.69)$$

where the peak-to-peak height is a, and the width measured between the two peaks is $2b$. The slope at $u=0$ is a/b. Comparing Eq. (10.67) with the functional form (10.69), we find that the peak-to-peak height is

$$\ell_0 \frac{d\alpha}{d\ell}\bigg|_{\text{p-p}} = \frac{\kappa_2(\kappa_2+2)}{8\sqrt{\left(v^2+1+\frac{\kappa_2}{2}\right)\left(v^2+1+\frac{\kappa_2}{6}\right)\left(v^2+1+\frac{2\kappa_2}{3}\right)\left[v^2+(1+\frac{\kappa_2}{2})^2\right]}}, \qquad (10.70)$$

and the power-broadened width is

$$\Delta u|_{\text{p-p}} = 2\sqrt{\frac{\left(v^2+1+\frac{2\kappa_2}{3}\right)\left[v^2+(1+\frac{\kappa_2}{2})^2\right]}{\left(v^2+1+\frac{\kappa_2}{2}\right)\left(v^2+1+\frac{\kappa_2}{6}\right)}}. \qquad (10.71)$$

These quantities are plotted for $v=0$ in Fig. 10.9.

The sharp dependence of the nonlinear Faraday rotation signal on the magnetic field finds application in sensitive *atomic magnetometers*. To increase sensitivity to magnetic fields, it is generally desirable to maximize the slope of the signal as a function of magnetic field. For small fields near zero, the slope is given by

$$\frac{d}{du}\left(\ell_0 \frac{d\alpha}{d\ell}\right)_{u=0} = \frac{\kappa_2(\kappa_2+2)}{8\left(v^2+1+\frac{2\kappa_2}{3}\right)\left[v^2+(1+\frac{\kappa_2}{2})^2\right]}. \qquad (10.72)$$

By maximizing with respect to κ_2, or by looking at Fig. 10.9, we see that to achieve the highest slope the saturation parameter should be of order unity.

10.2.8 Doppler broadening

So far we have assumed that the atoms are motionless, or at least have no velocity component along the light propagation direction. This will not be the case in a real experiment unless cooling and trapping techniques are employed. In a vapor cell, atomic velocities will be distributed according to the Maxwell–Boltzmann distribution, which, for a component v of the velocity, takes the form

$$f_v(v)dv = \left(\frac{m}{2\pi k_B T}\right)^{1/2} e^{-mv^2/(2k_B T)} dv, \qquad (10.73)$$

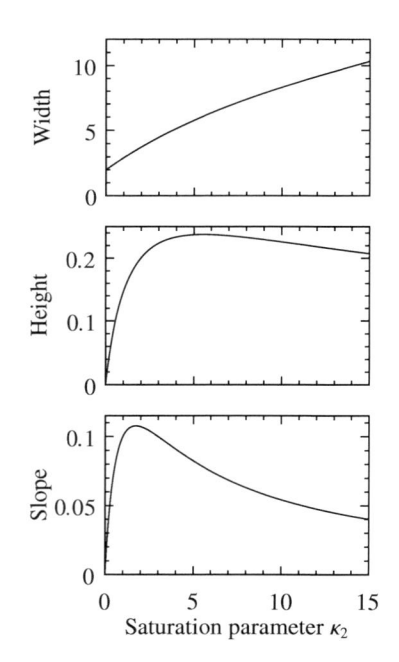

Fig. 10.9 Peak-to-peak width $\Delta u|_{\text{p-p}}$, height $\ell_0 \frac{d\alpha}{d\ell}\big|_{\text{p-p}}$ and small-field slope $\frac{d}{du}\left(\ell_0 \frac{d\alpha}{d\ell}\right)_{u=0}$ of NMOR for zero detuning as a function of saturation parameter κ_2.

where m is the atomic mass, k_B is the Boltzmann constant, and T is the temperature.

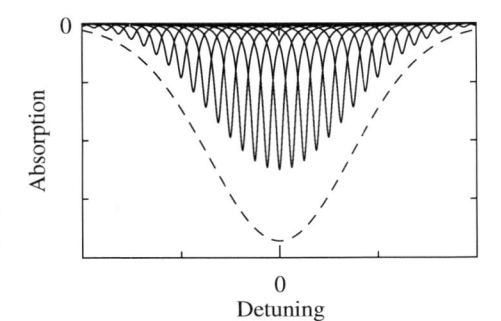

Fig. 10.10 Convolution of the Doppler-free linear absorption spectrum with the Gaussian Doppler profile. The solid lines show the contributions of individual velocity groups to the Doppler-broadened spectrum (dashed line).

We now determine the effect of the atomic velocity on the optical signals. Suppose that an atom is stationary at position $z = z_0$, with the light propagating along the z-axis. The optical electric field viewed at the location of the atom has time dependence $\sin(\omega t - kz_0)$, neglecting an overall phase, where ω is the light frequency and k is the light wave number. This is the dependence used in our previous calculation, where we have set $z_0 = 0$. Now suppose the atom is not stationary, but rather moves with a constant velocity v, so that its longitudinal position is $z(t) = z_0 + vt$. The electric field of the light as viewed at the position of the atom now has time dependence $\sin[\omega t - kz(t)] = \sin[(\omega - kv)t - kz_0]$. This shows that the light frequency has been replaced by an effective light frequency $\omega - kv$.

The shift $\Delta_v = -kv$ is the *non-relativistic Doppler shift* due to the atomic velocity. The calculation of the optical signals for an atom of velocity v goes through exactly as for a stationary atom, with the replacement $\Delta \to \Delta + \Delta_v$. The Doppler-broadened signals can be found by integrating the Doppler-shifted signal over velocity, after weighting with the Maxwell–Boltzmann distribution. This procedure relies on the assumption that the vapor cell does not contain buffer gas and is not antirelaxation coated, so that atoms cannot change their velocity without losing their polarization. (Velocity-mixing effects are considered in Sec. 11.5.) In addition, we neglect subtle effects such as the correlation of transit rate with velocity.

It is convenient to write the Maxwell–Boltzmann distribution in terms of the Doppler shift. This gives

$$f_{\Delta_v}(\Delta_v)d\Delta_v = \left(\frac{m}{2\pi k^2 k_B T}\right)^{1/2} e^{-m\Delta_v^2/(2k^2 k_B T)} d\Delta_v$$
$$= \frac{1}{\Gamma_D \sqrt{\pi}} e^{-\Delta_v^2/\Gamma_D^2} d\Delta_v, \quad (10.74)$$

where $\Gamma_D = k\sqrt{2k_B T/m}$ is the *Doppler width* (note that there are other definitions of the Doppler width that differ by numerical factors from this one). The Doppler width is the characteristic spectral width of optical signals for Doppler-broadened media.

If we designate one of the optical signals (10.60) by $S(\Delta)$, the Doppler-broadened signal is given by the integral

$$S_{\text{DB}}(\Delta) = \int_{-\infty}^{\infty} S(\Delta + \Delta_v) f_{\Delta_v}(\Delta_v) d\Delta_v, \quad (10.75)$$

i.e., the *convolution* of S and f_{Δ_v}. The procedure of convolution is illustrated for the case of linear absorption in Fig. 10.10. The magnetic field is set to zero, and the Doppler width is assumed to be 10 times the natural width. The signal at each detuning Δ is the total absorption

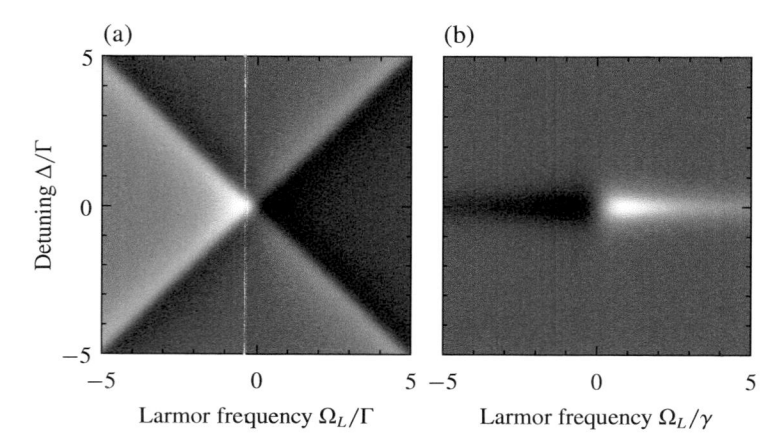

212 Light–atom interaction observed in transmitted light

Fig. 10.11 Density plots of the (a) linear (Eq. 10.64c) and (b) nonlinear (Eq. 10.67) Doppler-free magneto-optical rotation signals as a function of detuning and Larmor frequency. Note the different horizontal scales. In part (b), the saturation parameter has been chosen as $\kappa_2 = 2$.

from atoms of all velocities (velocity groups). The contributions of a representative sample of velocity groups are shown as solid curves, and the integral is shown as a dashed line.

It is clear that the main effect of Doppler broadening is simply to increase the spectral linewidth from $\sim\Gamma$ to $\sim\Gamma_D$. There are some additional consequences of Doppler broadening for the linear effects, however. The reason for this can be illustrated by 2D plots of the optical signals, such as those in Fig. 10.11, which compares magneto-optical rotation in the linear and nonlinear cases.

In Fig. 10.11(a), linear optical rotation as a function of detuning and Larmor frequency is displayed as a density plot, with white representing the largest positive signal and black the most negative signal. Because of the symmetry between detuning and the Larmor frequency for the linear effects discussed in Sec. 10.2.6, the signal peaks appear as diagonal features in the plot. Spectral convolution with the Doppler distribution will blur this plot in the vertical direction. However, because the features lie along a diagonal, blurring in the vertical direction will also effectively blur the features in the horizontal direction. As a result, not only is the spectral linewidth broadened, but the *magnetic* resonance linewidth is also broadened from Γ to Γ_D for the linear effect. This makes sense, as the original explanation for the linear effects involved splitting of the spectral resonances—if the spectral linewidths are broader, a larger magnetic field is needed in order to split them.

Another aspect of the linear signal plot to note is that there are positive (light) and negative (dark) signals along any given vertical axis. When blurring is performed along the vertical direction, the light and dark areas will overlap and cancel, reducing the magnitude of the signal. In general, if the integral of the Doppler-free spectrum is zero, the magnitude of the Doppler-broadened signal will be suppressed by Γ/Γ_D compared to a signal whose Doppler-free spectrum is all of one sign.

For comparison, nonlinear magneto-optical rotation is plotted in Fig. 10.11(b). In this case, the features lie along the horizontal and vertical axes, so that when the plot is blurred in the vertical direction, the magnetic-resonance feature in the horizontal direction is not broadened. In addition, each spectrum (along a vertical axis) contains only a positive or negative signal, so there is no cancelation induced by the blurring. This is one of the factors that makes NMOR an attractive measurement tool for thermal gases. Note, however, that in more complicated systems, such as those with hyperfine structure, there can be cancelations in the Doppler-broadened signal. The character of the Doppler-broadened spectrum can also be strongly affected by velocity-changing atomic collisions. These phenomena will be discussed in detail in Chapter 14.

The brief discussion of the nonlinear magneto-optical effects that we have given here will be continued and further developed in Part II of the book.

10.3 Perturbative approach

In Sec. 10.2 we were able to distinguish between linear and nonlinear effects based on their dependence on the magnetic field. We saw that the linear effects result in magnetic-field resonances with widths determined by the upper-state decay rate, whereas the resonance widths for the nonlinear effects are determined by the ground-state decay rate. We may, however, be interested in a more systematic classification of the linear and nonlinear effects produced by the system, one that does not depend on the knowledge of the resonance widths and that can distinguish between nonlinear effects of various orders.

In Sec. 9.1 we differentiated between linear and nonlinear effects based on the response of the optical properties of the atomic medium to the electric field of the light. This idea is the basis of the perturbative method for solving the density-matrix equations, which we discuss here. This method allows the effects that appear at any particular order in the electric field to be determined, the lowest-order of which are the linear effects. A practical motivation for using the perturbative method is that it allows low-order analytic solutions to be obtained for systems for which solving the complete nonperturbative density-matrix equations is not feasible. (As the atomic system is increased in size beyond the $F = 1 \to F' = 0$ system considered in Sec. 10.2, the solution of the full evolution equations rapidly increases in complexity.)

As we saw in Sec. 10.1, the optical properties of an atomic medium are determined by the atomic density matrix ρ, through the medium polarization $\mathbf{P} = n \operatorname{Tr} \rho \mathbf{d}$. When light interacts with the medium, the density matrix acquires a dependence on the optical field: $\rho = \rho(\mathcal{E})$. To analyze this dependence, we expand the density matrix in a power series[3]

$$\rho(\mathcal{E}) = \sum_{n=0}^{\infty} \rho^{(n)} \mathcal{E}_0^n, \qquad (10.76)$$

where the $\rho^{(n)}$ are independent of the magnitude $\mathcal{E}_0$ of the electric field. The zeroth-order term $\rho^{(0)}$ describes the density matrix in the absence of light and is not responsible for any optical effects. What about the first-order term $\rho^{(1)} \mathcal{E}_0$? Examination of the formulas (10.10) for the change in the optical parameters in the medium shows that, because of the factor of $\mathcal{E}_0$ in the denominators of these expressions, the dependence on the electric-field amplitude cancels when the first-order term is substituted in. Thus $\rho^{(1)}$ is responsible for optical effects that are independent of the magnitude of the electric field, i.e., linear effects. The higher-order terms are responsible for nonlinear effects.

---

[3] A detailed discussion of this expansion can be found, for example, in the books by Stenholm (2005) and Boyd (2008).

214 *Light–atom interaction observed in transmitted light*

To see how the various orders of ρ are coupled, we substitute the expansion (10.76) into the Liouville equation

$$\dot{\rho} = -\frac{1}{2}\left(\widehat{\Gamma}\rho + \rho\widehat{\Gamma}\right) - \frac{i}{\hbar}[H,\rho] + \Lambda, \qquad (10.77)$$

where the Hamiltonian $H = H_0 + H_l$ is assumed to consist of a diagonal unperturbed Hamiltonian H_0 in addition to the optical perturbation $H_l = -\mathbf{d}\cdot\mathcal{E}$.

> For simplicity, we will neglect the fact that the repopulation matrix Λ is ordinarily a function of ρ due to the term describing spontaneous decay. (Other processes, such as spin-exchange collisions, can also cause Λ to depend on ρ.) In Sec. 12.2 a perturbative calculation is performed that includes the effect of repopulation due to spontaneous decay.

If we consider the time evolution of the n-th order term $\rho^{(n)}$, we see that it must depend only on terms in the right-hand side that are also n-th order in the electric field. Since the light–atom interaction Hamiltonian is linear in $\mathcal{E}_0$ and all the other quantities are independent of $\mathcal{E}_0$, we have

$$\dot{\rho}^{(n)} = -\frac{1}{2}\left(\widehat{\Gamma}\rho^{(n)} + \rho^{(n)}\widehat{\Gamma}\right) - \frac{i}{\hbar}\left[H_0, \rho^{(n)}\right] - \frac{i}{\hbar}\left[-\mathbf{d}\cdot\hat{\mathcal{E}}, \rho^{(n-1)}\right] + \delta_{n,0}\Lambda, \qquad (10.78)$$

where $\hat{\mathcal{E}} = \mathcal{E}/\mathcal{E}_0$ and Λ is only present in the equation for $\dot{\rho}^{(0)}$.

Taking the matrix element $\dot{\rho}^{(n)}_{ab} = \langle a|\dot{\rho}^{(n)}|b\rangle$ and using the fact that H_0 and $\widehat{\Gamma}$ are diagonal, we have

$$\dot{\rho}^{(n)}_{ab} = -(i\omega_{ab} + \gamma_{ab})\rho^{(n)}_{ab} - \frac{i}{\hbar}\left[-\mathbf{d}\cdot\hat{\mathcal{E}}, \rho^{(n-1)}\right]_{ab} + \delta_{n,0}\Lambda_{ab}, \qquad (10.79)$$

where $\hbar\omega_{ab} = \langle a|H_0|a\rangle - \langle b|H_0|b\rangle$ is the energy difference between levels a and b, and $\gamma_{ab} = (\Gamma_a + \Gamma_b)/2$ is the average of the relaxation rates of the levels a and b. The density matrix ρ of a given order $\rho^{(n)}$ can thus be expressed in terms of the elements $\rho^{(n-1)}$ of the next lower order.

To further characterize each order of ρ, consider the matrix element of the commutator,

$$\left[-\mathbf{d}\cdot\hat{\mathcal{E}}, \rho^{(n-1)}\right]_{ab} = -\sum_c \left(\mathbf{d}_{ac}\cdot\hat{\mathcal{E}}\rho^{(n-1)}_{cb} - \rho^{(n-1)}_{ac}\mathbf{d}_{cb}\cdot\hat{\mathcal{E}}\right). \qquad (10.80)$$

Suppose that $\rho^{(n-1)}$ does not contain any optical coherences (i.e., coherences between the opposite-parity lower and upper states of the optical transition) and only contains populations and coherences between states of the same parity. Because $\mathbf{d}$ only couples opposite-parity states, we see that this term of the Liouville equation is zero unless $|a\rangle$ and $|b\rangle$ are of opposite parity (ρ_{ab} represents an optical coherence). Conversely, if $\rho^{(n-1)}$ consists entirely of optical coherences, the term is zero unless ρ_{ab} is not an optical coherence.

The source term for the zeroth-order expansion coefficient $\rho^{(0)}$ is Λ, which we can assume does not contain any optical coherences. From the above reasoning, $\rho^{(0)}$ is the source term for the first-order term $\rho^{(1)}$, which consists entirely of optical coherences, $\rho^{(1)}$ is the source for $\rho^{(2)}$, which has no optical coherences, and so on. In general, odd orders of ρ describe optical coherences and even orders describe density-matrix elements between states with the same parity (for an $F \to F'$ transition these are level populations and Zeeman coherences.

The situation is diagrammed in Fig. 10.12 for the lowest terms in the perturbative expansion of the density matrix. We assume that the atoms are in the ground state in the absence of light. This is indicated by using unprimed indices a, b for the zeroth-order term $\rho^{(0)}$. The density

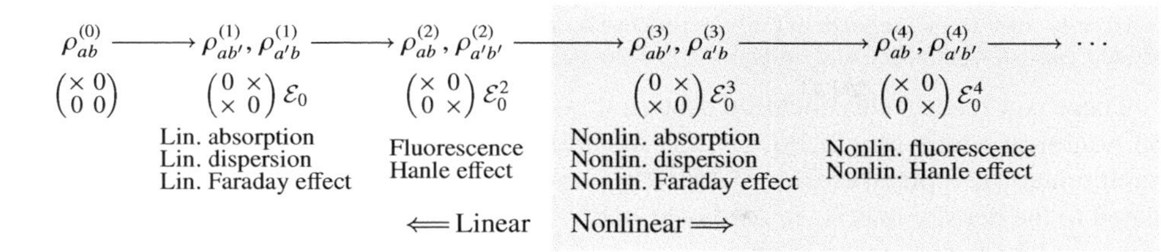

Perturbative approach 215

$$\rho_{ab}^{(0)} \longrightarrow \rho_{ab'}^{(1)}, \rho_{a'b}^{(1)} \longrightarrow \rho_{ab}^{(2)}, \rho_{a'b'}^{(2)} \longrightarrow \rho_{ab'}^{(3)}, \rho_{a'b}^{(3)} \longrightarrow \rho_{ab}^{(4)}, \rho_{a'b'}^{(4)} \longrightarrow \cdots$$

$$\begin{pmatrix} \times & 0 \\ 0 & 0 \end{pmatrix} \quad \begin{pmatrix} 0 & \times \\ \times & 0 \end{pmatrix} \mathcal{E}_0 \quad \begin{pmatrix} \times & 0 \\ 0 & \times \end{pmatrix} \mathcal{E}_0^2 \quad \begin{pmatrix} 0 & \times \\ \times & 0 \end{pmatrix} \mathcal{E}_0^3 \quad \begin{pmatrix} \times & 0 \\ 0 & \times \end{pmatrix} \mathcal{E}_0^4$$

Lin. absorption
Lin. dispersion   Fluorescence   Nonlin. absorption   Nonlin. fluorescence
Lin. Faraday effect Hanle effect   Nonlin. dispersion   Nonlin. Hanle effect
                                         Nonlin. Faraday effect

⇐ Linear   Nonlinear ⇒

Fig. 10.12 Terms in the perturbative expansion of the density matrix for atoms interacting with an optical field.

matrix corresponding to this term is also shown; the cross indicates a nonzero matrix element for the ground state. In general, the ground state has sublevel structure—in this case the cross can be interpreted as representing a submatrix with at least one nonzero entry.

The lowest-order interaction of the light field with the zeroth-order density matrix gives rise to the first-order density matrix $\rho^{(1)}$. The matrix elements here are written with primed (indicating the upper state) and unprimed indices to show that $\rho^{(1)}$ consists of optical coherences. This is also shown by the schematic first-order density matrix (with explicit electric-field dependence). These optical coherences correspond to dipole polarization that is linear in the electric field. As discussed above, this polarization is responsible for the linear effects described in Sec. 10.2.6, some of which are listed in the figure.

In the next order of perturbation, $\rho^{(1)}$ is the source for the second-order term $\rho^{(2)}$, consisting of populations and coherences within the ground and excited states. This is the lowest order at which atoms appear in the upper state and angular-momentum (Zeeman) polarization is created in the ground state. This term is responsible for fluorescence and quantum interference effects (e.g., quantum beats), excited-state level crossings and the Hanle effect. (Some of these effects were discussed in Chapter 8 in the context of experiments with two light fields—in that case, the effects are generally considered to be of higher order). While the density matrix due to this term is quadratic in the electric field, the optical coherences through this order are still linear in the electric field, because they arise entirely from $\rho^{(1)}$. Because the optical properties of the medium are determined by the optical coherences, the observable effects generated at the second order of perturbation theory are still generally considered linear.

The next order, $\rho^{(3)}$, again consists of optical coherences and is the lowest order in which a *nonlinear polarization* at the optical frequency appears. It is responsible for the onset of the nonlinear effects discussed in Sec. 10.2.7, such as the nonlinear Faraday effect. Because the optical coherences are proportional to $\mathcal{E}_0^3$, the optical signals are quadratic in the optical field, or linear in the light intensity. This explains the possibly counter-intuitive fact that the lowest-order nonlinear effects are linear in light intensity. Finally, the appearance of $\rho^{(4)}$ marks the onset of nonlinear fluorescence and the nonlinear Hanle effect.

To summarize, we have adopted the following commonly used classification system: processes in which the signals are proportional to the matrix elements of $\rho^{(1)}$ and $\rho^{(2)}$ are called *linear processes*, while all higher-order processes are referred to as *nonlinear processes*.

> The formalism of *nonlinear wave mixing* (Shen 1984; Boyd 2008) allows one to describe *linear* electro- and magneto-optical effects as three- or four-wave-mixing processes, in which at least one of the mixing waves is at zero frequency. For example, linear Faraday rotation is seen in this picture as mixing of the incident linearly polarized optical field with a zero-frequency magnetic field to produce an optical field of orthogonal polarization. While this is a perfectly consistent view, we prefer to adhere to a more common practice of calling these processes linear. In fact, in this book, the term *nonlinear*

will only refer to the dependence of the process on *optical* fields. Thus, we also consider as linear a broad class of optical-RF and optical-microwave double-resonance phenomena.

There is a related classification system that is often used to distinguish between linear and nonlinear processes. In Eq. (10.13) we introduced the electric susceptibility tensor $\overleftrightarrow{\chi}$, which relates the optical electric field to the complex electric polarization $\widetilde{\mathbf{P}}$, which is itself related to the density matrix. In nonlinear optics it is common to perform the expansion of $\widetilde{\mathbf{P}}$ corresponding to the expansion of ρ:

$$\widetilde{\mathbf{P}} = \sum_{n=1}^{\infty} \widetilde{\mathbf{P}}^{(n)} = \sum_{n=1}^{\infty} \overleftrightarrow{\chi}^{(n)} \cdot \widetilde{\mathcal{E}}^n, \tag{10.81}$$

where the $\overleftrightarrow{\chi}^{(n)}$ are the n-th order electric susceptibilities. From the above discussion, we find that only the odd orders of $\overleftrightarrow{\chi}^{(n)}$ are nonzero, and that $\overleftrightarrow{\chi}^{(1)}$ processes are linear, while $\overleftrightarrow{\chi}^{(3)}$ and higher-order processes are nonlinear.

The perturbative approach is useful for understanding the onset of a given effect with increasing light power. It works well in most cases involving nonresonant light fields. In the case of resonant laser excitation, however, caution must be used, since there is no way to make meaningful distinctions between different perturbative orders once the transition is saturated [the series (10.76) and (10.81) are non-converging and all orders of these expansions are coupled].

This discussion can be generalized to the case of experiments with multiple light beams. One arrangement that we will consider is a pump-probe measurement of nonlinear magneto-optical rotation using a weak probe field. In this case the pump beam creates ground-state polarization at second order ($\rho^{(2)}$), and this polarization influences the optical coherences produced by the probe at first order ($\rho^{(1)}$). Thus the observed medium polarization depends on two factors of pump and one factor of probe electric field, corresponding to a $\overleftrightarrow{\chi}^{(3)}$ process. This is the same order at which NMOR appears in the single-beam case. On the other hand, the two-beam fluorescence experiments discussed in Chapter 8 require two excitation steps, each of which is second order. Thus the observed effect is fourth order in the electric field, which is higher order than the lowest-order fluorescence and Hanle effects produced by a single beam.

Part II

Advanced topics

11
Nonlinear magneto-optical rotation

In Chapter 10 we discussed magneto-optical rotation as an example of a light–atom interaction effect that can be observed in transmitted light. We concentrated in particular on the linear effect and the low-power, polarization-dependent nonlinear effect. In this chapter, we extend this discussion to include the effect of light-induced *Bennett structures* in the atomic velocity distribution and the high-light-power process of alignment-to-orientation conversion. We also examine the effects of atomic collisions that change the atoms' velocity but preserve their polarization. We will see that such collisions have a significant effect of the spectra of the polarization-dependent effects; in order to account for such collisions, it is necessary to perform Doppler averaging in a manner different from that discussed in Chapter 10. Finally, we discuss experiments with optically thick atomic media, and experiments using the technique of modulated light.

11.1 Nested nonlinear magneto-optical rotation features

In the early days of NMOR studies, there was considerable confusion in the literature as to the physical origin of the nonlinear-optical-rotation effects observed in various experiments. In order to clarify the situation, Barkov *et al.* (1989) measured magneto-optical rotation on several low-angular-momentum transitions of atomic samarium (Sm). Three features of different widths centered at zero magnetic field were observed in the magnetic-field dependence of optical rotation, one nested inside the other (Fig. 11.1). The broadest feature, with a characteristic width determined by the Doppler width, was identified as the linear effect (Sec. 10.2.6), and

Fig. 11.1 Density-matrix calculation (Budker *et al.* 2002b) of light-polarization rotation as a function of magnetic field for the case of the samarium 571-nm ($J = 1 \rightarrow J' = 0$) line, closely reproducing experimental results described by Barkov *et al.* (1989); see also Zetie *et al.* (1992).

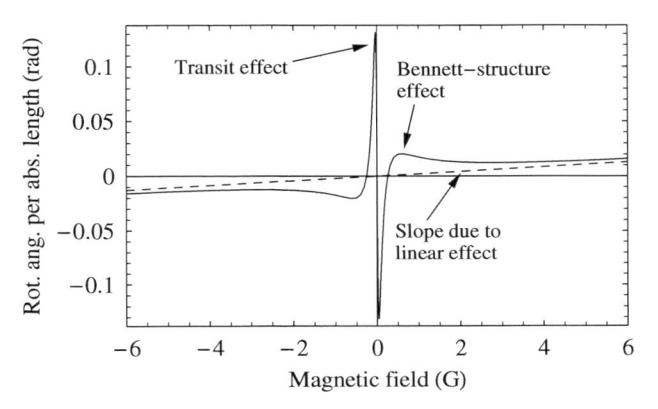

the narrowest, with characteristic width determined by the transit width, as the nonlinear atomic-alignment effect discussed in Sec. 10.2.7. Intermediate between the two, there was an additional nonlinear effect with characteristic width determined by the natural width of the transition. The mechanism responsible for this feature was identified as Bennett-structure formation in the atomic velocity distribution, as will be discussed in the next section.

As the narrowest of the features, the atomic-alignment effect produces the largest rotation for small magnetic fields [Barkov *et al.* (1989) observed nonlinear rotation $\sim 10^4$ times larger than that due to the linear effect using a buffer-gas-free metal-vapor cell]. Note, however, that this is not the end of the story. As we will see later in this chapter, the nested-NMOR-feature plot takes a dramatic further twist when an antirelaxation coating is applied to the inner walls of an alkali-vapor cell: another feature appears, with width—determined by the ground-state-polarization decay rate—that can be many orders of magnitude narrower still. This effect will be discussed in Sec. 11.5.

> The experimental results of Barkov *et al.* (1989) were subsequently compared to theoretical predictions, showing good agreement (Kozlov 1989; Zetie *et al.* 1992). At about the same time, precision measurements and detailed calculations of the nonlinear Faraday effect on the cesium $D2$ line were performed (Chen *et al.* 1990; Weis *et al.* 1993; Kanorsky *et al.* 1993). Because Cs has hyperfine structure and high angular momentum, the situation was somewhat more complicated than for transitions between low-angular-momentum states of the nuclear-spinless isotopes of Sm. Nevertheless, excellent quantitative agreement between theory and experiment was obtained.

11.2 Bennett-structure effects

Consider an atomic vapor, and a particular transition from the atoms' ground state. The atoms have a Maxwell–Boltzmann thermal velocity distribution. Because of the Doppler shift, the transition frequency is different for atoms with different velocities along a given axis. If a laser beam is directed through the vapor, only a particular set of atoms will have the correct velocity to put them in resonance with the narrow-band light. If the excited state decays to some state other than the initial one, then optical pumping will remove the atoms in the resonant velocity group from the ground state. If we plot the ground-state population as a function of velocity along the light propagation axis, we see a "hole" burned in the distribution, with width equal (neglecting power-broadening effects) to the resonance width of the transition, i.e., the natural width (Fig. 11.2a).

The hole in the velocity distribution, known as a *Bennett hole*, can be observed by its effect on the absorption and index of refraction of the medium. If the frequency of a probe laser beam is scanned across the resonance, the influence of the medium will clearly be reduced for probe light at the frequency corresponding to the hole. In Sec. 10.2.6 we saw that in the presence of a magnetic field, the spectra of the real part of the index of refraction for σ^+ and σ^- light are shifted in opposite directions, leading to optical rotation (the linear effect). For an unperturbed Doppler-broadened medium, the spectrum of the index of refraction has characteristic width Γ_D, leading to an optical rotation spectrum of the corresponding width. A narrow hole in the Doppler distribution, however, can be split with a much smaller magnetic field. Thus, in the magnetic-field spectrum, the magnetic-resonance feature due to the hole is nested inside that of the linear effect. Although the mechanism that produces the optical rotation is the same as that for the linear effect, the *hole-burning* effect requires optical pumping to produce the hole, and is therefore nonlinear.

Note that the hole-burning effect requires that the system be open. For a closed transition, excited atoms will decay back to the ground state, filling in the hole. Furthermore, if the same

The role of alignment-to-orientation conversion in nonlinear magneto-optical rotation   221

Fig. 11.2 (a) A hole burned in the ground-state-atom velocity distribution as a result of optical pumping on an open transition. The Doppler distribution is a Gaussian of width Γ_D. Atoms Doppler shifted into resonance with the light frequency ω_L are transferred out of the ground state, leaving a hole of width Γ, the natural width of the transition. (b) When a magnetic field is applied n_+ and n_- are shifted relative to each other (upper trace). The real part of the indices of refraction are shown, so that the features in plot (a) correspond to dispersive shapes in this plot. The centers of the narrow features are indicated by dashed lines. Polarization rotation of probe laser light is proportional to the difference Re $(n_+ - n_-)$ (lower trace). Features due to the Doppler distribution and the Bennett holes can be seen. Since the Bennett-related feature is caused by the removal of atoms from the Doppler distribution, the sign of rotation due to this effect is opposite to that of the linear rotation.

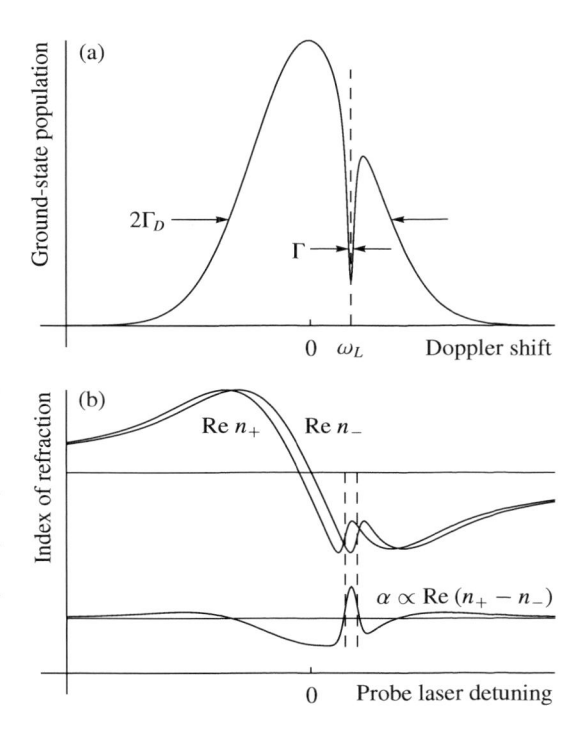

magnetic field is present during both optical pumping and probing, the Bennett holes will be created, but they will not produce optical rotation. This is because the effect requires that the σ^+ and σ^- features be shifted relative to each other after they are created.

It would seem that the hole-burning effect only occurs under a specialized set of circumstances. However, in the more common case of a closed transition *and* a magnetic field present during both pumping and probing, a related effect can occur, involving not holes but *peaks* in the velocity distribution. The peaks are created by excited atoms decaying to various Zeeman-shifted ground-state sublevels. The mechanism for producing optical rotation is the same, although the fact that the features are peaks, rather than holes, means that the rotation is of the opposite sign. The hole-burning and peak-building effects are together known as Bennett-structure effects, and are analyzed in more detail by Budker *et al.* (2002b).

Light can also modify the atomic velocity distribution by actually changing the atomic velocities through radiation pressure. Under certain conditions this can lead to redistribution of atoms among velocity groups, thus deforming the Bennett structures and modifying the nonlinear optical properties of the medium.

11.3 The role of alignment-to-orientation conversion in nonlinear magneto-optical rotation

In Sec. 5.6 we considered an electric field applied to an oriented atomic system. We saw that the electric field can induce quantum beats that produce alignment in the atomic system (Fig. 5.5). Because the polarization state oscillates, we can equally well say that if we begin with an aligned atomic system (e.g., that shown in the third panel of Fig. 5.5, for $t = \tau_S/4$), it can be converted to an oriented one. We refer to this process as alignment-to-orientation conversion

(AOC). Note that the orientation is produced in a direction that is perpendicular to both the electric field and the atomic-alignment axis, so if the electric field is along the alignment axis, the effect will not occur.

It turns out that, under certain conditions, a very similar process plays a crucially important role in NMOR. In a standard NMOR experiment, a static magnetic field and a linearly polarized optical field are applied to the atomic system. At first glance, it might seem that the electric field necessary to produce AOC is not present. However, the oscillating electric field present in the optical field can produce the same effect if the optical-field strength is high enough.

As indicated in the discussion of in Secs. 4.5 and 10.2.2, under the rotating-wave approximation the effect of an AC electric field is similar to that of a static field. The main difference is that level mixing due to an AC field depends on the detuning of the optical frequency from resonance, rather than the energy splitting between the levels; if the field is precisely on resonance, no shifts will occur. As discussed in detail in Chapter 19, the AC field produces the same kinds of tensor splitting of the ground-state Zeeman sublevels as does the static field. This means that this field can induce quantum beats that convert alignment to orientation.

In an NMOR experiment, the magnetic field, in addition to the electric field, is required for AOC to occur. This is because the linearly polarized light creates atomic alignment with an axis along the electric-field direction, and in the absence of the magnetic field, there is nothing to break the symmetry about this axis. However, when a magnetic field is applied along the light propagation axis, the alignment axis precesses away from the light polarization axis, and AOC can then take place. Since the magnetic field is perpendicular to both the light polarization axis and the alignment axis, orientation will be produced either along or opposite to the magnetic-field direction, depending on the sign of the g-factor and whether the light is tuned above or below resonance.

Figure 11.3(a) shows the ground-state atomic polarization produced in an NMOR experiment with low light power, as described in Sec. 10.2.7. A $F = 1 \to F' = 1$ transition subject to x-polarized light is assumed. The first frame shows the atomic alignment produced by the light prior to the application of the magnetic field. The z-directed magnetic field **B** is then turned on, and the resulting Larmor precession of the atomic polarization is shown in the second and third frames.

The situation for higher light power is shown in Fig. 11.3(b). Here the light detuning and intensity are chosen so that the AC-Stark frequency shifts due to the optical electric field **E** are comparable to the Zeeman splitting and the ground-state polarization relaxation rate. Prior to the application of the magnetic field, the atomic alignment axis is along the electric field axis, so quantum beats do not occur (first frame). When the magnetic field is applied, the atomic alignment begins to precess and gains a component along the y-axis. This allows AOC to occur, producing orientation in the z direction (second and third frames).

As discussed in Sec. 10.2.7, optical rotation caused by the low-power atomic alignment effect shown in Fig. 11.3(a) is due to linear dichroism, i.e., a difference in absorption for light of orthogonal linear polarizations. The atomic orientation produced in the high-power effect, on the other hand, is symmetric about the light propagation axis, and so does not produce linear dichroism. (Furthermore, as shown in Fig. 11.3(b), the precession of atomic alignment away from the light-polarization direction is suppressed.) However, an atomic ensemble oriented along the light-propagation direction interacts with the two *circular* light polarizations differently. This is clear if we remember that orientation in the z direction corresponds to different populations for the $m = \pm 1$ Zeeman sublevels (Fig. 11.4). As we saw for the linear

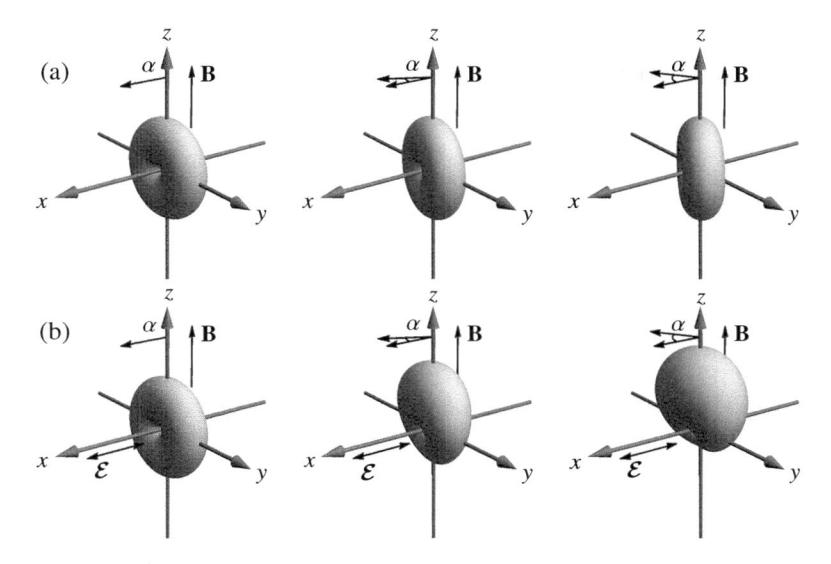

Fig. 11.3 (a) Sequence showing the evolution of the AMPS corresponding to optically pumped ground-state atomic alignment in a longitudinal magnetic field for an $F = 1 \rightarrow F' = 1$ transition at low light powers (time proceeds from left to right). In the first plot, the atoms have been optically pumped into an aligned state by x-polarized light, which then precesses in the magnetic field (second and third plots). This produces optical rotation by an angle α. (b) Sequence showing the evolution of the same ground-state atomic alignment in a longitudinal magnetic field for high light power. The light frequency is slightly detuned from resonance to allow for a nonzero light shift. The combined action of the magnetic and light fields produces orientation along the z-axis.

Fig. 11.4 A $F = 1 \rightarrow F' = 1$ atomic transition interacting with x-polarized light. The ground state is oriented along $\hat{z}$, so that the population of the $m = 1$ sublevel is higher than that of the $m = -1$ sublevel. This results in stronger interaction with σ^- light ($m \rightarrow m' = m - 1$ transitions) than with σ^+ light ($m \rightarrow m' = m + 1$ transitions).

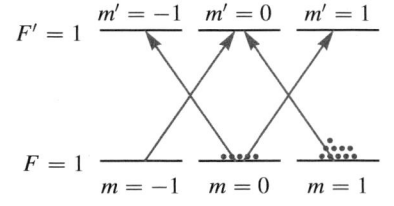

optical rotation effect (Sec. 10.2.6), a difference in the refractive indices for the two circular polarizations of light (circular birefringence) results in optical rotation.

The light power for which the highest sensitivity to magnetic fields is obtained is often in the range in which AOC is significant. Thus, an understanding of this effect is important for the analysis of magnetometers based on NMOR. From an experimental standpoint, there are a few key differences between the optical rotation effect due to AOC and the one due to pure atomic alignment. The first is the light-power dependence. In the pure alignment effect, the light participates in one nonlinear process—optical pumping—to produce the alignment that induces optical rotation. This effect is therefore linear in the light power at low power. In the AOC effect, on the other hand, the light fulfills two roles—it performs optical pumping and also induces the Stark shifts that convert the alignment to orientation. Thus the AOC effect is proportional to the square of the light power to lowest order. This is why the alignment effect is referred to as the low-power effect, and the AOC effect as the high-power effect. Second, the AC Stark shifts go to zero when the light is on resonance. Thus, in a Doppler-free experiment, the alignment effect always dominates when the light is on resonance, while the AOC effect becomes important when the light is detuned from resonance. (In a Doppler-broadened ensemble, the light is always detuned from resonance for some of the velocity groups, so the AOC effect does not exhibit this behavior.) Finally, for some transitions the AOC effect can produce optical rotation of the opposite sign to that produced by the low-power effect. For $F \to F' = F - 1$ and $F \to F' = F$ transitions the two effects produce rotation of the same sign, whereas for $F \to F' = F + 1$ transitions, the signals are of the opposite sign. This is in fact the clearest indication that AOC is occurring in an NMOR experiment: as the light power is increased, the signal on a $F \to F' = F + 1$ transition will flip sign.

Although the phenomenon of AOC has been studied in a variety of different contexts (Lombardi 1969; Pinard and Aminoff 1982; Hilborn *et al.* 1994; Auzinsh and Ferber 1995; Dovator and Okunevich 2001; Kuntz *et al.* 2002), its role in NMOR was only recognized recently (Budker *et al.* 2000a). However, as is often the case with "new" phenomena, a closely related discussion can be found in the classic literature (Cohen-Tannoudji and Dupont-Roc 1969).

11.4 Buffer-gas vapor cells

Noble gasses such as helium or argon are commonly added as *buffer gasses* to atomic vapor cells along with the subject species. Velocity-changing collisions with the buffer gas atoms slow the diffusion rate of the subject atoms through the cell. This may be intended to keep atoms in the interaction region of the cell (e.g., inside the light beam) longer, or to reduce the frequency of collisions with the cell walls.

If a Bennett hole or peak (Sec. 11.2) has been created in the atomic-velocity distribution, velocity-changing collisions will tend to wash it out. Thus, NMOR related to Bennett structures is suppressed in the presence of buffer gas, as was seen in early studies of NMOR in Sm (Davies *et al.* 1987; Baird *et al.* 1989; Barkov *et al.* 1989).

In addition to changing the atomic velocity, collisions with buffer gas also generally destroy atomic polarization, and thus also suppress optical rotation due to polarization effects such as the alignment effect described in Sec. 10.2.7 and the AOC effect described in Sec. 11.3. However, a very important exception is effects with atoms with total electronic angular momentum in the ground state $J = 1/2$, e.g., the alkali atoms. For such atoms, the cross-sections for depolarizing buffer-gas collisions are typically 4–10 orders of magnitude smaller than those for velocity-changing collisions (Walker *et al.* 1997; Walker 1989). This means that buffer gas can be used to increase the lifetime of ground-state polarization of alkali atoms by reducing the rate of depolarizing wall collisions, with a smaller increase in depolarization rate due to buffer-gas collisions.[1] As discussed in Sec. 10.2.7, the low-power width of NMOR features related to atomic polarization effects is determined by the ground-state polarization relaxation rate. Thus longer polarization lifetimes lead to a higher slope of the optical rotation signal as a function of magnetic field for small fields. As mentioned at the end of Sec. 10.2.7, this generally results in higher sensitivity to small magnetic fields. In the next section we will discuss an even more effective way of increasing the ground-state polarization lifetimes of alkali atoms in vapor cells.

---

[1] A systematic study of NMOR in Rb cells with buffer gas of different pressures, and a comparative analysis of the magnetometric sensitivity with such cells and an antirelaxation-coated cell of the same dimensions, was conducted by Park *et al.* (2010).

Fig. 11.5 Longitudinal magnetic-field dependence of optical rotation in a paraffin-coated ^{85}Rb vapor cell (Budker et al. 1998). The background slope is due to the Bennett-structure effect. The dispersion-like structure is due to the transit effect. The inset shows the near-zero B_z-field behavior at a 2×10^5 magnification of the magnetic-field scale. Light intensity is ~ 100 μW/cm^2. The laser is tuned ~ 150 MHz to the high frequency side of the $F = 3 \rightarrow F'$ absorption peak.

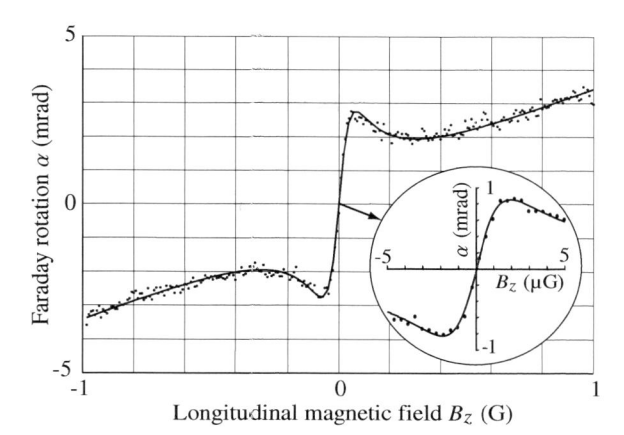

The redistribution of atomic velocities due to velocity-changing collisions, sometimes referred to as "velocity mixing," can also have a significant effect on the spectra of nonlinear effects in atoms with hyperfine structure. This effect is described in the next section and in chapters 13 and 14 in the context of antirelaxation-coated vapor cells, in which it also occurs.

11.5 Antirelaxation-coated cells

When an alkali atom strikes the glass wall of a vapor cell and then returns to the interior of the cell, its polarization state is essentially completely randomized. This means that for a dilute alkali vapor contained in a buffer-gas-free cell, the ground-state polarization lifetime is given by the average time of flight of an atom across the cell. If the interaction region (light beam) takes up only a fraction of the cell, the useful polarization lifetime is even shorter, because an atom is unlikely to leave the interaction region and then return to it before making contact with the cell walls.

There are surfaces that are much more gentle to the polarization of alkali atoms than glass. In particular, if the inner walls of the vapor cell are coated with a chemically inert substance like paraffin (chemical formula $C_n H_{2n+2}$), collisions with the cell walls have a much reduced effect on the atomic polarization. In fact, with specialized paraffin coatings the atoms can make up to 10^4 collisions with the cell walls before depolarizing (Bouchiat and Brossel 1966; Alexandrov and Bonch-Bruevich 1992; Alexandrov et al. 1996). Other materials recently discovered to be effective as antirelaxation coatings can allow as many as 10^6 collisions.

Atomic polarization, then, can have a much longer lifetime in a coated than an uncoated cell, leading to much narrower features due to nonlinear magneto-optical effects. Figure 11.5 shows a measurement of NMOR in a coated cell as a function of magnetic field. The widest feature, seen here as a background slope, is due to the Bennett-structure effect (Sec. 11.2). A narrower feature, with width on the order of 0.1 G, is due to the polarization evolution effect (precession of alignment or AOC). The width of this feature is determined by the transit rate of atoms through the light beam, so this effect is often called the *transit effect*. We have seen both the Bennett-structure effect and the transit effect in uncoated cells. However, in measurements using coated cells, there is another feature hiding in the center of the plot.

If the magnetic-field range around zero is magnified by several orders of magnitude, a new effect can be observed with width on the order of 1 μG. The mechanism for this effect is also polarization evolution, but evolution over a much longer time scale. Rather than being

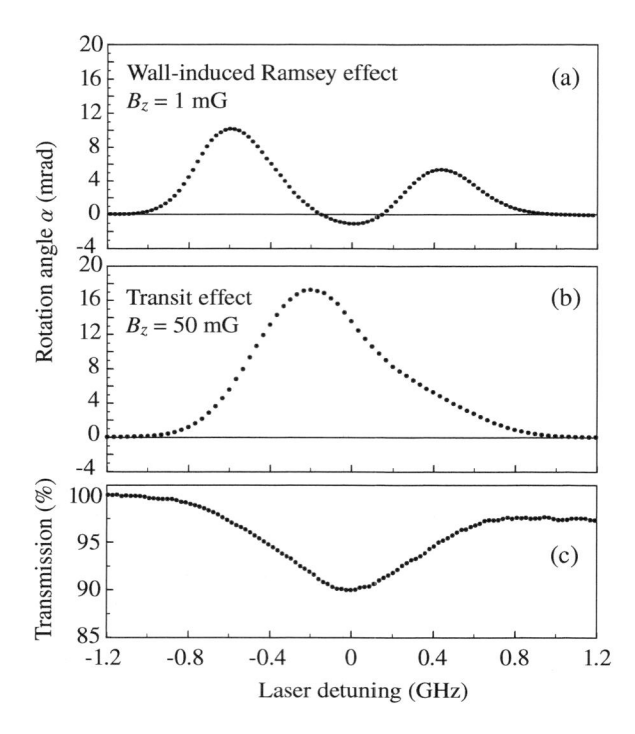

Fig. 11.6 (a) Wall-induced Ramsey rotation spectrum for the $F = 3 \to F'$ component of the $D1$ line of ^{85}Rb obtained by Budker *et al.* (2000b) for light intensity 1.2 mW/cm^2 and beam diameter $\sim$ 3 mm. (b) Transit effect rotation spectrum, for light intensity 0.6 mW/cm^2. (c) Light transmission spectrum for light intensity 1.2 mW/cm^2. The background slope in light transmission is due to the change in incident laser power during the frequency scan.

optically pumped and then probed in one transit through the light beam, in this case the atoms responsible for the effect are pumped and then fly out of the light beam, collide with the cell walls numerous times, and then fly through the beam again, where they induce optical rotation. The time over which this process may occur is the ground-state polarization lifetime, so the polarization relaxation rate determines the width of the feature due to this effect.

This coated-cell effect involves a mechanism quite similar to that found in *Ramsey's method of separated oscillatory fields*. Norman Ramsey received the Nobel prize for the invention of this method, which involves separated fields used to prepare and then detect polarization in a molecular beam. The separation between the fields allows the accumulation of precession angle due to a static magnetic field—the farther apart the two fields, the longer the polarization can precess. In the coated vapor-cell case, optically pumped atoms return to the light field due to wall collisions rather than linear motion, so the effect is sometimes known as the *wall-induced Ramsey effect*, or just as the "wall effect."

> It is interesting to note that Ramsey himself (Kleppner, Ramsey and Fjelstadt 1958), in order to decrease the resonance widths in experiments with separated oscillatory fields, constructed a "storage box" with Teflon-coated walls in which atoms would bounce around for a period of time before emerging to pass through the second oscillatory field. Teflon worked very well with hydrogen atoms, but was found to be incompatible with the chemically aggressive alkali vapors. On the other hand, paraffin coating—which works wonders with the alkalis—does not work with hydrogen.

The NMOR spectra for the wall-induced Ramsey effect can be quite different from those for the transit effect (Fig. 11.6). In the wall-induced Ramsey effect, atoms undergo many velocity-changing collisions between pump and probe interactions (velocity mixing). This means that the Doppler shift for an atom during pumping may be entirely different than the one during probing. If there are hyperfine transitions whose frequency separation is smaller than then Doppler width, it is possible that the light could be resonant with one hyperfine transition when the atom is pumped, and then, due to a different Doppler shift, be resonant with another transition when it is probed. Both the ground-state polarization produced by optical pumping and the effect on the light of the atomic polarization that has evolved in the magnetic field depend on the nature of the transition. For the $F = 3 \to F' = 2, 3$ component of the ^{85}Rb $D1$ line, the contribution to optical rotation from atoms pumped and probed on different transitions has opposite sign to that from atoms pumped and probed on the same transition. Thus, when the laser is tuned near resonance with one of the transitions, atoms will tend to be pumped and probed on that same resonance, and produce rotation of one sign. On the other hand, when the laser is tuned between the two resonances, atoms are equally likely to be pumped and probed on either the same or different resonances, and so there are two contributions of the opposite sign which cancel. Therefore, the wall-induced Ramsey NMOR spectrum consists of two peaks, as seen in Fig. 11.6(a). In the transit effect for buffer-gas-free cells, in contrast, atoms remain in a particular velocity group during both optical pumping and probing. The transit-effect spectrum has a single peak because for each atom light is resonant with the same transition during both pumping and probing (Fig. 11.6b).

The cancelation of the optical rotation signal when Doppler-broadened hyperfine transitions overlap may also be thought of in terms of the polarization moments that may be created in the ground state. When there is overlap of the hyperfine resonances and atomic velocity mixing, the effect of the hyperfine structure on optical pumping may be effectively negated. This means that light may only create polarization moments that can be supported by the electronic angular momentum, regardless of the nuclear spin. In the case of the alkali atoms, the ground-state electronic angular momentum is $J = 1/2$, so, as discussed in Sec. 5.7, the highest-rank multipole moment that can be supported is $\kappa = 1$, i.e., orientation. Because the NMOR effect requires the creation of alignment ($\kappa = 2$) in the ground state, the signal is suppressed. This discussion will be returned to in more detail in chapters 13 and 14.

In order to extend the theoretical treatment of Chapter 10 to describe antirelaxation-coated cells, the velocity-mixing mechanism must be taken into account. In the low-light-power limit, this can be done by modifying the convolution procedure for Doppler broadening presented in Sec. 10.2.8. That procedure, as originally presented, assumes that the atomic velocities do not change, as for the transit effect. Then the signal due to each velocity group can be found independently and all velocity groups integrated over. If we assume, rather, that there is complete velocity mixing between pumping and probing, we must integrate over the density matrix found for each velocity group after optical pumping. Assuming that the atoms in each velocity group possess this aggregate density matrix, we can then integrate the optical rotation over atomic velocity to find the observed signal. This approach will be taken in chapters 13 and 14.

As the light power is increased, it becomes important that, in general, only part of the vapor cell is illuminated. This means that the density matrices for atoms inside and outside the laser beam must be kept track of separately. Furthermore, atoms usually do not undergo complete velocity mixing while they are inside the laser beam, so density matrices for each atomic velocity group inside the beam must be considered. These density matrices are all coupled by the transfer of atoms into and out of the light beam, and the evolution equations for each must be solved simultaneously. In this book we will avoid this approach and concentrate our theoretical analysis instead on the low-power case.

An additional complication in the treatment of antirelaxation-coated cells is that the polarization lifetime is long enough so that the effects of collisions between alkali atoms become significant. Alkali-alkali collisions can relax atomic polarization and also transfer it between the colliding atoms. These so-called *spin-exchange collisions* thus represent a relaxation mechanism with a more complex character than that of the normally isotropic relaxation due to collisions with the walls (in fact, due to the conservation of total angular momentum, the relaxation itself can be nonlinear). Expressions for the effect of spin-exchange collisions on the density matrix have been obtained; however, we will generally use the simpler isotropic model for relaxation in our calculations.

> The following references provide further reading.
>
> In their early work on optical pumping, Robinson et al. (1958) showed that paraffin coatings could reduce the relaxation of atomic polarization due to wall collisions. Later on, working with a paraffin-coated Cs vapor cell, Kanorskii et al. (1995) discovered a narrow feature (of width ~ 1 mG) in the magnetic-field dependence of Faraday rotation. Kanorskii et al. (1995) described the feature as a Ramsey resonance induced by multiple wall collisions. Bouchiat and Brossel (1966), Alexandrov and Bonch-Bruevich (1992), and Alexandrov et al. (1996) investigated the properties of paraffin coatings. Budker et al. (1998) performed an investigation of the wall-induced Ramsey effect in NMOR using Rb atoms contained in paraffin-coated cells and observed magnetic resonances of width ~ 1 µG (Fig. 11.5). Coatings leading to longer lifetimes are being developed by Balabas et al. (2010).
>
> Ramsey's method is discussed in his Nobel lecture (Ramsey 1990). Skalla and Waeckerle (1997) studied the wall-induced Ramsey effect in cells with various geometries (cylindrical, spherical, and toroidal), and used spatially separated pump and probe fields to measure *Berry's topological phase* (Berry 1984). Pustelny et al. (2006) investigated the possibilities of applying the separated optical field method to improve the sensitivity of NMOR-based magnetometers.
>
> Equations describing density-matrix evolution due to spin-exchange in alkali–alkali collisions have been obtained by Okunevich (1994; 1995) and Vallés and Alvarez (1994; 1996), following work of Grossetête (1964).

11.6 Optically thick media

One of the most important applications of nonlinear magneto-optical rotation is magnetometry, whose sensitivity tends to improve with the number of atoms involved in the measurement. As the number of atoms is increased, however, the medium becomes optically thick and transmits less light, which has the effect of decreasing sensitivity. Thus, in order to find the optimal conditions for magnetometry, one must study NMOR in optically thick media.

> Some experimental investigation on this topic was performed by Novikova et al. (2001b), who used a 5-cm long uncoated buffer-gas-free vapor cell containing ^{87}Rb and a laser beam tuned to the maximum of the NMOR spectrum near the D1 line (laser power was 2.5 mW, beam diameter ~ 2 mm). They measured the maximum (with respect to laser frequency and magnetic field) polarization rotation φ_{max} as a function of atomic density. It was found that φ_{max} increases essentially linearly up to $n \approx 3.5 \times 10^{12}$ cm^{-3}. At higher densities, $d\varphi_{max}/dn$ decreases and eventually becomes negative. The maximum observed rotation was ≈ 10 rad with the applied magnetic field of ≈ 0.6 G (this value of magnetic field corresponds to about an order of magnitude power broadening compared to low-power NMOR).

Let us discuss some salient features of the interaction of resonant light with an optically thick medium, following the work of Rochester and Budker (2002). Note that the primary goal of that work was to derive the scaling of optimal magnetometric sensitivity with respect to the optical thickness of the sample. We will not discuss the sensitivity here, referring the reader to the original paper and to a general review of all-optical magnetometers by Budker and Romalis (2007).

We analyze NMOR for an isolated $F = 1 \to F = 0$ transition, for which analytical solutions for the density matrix were obtained in chapter 10. We assume that the transverse thickness of the sample is small, so that the reabsorption of spontaneously emitted radiation (radiation trapping) can be neglected. (This assumption may not be justified in actual experiments with optically thick media.) We also neglect mixing between different velocity groups due to velocity-changing collisions. Such mixing is important in the presence of a buffer gas, or for antirelaxation-coated cells.

11.6.1 Absorption and optical rotation in a thick medium on resonance

We consider the system discussed in Sec. 10.2: a $F = 1 \to F = 0$ transition subject to linearly-polarized incident light, with a magnetic field directed along the light propagation direction (the Faraday geometry). In that discussion the absorption and optical-rotation signals were obtained for a Doppler-free, optically thin medium under the assumption that $\gamma \ll \Gamma$, where γ is the transit relaxation rate and Γ is the natural linewidth. We now further assume that the optical-pumping saturation parameter $\kappa_2 = \Omega_R^2/(\Gamma\gamma)$ is much greater than unity. Here $\Omega_R = \|d\|\mathcal{E}_0/(\sqrt{3}\hbar)$ is the optical Rabi frequency, where $\|d\|$ is the reduced dipole matrix element and $\mathcal{E}_0$ is the optical field amplitude. (We will see from the final result that $\kappa_2 \gg 1$ everywhere in the medium when the input light power is optimal.) Finally, as we are interested in the small-field slope of the magnetic resonance, we take the case in which the magnetic field is arbitrarily small.

We now apply these additional assumptions to Eqs. (10.56), (10.60a), and (10.60c). We find that the absorption coefficient per unit length, α, is given by

$$\alpha = -\frac{2}{\mathcal{E}_0}\frac{d\mathcal{E}_0}{d\ell} \approx \frac{\alpha_0}{(2\Delta/\Gamma)^2 + 2\kappa_2/3}, \tag{11.1}$$

where Δ is the light-frequency detuning from resonance, and $\alpha_0 = 1/\ell_0$, the unsaturated absorption coefficient on resonance, is obtained from Eq. (10.61):

$$\alpha_0 \approx \frac{1}{6\pi}\lambda^2 n, \tag{11.2}$$

where λ is the transition wavelength and n is the atomic density.

In this section we denote the optical-rotation angle by β to avoid confusion with the absorption coefficient. The slope of optical rotation per unit optical path length, $d\beta/(d\Omega_L dx)$, where $\Omega_L = g\mu_B B/\hbar$ is the Larmor frequency (g is the Landé factor, and μ_B is the Bohr magneton), is found to be

$$\frac{\gamma d\beta}{d\Omega_L dx} \approx \frac{\alpha_0 \left[4\gamma(2\Delta/\Gamma)^2 + \Gamma\kappa_2^2\right]}{\Gamma\left[(2\Delta/\Gamma)^2 + 2\kappa_2/3\right]\left[4(2\Delta/\Gamma)^2 + \kappa_2^2\right]}. \tag{11.3}$$

Noting that for large κ_2, the power-broadened width of the optical-rotation spectrum goes as $\Gamma\sqrt{\kappa_2}$, we can assume that $\Delta \lesssim \Gamma\sqrt{\kappa_2}$. Equation (11.3) then reduces to

$$\frac{\gamma d\beta}{d\Omega_L dx} \approx \frac{\alpha_0}{(2\Delta/\Gamma)^2 + 2\kappa_2/3}. \tag{11.4}$$

The above results apply to an optically thin medium in the Doppler-free case, or equivalently the case in which the power-broadened linewidth is much greater than the Doppler width. We

230   *Nonlinear magneto-optical rotation*

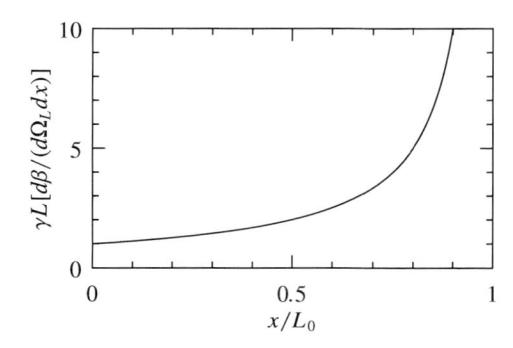

Fig. 11.7 Normalized contribution to small-field slope (in radians) of a slice of the medium of width dx at position x, given by Eq. (11.7).

will now generalize these results to describe an optically thick medium. We initially assume $\Delta = 0$ (we will have to include nonzero detunings in the discussion of the Doppler-broadened case below). As the light passes through the medium, its intensity and polarization angle change according to Eqs. (11.1) and (11.4). (Generally, the medium in the presence of a magnetic field also produces light ellipticity; however, the ellipticity is an odd function of detuning and is zero on resonance.) Writing the light intensity in terms of the saturation parameter κ_2, the rate of change of the saturation parameter as light travels through the medium is given by

$$\frac{d\kappa_2(x)}{dx} = -\alpha \kappa_2(x) \approx -\frac{3}{2}\alpha_0, \quad (11.5)$$

so solving for the saturation parameter as a function of position, we obtain

$$\kappa_2(x) \approx \kappa_2(0) - \frac{3}{2}\alpha_0 x. \quad (11.6)$$

This means that the light can penetrate a distance $L_0 = 2\kappa_2(0)/(3\alpha_0)$ into the medium before being extinguished.

Note the remarkable difference between the way that absorption occurs in the case of low light power (no saturation) and the case of high light power (saturated absorption) described by Eq. (11.6). In the low-power case, the absorption coefficient is independent of the light intensity. Solving $d\kappa_2/dx = -\alpha\kappa_2$ for constant α gives $\kappa_2(x) = \kappa_2(0)\exp(-\alpha_0 x)$ for a thick medium—i.e., the $d\kappa_2/dx = -\alpha\kappa_2$ for constant α gives $\kappa_2(x) = \kappa_2(0)\exp(-\alpha_0 x)$ for a thick attenuation is linear.

The contribution to the small-field optical rotation of the "slice" of the medium at position x is found by substituting Eq. (11.6) into Eq. (11.4):

$$\frac{\gamma d\beta}{d\Omega_L dx}(x) \approx \frac{\alpha_0}{2\kappa_2(0)/3 - \alpha_0 x} = \frac{1}{L_0 - x}. \quad (11.7)$$

This contribution is plotted as a function of depth in the medium in Fig. 11.7. The plot illustrates that the part of the medium near the end of the optical path length contributes more to the overall rotation than the part near its beginning. This is because light power, and correspondingly, the light broadening of the resonance, is lower at the end of the medium.

Choosing a length $L < L_0$ of the medium, we can integrate over the path length to find the total slope of optical rotation:

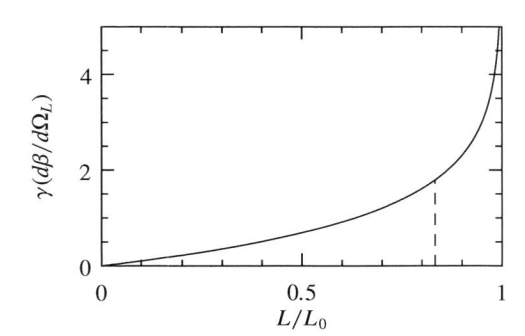

Fig. 11.8 Normalized slope (in radians) of NMOR as a function of normalized length L/L_0 of the medium. The value of L/L_0 that produces the greatest magnetometric sensitivity is shown with a dashed line.

$$\begin{aligned}\frac{\gamma d\beta}{d\Omega_L} &= \int_0^L \frac{\gamma d\beta}{d\Omega_L dx}(x) dx \\ &\approx \int_0^L \frac{1}{L_0 - x} dx \\ &= \ln\left(\frac{1}{1 - L/L_0}\right).\end{aligned} \quad (11.8)$$

The slope (11.8) is plotted as a function of L/L_0 in Fig. 11.8. We see that if we make the medium very thick, the NMOR angle can be made correspondingly large; however, the intensity of the light transmitted through the sample is low.

11.6.2 The Doppler-broadened case

Now we consider the case where the Doppler width Γ_D is much greater than the power-broadened linewidth. In this case, we need to convolve the Doppler-free spectra with the Maxwellian velocity distribution, as discussed in Sec. 10.2.8. In the limit of large Doppler width, we can approximate the Doppler distribution (Eq. 10.74) by its central value, i.e.,

$$f_{\Delta_v}(\Delta_v) d\Delta_v \approx f_{\Delta_v}(0) d\Delta_v \approx \frac{1}{\Gamma_D \sqrt{\pi}} d\Delta_v. \quad (11.9)$$

The Doppler-broadened spectrum is then proportional to the integral over the Doppler-free spectrum. The Doppler-broadened absorption coefficient is given by (see Eq. 11.1)

$$\begin{aligned}\alpha_{\text{DB}} &\approx \frac{1}{\sqrt{\pi}\Gamma_D} \int_{-\infty}^{\infty} \alpha(\Delta) d\Delta \\ &\approx \frac{\alpha_0}{\sqrt{\pi}\Gamma_D} \int_{-\infty}^{\infty} \frac{d\Delta}{(2\Delta/\Gamma)^2 + 2\kappa_2/3} \\ &= \sqrt{\frac{3\pi}{8}} \frac{\Gamma}{\Gamma_D} \frac{\alpha_0}{\sqrt{\kappa_2}}.\end{aligned} \quad (11.10)$$

Comparing Eqs. (11.10) and (11.1), we see that we have reproduced a well-known result (see, for example, Demtröder 1996, Sec. 7.2.1) that resonant absorption falls as $1/\kappa_2$ for Doppler-free media and as $1/\sqrt{\kappa_2}$ for Doppler-broadened media when $\kappa_2 \gg 1$. The change in κ_2 per unit length is

$$\frac{d\kappa_2(x)}{dx} = -\alpha_{DB}\kappa_2(x)$$
$$\approx -\frac{\sqrt{6\pi}}{4}\frac{\Gamma}{\Gamma_D}\alpha_0\sqrt{\kappa_2(x)}, \tag{11.11}$$

and solving for κ_2 as a function of position, we obtain

$$\kappa_2(x) \approx \left(\sqrt{\kappa_2(0)} - \frac{\sqrt{6\pi}}{8}\frac{\Gamma}{\Gamma_D}\alpha_0 x\right)^2 = \kappa_2(0)\left(1 - x/L_0'\right)^2. \tag{11.12}$$

where

$$L_0' = \frac{8}{\alpha_0\sqrt{6\pi}}\frac{\Gamma_D}{\Gamma}\sqrt{\kappa_2(0)} \tag{11.13}$$

is the maximum length of the medium in the Doppler-broadened case. Note that the behavior of the saturation parameter as a function of distance is different in the Doppler-broadened case compared to the Doppler-free case (cf. Eq. 11.6) in which the saturation parameter falls approximately linearly with distance.

Performing the integral to find the Doppler-broadened spectrum of small-field optical rotation gives

$$\left.\frac{\gamma d\beta}{d\Omega dx}\right|_{DB} \approx \frac{1}{\sqrt{\pi}\Gamma_D}\int_{-\infty}^{\infty}\frac{\gamma d\beta}{d\Omega dx}(\Delta)\,d\Delta$$
$$\approx \frac{\alpha_0}{\sqrt{\pi}\Gamma_D}\int_{-\infty}^{\infty}\frac{1}{(2\Delta/\Gamma)^2 + 2\kappa_2/3}d\Delta \tag{11.14}$$
$$\approx \frac{\sqrt{6\pi}}{4}\frac{\Gamma}{\Gamma_D}\frac{\alpha_0}{\sqrt{\kappa_2}}.$$

We see that the rotation per unit length scales as $1/\sqrt{\kappa_2}$, in contrast to the $1/\kappa_2$ scaling for the Doppler-free case [similar to the situation with absorption; cf. Eqs. (11.1), (11.10)]. This is because the number of atoms producing the effect is, in a sense, not fixed; with increasing light power, a larger fraction of the Doppler distribution is involved.

Substituting Eq. (11.12) into Eq. (11.14) to find the contribution to the slope as a function of position gives

$$\left.\frac{\gamma d\beta}{d\Omega dx}(x)\right|_{DB} \approx \frac{2}{L_0' - x}. \tag{11.15}$$

It is interesting to note that while the light power and the rotation slope per unit length behave differently in the Doppler-broadened case compared to the Doppler-free case, Eq. (11.15) has the same functional form as for the Doppler-free case (Eq. 11.7).

Integrating over a medium of length $L < L_0'$, we obtain

$$\left.\frac{\gamma d\beta}{d\Omega}\right|_{DB} \approx \int_0^L \left.\frac{\gamma d\beta}{d\Omega dx}(x)\right|_{DB} dx$$
$$\approx 2\ln\left(\frac{1}{1 - L/L_0'}\right). \tag{11.16}$$

The behavior of Eq. (11.16) as a function of L is qualitatively similar to that of Eq. (11.8) shown in Fig. 11.8. However, the dependence on $\kappa_2(0)$ (through the length L_0') is different.

Optically thick media 233

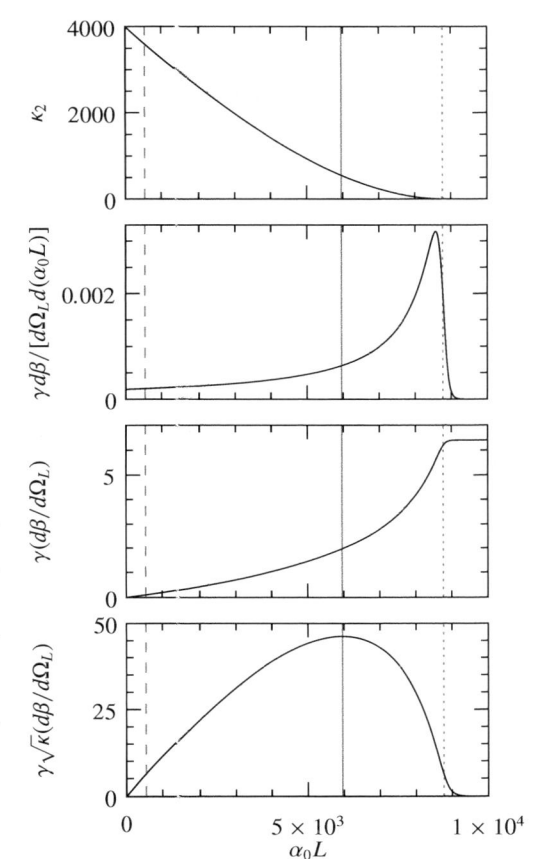

Fig. 11.9 Normalized (a) saturation parameter, (b) differential small-field rotation, (c) total small-field rotation, and (d) inverse magnetometric sensitivity as a function of optical depth with initial saturation parameter $\kappa_2(0) = 4 \times 10^3$ and $\Gamma_D/\Gamma = 60$. Plots (b), (c), and (d) are in units of radians. The dashed line indicates the transition from the Doppler-free to the Doppler-broadened regime [$\kappa_2 = (\Gamma_D/\Gamma)^2$], and the dotted line indicates the point at which nonlinear effects begin to turn off ($\kappa_2 = 1$). Linear optical rotation is neglected in this plot. The solid line indicates the optical depth at which maximum sensitivity is achieved.

11.6.3 The general case

The Doppler-broadened case is realized when the power-broadened width $\sim \Gamma\sqrt{\kappa_2(x)}$ is smaller than the Doppler width for all x within the sample. As light power is increased, a gradual transition to the Doppler-free case occurs, where power broadening at the front of the sample starts to exceed the Doppler width.

A numerical result can be obtained for the general case. In a typical experiment, light power is ~ 1 mW, the laser beam diameter is ~ 0.1 cm, $\lambda \approx 800$ nm, and $\Gamma_D/\Gamma \approx 60$. Thus the effective ground state relaxation rate due to the transit of atoms through the laser beam is $\gamma \approx 2\pi \times 50$ kHz and the initial saturation parameter is $\kappa_2(0) \approx 4 \times 10^3$.

We hold the laser power constant and vary the nominal optical depth $\alpha_0 L$ by changing the atomic density or the optical path length. Normalized transmission, differential small-field rotation, total small-field rotation, and magnetometric sensitivity (proportional to the product of differential small-field rotation and the square root of the transmitted light power, see Rochester and Budker 2002) are plotted in Fig. 11.9 as a function of optical depth. For small optical depth, $\kappa_2 > (\Gamma_D/\Gamma)^2$ and the medium is effectively Doppler-free. Transmission falls linearly until the transition to the Doppler-broadened case is made (dashed line). Then transmission falls quadratically until the linear optics regime is reached (dotted line), after which it falls exponentially. Differential small-field rotation initially rises, as κ_2 falls and power broadening is reduced, until nonlinear effects begin to turn off. (Linear optical rotation is neglected in this plot.) Since magnetometric sensitivity depends both on total optical rotation and transmission, an intermediate value for the optical depth produces the greatest sensitivity (solid line).

11.7 Nonlinear magneto-optical rotation with modulated light

In each of the magneto-optical effects that we have discussed in this chapter and in Chapter 10, the magnetic-field dependence of optical rotation is either a dispersive Lorentzian or is qualitatively similar to one: the rotation angle is linear in magnetic field at low fields and drops off with the field once the Larmor frequency exceeds a characteristic rate or width. For the linear effect in a Doppler-broadened medium the nominal width is the Doppler width, for the Bennett-structure effects it is the natural width, and for the polarization-dependent effects it is the polarization relaxation rate. From a practical point of view, this behavior means that there is a trade-off between the sensitivity and *dynamic range* of magnetometers based on these effects—narrower widths mean steeper slopes and higher sensitivity, but smaller cut-off fields above which the rotation goes rapidly to zero, narrowing the range of fields over which the magnetometer is sensitive.

The polarization-dependent nonlinear magneto-optical rotation effects, having the narrowest widths, exhibit this problem most strongly. For example, as noted in Sec. 11.5, the wall-induced Ramsey effect has a typical width of 1 µG, so that magnetometers based on this effect are sensitive only to magnetic fields of this strength or weaker. It turns out, however, that a simple modification of the setup—adding modulation to the pump light and detecting optical rotation synchronously with modulation—can create additional resonances with comparably small widths, but centered at nonzero values of the magnetic field, as opposed to the "zero-field" resonances discussed above. These resonances can extend the dynamic range of magnetometers while maintaining high sensitivity. Below, we briefly describe nonlinear optical rotation with amplitude and frequency modulated light. A discussion of other types of modulation in the context of nonlinear magneto-optics along with an extensive bibliography can be found in a review by Alexandrov *et al.* (2005).

11.7.1 Amplitude modulation

Consider a group of atoms polarized by a short pulse of linearly polarized pump light. If the atoms are in a magnetic field they will begin to precess, and may induce optical rotation in a probe light beam. Contrary to the case with cw pump light in which steady-state signals are observed, here we observe a time-dependent, transient signal: the optical rotation oscillates as the polarization performs complete revolutions, and the signal dies away as the polarization relaxes (Fig. 11.10a). We will assume the Faraday geometry, in which the pump light and the probe light both propagate along the magnetic-field direction. Note that linearly polarized light produces polarization (a component of rank-two alignment with $q = 2$) with two-fold symmetry about the light propagation direction (for example, see Figs. 9.6 and 9.11, where the light propagation direction lies somewhere in the *x-y* plane). This means that it takes one-half of a Larmor period for the polarization to return to its original state, i.e., the quantum-beat frequency is twice the Larmor frequency. Consequently, the oscillation frequency of the rotation signal is also twice the Larmor frequency Ω_L. In general, a polarization moment ρ_q^κ has q-fold symmetry about the reference axis (Fig. 11.11), and so would produce optical rotation that oscillates at $q\Omega_L$.

In order to optically pump more atoms, we may wish to use more pulses. If a train of light pulses is used with some arbitrary repetition rate, the atoms polarized in each pulse will, in general, precess out of phase with the atoms polarized in the previous pulses, tending to wash out the medium polarization and the time-dependent signal (Fig. 11.10b). However, if the pump pulses are synchronized with the oscillation period of the signal, the signal due to each successive group of polarized atoms will reinforce the signal due to the previously pumped atoms, producing a larger time-dependent aggregate signal (Fig. 11.10c). Thus we see that a resonance is achieved when the signal oscillation frequency $2\Omega_L$ is equal to the pulse repetition frequency Ω_m.

Actually, reinforcement of the signal occurs when a pulse occurs every two oscillation periods, as well (Fig. 11.10d). In fact, we can see that a resonance condition is achieved whenever $2\Omega_L = n\Omega_m$, where n is an integer, i.e., whenever twice the Larmor frequency is a harmonic of the pulse repetition rate. Furthermore, it is not necessary to use short pump pulses—any modulation of the light amplitude at the frequency Ω_m will produce a similar result.

Under resonance conditions, the polarization produced by successive pumping cycles adds constructively, producing net rotating polarization in the medium. This effect, introduced by Bell and Bloom (1961), is known as *synchronous optical pumping* or *optically driven spin precession*. This is a particular case of a general class of phenomena known as *beat resonances* exhibited when a parameter of the pump light or external field is modulated, as discussed below. In nonlinear magneto-optical experiments, such resonances are generally observed using *lock-in detection* of the transmission or polarization signal of the probe light, at a harmonic of the modulation frequency.

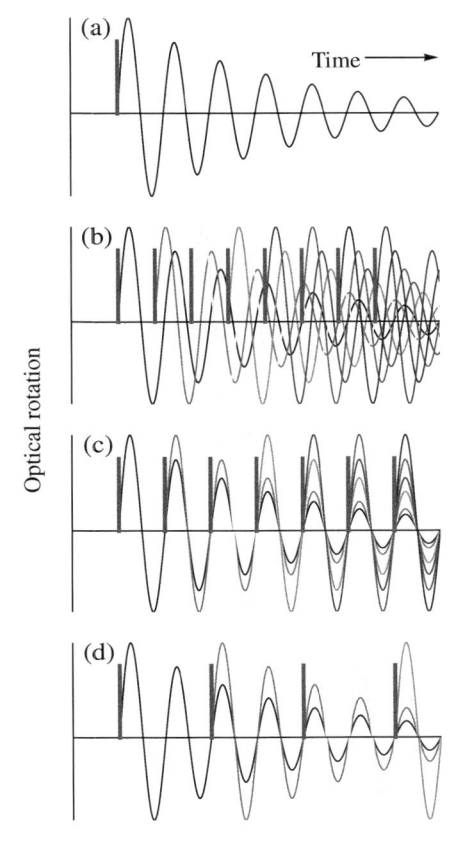

Fig. 11.10 Optical rotation signal in response to (a) a single pump pulse (vertical bar), (b) a train of pulses, (c) pulses at twice the Larmor frequency, (d) pulses at the Larmor frequency. In (c) and (d) the response to successive pulses is additive, resulting in a resonance in the oscillating signal.

It is very common in experimental practice that some experimental parameter is modulated, for example, sinusoidally, and a signal is detected either at the frequency of the modulation or its harmonic. Typically, *phase sensitive detection* is performed with a device known as *lock-in amplifier* or *lock-in detector*, which has two inputs: "signal" and "reference."

Let us say that the parameter that is modulated goes as $\cos\Omega_m t$, where Ω_m is the modulation frequency. The reference should be fixed-amplitude signal oscillating at the frequency of the signal that we would like to detect, i.e., a harmonic of Ω_m. The phase of the reference can be chosen as desired; let us say, that the reference is in phase with the modulation. Given an input signal $S(t)$ that, generally, contains noise in addition to the useful signal we are looking for, a lock-in detector multiplies $S(t)$ by

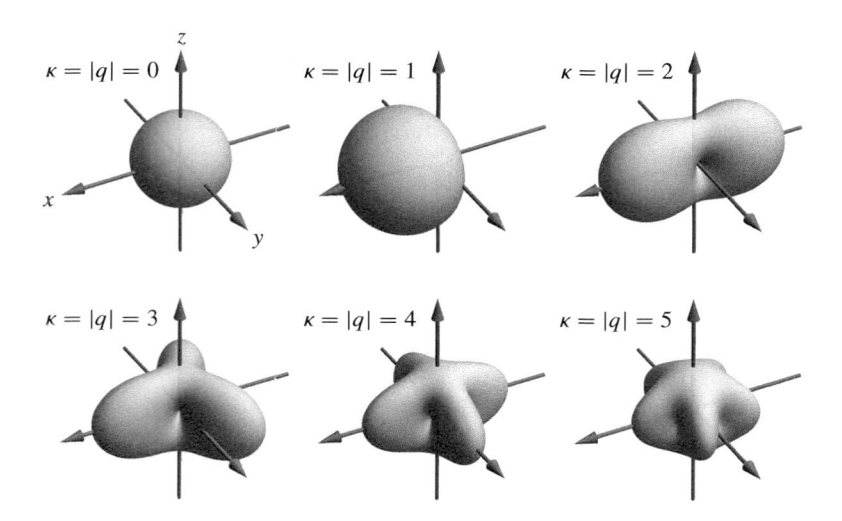

Fig. 11.11 Angular-momentum polarization surfaces for $F = 4$ states composed only of population (ρ_0^0) and the maximum possible values of the components $\rho_{\pm\kappa}^\kappa$ for a particular κ. $\kappa = 0$: monopole moment (isotropic state with population only); $\kappa = 1$: dipole moment (oriented state); $\kappa = 2$: quadrupole moment (aligned state); $\kappa = 3$: octupole moment; $\kappa = 4$: hexadecapole moment; $\kappa = 5$: triakontadipole moment. Polarization moments with index q have $|q|$-fold symmetry about the z-axis.

the reference signal and integrates the resulting signal with a time constant τ. It also does the same with the reference shifted by $\pi/2$. This provides the in-phase and quadrature signals:

$$\text{In-phase output}(t) \propto \int_{-\infty}^{t} S(t') \cos n\Omega_m t' e^{-(t-t')/\tau} dt', \tag{11.17}$$

$$\text{Quadrature output}(t) \propto \int_{-\infty}^{t} S(t') \sin n\Omega_m t' e^{-(t-t')/\tau} dt'. \tag{11.18}$$

Here it is assumed that the n-th harmonic is detected. If τ is much longer than the modulation period, this procedure averages out all the contributions to $S(t)$, except those at the "correct" frequency and phase.

Lock-in detection is extremely useful for discriminating the useful signal from noise and other spurious signals occurring at different frequency and/or not phased with the reference. It can be thought of as phase-sensitive spectral filtering of the signal with the central frequency $n\Omega_m$ and the bandwidth given by $1/\tau$.

11.7.2 Frequency modulation

Frequency (rather than amplitude) modulation of the pump light can be used to produce an effect similar to that discussed in Sec. 11.7.1. Here, the optical pumping rate is modulated as a result of its frequency dependence. One example of this is *nonlinear magneto-optical (Faraday) rotation with frequency-modulated light (FM NMOR)*. In this technique, linearly polarized light near-resonant with an atomic transition is directed parallel to the magnetic field. The frequency of the light is modulated, moving it closer and farther from resonance. This causes the rates of optical pumping and probing to acquire a periodic time dependence. Similarly to the case of amplitude modulation, a resonance occurs when the quantum-beat

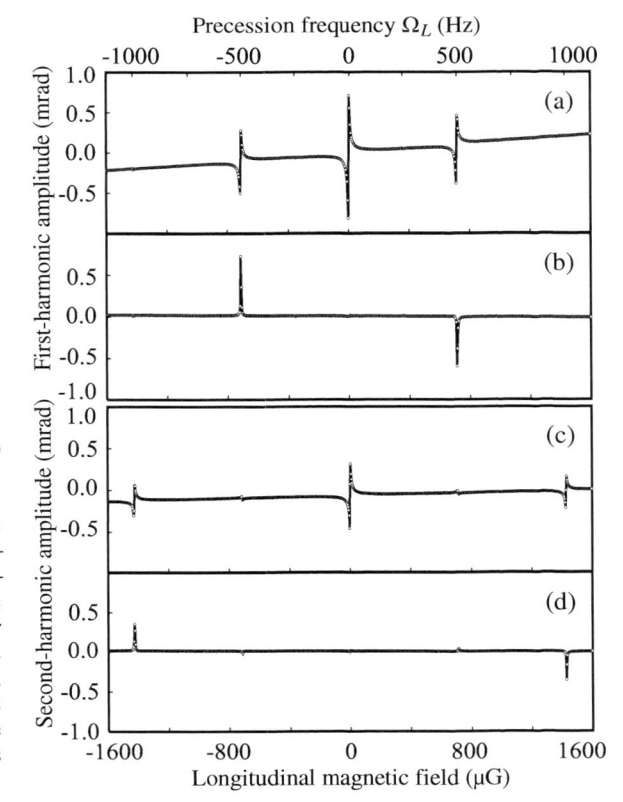

Fig. 11.12 (From Budker et al. 2002c.) Signals detected at the first harmonic (a,b) and second harmonic (c,d) of Ω_m as a function of longitudinal magnetic field. This experiment employed buffer-gas-free, paraffin-coated vapor cells containing isotopically enriched ^{87}Rb. The laser was tuned near the $D1$ line, laser power was 15 μW, beam diameter ~2 mm, $\Omega_m = 1$ kHz, and the modulation amplitude $\Delta\omega_m = 220$ MHz. Traces (a,c) and (b,d) correspond to the in-phase and the quadrature outputs of the signals from the lock-in detector, respectively.

frequency $q\Omega_L$ for a polarization moment ρ_q^κ equals a harmonic of the modulation frequency $n\Omega_m$ (the lowest-order polarization moment here has $\kappa = 2$, $q = 2$; we will discuss the case of higher κ, which can support higher q, in Chapter 18). The atomic sample is pumped into a macroscopic rotating polarized state that causes a periodic modulation of the plane of light polarization at the output of the medium. The amplitude of time-dependent optical rotation at various harmonics of Ω_m can be measured with a phase-sensitive lock-in detector (Fig. 11.12).

Consider Fig. 11.12(a,b), which show the in-phase and quadrature signals detected at the first harmonic. At the center of the in-phase plot is the zero-field resonance, the same feature seen in nonlinear Faraday rotation with cw light. Near zero magnetic field, the modulation frequency is much larger than the Larmor frequency, so the fact that the light is modulated does not significantly affect the optical pumping. Thus the medium polarization reaches a steady state, as it would for cw light (Sec. 10.2.7), with its alignment axis tilted with respect to the light polarization. Considering the light now as a probe, the atomic alignment induces optical rotation, with the amount of rotation depending on how close the probe light is to optical resonance. Thus, as the probe frequency is modulated, the optical rotation signal acquires time dependence at harmonics of the modulation frequency, resulting in the zero-field resonance. If we define the beginning of the light modulation cycle to be the point in the cycle at which the light interacts most strongly with the atoms, the zero-field optical rotation signal will be in phase with the light modulation, and thus is largely suppressed in the quadrature component of the signal (Fig. 11.12b).

At $\Omega_L = 500$ Hz, the $2\Omega_L = \Omega_m$ resonance discussed above is seen. Precisely on resonance, a large rotating polarization is set up in the medium; this results in a peak in the quadrature component of the signal. The reason that this signal appears in the quadrature component is that the atomic polarization must rotate away from the light polarization axis in order for it to induce optical rotation—the maximum rotation is produced at an angle $\pi/4$. This results in a phase lag between optical pumping and the detection of optical rotation. As the magnetic field is moved out of the resonance condition, the net atomic polarization begins to wash out, as discussed above. This reduces the signal in the quadrature component. However, at the same time, an effective phase shift is introduced between pumping and probing, due to the difference between the modulation frequency and the quantum-beat frequency. This allows the signal to appear in the in-phase component as a dispersive-Lorentzian-shaped feature.

At $\Omega_L = 500$ Hz, very small features corresponding to the $2\Omega_L = 2\Omega_m$ resonance can be observed. The background slope seen in the in-phase component of the signal is due to the zero-field resonance for the transit effect (Sec. 11.5).

The FM NMOR technique is useful for increasing the dynamic range of NMOR-based magnetometers. The beat resonances have width comparable to that of the zero-field resonance (since the dominant relaxation mechanisms are the same), but in principle can be centered at any desired magnetic field.

12

Perturbative and approximate methods for light–atom interactions

The idea of the perturbative solution of the density matrix evolution equations was introduced in Sec. 10.3. This technique allows a systematic classification of light–atom interactions of various orders and, in the low-light-power limit, general analytic formulas that are useful as building blocks for obtaining further results.

In this chapter we first discuss the operator that describes the transfer of atoms from upper to lower states via spontaneous decay. This operator appears in the Liouville equation for a general atomic system. Using the perturbative method, we solve the Liouville equation to second order in the optical electric field. This solution allows us to obtain lowest-order formulas for optical pumping (depopulation and repopulation), the excitation of atoms to the upper state (excitation matrix), and observable effects in transmitted light. These problems were solved in a general form in the 1960s, most notably, in the work of Barrat and Cohen-Tannoudji (1961a), Cohen-Tannoudji (1962a;b), Dyakonov (1965), and Happer and Mathur (1967).

Note that, as discussed in Sec. 10.3, the lowest-order nonlinear effects are *third* order in the electric field. However, by examining pump-probe arrangements in which the interaction with each light field is second order or below, we can use the second-order perturbative solution to study higher-order effects.

12.1 Polarization transfer in spontaneous decay

Barrat and Cohen-Tannoudji (1961a) have shown that, given a density matrix ρ_{rs} in an upper state, the rate of change of the lower-state density matrix ρ_{mn} due to spontaneous emission is given by

$$\dot{\rho}_{mn} = \sum_{r,s} \frac{4\omega_{rm}^3}{3\hbar c^3} \mathbf{d}_{mr} \cdot \mathbf{d}_{sn} \rho_{rs} = \sum_{r,s} F_{mn}^{sr} \rho_{rs}, \qquad (12.1)$$

where $F_{mn}^{sr} \rho_{rs}$, given by

$$F_{mn}^{sr} = \frac{4\omega_{rm}^3}{3\hbar c^3} \mathbf{d}_{mr} \cdot \mathbf{d}_{sn}, \qquad (12.2)$$

is the *spontaneous-emission operator*. The sum over the upper pair of indices can also be written as a trace, which results in the compact form

$$\dot{\rho} = \mathrm{Tr}\,\rho F. \qquad (12.3)$$

The rigorous derivation of this result requires quantum electrodynamics. As an alternative, let us present a heuristic explanation for the various factors in Eq. (12.1). A well-known classical electrodynamics formula tells us that the rate of emission of radiation from an oscillating dipole moment $\mathbf{d}(t) = \mathbf{d} \sin \omega t$ is

$$\mathcal{I} = \frac{2}{3c^3}\ddot{\mathbf{d}}(t)^2 = \frac{2\omega^4}{3c^3}\mathbf{d} \cdot \mathbf{d}. \quad (12.4)$$

To convert this expression into a rate for the emission of photons, we divide by the energy of a photon, $\hbar\omega$, to find

$$\frac{2\omega^3}{3\hbar c^3}\mathbf{d} \cdot \mathbf{d}. \quad (12.5)$$

This formula resembles Eq. (12.1), which can be considered its quantum-mechanical analog. In the quantum-mechanical case, a sum must be performed over all possible dipole moments that connect the upper state to a particular lower-state population or coherence.

Another way to justify these formulas comes from symmetry considerations. We have seen that the dipole transition rate induced by an optical field $\mathcal{E}$ is given by $(\mathbf{d} \cdot \mathcal{E})^2$, i.e., the square of the transition matrix element. Just keeping the directional information from the electric field, we can write this in terms of the polarization vector $\hat{\varepsilon}$: $(\mathbf{d} \cdot \hat{\varepsilon})^2$. In spontaneous decay, the electric field inducing the transitions (produced by the quantum vacuum fluctuations) has all possible polarizations. The total transition probability is the sum over the probabilities for all these polarizations; since there is no particular polarization of the electric field, no polarization information can remain in the formula for the total spontaneous decay rate. In this case, the sum over the terms $(\mathbf{d} \cdot \hat{\varepsilon})^2$ must not contain $\hat{\varepsilon}$, but must still be a scalar that is quadratic in $\mathbf{d}$. The only possibility then is $\mathbf{d} \cdot \mathbf{d}$.

The rate for emission into a particular electromagnetic mode with polarization $\hat{\varepsilon}$ while generating a particular ground-state coherence ρ_{mn} is

$$\frac{d\Phi}{d\Omega} = \frac{\omega^3}{2\pi\hbar c^3}\sum_{rs}\hat{\varepsilon}^* \cdot \mathbf{d}_{mr}\rho_{rs}\mathbf{d}_{sn} \cdot \hat{\varepsilon}, \quad (12.6)$$

where $d\Omega$ is the differential solid angle in the propagation direction of the mode. Summing over two transverse polarizations and integrating over angles introduces a factor $8\pi/3$ and yields Eq. (12.1). If we take the trace over the ground state and multiply by $\hbar\omega$ to convert to intensity, on the other hand, we find the fluorescence intensity of a particular polarization emitted into a solid angle:

$$\frac{d\mathcal{I}}{d\Omega} = \frac{\omega^4}{2\pi c^3}\sum_{mrs}\hat{\varepsilon}^* \cdot \mathbf{d}_{mr}\rho_{rs}\mathbf{d}_{sm} \cdot \hat{\varepsilon}. \quad (12.7)$$

We now examine what kinds of atomic polarization can be transferred from the upper to the lower state. The explicit form of the spontaneous-emission operator (Eq. 12.1) tells us that, in general, coherences can be transferred from the upper to the lower state via spontaneous emission. This is not surprising, because we can quite easily create a situation in which it is clear that only populations should be transferred when viewed in one basis, but must involve coherences when the basis is rotated.

Consider a closed $J = 1/2 \to J' = 1/2$ transition. Suppose that initially we have only the $m' = 1/2$ sublevel populated, which can then decay to the ground state. The initial excited-state density matrix is

Polarization transfer in spontaneous decay 241

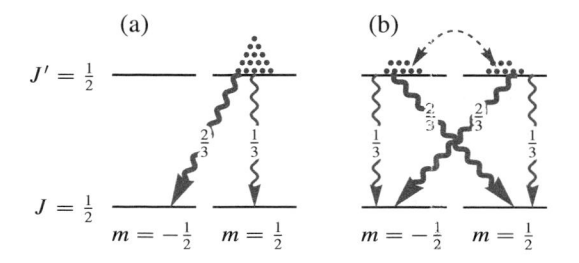

Fig. 12.1 Spontaneous decay to a $J = 1/2$ ground state of (a) a $|J' = 1/2, m' = 1/2\rangle$ state and (b) the same state represented in a basis in which it is polarized along $\hat{\mathbf{x}}$. Arrows representing spontaneous decay transitions are labeled with the corresponding branching ratios; coherence between atomic states is indicated with a dashed arrow. In (a) only populations are transferred to the ground state; in (b) both populations and coherences are transferred.

$$\rho' = \begin{pmatrix} 1 & 0 \\ 0 & 0 \end{pmatrix}. \tag{12.8}$$

The final populations of the ground $m = 1/2$ and $m = -1/2$ sublevel are proportional to the square of the corresponding $3j$ symbols (or Clebsch–Gordan coefficients) and are found to be $1/3$ and $2/3$, respectively (Fig. 12.1a). Since the initial state (spins "pointing" along $\hat{\mathbf{z}}$, in the quantum-mechanical sense) is symmetric about the z-axis, and the process of spontaneous decay has no preferred direction, the final state must also be symmetric about the z-axis. Therefore there can be no coherences in the final state ground-state density matrix, and it is given by

$$\rho = \begin{pmatrix} \frac{1}{3} & 0 \\ 0 & \frac{2}{3} \end{pmatrix}. \tag{12.9}$$

In this basis only populations have been transferred by spontaneous decay.

Now, suppose that we rotate the basis so that in the new basis the spins in the upper state "point" along $\hat{\mathbf{x}}$. By symmetry, a state that points along $\hat{\mathbf{x}}$ has equal projections on the $m' = \pm 1/2$ states, so the initial-state populations must be the same and equal to $1/2$. Also, because the initial state is a pure state, the coherences between the two basis states must be maximal, i.e., their magnitudes must also be $1/2$. It turns out that the coherences are real (real coherences are generally associated with polarization along x, while imaginary coherences are associated with polarization along y). This can be verified by actually performing the rotation using Eq. (5.35). Therefore, in this basis, the density matrix for the same initial state takes the form

$$\tilde{\rho}' = \frac{1}{2} \begin{pmatrix} 1 & 1 \\ 1 & 1 \end{pmatrix}. \tag{12.10}$$

Applying the same rotation to the ground-state density matrix ρ gives

$$\tilde{\rho} = \frac{1}{2} \begin{pmatrix} 1 & -\frac{1}{3} \\ -\frac{1}{3} & 1 \end{pmatrix}. \tag{12.11}$$

This leads us to the conclusion that coherences in the initial density matrix $\tilde{\rho}'$ have been transferred to the ground-state density matrix $\tilde{\rho}$ by spontaneous decay. The coherences have, however, been "diluted" somewhat—their magnitude is $1/3$ of that found in the initial, fully polarized state. This corresponds to the dilution of the longitudinal polarization that occurs between Eqs. (12.8) and (12.9): ρ' has the maximum possible difference between the two upper-state populations, and in ρ the difference between the ground-state populations is smaller. It can also be understood by examining the diagram of the processes in Fig. 12.1(b). Since the different modes of the quantum vacuum fluctuations are uncorrelated, coherence between two atomic states can only be transferred to ground states if both the states decay into the same mode, i.e., both transitions have the same Δm. The only pair of transitions that satisfy this in this case are the two $\Delta m = 0$ transitions. But these transitions have a branching ratio of $1/3$, so only that fraction of the coherence is transferred.

> Note that not only is the polarization diluted upon transfer to the ground state, but it also changes direction—the excited state (12.8) is oriented in the positive $\hat{z}$ direction, while the resulting ground state (12.9) has orientation along $-\hat{z}$. Such a reversal of orientation direction in the decay of a stretched state is a feature of $J = 1/2 \to J' = 1/2$ transitions and does not occur for states with higher angular momenta. A more detailed discussion of the dilution of polarization in spontaneous emission can be found in Chapter 5 of the book by Auzinsh and Ferber (1995).

Using Eq. (12.1) and the Wigner–Eckart theorem (Eq. 3.121), we can find the result directly in any basis. We expand the dot product in the spherical basis and use the relation $d_q^\dagger = (-1)^q d_{-q}$ to obtain for a $J \to J'$ transition

$$F_{m_1 m_2}^{m'_1 m'_2} = \frac{4\omega^3}{3\hbar c^3} \sum_q (-1)^q \langle J m_1 | d_q | J' m'_2 \rangle \langle J' m'_1 | d_{-q} | J m_2 \rangle$$

$$= \frac{4\omega^3}{3\hbar c^3} \sum_q \langle J m_1 | d_q | J' m'_2 \rangle \langle J m_2 | d_q | J' m'_1 \rangle^* \qquad (12.12)$$

$$= (-1)^{2J - m_1 - m_2} (2J' + 1) \Gamma \sum_q \begin{pmatrix} J & 1 & J' \\ -m_1 & q & m'_2 \end{pmatrix} \begin{pmatrix} J & 1 & J' \\ -m_2 & q & m'_1 \end{pmatrix},$$

where in the last line we have used (Eq. 7.62)

$$\frac{4\omega^3}{3\hbar c^3} \frac{1}{2J' + 1} |\langle J \| d \| J' \rangle|^2 = \Gamma, \qquad (12.13)$$

where Γ is the spontaneous decay rate. Using this, we can write the matrix elements of the spontaneous-emission operator for the $J = 1/2 \to J' = 1/2$ system as

$$F = \begin{pmatrix} F_{1/2,1/2} = \Gamma \begin{pmatrix} \frac{1}{3} & 0 \\ 0 & \frac{2}{3} \end{pmatrix} & F_{1/2,-1/2} = \Gamma \begin{pmatrix} 0 & 0 \\ -\frac{1}{3} & 0 \end{pmatrix} \\ F_{-1/2,1/2} = \Gamma \begin{pmatrix} 0 & -\frac{1}{3} \\ 0 & 0 \end{pmatrix} & F_{-1/2,-1/2} = \Gamma \begin{pmatrix} \frac{2}{3} & 0 \\ 0 & \frac{1}{3} \end{pmatrix} \end{pmatrix}. \qquad (12.14)$$

Here the outer matrix corresponds to the lower indices of F_{mn}^{sr} and is in the basis of the ground states; each of its matrix elements is a matrix in the basis of the upper state corresponding to the upper indices.

To obtain the rate of change of a ground-state density-matrix element ρ_{mn}, we take the trace $\mathrm{Tr}\, F_{mn}\rho'$. Performing this procedure with the initial states (12.8) and (12.10), we obtain

$$\dot{\rho} = \Gamma \begin{pmatrix} \frac{1}{3} & 0 \\ 0 & \frac{2}{3} \end{pmatrix} \quad \text{and} \quad \dot{\rho} = \frac{\Gamma}{2} \begin{pmatrix} 1 & -\frac{1}{3} \\ -\frac{1}{3} & 1 \end{pmatrix}, \qquad (12.15)$$

as expected. These are the *initial* rates for the repopulation of the ground state—as the upper-state populations and coherences exponentially decay at the rate Γ, the ground-state repopulation rate will decrease correspondingly.

12.2 Perturbative solution of the steady-state density matrix

In Sec. 10.2 we solved the Liouville equation for a $F = 1 \to F' = 0$ system subject to light with arbitrarily large intensity. In this section we will find solutions for a general system in the limit of low light intensity.

We write the Liouville equation (Sec. 5.5) as before, except that we now separate from the repopulation matrix Λ the term describing repopulation due to spontaneous decay discussed in the previous section. This allows all of the dependence on ρ to be written explicitly:

$$\dot{\rho} = \frac{1}{i\hbar}[H, \rho] - \frac{1}{2}\{\widehat{\Gamma}, \rho\} + \Lambda + \mathrm{Tr}\,(F\rho), \tag{12.16}$$

where H is the total Hamiltonian, $\widehat{\Gamma}$ is the relaxation matrix, Λ is the repopulation matrix, and F is the spontaneous emission operator. Here we have used the *anticommutator* $\{A, B\} = AB + BA$. If we neglect relaxation mechanisms, such as spin-exchange collisions, that depend on the density matrix, we can assume that Λ is independent of ρ.

We now take the total Hamiltonian to be composed of a diagonal part H_0 and a time-independent perturbation $\hbar V$, so that $H = H_0 + \hbar V$, where we write V in frequency units for notational convenience. The unperturbed Hamiltonian H_0 describes the level structure of the atom, and can also account for static external fields directed along the z-axis, so that their contribution to the Hamiltonian is diagonal. (The method is also applicable to a non-diagonal H_0 if it can first be diagonalized.) We will later take the perturbation V to be the Hamiltonian for light interaction under the rotating-wave approximation.

Taking the matrix element $\dot{\rho}_{mn} \equiv \langle m|\dot{\rho}|n\rangle$, where $|m\rangle$ and $|n\rangle$ are eigenstates of H_0 with eigenenergies E_m and E_n, we obtain

$$\begin{aligned}\dot{\rho}_{mn} &= -\frac{i}{\hbar}\langle m|[H_0, \rho]|n\rangle - i\langle m|[\hbar V, \rho]|n\rangle - \frac{1}{2}\langle m|\{\widehat{\Gamma}, \rho\}|n\rangle \\ &\quad + \langle m|\Lambda|n\rangle + \langle m|\mathrm{Tr}\,(F\rho)|n\rangle \\ &= -i\widetilde{\omega}_{mn}\rho_{mn} - i\sum_p \left[V_{mp}\rho_{pn} - \rho_{mp}V_{pn}\right] + \Lambda_{mn} + \sum_{rs} F^{sr}_{mn}\rho_{rs},\end{aligned} \tag{12.17}$$

where we define the complex frequency splitting $\widetilde{\omega}_{mn} = (E_m - E_n)/\hbar - i(\Gamma_m + \Gamma_n)/2$, with $\Gamma_m = \widehat{\Gamma}_{mm}$ ($\widehat{\Gamma}$ is assumed to be diagonal). If $\Gamma_m \ne 0$ for all m, the density matrix evolution is damped, and the density matrix reaches a steady state at large t. We can then set $\dot{\rho}_{mn} = 0$ on the left side of Eq. (12.17). We can then move the first term on the right side to the left to find an implicit steady-state solution for ρ_{mn}:

$$\rho_{mn} = \frac{1}{i\widetilde{\omega}_{mn}}\left(\Lambda_{mn} + \sum_{rs} F^{sr}_{mn}\rho_{rs} - i\sum_p \left(V_{mp}\rho_{pn} - \rho_{mp}V_{pn}\right)\right). \tag{12.18}$$

We can find a series expansion for ρ_{mn} by recursively substituting this expression into itself: every instance of a matrix element of ρ on the right side is replaced using the equality (12.18), and this procedure is repeated on the result. We assume that the atomic structure is such that there are no cascade decays, i.e., quantities such as $F_{sr}^{pk} F_{mn}^{sr}$ are zero. We also assume that Λ repopulates only the ground state, so that, for example, $F_{mn}^{sr} \Lambda_{rp}$ is zero (since $|r\rangle$ must be an excited state for F_{mn}^{sr} to be nonzero). If we drop all terms containing three or more factors of V, the process terminates, yielding

$$\rho_{mn} \approx \frac{1}{i\widetilde{\omega}_{mn}} \Bigg\{ \Lambda_{mn} + \sum_{p} \left(\frac{\Lambda_{mp} V_{pn}}{\widetilde{\omega}_{mp}} - \frac{V_{mp} \Lambda_{pn}}{\widetilde{\omega}_{pn}} \right) \\
+ \sum_{pk} \left[\frac{\Lambda_{mp} V_{pk} V_{kn}}{\widetilde{\omega}_{mp} \widetilde{\omega}_{mk}} - \frac{V_{mp} \Lambda_{pk} V_{kn}}{\widetilde{\omega}_{pk}} \left(\frac{1}{\widetilde{\omega}_{mk}} + \frac{1}{\widetilde{\omega}_{pn}} \right) + \frac{V_{mp} V_{pk} \Lambda_{kn}}{\widetilde{\omega}_{pn} \widetilde{\omega}_{kn}} \right] \\
+ i \sum_{rs} \frac{F_{mn}^{sr}}{\widetilde{\omega}_{rs}} \sum_{pk} \frac{V_{rp} \Lambda_{pk} V_{ks}}{\widetilde{\omega}_{pk}} \left(\frac{1}{\widetilde{\omega}_{rk}} + \frac{1}{\widetilde{\omega}_{ps}} \right) \Bigg\}. \quad (12.19)$$

This is the steady-state solution to the density matrix valid to second order. This expression can be simplified. In the case in which V represents the electric-dipole interaction Hamiltonian, it only mixes ground and excited states and does not mix ground states with each other. If we additionally assume that Λ represents initially unpolarized atoms and we restrict our attention to ground-state density-matrix elements ρ_{mn}, Eq. (12.19) reduces to

$$\rho_{mn} \approx \frac{\Lambda_{mm}}{i\widetilde{\omega}_{mn}} \Bigg[\delta_{mn} + \sum_{p} V_{mp} V_{pn} \left(\frac{1}{\widetilde{\omega}_{mm} \widetilde{\omega}_{mp}} + \frac{1}{\widetilde{\omega}_{nn} \widetilde{\omega}_{pn}} \right) \\
+ i \sum_{rs} \frac{F_{mn}^{sr}}{\widetilde{\omega}_{rs}} \sum_{k} \frac{V_{rk} V_{ks}}{\widetilde{\omega}_{kk}} \left(\frac{1}{\widetilde{\omega}_{rk}} + \frac{1}{\widetilde{\omega}_{ks}} \right) \Bigg], \quad (12.20)$$

where k runs over the ground states and p, r, and s run over the excited states. In Eq. (12.20) the first term in the square brackets represents the initial unpolarized ground state, the second term accounts for the loss of atoms that have been pumped out of the ground state, and the last term accounts for atoms that have returned to the ground state through spontaneous decay.

12.3 The optical-field case

We now choose the perturbation V to represent the interaction of the atoms with a light field. As we have seen previously, the light–atom interaction Hamiltonian in the dipole approximation is given by $H_l = -\mathbf{d} \cdot \mathcal{E}$. Here we write the electric field $\mathcal{E}$ in terms of a complex polarization vector $\hat{\varepsilon}$ as

$$\mathcal{E} = \mathcal{E}_0 \operatorname{Re} \hat{\varepsilon} e^{i(\mathbf{k} \cdot \mathbf{r} - \omega t)}, \quad (12.21)$$

where ω is the light frequency, $\mathbf{k}$ is the wave vector, and $\mathcal{E}_0$ is the real electric field amplitude. We assume that a particular atom is in uniform motion, so that the position vector $\mathbf{r}$ is given by $\mathbf{r} = \mathbf{r}_0 + \mathbf{v}t$, where $\mathbf{r}_0$ is the initial position and $\mathbf{v}$ is the velocity of the atom.

We now employ the rotating-wave approximation to remove the optical-frequency time dependence from the Hamiltonian. We apply a unitary transformation via the diagonal matrix U with matrix elements

$$U_{pp} = e^{i[\mathbf{k} \cdot (\mathbf{r}_0 + \mathbf{v}t) - \omega t]} \quad (12.22)$$

when $|p\rangle$ is an excited state, and $U_{pp} = 1$ when it is a ground state. Proceeding as in Sec. 10.2.2, we find that in the rotating frame (denoted here by a prime) the upper-state eigenenergies (diagonal terms of the Hamiltonian) are modified to

$$E'_p = E_p - \hbar(\omega - \mathbf{k} \cdot \mathbf{v}), \tag{12.23}$$

and the matrix elements of H_l become

$$\begin{aligned}(H'_l)_{pk} &= -\frac{1}{2}\mathcal{E}_0 \hat{\boldsymbol{\varepsilon}} \cdot \mathbf{d}_{pk} \\ (H'_l)_{kp} &= -\frac{1}{2}\mathcal{E}_0 \hat{\boldsymbol{\varepsilon}}^* \cdot \mathbf{d}_{kp}\end{aligned} \tag{12.24}$$

for $|p\rangle$ an excited state and $|k\rangle$ a ground state.

Thus Eq. (12.20) can be written in the rotating frame by setting $V = H'_l/\hbar$ and replacing transition frequencies ω_{pk} between the excited and ground states with

$$\omega'_{pk} = \omega_{pk} - \omega + \mathbf{k} \cdot \mathbf{v}. \tag{12.25}$$

This gives

$$\begin{aligned}\rho_{mn} \approx \frac{\Lambda_{mm}}{i\widetilde{\omega}_{mn}}\bigg[&\delta_{mn} + \frac{\mathcal{E}_0^2}{4\hbar^2}\sum_p \hat{\boldsymbol{\varepsilon}}^* \cdot \mathbf{d}_{mp}\hat{\boldsymbol{\varepsilon}} \cdot \mathbf{d}_{pn}\left(\frac{1}{\widetilde{\omega}_{nn}\widetilde{\omega}'_{pn}} - \frac{1}{\widetilde{\omega}_{mm}(\widetilde{\omega}'_{pm})^*}\right) \\ &+ i\frac{\mathcal{E}_0^2}{4\hbar^2}\sum_{rs}\frac{F^{sr}_{mn}}{\widetilde{\omega}_{rs}}\sum_k\frac{\hat{\boldsymbol{\varepsilon}} \cdot \mathbf{d}_{rk}\hat{\boldsymbol{\varepsilon}}^* \cdot \mathbf{d}_{ks}}{\widetilde{\omega}_{kk}}\left(\frac{1}{\widetilde{\omega}'_{rk}} - \frac{1}{(\widetilde{\omega}'_{sk})^*}\right)\bigg].\end{aligned} \tag{12.26}$$

We assume that the excited states relax at a rate Γ and the ground states at a rate $\gamma \ll \Gamma$, so $\widetilde{\omega}_{mm} = \widetilde{\omega}_{nn} = -i\gamma$. A common situation, and the one that will be of interest to us in Chapter 13, is when coherences only develop between nearly degenerate ground or upper states, i.e., for which $\omega_{mn}, \omega_{rs} \ll \Gamma$. Note that this is the case for the nonlinear effects studied in Chapter 10. Making this assumption, we find

$$\rho_{mn} \approx \frac{\Lambda_{mm}}{i\widetilde{\omega}_{mn}}\bigg[\delta_{mn} - \frac{\mathcal{E}_0^2}{4\hbar^2}\frac{\Gamma}{\gamma}\bigg(\sum_p \frac{\hat{\boldsymbol{\varepsilon}}^* \cdot \mathbf{d}_{mp}\hat{\boldsymbol{\varepsilon}} \cdot \mathbf{d}_{pn}}{(\omega'_{pm})^2 + (\Gamma/2)^2} - \sum_{rs}\frac{F^{sr}_{mn}}{\Gamma}\sum_k\frac{\hat{\boldsymbol{\varepsilon}} \cdot \mathbf{d}_{rk}\hat{\boldsymbol{\varepsilon}}^* \cdot \mathbf{d}_{ks}}{(\omega'_{rk})^2 + (\Gamma/2)^2}\bigg)\bigg]. \tag{12.27}$$

The density matrix, as determined by Eq. (12.27), is a function of the atomic velocity $\mathbf{v}$ through the velocity dependence of the Doppler-shifted transition frequencies. Equation (12.27) can be applied directly to the individual velocity groups in a Doppler-broadened ensemble, assuming that the atomic velocities do not change during the polarization relaxation time. This is the case for the transit effect studied in Chapter 10. When a polarization-preserving technique like buffer gas or an antirelaxation coating is used, however, atoms undergo collisions that change their velocities without destroying their polarization. This means that we must account for atomic polarization in one velocity group that originated in another velocity group. In the limit in which the collision rate is much higher than the polarization relaxation rate, the polarization in all of the velocity groups becomes the same. This is the *complete-mixing* approximation.

We can find the mixed density matrix by taking a weighted average of the density matrices for each velocity group. The weighting function is the Maxwellian velocity distribution, written in frequency space as (Eq. 10.74)

$$f_{\Delta_v}(\Delta_v)d\Delta_v = \frac{1}{\Gamma_D\sqrt{\pi}}e^{-\Delta_v^2/\Gamma_D^2}d\Delta_v. \tag{12.28}$$

Assuming that the Doppler width Γ_D is much greater than Γ, performing the integral over atomic velocity amounts to replacing the Lorentzian spectrum with a Gaussian one, i.e., making the replacement

$$\frac{\Gamma}{(\omega_{rk} - \omega + \mathbf{k}\cdot\mathbf{v})^2 + (\Gamma/2)^2} \to \frac{2\sqrt{\pi}}{\Gamma_D}e^{-(\omega-\omega_{rk})^2/\Gamma_D^2}. \tag{12.29}$$

We then have for the velocity-averaged ground-state density matrix

$$\rho_{mn} \approx \frac{\Lambda_{mm}}{i\tilde{\omega}_{mn}}\left[\delta_{mn} - \frac{\sqrt{\pi}\mathcal{E}_0^2}{2\hbar^2\gamma\Gamma_D}\left(\sum_p \hat{\boldsymbol{\varepsilon}}^*\cdot\mathbf{d}_{mp}\hat{\boldsymbol{\varepsilon}}\cdot\mathbf{d}_{pn}e^{-(\omega-\omega_{pm})^2/\Gamma_D^2}\right.\right.$$
$$\left.\left. - \sum_{krs}\frac{4\omega_{rm}^3}{3\hbar c^3 \Gamma}\mathbf{d}_{mr}\cdot\mathbf{d}_{sn}\hat{\boldsymbol{\varepsilon}}\cdot\mathbf{d}_{rk}\hat{\boldsymbol{\varepsilon}}^*\cdot\mathbf{d}_{ks}e^{-(\omega-\omega_{rk})^2/\Gamma_D^2}\right)\right], \tag{12.30}$$

where we have used the explicit form (12.2) of the spontaneous emission operator.

12.4 Repopulation and depopulation

The first term in Eqs. (12.27) and (12.30) describes the unperturbed ground-state density matrix, while the second term of each of these equations describes the perturbation induced by the light. There are two contributions—the first describes the effect of depopulation pumping, i.e., atoms that have been removed from the ground state by optical pumping. Defining $\Lambda_{mm} = \gamma/w$, where w is the multiplicity of the ground state, this contribution takes the form

$$\rho_{mn}^{(\text{depop.})} \approx -\frac{\pi}{2}\frac{\mathcal{E}_0^2}{w\hbar^2(i\omega_{mn}+\gamma)}\sum_p \hat{\boldsymbol{\varepsilon}}^*\cdot\mathbf{d}_{mp}\hat{\boldsymbol{\varepsilon}}\cdot\mathbf{d}_{pn}G(\omega'_{pm}), \tag{12.31}$$

where the lineshape function $G(\Delta)$ is given by

$$G(\Delta) = \frac{1}{\pi}\frac{\Gamma/2}{\Delta^2 + (\Gamma/2)^2} \tag{12.32}$$

when there is no velocity mixing, and

$$G(\Delta) = \frac{1}{\Gamma_D\sqrt{\pi}}e^{-\Delta^2/\Gamma_D^2} \tag{12.33}$$

when there is complete mixing. (In the case with no mixing, the argument to the lineshape function is the Doppler-shifted light detuning, while in the complete-mixing case it is the unshifted detuning.)

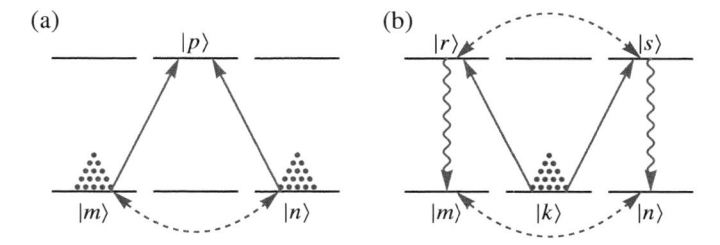

Fig. 12.2 Generation of coherence in the ground state of an atomic transition. (a) Two ground-state sublevels $|m\rangle$, $|n\rangle$ connected to the same upper-state sublevel $|p\rangle$. (b) A ground-state sublevel $|k\rangle$ connected to two upper-state sublevels $|r\rangle$, $|s\rangle$, generating coherence between them. The coherence is then transferred to a pair of ground-state sublevels $|m\rangle$, $|n\rangle$ via spontaneous decay.

The second contribution describes repopulation pumping—atoms that have been returned to the ground state through spontaneous decay. It is given by

$$\rho_{mn}^{(\text{repop.})} \approx \frac{\pi}{2} \frac{\mathcal{E}_0^2}{w\hbar^2 (i\omega_{mn} + \gamma)} \sum_{krs} \frac{4\omega_{rm}^3}{3\hbar c^3 \Gamma} \mathbf{d}_{mr} \cdot \mathbf{d}_{sn} \hat{\boldsymbol{\varepsilon}} \cdot \mathbf{d}_{rk} \hat{\boldsymbol{\varepsilon}}^* \cdot \mathbf{d}_{ks} G(\omega_{rk}'). \tag{12.34}$$

In a perturbative analysis, the various effects of the light can be assumed to be acting independently. Therefore, these terms can be considered separately in order to analyze the various processes occurring during light–atom interactions. We will do this in Chapter 13.

The terms describing depopulation and repopulation pumping can be converted into rates by noting that the characteristic time for the ground state to reach equilibrium is $1/\gamma$. Because Eqs. (12.31) and (12.34) give the steady-state value for each process, the rate of change for each process must be obtained by multiplying by γ:

$$\dot{\rho}_{mn}^{(\text{depop.})} \approx -\frac{\pi}{2} \frac{\gamma \mathcal{E}_0^2}{w\hbar^2 (i\omega_{mn} + \gamma)} \sum_p \hat{\boldsymbol{\varepsilon}}^* \cdot \mathbf{d}_{mp} \hat{\boldsymbol{\varepsilon}} \cdot \mathbf{d}_{pn} G(\omega_{pm}'), \tag{12.35}$$

and

$$\dot{\rho}_{mn}^{(\text{repop.})} \approx \frac{\pi}{2} \frac{\gamma \mathcal{E}_0^2}{w\hbar^2 (i\omega_{mn} + \gamma)} \sum_{krs} \frac{4\omega_{rm}^3}{3\hbar c^3 \Gamma} \mathbf{d}_{mr} \cdot \mathbf{d}_{sn} \hat{\boldsymbol{\varepsilon}} \cdot \mathbf{d}_{rk} \hat{\boldsymbol{\varepsilon}}^* \cdot \mathbf{d}_{ks} G(\omega_{rk}'). \tag{12.36}$$

By examining the formulas for repopulation and depopulation pumping, we can see how coherences can be generated in the ground state by the light. The term for depopulation pumping on a coherence ρ_{mn} is nonzero only if $|m\rangle$ and $|n\rangle$ are both connected to a single upper state $|p\rangle$ (Fig. 12.2a). Coherence is created by repopulation pumping, on the other hand, when the light connects the states via the chain $|m\rangle \leftrightarrow |r\rangle \leftrightarrow |k\rangle \leftrightarrow |s\rangle \leftrightarrow |n\rangle$. The ground state $|k\rangle$ is connected to two upper states $|r\rangle$ and $|s\rangle$, generating coherence between them; the coherence is then transferred to the ground state by spontaneous decay (Fig. 12.2b). Note that if there is initially coherence in the ground state, it can be transferred to the upper state by the light, allowing for another path by which repopulation pumping can create coherence in the ground state (see Sec. 12.5).

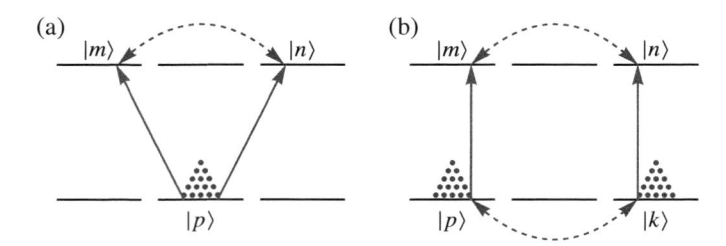

Fig. 12.3 Generation by excitation light of coherence in the upper state of an atomic transition. (a) A ground-state sublevel connected to two upper-state sublevels. (b) A pair of ground-state sublevels with coherence between them connected to a pair of upper-state sublevels.

12.5 Optical excitation

We can also use Eq. (12.19) to examine the effect of excitation light on the upper state. Assuming now that $|m\rangle$ and $|n\rangle$ are upper states, Eq. (12.19) reduces to

$$\rho_{mn} \approx -\frac{1}{i\widetilde{\omega}_{mn}} \sum_{pk} \frac{V_{mp}\Lambda_{pk}V_{kn}}{\widetilde{\omega}_{pk}} \left(\frac{1}{\widetilde{\omega}_{mk}} + \frac{1}{\widetilde{\omega}_{pn}} \right), \qquad (12.37)$$

where $|p\rangle$ and $|k\rangle$ now represent ground states. Under the assumption that ground- and upper-state coherences only develop between states that are split by much less than Γ, this formula becomes

$$\rho_{mn} \approx \sum_{pk} \frac{V_{mp}\Lambda_{pk}V_{kn}}{(i\omega_{pk}+\gamma)\left[\omega_{pn}^2 + (\Gamma/2)^2\right]}. \qquad (12.38)$$

Equation (12.38) indicates the ways in which coherences can be produced in the upper state. If Λ is diagonal, as would be the case for an initially unpolarized ground state, then we must have $p = k$. In that case, the only way that an upper-state coherence ρ_{mn} can be nonzero is if V_{mp} and V_{pn} are both nonzero; i.e., the light connects $|p\rangle$ to both $|m\rangle$ and $|n\rangle$ (Fig. 12.3a). If Λ has off-diagonal elements Λ_{pk}, representing initial ground-state coherences ρ_{pk}, then ρ_{mn} can be nonzero if the light connects $|p\rangle$ to $|m\rangle$ and $|k\rangle$ to $|n\rangle$ (Fig. 12.3b).

Substituting the optical Hamiltonian into Eq. (12.38) and using $\Lambda = \gamma \rho^{(0)}$, we have

$$\begin{aligned}\rho_{mn} &\approx \frac{\mathcal{E}_0^2}{4\hbar^2} \sum_{pk} \frac{\gamma}{i\omega_{pk}+\gamma} \frac{\hat{\varepsilon}\cdot\mathbf{d}_{mp}\rho_{pk}^{(0)}\hat{\varepsilon}^*\cdot\mathbf{d}_{kn}}{(\omega'_{pn})^2 + (\Gamma/2)^2} \\ &= \frac{\pi\mathcal{E}_0^2}{2\Gamma\hbar^2} \sum_{pk} \hat{\varepsilon}\cdot\mathbf{d}_{mp}\rho_{pk}^{(0)}\hat{\varepsilon}^*\cdot\mathbf{d}_{kn} G\left(\omega'_{pn}\right),\end{aligned} \qquad (12.39)$$

where in the last line we have assumed that $|p\rangle$ and $|k\rangle$ are degenerate.

> Additional discussion, including an examination of a set of monochromatic light fields of different frequencies and the transition to the broad-line approximation (Sec. 12.8), can be found in the book by Aleksandrov *et al.* (1993).

12.6 Absorption and optical rotation signals

In Sec. 10.1 we found expressions for optical signals by using the wave equation to relate the medium polarization to the changes in optical parameters. In that section we primarily used the

α–ϵ parametrization of the light field—it will also be convenient to write these expressions using the complex polarization vector $\hat{\varepsilon}$ to characterize the polarization of the light field. After obtaining these formulas, we will apply the perturbative solution of the density matrix to them. This will supply expressions for the optical signals in terms of the ground-state density matrix, valid to lowest order in the probe light field.

Equation (12.21) describes the electric field of the light. The complex polarization vector $\hat{\varepsilon}$ is normalized so that $\hat{\varepsilon} \cdot \hat{\varepsilon}^* = 1$. We are free to choose a phase convention for $\hat{\varepsilon}$; a convenient choice is to assume that $\hat{\varepsilon} \cdot \hat{\varepsilon}$ is real and nonnegative. Then $0 \le \hat{\varepsilon} \cdot \hat{\varepsilon} \le 1$, with $\hat{\varepsilon} \cdot \hat{\varepsilon} = 1$ for linear polarization and $\hat{\varepsilon} \cdot \hat{\varepsilon} = 0$ for circular polarization. The dot product between the real and imaginary parts of $\hat{\varepsilon}$ is then given by

$$(\operatorname{Re}\hat{\varepsilon}) \cdot (\operatorname{Im}\hat{\varepsilon}) = -\frac{i}{4}\left(\hat{\varepsilon} + \hat{\varepsilon}^*\right) \cdot \left(\hat{\varepsilon} - \hat{\varepsilon}^*\right)$$
$$= -\frac{i}{4}\left[\hat{\varepsilon} \cdot \hat{\varepsilon} - (\hat{\varepsilon} \cdot \hat{\varepsilon})^*\right] \tag{12.40}$$
$$= 0,$$

so we can define a pair of orthogonal real unit vectors by

$$\hat{\mathbf{e}}_1 = \frac{\operatorname{Re}\hat{\varepsilon}}{|\operatorname{Re}\hat{\varepsilon}|}, \tag{12.41a}$$

$$\hat{\mathbf{e}}_2 = \hat{\mathbf{k}} \times \hat{\mathbf{e}}_1 = \pm \frac{\operatorname{Im}\hat{\varepsilon}}{|\operatorname{Im}\hat{\varepsilon}|}. \tag{12.41b}$$

The magnitudes of the real and imaginary parts of $\hat{\varepsilon}$ are given by

$$|\operatorname{Re}\hat{\varepsilon}| = \frac{1}{2}|\hat{\varepsilon}^* + \hat{\varepsilon}| = \sqrt{\frac{1 + \hat{\varepsilon} \cdot \hat{\varepsilon}}{2}}, \tag{12.42a}$$

$$|\operatorname{Im}\hat{\varepsilon}| = \frac{1}{2}|i\left(\hat{\varepsilon}^* - \hat{\varepsilon}\right)| = \sqrt{\frac{1 - \hat{\varepsilon} \cdot \hat{\varepsilon}}{2}}. \tag{12.42b}$$

We can rewrite the optical electric field (12.21) in terms of $\hat{\mathbf{e}}_1$ and $\hat{\mathbf{e}}_2$ as

$$\mathcal{E} = \mathcal{E}_0 \operatorname{Re}\left(e^{i(\mathbf{k}\cdot\mathbf{r} - \omega t + \varphi)}\left(|\operatorname{Re}\hat{\varepsilon}|\hat{\mathbf{e}}_1 \pm i|\operatorname{Im}\hat{\varepsilon}|\hat{\mathbf{e}}_2\right)\right). \tag{12.43}$$

We can also parametrize the optical field as we have done previously, in terms of the rotation angle φ and ellipticity ϵ:

$$\mathcal{E} = \operatorname{Re}\left\{\mathcal{E}_0 e^{i(\mathbf{k}\cdot\mathbf{r}-\omega t+\varphi)}\left[(\cos\alpha\cos\epsilon - i\sin\alpha\sin\epsilon)\hat{\mathbf{e}}_1 + (\sin\alpha\cos\epsilon + i\cos\alpha\sin\epsilon)\hat{\mathbf{e}}_2\right]\right\}. \tag{12.44}$$

With the basis vectors defined in terms of the polarization vector by Eq. (12.41), we can compare Eqs. (12.43) and (12.44) to see that $\alpha = 0$. In addition, using Eq. (12.42), we can relate ϵ to $\hat{\varepsilon}$:

$$\cos\epsilon = \sqrt{\frac{1 + \hat{\varepsilon} \cdot \hat{\varepsilon}}{2}}, \tag{12.45a}$$

$$\sin\epsilon = \pm\sqrt{\frac{1 - \hat{\varepsilon} \cdot \hat{\varepsilon}}{2}}, \tag{12.45b}$$

which gives
$$\epsilon = \pm \arctan\sqrt{\frac{1-\hat{\boldsymbol{\varepsilon}}\cdot\hat{\boldsymbol{\varepsilon}}}{1+\hat{\boldsymbol{\varepsilon}}\cdot\hat{\boldsymbol{\varepsilon}}}}. \quad (12.46)$$

In Sec. 10.1, we found that the change in the optical parameters over an infinitesimal path length $d\ell$ is given by (assuming $\alpha = 0$)

$$\frac{1}{\mathcal{E}_0}\frac{d\mathcal{E}_0}{d\ell} = \frac{d\ln\mathcal{E}_0}{d\ell} = \frac{2\pi\omega}{\mathcal{E}_0 c}(P_2\cos\epsilon + P_3\sin\epsilon), \quad (12.47\text{a})$$

$$\frac{d\varphi}{d\ell} = \frac{2\pi\omega}{\mathcal{E}_0 c}\sec 2\epsilon\,(P_1\cos\epsilon + P_4\sin\epsilon), \quad (12.47\text{b})$$

$$\frac{d\alpha}{d\ell} = \frac{2\pi\omega}{\mathcal{E}_0 c}\sec 2\epsilon\,(P_1\sin\epsilon + P_4\cos\epsilon), \quad (12.47\text{c})$$

$$\frac{d\epsilon}{d\ell} = -\frac{2\pi\omega}{\mathcal{E}_0 c}(P_2\sin\epsilon - P_3\cos\epsilon). \quad (12.47\text{d})$$

Here the components P_i of the medium polarization $\mathbf{P} = n\langle\mathbf{d}\rangle$ are defined by

$$\mathbf{P} = \text{Re}\left\{e^{i(\mathbf{k}\cdot\mathbf{r}-\omega t+\varphi)}\left[(P_1 - iP_2)\,\hat{\mathbf{e}}_1 + (P_3 - iP_4)\,\hat{\mathbf{e}}_2\right]\right\}. \quad (12.48)$$

Evaluating the trace to find the medium polarization, we find

$$\begin{aligned}\mathbf{P} &= n\,\text{Tr}\,\rho\mathbf{d} \\ &= n\sum_{mp} 2\,\text{Re}\left(\rho_{pm}\mathbf{d}_{mp}\right) \\ &= n\sum_{mp} 2\,\text{Re}\left(\rho'_{pm}\mathbf{d}_{mp}e^{i(\mathbf{k}\cdot\mathbf{r}-\omega t+\varphi)}\right),\end{aligned} \quad (12.49)$$

where m runs over ground states, and p runs over excited states. In the last step we have written $\mathbf{P}$ in terms of the density matrix in the rotating frame, using $\rho = U\rho' U^\dagger$.

Comparing Eqs. (12.48) and (12.49), we find for the components of $\mathbf{P}$:

$$P_1 = 2n\sum \text{Re}\left[\rho'_{pm}\mathbf{d}_{mp}\cdot\hat{\mathbf{e}}_1\right], \quad (12.50\text{a})$$

$$P_2 = -2n\sum \text{Im}\left[\rho'_{pm}\mathbf{d}_{mp}\cdot\hat{\mathbf{e}}_1\right], \quad (12.50\text{b})$$

$$P_3 = 2n\sum \text{Re}\left[\rho'_{pm}\mathbf{d}_{mp}\cdot\hat{\mathbf{e}}_2\right], \quad (12.50\text{c})$$

$$P_4 = -2n\sum \text{Im}\left[\rho'_{pm}\mathbf{d}_{mp}\cdot\hat{\mathbf{e}}_2\right]. \quad (12.50\text{d})$$

We can write Eqs. (12.47) in terms of $\hat{\boldsymbol{\varepsilon}}$ and ρ' using Eqs. (12.41), (12.42), and (12.50). We find that we can write for an observable $\mathcal{O}$:

$$\frac{d\mathcal{O}}{d\ell} = -\frac{4\pi\omega n}{\mathcal{E}_0 c}\sum \text{Im}\left[\rho'_{pm}\mathbf{d}_{mp}\cdot\mathbf{u}_\mathcal{O}\right], \quad (12.51)$$

where

Absorption and optical rotation signals 251

$$\mathbf{u}_{\ln(\mathcal{E}_0)} = \hat{\boldsymbol{\varepsilon}}^*, \tag{12.52a}$$

$$\mathbf{u}_\varphi = -i\frac{\hat{\boldsymbol{\varepsilon}}^*}{|\hat{\boldsymbol{\varepsilon}} \cdot \hat{\boldsymbol{\varepsilon}}|}, \tag{12.52b}$$

$$\mathbf{u}_\alpha = \hat{\mathbf{k}} \times \frac{\operatorname{Re}\hat{\boldsymbol{\varepsilon}}}{|\hat{\boldsymbol{\varepsilon}} \cdot \hat{\boldsymbol{\varepsilon}}|} - i\frac{\operatorname{Re}\hat{\boldsymbol{\varepsilon}}}{|\hat{\boldsymbol{\varepsilon}} \cdot \hat{\boldsymbol{\varepsilon}}|}\sqrt{\frac{1-\hat{\boldsymbol{\varepsilon}}\cdot\hat{\boldsymbol{\varepsilon}}}{1+\hat{\boldsymbol{\varepsilon}}\cdot\hat{\boldsymbol{\varepsilon}}}}, \tag{12.52c}$$

$$\mathbf{u}_\epsilon = -i\left(\hat{\mathbf{k}} \times \operatorname{Re}\hat{\boldsymbol{\varepsilon}} \mp i\operatorname{Re}\hat{\boldsymbol{\varepsilon}}\sqrt{\frac{1-\hat{\boldsymbol{\varepsilon}}\cdot\hat{\boldsymbol{\varepsilon}}}{1+\hat{\boldsymbol{\varepsilon}}\cdot\hat{\boldsymbol{\varepsilon}}}}\right). \tag{12.52d}$$

For linear polarization, this reduces to

$$\mathbf{u}_{\ln(\mathcal{E}_0)} = \hat{\boldsymbol{\varepsilon}}, \tag{12.53a}$$

$$\mathbf{u}_\varphi = -i\hat{\boldsymbol{\varepsilon}}, \tag{12.53b}$$

$$\mathbf{u}_\alpha = \hat{\mathbf{k}} \times \hat{\boldsymbol{\varepsilon}}, \tag{12.53c}$$

$$\mathbf{u}_\epsilon = -i\hat{\mathbf{k}} \times \hat{\boldsymbol{\varepsilon}}. \tag{12.53d}$$

We can use the perturbation expansion obtained in Sec. 12.2 to find a lowest-order expression for Eq. (12.51) in terms of the ground-state density matrix elements. As discussed in Sec. 10.3, optical polarization first appears at first order in the expansion in the electric field. From Eq. (12.19) we have for the first-order optical coherences

$$\rho'_{pm} \approx \sum_n \frac{iV_{pn}\Lambda_{nm}}{\widetilde{\omega}'_{pm}\widetilde{\omega}_{mn}}. \tag{12.54}$$

We assume here that Λ represents the influx of ground-state atoms that have been polarized by some prior pump-light interaction. If these atoms have density matrix $\rho^{(0)}$ then $\Lambda_{nm} = \gamma\rho^{(0)}_{nm}$ and we have

$$\rho'_{pm} \approx -\frac{i}{2}\frac{\gamma\mathcal{E}_0}{\hbar}\sum_n \frac{\hat{\boldsymbol{\varepsilon}}\cdot\mathbf{d}_{pn}\rho^{(0)}_{nm}}{\widetilde{\omega}'_{pm}\widetilde{\omega}_{mn}}. \tag{12.55}$$

Substituting this expression into Eq. (12.51), we find

$$\frac{d\mathcal{O}}{d\ell} = \frac{2\pi\omega\gamma n}{\hbar c}\sum_{pmn}\operatorname{Re}\left[\frac{\hat{\boldsymbol{\varepsilon}}\cdot\mathbf{d}_{pn}\rho^{(0)}_{nm}\mathbf{d}_{mp}\cdot\mathbf{u}_\mathcal{O}}{\widetilde{\omega}'_{pm}\widetilde{\omega}_{mn}}\right]. \tag{12.56}$$

Neglecting coherences between nondegenerate ground states, we find

$$\frac{d\mathcal{O}}{d\ell} = -\frac{2\pi\omega n}{\hbar c}\sum_{pmn}\operatorname{Im}\left[\frac{\hat{\boldsymbol{\varepsilon}}\cdot\mathbf{d}_{pn}\rho^{(0)}_{nm}\mathbf{d}_{mp}\cdot\mathbf{u}_\mathcal{O}}{\widetilde{\omega}'_{pm}}\right]$$
$$= -\frac{2\pi\omega n}{\hbar c}\operatorname{Im}\left[\hat{\boldsymbol{\varepsilon}}\cdot\overleftrightarrow{\boldsymbol{\beta}}\cdot\mathbf{u}_\mathcal{O}\right], \tag{12.57}$$

where we have defined

$$\overleftrightarrow{\boldsymbol{\beta}} = \sum_{pmn}\frac{\mathbf{d}_{pn}\rho^{(0)}_{nm}\mathbf{d}_{mp}}{\widetilde{\omega}'_{pm}}. \tag{12.58}$$

In Chapter 13 we calculate the light absorption under the assumption that $\rho_{nm}^{(0)}$ is independent of $\mathbf{v}$ (complete-mixing approximation). From Eqs. (12.25), (12.52a), and (12.57) we obtain

$$\frac{1}{\mathcal{E}_0}\frac{d\mathcal{E}_0}{d\ell} = -\frac{2\pi\omega n}{\hbar c}\sum_{pmn}\text{Im}\left(\frac{\hat{\boldsymbol{\varepsilon}}\cdot\mathbf{d}_{pn}\rho_{nm}^{(0)}\mathbf{d}_{mp}\cdot\hat{\boldsymbol{\varepsilon}}^*}{\omega_{pm}-\omega+\mathbf{k}\cdot\mathbf{v}-i\Gamma/2}\right)$$
$$= -\frac{\pi\omega n\Gamma}{\hbar c}\sum_{pmn}\frac{\hat{\boldsymbol{\varepsilon}}\cdot\mathbf{d}_{pn}\rho_{nm}^{(0)}\mathbf{d}_{mp}\cdot\hat{\boldsymbol{\varepsilon}}^*}{(\omega_{pm}-\omega+\mathbf{k}\cdot\mathbf{v})^2+(\Gamma/2)^2}. \quad (12.59)$$

The total light absorption is given by the average of Eq. (12.59) over atomic velocities. Weighting by a Maxwellian Doppler distribution and integrating as in Sec. 12.2, we find for the fractional change in electric field

$$\frac{1}{\mathcal{E}_0}\frac{d\mathcal{E}_0}{d\ell} = -\frac{2\pi^2\omega n}{\hbar c}\sum_{pmn}\hat{\boldsymbol{\varepsilon}}\cdot\mathbf{d}_{pn}\rho_{nm}^{(0)}\mathbf{d}_{mp}\cdot\hat{\boldsymbol{\varepsilon}}^*G\left(\omega_{pm}\right), \quad (12.60)$$

where the lineshape function G is as defined in Eq. (12.33).

12.7 What kind of atomic polarization can influence the absorption and emission of light?

Say we have a polarized light beam that interacts with an ensemble of atoms, and we measure absorption. What ground-state atomic polarization moments can influence absorption? A related question is—what kind of polarization in the excited state can influence emission?

The answer is given by the form of the equations for absorption (Eq. 12.60) and emission (Eq. 12.7). Consider absorption from a state with total angular momentum F. The left side of Eq. (12.60) does not refer to atomic wave functions. Correspondingly, it must be a rank-zero quantity. Therefore, the right side must be rank zero as well. On the right, there are two occurrences of the rank-one electric-dipole operator; these are contracted with the density matrix, which may have any polarization moments of rank up to $2F$. The tensor contraction rules say that when contracting two tensors of rank κ and κ', the rank of the result is in the range $|\kappa-\kappa'|$ to $\kappa+\kappa'$. This means that when the two dipole operators are combined the maximum possible rank is two, and only the polarization moments of ρ with rank up to two can contribute to the absorption. A similar conclusion holds for emission, as well as, in fact, other types of light–atom interactions such as optical polarization rotation. We discuss specific examples of this in Chapter 13.

Ultimately, these results are related to the fact that the photon is a spin-one particle, and we are considering low-power perturbative processes which correspond to a single photon participating in absorption or emission. Using light power such that multi-photon processes occur, higher-rank polarization moments can be created and detected (Chapter 18).

12.8 The broad-line approximation

Besides the case of low light power, another situation in which useful approximate solutions for the density matrix can be found is when the incident light has a broad spectral profile. Under the *broad-line approximation* (BLA), this broad linewidth is assumed to be the result of fast phase fluctuations of the light field, corresponding to a short coherence lifetime and fast relaxation of optical coherences, faster than the other relaxation rates and pumping rates in the problem. Under this approximation, the optical coherences can be eliminated from the time-dependent density-matrix-evolution equations, producing a smaller set of rate equations for the ground- and excited-state populations and coherences.

At first glance, it would seem that this is an approximation that would best describe spectroscopic experiments of the past, when narrow-band lasers were not available, and physicists had to contend with using spectrally broad light sources such as discharge lamps. While this is certainly the case (the broad-line approximation was introduced in the 1960s; see the original paper reprinted in the book by Cohen-Tannoudji, 1994), this approximation remains a useful tool. For example, it can be applied to short pulses (Sec. 15.2.2), and can be valuable in some situations with broadened laser light (Auzinsh *et al.* 2009b). It can also be useful for theoretical analysis of a system, as rate equations are often more intuitive than the complete Liouville equation.

We first take a detour and describe how the optical coherences can be eliminated from the equations for the steady-state density matrix, without making any approximations. We then show how this approach can be extended to the time-dependent evolution equations under the broad-line approximation.

12.8.1 Reduction of steady-state density-matrix equations

Consider the Liouville equation written in the form (12.17):

$$\dot{\rho}_{mn} = -i\widetilde{\omega}_{mn}\rho_{mn} + \Lambda_{mn} + \sum_{rs} F^{sr}_{mn}\rho_{rs} - i\sum_{p}\left[V_{mp}\rho_{pn} - \rho_{mp}V_{pn}\right]. \quad (12.61)$$

Here V describes an interaction that connects lower and upper states. Previously in this chapter we assumed that this interaction was weak, but in this section we take it to have arbitrary strength, so that the analysis is nonperturbative. We will consider the optical-field case under the rotating-wave approximation, in which V is given by (Eq. 12.24)

$$V_{pk} = -\frac{\mathcal{E}_0}{2\hbar}\hat{\varepsilon}\cdot\mathbf{d}_{pk}, \quad (12.62\mathrm{a})$$

$$V_{kp} = -\frac{\mathcal{E}_0}{2\hbar}\hat{\varepsilon}^*\cdot\mathbf{d}_{kp}, \quad (12.62\mathrm{b})$$

for $|p\rangle$ an excited state and $|k\rangle$ a ground state. Also, the complex frequency splitting $\widetilde{\omega}$ is given for the optical transitions by (Eq. 12.25)

$$\widetilde{\omega}_{pk} = \omega_{pk} - \omega + \mathbf{k}\cdot\mathbf{v} - \frac{i}{2}(\Gamma + \gamma), \quad (12.63\mathrm{a})$$

$$\widetilde{\omega}_{kp} = -(\omega_{pk} - \omega + \mathbf{k}\cdot\mathbf{v}) - \frac{i}{2}(\Gamma + \gamma), \quad (12.63\mathrm{b})$$

where we omit the prime previously used to indicate the rotating basis.

In the steady-state condition, the time derivatives are zero, and we can solve the Liouville equation implicitly as in Eq. (12.18). Now consider an optical coherence ρ_{mp}, where m is a ground state and p is an excited state, or vice versa. Such a coherence is not replenished by atomic transit or spontaneous decay, so Eq. (12.18) reduces to

$$\rho_{mp} = \frac{1}{\widetilde{\omega}_{mp}}\sum_{r}\left(\rho_{mr}V_{rp} - V_{mr}\rho_{rp}\right), \quad (12.64)$$

Note that the right-hand side of Eq. (12.64) is free of optical coherences—ρ_{mr} and ρ_{rp} are either ground- or excited-state density-matrix elements. We can therefore substitute Eq. (12.64) back into the steady-state version of (12.61) to eliminate the optical coherences:

$$0 = -i\widetilde{\omega}_{mn}\rho_{mn} + \Lambda_{mn} + \sum_{rs} F^{sr}_{mn}\rho_{rs}$$

$$-i\sum_{pr}\left[\left(\frac{1}{\widetilde{\omega}_{pn}} + \frac{1}{\widetilde{\omega}_{mr}}\right)V_{mp}\rho_{pr}V_{rn} - \frac{V_{mp}V_{pr}\rho_{rn}}{\widetilde{\omega}_{pn}} - \frac{\rho_{mp}V_{pr}V_{rn}}{\widetilde{\omega}_{mr}}\right]. \quad (12.65)$$

This is now a complete set of equations for the lower- and upper-state density-matrix elements, a much smaller set of variables than the full system including the optical coherences.

Equation (12.65) also has the advantage that it manifestly displays the resonant character of the interaction. Assuming for simplicity that both the lower states and the upper states are nearly degenerate, we can write the Doppler-shifted light detuning of Eq. (12.63) as $-(\omega_{pk} - \omega + \mathbf{k}\cdot\mathbf{v}) = \Delta$, giving

$$\frac{1}{\widetilde{\omega}_{pn}} = -\frac{1}{\pm\Delta + \frac{i}{2}(\Gamma+\gamma)} = i\left(\frac{\frac{1}{2}(\Gamma+\gamma)}{\Delta^2 + \frac{1}{4}(\Gamma+\gamma)^2} \pm \frac{i\Delta}{\Delta^2 + \frac{1}{4}(\Gamma+\gamma)^2}\right), \quad (12.66)$$

where we take the plus sign if $|n\rangle$ is an upper state, and the minus sign if it is a lower state. Each of the terms in the parentheses are resonant lineshape factors, the first term describing atomic transitions, and the second describing AC Stark shifts (Sec. 4.5.3). The terms in the square brackets in Eq. (12.65) can also be assigned physical meanings: for a lower state, the first term describes transitions from the upper states due to stimulated emission, while the second and third terms describe transfer of atoms between the lower states (optical pumping) as well as Stark shifts; for an upper state, the first term describes transitions from the ground state, while the second and third describe transfer between the upper states.

Note that we have not used any approximations in deriving Eq. (12.65). The derivation does, however, depend crucially on the steady-state condition—otherwise, we would not be able to solve for the optical coherences to obtain Eq. (12.64). In the next section we discuss the broad-line approximation, under which it is possible to perform this step even when solving the time-dependent equations.

12.8.2 Adiabatic elimination

Although the broad-line approximation for light–atom interactions can be introduced into the density-matrix formalism phenomenologically (Cohen-Tannoudji 1994), a more systematic approach allows a derivation beginning from the Liouville equation. We first give some physical justification for the technique, and then outline a rigorous method.

As mentioned above, the condition for the broad-line approximation is that the spectral width of the light is broader than any other relaxation or transition rate in the problem. Because spectrally broad light can be viewed as radiation with rapidly and chaotically fluctuating phase, this means that the coherence lifetime τ_l of the light is much shorter than any other relaxation time. This light generates optical coherences, which are subject to the same relaxation lifetime τ_l. Now consider a time period that is longer than τ_l, but shorter than any other relaxation lifetime, so that the lower- and upper-state density-matrix elements are essentially constant over this time. The optical coherences will quickly relax to the steady state determined by the other density-matrix elements.[1] Subsequent light-phase fluctuations will alter the optical coherences, but the new values will quickly relax also. Thus the optical coherences will fluctuate, but their mean values will be those of the steady state. If we neglect the fluctuations, we can set the time derivatives of the optical coherences to zero in the Liouville equation. This procedure is known as *adiabatic elimination* (Stenholm 2005). Note that taking this step means that the equations are valid only over timescales long enough that the fluctuations can be averaged over. On even longer timescales, we see that the mean values of the optical coherences depend on the current values of the other density-matrix elements, but not on the history of the optical coherences themselves, due to the randomization caused by the fluctuations. In other words, from the point of view of solving the system of differential equations for the density-matrix evolution, the optical coherences are now dependent variables, rather than independent variables, and can be eliminated from the equations, as was done in the previous section under steady-state conditions. This provides a set of rate equations for the density-matrix elements, identical to those obtained in the previous section, except that the time derivatives for the ground- and excited-state density-matrix elements have not been set to zero:

$$\dot{\rho}_{mn} = -i\widetilde{\omega}_{mn}\rho_{mn} + \Lambda_{mn} + \sum_{rs} F_{mn}^{sr}\rho_{rs}$$
$$- i\sum_{pr}\left[\left(\frac{1}{\widetilde{\omega}_{pn}} + \frac{1}{\widetilde{\omega}_{mr}}\right)V_{mp}\rho_{pr}V_{rn} - \frac{V_{mp}V_{pr}\rho_{rn}}{\widetilde{\omega}_{pn}} - \frac{\rho_{mp}V_{pr}V_{rn}}{\widetilde{\omega}_{mr}}\right]. \quad (12.67)$$

This is the set of approximate equations we have set out to obtain, describing the evolution of the density matrix in the presence of broad-band light. However, our derivation has not been completely mathematically justified. In particular, we have retained the information about the mean values of the optical coherences, but discarded the effects of fluctuations about the mean. A more rigorous derivation of the rate equations under the broad-line approximation was obtained by Blushs and Auzinsh (2004), using a procedure called *decoherence analysis*. In this derivation, the fluctuating phase of the light field is represented explicitly as a time-dependent light phase $\varphi(t)$. There are two models commonly used to describe the phase instabilities of the light, in which the phase either randomly jumps or undergoes random continuous drift (*phase diffusion*). Each of these models results in a light-frequency distribution with a Lorentzian spectral profile, and to our level of approximation, they each lead to the same set of evolution equations.

To find the desired rate equations, the differential equations for the optical coherences obtained from the Liouville equation are converted to integral form, and an average is taken over all possible fluctuations of $\varphi(t)$. Using one of the above models for the light-phase fluctuations, and employing the *decorrelation approximation*, in which the correlations between the fluctuations of the ground- and excited-state density-matrix elements and the optical coherences are neglected, the averages can be evaluated to find evolution equations for the mean values of the density-matrix elements. This gives solutions for the optical coherences that can be substituted back into the Liouville equation, as before. The result is equations for the evolution of the average values of the density-matrix elements that are identical to Eq. (12.67) except for a modification made to the effective linewidth of the transition. Equation (12.66) becomes

---

[1] Here we consider the density matrix under the rotating-wave approximation, in which the optical coherences are constant in the steady state.

$$\frac{1}{\widetilde{\omega}_{pn}} = -\frac{1}{\pm\Delta + \frac{i}{2}(\Gamma + \gamma + \Delta\omega)}$$
$$= i\left(\frac{\frac{1}{2}(\Gamma + \gamma + \Delta\omega)}{\Delta^2 + \frac{1}{4}(\Gamma + \gamma + \Delta\omega)^2} \pm \frac{i\Delta}{\Delta^2 + \frac{1}{4}(\Gamma + \gamma + \Delta\omega)^2}\right), \quad (12.68)$$

where $\Delta\omega$ is the full width at half maximum of the light spectral profile, and Δ is now the Doppler-shifted detuning of the center of the laser line from resonance. Thus the effective linewidth of the transition is broadened by the laser linewidth, as one would expect on physical grounds.

We will use the broad-line approximation to calculate optical pumping by short pulses in Sec. 15.2.2.

13

Polarization effects in transitions with partially resolved hyperfine structure

Atomic polarization, created in a medium by polarized light, can modify the optical response of the medium, affecting the light field. As we have already mentioned, the absorption of light of a particular polarization by atoms in a polarized state can be reduced [*electromagnetically induced transparency* (Fleischhauer *et al.* 2005)] or increased [*electromagnetically induced absorption* (Lezama *et al.* 1999)] compared to that for an unpolarized state. *Coherent population trapping* (Nasyrov *et al.* 2006) is a closely related phenomenon, the study of which led to the discovery of an interesting effect that is also a powerful tool for the manipulation of atomic states: coherent population transfer between atomic states, known as *STIRAP* [*stimulated Raman adiabatic passage* (Bergmann *et al.* 1998)]. *"Lasing without inversion"* (Kocharovskaya and Khanin Ya 1988; Scully *et al.* 1989) is another related effect.

Additional effects are encountered when atoms interact with coherent light in the presence of a magnetic field (Budker *et al.* 2002a; Alexandrov *et al.* 2005).[1] These magneto-optical effects—especially those involving magnetic-field-induced evolution of long-lived ground-state polarization—can be used to perform sensitive magnetometry (Budker and Romalis 2007).

> These effects are also often referred to as "coherence effects," although this is something of a misnomer, as in some cases the effects can be described using a basis in which there are no ground-state coherences (Kanorsky *et al.* 1993).

The atomic polarization responsible for specific effects, such as nonlinear magneto-optical rotation (NMOR), can be described in terms of the polarization moments (PM) in the multipole expansion of the density matrix (Sec. 5.7). The lowest-rank multipole moments correspond to population, described by a rank $\kappa = 0$ tensor, orientation, described by a rank $\kappa = 1$ tensor, and alignment, described by a rank $\kappa = 2$ tensor. It is these three lowest-rank multipole moments that can directly affect light absorption and laser-induced fluorescence (Dyakonov 1965; Auzinsh and Ferber 1995), and thus can be created and detected through single-photon interactions. An atomic state with total angular momentum F can support multipole moments with rank up to $\kappa = 2F$; multi-photon interactions and multipole transitions higher than dipole allow the higher-order moments to be created and detected. Magneto-optical techniques can be used to selectively address individual high-rank multipoles as discussed in Chapter 18.

---

[1] Budker and Rochester (2004) discuss a relationship between these effects and electromagnetically induced absorption.

Magneto-optical coherence effects that involve linearly polarized light generally require the production and detection of polarization corresponding to atomic alignment. [There are multi-field, high-light-power effects in which alignment is converted to orientation, which is then detected (Sec. 11.3); these effects still depend on the creation of alignment by the light]. Thus, for ground-state coherence effects, the ground state in question must have angular momentum of at least $F = 1$ in order to support a rank-two polarization moment. The alkali atoms K, Rb, and Cs—commonly used for magneto-optical experiments—each have ground-state hyperfine sublevels with $F \geq 1$. If light is tuned to a suitable transition between a ground-state and an excited-state hyperfine sublevel, alignment can be created and detected in the ground state.

The situation changes, however, if the hyperfine structure is not resolved. If the hyperfine transitions are completely unresolved (as was the case in early work that used broad-band light sources such as electrodeless discharge lamps to excite atoms), then it is the fine-structure transition that is effectively excited—the $D1$ line ($n^2 S_{1/2} \to n^2 P_{1/2}$) or the $D2$ line ($n^2 S_{1/2} \to n^2 P_{3/2}$). In this case, the effects related to the excitation of a particular hyperfine transition are averaged out when all transitions are summed over. Thus the effect of the nuclear spin is removed, and the states have effective total angular momentum $J = 1/2$ for the ground state and $J = 1/2$ or $3/2$ for the excited state. In this case the highest rank multipole moment that can be supported by the ground state is orientation ($\kappa = 2J = 1$), and effects depending on atomic ground-state alignment will not be apparent.

In practical experiments with alkali atoms in vapor cells, even when narrow-band laser excitation is used, the hyperfine structure is in general only partially resolved, due to Doppler broadening. At room temperature, the Doppler widths of the atomic transitions in K, Rb, and Cs range from 463 MHz for K to 226 MHz for Cs. The ground-state hyperfine splittings, ranging from 462 MHz for K to 9.192 GHz for Cs, are on the order of or greater than the Doppler widths, while the excited-state hyperfine splittings, ranging from 8 MHz to 1.167 GHz, are generally on the order of or smaller than the Doppler width. Thus the question arises: how do coherence effects depend on the ground- and excited-state hyperfine splitting when the hyperfine structure is neither completely resolved nor completely unresolved?

In this chapter we discuss transitions for which one or the other of the excited- or ground-state hyperfine structure (hfs) is completely unresolved. We determine which polarization moments can be created in the ground state via single-photon interactions, and which moments can be detected through their influence on light absorption. We find that the two contributions to the ground-state polarization—absorption and polarization transfer through spontaneous decay—depend differently on the ground- and excited-state hyperfine structure.

In Chapter 14, we choose a particular system and investigate the detailed dependence of NMOR signals on the excited- and ground-state hyperfine splitting. We consider three cases: systems in which the atomic Doppler distribution can be neglected, and systems in which the Doppler distribution is broad compared to the natural linewidth and in which the rate of velocity-changing collisions is either much slower than or much faster than the ground-state polarization relaxation rate.

Throughout the discussion we use the low-light-intensity approximation in order to simplify the calculations and obtain analytic results. It can be shown, using higher-order perturbation theory and numerical calculations, that the essential results presented here hold for arbitrary light intensity, as well. Previous work that discusses the dependence of optical pumping on whether or not hyperfine structure (hfs) is resolved includes that of Happer and Mathur (1967); Happer (1972); and Lehmann (1967).

In this section, we discuss the creation and detection of atomic polarization in systems for which either the ground- or excited-state hyperfine structure is unresolved. This section deals with systems that can be described using the complete-mixing approximation, i.e., the assumption that atomic velocities are completely rethermalized in between optical pumping and probing. This is the case for experiments using buffer-gas or antirelaxation-coated vapor cells, in which atoms undergo frequent velocity-changing collisions during the ground-state polarization lifetime. The consequences of the complete-mixing approximation are similar to those of the broad-line approximation (Barrat and Cohen-Tannoudji 1961a;b; Cohen-Tannoudji 1962a;b), which takes the spectrum of the pump light to be broader than the Doppler width of the ensemble. In the complete-mixing case, narrow-band light produces polarization in a single velocity group in the Doppler distribution, and the polarization is averaged over all velocity groups through rethermalization, while in the broad-line case, the entire Doppler distribution is pumped directly.

13.1 Depopulation pumping

We consider an ensemble of atoms with nuclear spin I, a ground state with electronic angular momentum J_g, and an excited state with angular momentum J_e. The various ground- and excited-state hyperfine levels are labeled by F_g and F_e, respectively. The atoms are subject to weak monochromatic light with complex polarization vector $\hat{\varepsilon}$ and frequency ω, near-resonant with the atomic transition frequency $\omega_{J_g J_e}$. We assume that the atoms undergo collisions that mix different components of the Doppler distribution (or, equivalently, use the broad-line approximation). We also neglect coherences between different ground-state or different excited-state hyperfine levels (these coherences will not develop for low light power as long as the hyperfine splittings are larger than the natural width of the excited state). We first consider polarization produced in the ground state due to atoms absorbing light and being transferred to the excited state (depopulation pumping). The general form of the contribution to the ground-state density matrix due to this effect was found in Chapter 12:

$$\rho_{mn}^{(depop)} \propto \sum_r \hat{\varepsilon}^* \cdot \mathbf{d}_{mr} \hat{\varepsilon} \cdot \mathbf{d}_{rn} G(\omega - \omega_{rn}), \tag{13.1}$$

where m and n are degenerate ground states, r is an excited state, ω_{rn} is the transition frequency between r and n, and G is a function describing the spectral lineshape. If the natural width of the excited state is much smaller than the Doppler width Γ_D, G is approximately a Gaussian of the Doppler width. For the system described above, this takes the form

$$\begin{aligned}\rho_{F_g m, F_g m'}^{(depop)} &\propto \sum_{F_e m''} \langle F_g m | \hat{\varepsilon}^* \cdot \mathbf{d} | F_e m'' \rangle \\ &\quad \times \langle F_e m'' | \hat{\varepsilon} \cdot \mathbf{d} | F_g m' \rangle G(\omega - \omega_{F_e F_g}).\end{aligned} \tag{13.2}$$

Now suppose that the light frequency is tuned so that it is close, compared with the Doppler width, to an unresolved group of transition frequencies, and far from every other transition frequency (Fig. 13.1). We employ the simplest approximation that $G(\omega - \omega_{F_e F_g})$ takes the same value for each transition in the unresolved group, and is zero for all other transitions. With these approximations, Eq. (13.2) becomes

$$\rho_{F_g m, F_g m'}^{(depop)} \propto \sum_{F_e m''} \langle F_g m | \hat{\varepsilon}^* \cdot \mathbf{d} | F_e m'' \rangle \langle F_e m'' | \hat{\varepsilon} \cdot \mathbf{d} | F_g m' \rangle, \tag{13.3}$$

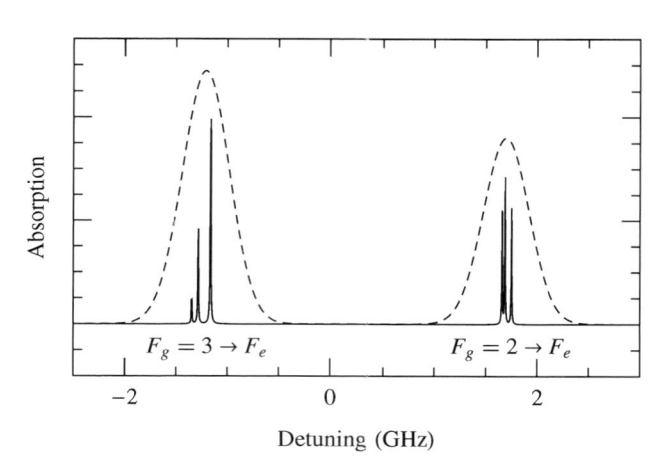

Fig. 13.1 (From Auzinsh et al. 2009a.) Doppler-free (solid line) and Doppler-broadened (dashed line) absorption spectra for the ^{85}Rb D2 line. A Maxwellian velocity distribution at room temperature is assumed. If the incident light frequency is tuned near the center of the $F_g = 2 \to F_e$ transition group, the condition discussed in the text is fulfilled. Namely, the light detuning from each resonance frequency is either much less than or much greater than the Doppler width. The condition holds somewhat less rigorously for light tuned to the center of the $F_g = 3 \to F_e$ transition group.

where the sum now runs over only those excited states F_e that connect via one of the unresolved resonant transitions to the ground state F_g in question. (This sum also arises in the broad-line approximation.)

We now investigate which coherences can be created in the ground state by the light. As we will see, this will determine which polarization moments can be created. Suppose first that the excited-state hfs is entirely unresolved. Then the sum over $|F_e m''\rangle\langle F_e m''|$ in Eq. (13.3) runs over all excited states, so that it is equivalent to the identity. We replace this sum with the sum over the eigenstates in the uncoupled basis $\sum_{m_I'' m_J''} |I m_I'' J_e m_J''\rangle\langle I m_I'' J_e m_J''|$. Further, we insert additional sums to expand the ground-state coupled-basis eigenstates in terms of the uncoupled basis. We also expand $\hat{\varepsilon}$ and $\mathbf{d}$ in terms of their spherical components. Equation (13.3) becomes

$$\rho_{F_g m, F_g m'}^{(depop)} \propto \sum (-1)^{q'+q''} (\varepsilon^*)_{q'} \varepsilon_{q''} \langle F_g m | I m_I J_g m_J \rangle \langle I m_I J_g m_J | d_{-q'} | I m_I'' J_e m_J'' \rangle$$
$$\times \langle I m_I'' J_e m_J'' | d_{-q''} | I m_I' J_g m_J' \rangle \langle I m_I' J_g m_J' | F_g m' \rangle$$
$$= \sum (-1)^{q'+q''} (\varepsilon^*)_{q'} \varepsilon_{q''} \langle F_g m | I m_I J_g m_J \rangle \langle J_g m_J | d_{-q'} | J_e m_J'' \rangle \qquad (13.4)$$
$$\times \langle J_e m_J'' | d_{-q''} | J_g m_J' \rangle \langle I m_I J_g m_J' | F_g m' \rangle,$$

where the inner products $\langle \cdots | \cdots \rangle$ are given by the Clebsch–Gordan coefficients, with

$$\langle J_3 m_3 | J_1 m_1 J_2 m_2 \rangle = \langle J_1 m_1 J_2 m_2 | J_3 m_3 \rangle. \qquad (13.5)$$

In the second line of Eq. (13.4) we have used the fact that the electric-dipole operator is diagonal in the nuclear-spin states.

We now use the Clebsch–Gordan condition $m_1 + m_2 = m_3$, as well as the related electric-dipole selection rule

$$\langle J_1 m_1 | d_q | J_2 m_2 \rangle = 0 \text{ unless } m_1 = m_2 + q, \qquad (13.6)$$

to determine which coherences $\rho_{F_g m, F_g m'}^{(depop)}$ can be nonzero in Eq. (13.4). Traversing the factors in the last line of Eq. (13.4) from left to right, we find that a term in the sum is zero unless

$$m = m_I + m_J, \quad m_J'' = m_J + q',$$
$$m_J' = m_J'' + q'', \quad m' = m_I + m_J'. \qquad (13.7)$$

From this we find that
$$|m' - m| = |q' + q''| \leq 2. \tag{13.8}$$

We can translate a limit on $|\Delta m|$ directly into a limit on the rank κ of polarization moments that can be created as follows. The polarization moments are the coefficients of the expansion of the density matrix into a sum of irreducible tensor operators (a set of operators with the rotational symmetries of the spherical harmonics). A polarization moment of rank κ has $2\kappa + 1$ components with projections $q = -\kappa, \ldots, \kappa$, which are related to the Zeeman-basis density-matrix elements by (Eq. 5.72)

$$\rho_q^\kappa = \sum_{mm'} (-1)^{F-m'} \langle FmF, -m'|\kappa q\rangle \rho_{m'm}, \tag{13.9}$$

From Eq. (13.9), a ground-state PM ρ_q^κ with a given value of $|q|$ can exist if and only if there is a $|\Delta m| = |q|$ coherence in the ground-state density matrix. A limit on $|q|$ is not by itself a limit on κ, because any polarization moment with rank $\kappa \geq |q|$ can have a component with projection q. However, if such a high-rank moment exists, we can always find a rotated basis such that the component with projection q in the original basis manifests itself as a component with projection κ in the rotated basis. Because Eq. (13.4) holds for arbitrary light polarization, it holds in the rotated basis, so we can conclude that no polarization moment ρ_q^κ with rank κ greater than the limit on $|\Delta m|$ can be created, regardless of the value of q.

For the case under consideration, this analysis reveals that only polarization moments with $\kappa \leq 2$ are present. This is a consequence of the fact that we are considering the lowest-order contribution to optical pumping (namely, second order in the incident light field), so that multi-photon effects are not taken into account. A single photon is a spin-one particle, so it can support polarization moments up to $\kappa_\gamma = 2$. For a polarization moment (PM) of rank κ to be created, the unpolarized (rank 0) density matrix must be coupled to a rank-κ PM by the rank $\kappa_\gamma \leq 2$ photon. The triangle condition for tensor products implies that $\kappa \leq \kappa_\gamma + 0 \leq 2$.

An additional condition on $|\Delta m|$ can be found from Eq. (13.7), using the fact that m_J and m_J' are projections of the ground-state electronic angular momentum, so that their absolute values are less than or equal to J_g. From the first and last conditions of Eq. (13.7) we find

$$|m' - m| = |m_J' - m_J| \leq 2J_g. \tag{13.10}$$

Thus the coherences that can be created within a ground-state hyperfine level F_g are limited to twice the ground-state electronic angular momentum J_g, even if $F_g > J_g$. As a consequence, polarization moments in the ground state are limited to rank $\kappa \leq 2J_g$. We can understand this restriction by examining Eq. (13.4). Because the excited-state hyperfine shifts have been eliminated from the expression and the electric-dipole operator does not act on the nuclear spin space, all traces of the hyperfine interaction in the excited state have been removed from Eq. (13.4). This is indicated by the fact that, in the last line of the equation, the nuclear spin does not appear in the state vectors describing excited states. Thus the excited state only couples to the electronic spin of the ground state, so that there is no mechanism for coupling two ground-state nuclear spin states. This means that any polarization moment present in the ground state must be supported by the electronic spin only.

Considering now the case in which the excited-state hfs is resolved and the ground-state hfs is unresolved, opposite to the case considered so far, we find no similar restriction. It is clear from Eq. (13.3) that the polarization produced in a ground-state hyperfine level is independent

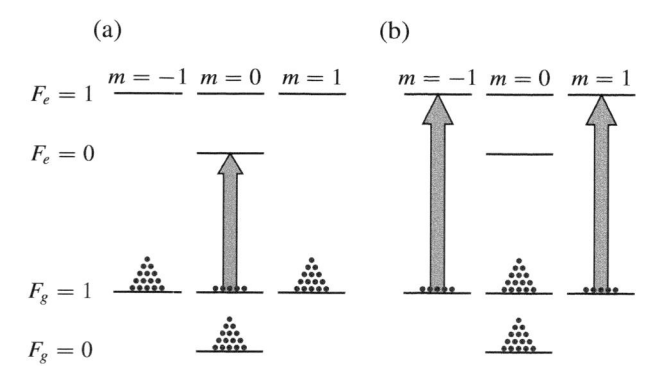

Fig. 13.2 (From Auzinsh *et al.* 2009a.) Excitation with *z*-polarized light on the (a) $1 \to 0$ and (b) $1 \to 1$ transitions of a totally resolved $J_g = 1/2 \to J_e = 1/2$ transition with $I = 1/2$. Alignment is produced in the $F_g = 1$ hyperfine level in both cases. Relative atomic populations are indicated by the number of dots displayed above each ground-state level. Relative transition strengths are indicated by the widths of the arrows—here the transition strengths are all the same.

of all of the other ground-state levels—only one ground-state level F_g appears in the equation. If the excited-state hfs is resolved, then likewise only one excited-state level F_e appears. Thus pumping on a transition $F_g \to F_e$ produces the same polarization in the level F_g as pumping on a completely isolated $F_g \to F_e$ transition, regardless of any nearby (unresolved) ground-state hyperfine levels. Any polarization moment up to rank $\kappa = 2F_g$ can be produced, subject to the restriction $\kappa \leq 2$ in the lowest-order approximation.

In fact, these results can be obtained without the need for any calculations. It is clear that if all the hyperfine splittings are set to zero, the nuclear spin is effectively noninteracting, and can be ignored. In this case, only polarization moments that can be supported by the electronic spin J_g can be produced in the ground state. In particular, if we consider polarization of a given (degenerate) ground-state hyperfine level, we must have $\kappa \leq 2J_g$. If the ground-state hyperfine splitting is increased, this conclusion must remain unchanged, because the light only couples the ground states to the excited states; to lowest order it does not make any difference what is going on in the other ground-state hyperfine levels. If the excited-state hyperfine splitting is then increased, the various $F_g \to F_e$ hyperfine transitions become isolated; for an isolated transition the limit on the ground-state polarization moments is $\kappa \leq 2F_g$. Thus we see that the limit $\kappa \leq 2J_g$ on the ground-state polarization moments occurs when the excited-state hfs is unresolved, and this limit does not depend on whether or not the ground-state hfs is resolved.

The total angular momentum F_g can be significantly larger than J_g. For example, Cs has $I = 7/2$ and $J_g = 1/2$, so that the maximum value of F_g is 4. Thus polarization moments up to rank eight can be produced in the ground state by depopulation pumping if the excited-state hfs is resolved, but only up to rank one if it is unresolved. To second order in the light field the ground-state polarization that can be created is limited to at most rank two in any case. However, the question of whether rank-two polarization can be created is an important one: ground-state alignment is crucial for nonlinear magneto-optical effects with linearly polarized light, as we discuss in Chapter 14.

This situation is illustrated for linearly polarized light resonant with an alkali $D1$ line ($J_g = J_e = 1/2$) in Figs. 13.2 and 13.3. We choose $I = 1/2$ for simplicity, and the quantization axis is taken along the direction of the light polarization. In Fig. 13.2 the hfs is completely resolved. Part (a) of the figure shows light resonant with the $F_g = 1 \rightarrow F_e = 0$ transition. Atoms are pumped out of the $|F_g = 1, m = 0\rangle$ sublevel, producing alignment in the $F_g = 1$ state. (Linearly polarized light in the absence of other fields can only produce even-rank moments, and an $F = 1$ state can only support polarization moments up to rank two; therefore, the anisotropy shown in Fig. 13.2 must correspond to alignment.)

If light is resonant with the $F_g = 1 \rightarrow F_e = 1$ transition, as in part (b), the $m = \pm 1$ sublevels of the $F_g = 1$ state are depleted, producing alignment with sign opposite to that in Fig. 13.2(a). This can be contrasted with the case in which the excited-state hyperfine structure is completely unresolved, shown in Fig. 13.3. Here, all the Zeeman sublevels of the $F_g = 1$ state are pumped out equally—the $|F_g = 1, m = 0\rangle$ sublevel on the $F_g = 1 \rightarrow F_e = 0$ transition, and the $|F_g = 1, m = \pm 1\rangle$ sublevels on the $F_g = 1 \rightarrow F_e = 1$ transition. [The relative pumping rates, which can be found from terms of the sum in Eq. (13.3), are all the same.] Thus no imbalance is created in the $F_g = 1$ sublevel populations, and no polarization is created in this state.

The same principle is illustrated for nuclear spin $I = 3/2$ in Fig. 13.4. The excited-state hfs is unresolved, and light is resonant with the $F_g = 2 \rightarrow F_e$ transitions. In this case, the $m = \pm 1$ ground-state sublevels are pumped on two different transitions. The total transition strength connecting each $F_g = 2$ sublevel to the excited state is the same, and so no polarization is produced in the $F_g = 2$ state.

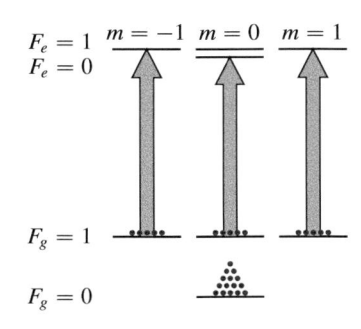

Fig. 13.3 (From Auzinsh et al. 2009a.) As Fig. 13.2 but with excited-state hfs unresolved; light is resonant with the $F_g = 1 \rightarrow F_e$ transition group. No alignment is produced in the $F_g = 1$ ground state.

The conclusions of this section must be modified when polarization produced in the ground state by spontaneous emission from the excited state is taken into account. We now consider the effect of this mechanism on the ground-state polarization (Sec. 13.2).

13.2 Excited state and repopulation pumping

Through second order in the incident light field (first order in light intensity), there is one additional contribution to the ground-state polarization besides the one considered in Sec. 13.1: that due to atoms being pumped to the excited state and then returning to the ground state via spontaneous emission (repopulation pumping). We first consider polarization produced in the excited state. The general form of the excited-state density matrix is (Happer 1972)

Fig. 13.4 (From Auzinsh et al. 2009a.) Excitation with z-polarized light on the $F_g = 2 \rightarrow F_e$ transition group of a $D1$ transition with unresolved excited-state hfs. The nuclear spin is $I = 3/2$. No alignment is produced in the $F_g = 2$ hyperfine level. The width of each arrow represents the relative transition strength, which can be obtained from terms of the sum in Eq. (13.3).

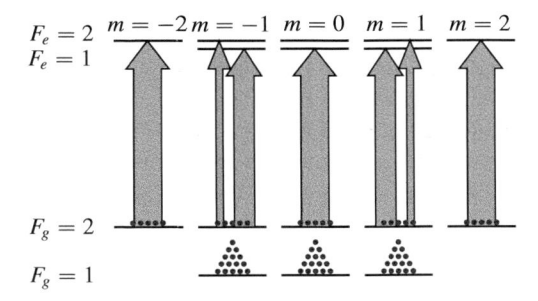

$$\rho_{rs} \propto \sum_k \hat{\varepsilon} \cdot \mathbf{d}_{rk}\hat{\varepsilon}^* \cdot \mathbf{d}_{ks} G'(\omega - \omega_{rk}), \tag{13.11}$$

where r and s are excited states and k is a ground state. Unlike the ground-state polarization, that of the excited state generally decays before it can be mixed by collisions. Thus, the lineshape function $G'(\omega - \omega_{rk})$ in this case describes a feature with a width equal to the (possibly power-broadened) natural width of the transition. Comparing this expression to the formula for ground-state depopulation pumping (Eq. 13.1), we find that, as one would expect, the roles of the ground-state and excited state have been reversed. This means that the results of Sec. 13.1, with F_g and F_e interchanged, can be applied to the excited-state polarization. In this case, there is a limit $\kappa \leq 2J_e$ on the polarization moments that can be produced in the excited state, that occurs only when the *ground state* hfs is unresolved. The restriction does not depend on whether or not the excited-state hfs is resolved. There is the additional limit $\kappa \leq 2$ for low light power.

When the polarized atoms in the exited state decay due to spontaneous emission, the polarization can be transferred to the ground state. This contribution to the ground-state density matrix is given by (Eq. 12.39)

$$\rho_{mn}^{(repop)} \propto \sum_{sr} \mathbf{d}_{mr} \cdot \mathbf{d}_{sn} \rho_{rs}, \tag{13.12}$$

with ρ_{rs} as given above. (Note that after the atomic polarization is transferred to the ground state, it is averaged over velocity groups by collisions, so that the lineshape again has the Doppler width.) The fact that this formula has no reference to individual transition frequencies leads us to expect that the polarization transfer should be independent of the hyperfine splittings. Indeed, writing this expression out for the case under consideration gives

$$\rho_{F_g m, F_g m'}^{(repop)} \propto \sum (-1)^p \langle F_g m | d_p | F_e m'' \rangle \langle F_e m'' | \rho | F_e m''' \rangle \\ \times \langle F_e m''' | d_{-p} | F_g m' \rangle, \tag{13.13}$$

and the only restriction to be obtained is $m' - m = m''' - m''$ (excited-state Δm equals ground-state Δm), while transforming to the uncoupled basis does not result in any additional limits. In other words, if the polarization moment can be supported in the ground state, it can be transferred from the excited state via spontaneous emission.

Combining these results, we see that there is a similar restriction on polarization created in the ground state by repopulation pumping as the one on polarization created by depopulation pumping. However, the restriction occurs in the opposite case. When the ground state is unresolved the polarization produced by repopulation pumping must have $\kappa \leq 2J_e$. This limit does not depend on whether the excited-state hfs is resolved.

We now illustrate the foregoing for a system with $J_g = J_e = I = 1/2$ pumped with linearly polarized light. In Fig. 13.5 both the ground- and excited-state hfs is resolved, and light is tuned to the $F_g = 1 \to F_e = 1$ transition. In part (a) of the figure, the pump light produces polarization in the $F_e = 1$ excited state. In part (b) the excited atoms spontaneously decay. This creates polarization in the $F_g = 1$ ground state, because more atoms are transferred to the $|F_g = 1, m = 0\rangle$ sublevel than to the $|F_g = 1, m = \pm 1\rangle$ sublevels. (In this and the following two figures, we do not show the atoms that decay to the $F_g = 0$ state.) Figure 13.6 is the same but with light tuned to the $F_g = 0 \to F_e = 1$ transition; polarization is also created in the $F_g = 1$ ground state in this case.

Fig. 13.5 (From Auzinsh et al. 2009a.) Level diagram for an $D1$ transition with resolved hfs for an atom with $I = 1/2$ showing (a) optical excitation and (b) spontaneous decay with linearly polarized light resonant with the $F_g = 1 \to F_e = 1$ transition. The branching ratio for each allowed decay is the same, leading to an excess of atoms in the $|F_g = 1, m = 0\rangle$ sublevels over the populations of the $|F_g = 1, m = \pm 1\rangle$ sublevels by a ratio of 2:1.

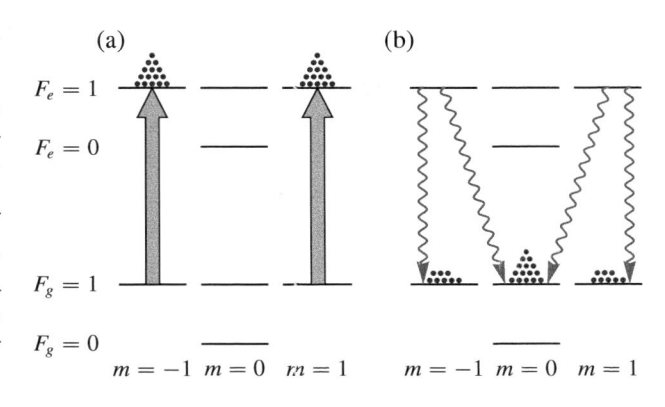

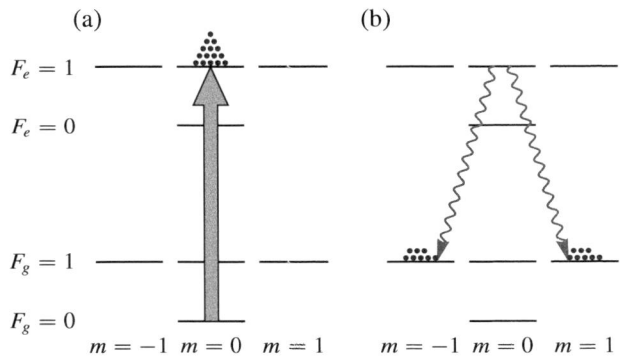

Fig. 13.6 (From Auzinsh et al. 2009a.) As Fig. 13.5, but with light tuned to the $F_g = 0 \to F_e = 1$ transition. In this case an excess of atoms results in the $m = \pm 1$ states, so that the polarization has the opposite sign to that in Fig. 13.5.

In Fig. 13.7 the ground-state hfs is now unresolved, while the excited-state hfs remains resolved. In this case, both ground-state hyperfine levels are pumped by the light, and equal populations are produced in the sublevels of the $F_e = 1$ state, as shown in part (a) of the figure. As seen in part (b), the excited-state atoms spontaneously decay in equal numbers to the $F_g = 1$ sublevels, so that no polarization is produced in the $F_g = 1$ state.

Note that in the opposite case, with unresolved excited-state hfs and resolved ground-state hfs, spontaneous decay is not prevented from producing polarization in the $F_g = 1$ ground state. In this case, atoms are pumped into the $F_e = 0$ state, as well as the $F_e = 1$ state, as shown in Fig. 13.8(a). Since the $|F_e = 0, m = 0\rangle$ state decays isotropically, the decay from this state does not cancel out the polarization created by decay from the $F_g = 1$ state (Fig. 13.8b). Thus we see that it is the ground-state hfs, and not the excited-state hfs, that needs to be resolved in order for polarization to be produced in the ground state due to spontaneous decay.

To summarize the results obtained so far, to lowest order in the excitation light, polarization can be either produced in the ground state directly through absorption, or transferred to the ground state by spontaneous emission. To this order, polarization moments due to both of these mechanisms must have rank $\kappa \leq 2$. In addition, if the excited-state hfs is unresolved, there is a limit $\kappa \leq 2J_g$ on the ground-state polarization due to depopulation, but no additional limit on the polarization due to repopulation. On the other hand, if the ground-state hfs is unresolved, there is a limit $\kappa \leq 2J_e$ on the ground-state polarization due to repopulation, but no additional limit on depopulation. Thus, unless both the excited-state and ground-state

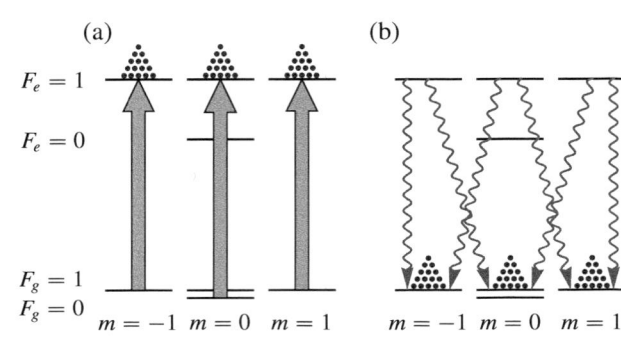

Fig. 13.7 (From Auzinsh et al. 2009a.) As Fig. 13.5 but with unresolved ground-state hfs; light is tuned to the $F_g \to F_e = 1$ transition group. The contributions to the ground-state polarization illustrated in Figs. 13.5 and 13.6 cancel, so that no ground-state polarization is produced.

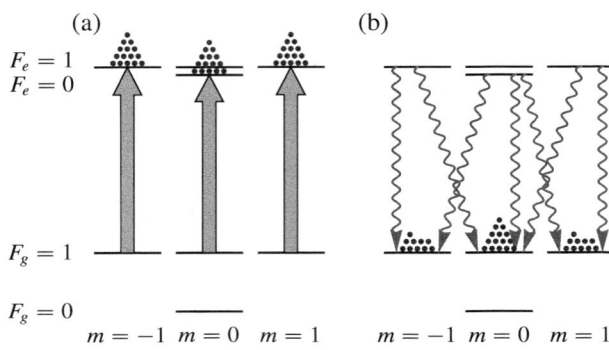

Fig. 13.8 (From Auzinsh et al. 2009a.) As Fig. 13.5 but with unresolved excited-state hfs; light is tuned to the $F_g = 1 \to F_e$ transition group. Three decay channels transfer atoms to the $|F_g = 1, m = 0\rangle$ sublevel, while the $|F_g = 1, m = \pm 1\rangle$ sublevels are each fed by two decay channels. Since all the branching ratios are the same, the resulting population imbalance is 3:2.

hyperfine structure is unresolved, one or the other of the mechanisms is capable of producing polarization of all ranks $\kappa \leq 2$.

13.3 Absorption

The absorption $\mathcal{A}$ of a weak probe light beam is given in terms of the ground-state density matrix by (Eq. 12.60)

$$\mathcal{A} \propto \sum_{mnr} \hat{\boldsymbol{\varepsilon}} \cdot \mathbf{d}_{rm} \rho_{mn} \hat{\boldsymbol{\varepsilon}}^* \cdot \mathbf{d}_{nr} G(\omega - \omega_{rm}), \tag{13.14}$$

or

$$\mathcal{A} \propto \sum \langle F_e m | \hat{\boldsymbol{\varepsilon}} \cdot \mathbf{d} | F_g m' \rangle \langle F_g m' | \rho | F_g m'' \rangle \langle F_g m'' | \mathbf{d} \cdot \hat{\boldsymbol{\varepsilon}}^* | F_e m \rangle G(\omega - \omega_{F_e F_g}), \tag{13.15}$$

where all quantities are as defined above. Using the approximation, as in Secs. 13.1 and 13.2, that the light is resonant with an unresolved transition group and far detuned from all other transitions, this formula reduces to

$$\mathcal{A} \propto \sum \langle F_e m | \hat{\boldsymbol{\varepsilon}} \cdot \mathbf{d} | F_g m' \rangle \langle F_g m' | \rho | F_g m'' \rangle \langle F_g m'' | \mathbf{d} \cdot \hat{\boldsymbol{\varepsilon}}^* | F_e m \rangle, \tag{13.16}$$

where the sum over F_g and F_e includes only those combinations that are in the unresolved resonant transition group.

We now investigate the dependence of the absorption on the ground-state polarization in various cases. Consider the case in which the ground-state hfs is completely resolved, and the excited-state structure is unresolved. The light is tuned to an unresolved transition group consisting of transitions between one ground-state hyperfine level F_g and all of the excited-state levels. The sum in Eq. (13.16) over the excited states is then a closure relation, and can be replaced with a sum over any complete basis for the excited state, in particular, the uncoupled basis. We also insert closure relations to expand the ground states $\langle F_g m''|$ and $|F_g m'\rangle$ in the uncoupled basis. We obtain

$$\begin{aligned}
\mathcal{A} &\propto \sum (-1)^{q'+q''} \varepsilon_{q'}(\varepsilon^*)_{q''} \langle I m_I J_e m_J | d_{-q'} | I m_I' J_g m_J'\rangle \langle I m_I' J_g m_J' | F_g m'\rangle \\
&\quad \times \langle F_g m' |\rho| F_g m''\rangle \langle F_g m'' | I m_I'' J_g m_J''\rangle \langle I m_I'' J_g m_J'' | d_{-q''} | I m_I J_e m_J\rangle \\
&= \sum (-1)^{q'+q''} \varepsilon_{q'}(\varepsilon^*)_{q''} \langle J_e m_J | d_{-q'} | J_g m_J'\rangle \langle I m_I' J_g m_J' | F_g m'\rangle \langle F_g m' |\rho| F_g m''\rangle \\
&\quad \times \langle F_g m'' | I m_I J_g m_J''\rangle \langle J_g m_J'' | d_{-q''} | J_e m_J\rangle,
\end{aligned}$$
(13.17)

where only one nuclear-spin summation variable remains in the last line. The dipole matrix element selection rules and Clebsch–Gordan conditions require that

$$m' = m_I + m_J', \quad m_J' = m_J + q',$$
$$m'' = m_I + m_J'', \quad m_J = m_J'' + q''$$
(13.18)

must be satisfied in order for a term in the sum to contribute to the absorption. These conditions can be combined to yield $|m' - m''| = |q' + q''| \leq 2$. Thus only coherences with $|\Delta m| \leq 2$ (and polarization moments with $\kappa \leq 2$) can affect the lowest-order absorption signal. The reason for this is analogous to the reason that polarization moments of maximum rank two can be created with a lowest-order interaction with the light. As discussed in Chapter 12, absorption occurs when an atom is transferred to the excited state, i.e., when population (rank zero polarization) is created in the excited state. Thus, to be observed in the signal, a ground-state atomic PM must be coupled to a $\kappa = 0$ excited-state PM by a spin-one photon, which can support polarization moments up to rank two. The triangle condition for tensor products then implies that the rank of the atomic polarization moment must be no greater than two.

Another restriction on the coherences that can affect absorption can be found from Eq. (13.18) by using the fact that $|m_J'| \leq J_g$ and $|m_J''| \leq J_g$. We find

$$|m' - m''| = |m_J' - m_J''| \leq 2J_g.$$
(13.19)

In other words, only polarization moments with $\kappa \leq 2J_g$ can affect the absorption signal, regardless of the value of F_g. Evidently, it is the excited-state hfs that determines which ground-state polarization moments can be detected in absorption, whether or not the ground-state hfs is resolved.

Considering the case in which both the excited- and ground-state hfs is entirely unresolved can lend some insight into this result. In this case, every combination of F_g and F_e enters in the sum in Eq. (13.16). If the ground-state hyperfine splitting is sent to zero, the sum must be extended to include matrix elements of ρ between different hyperfine levels. This means that all of the sums in Eq. (13.16) can be replaced with sums over uncoupled basis states, giving

$$\begin{aligned}
\mathcal{A} &\propto \sum (-1)^{q'+q''} \varepsilon_{q'}(\varepsilon^*)_{q''} \langle I m_I J_e m_J | d_{-q'} | I m_I' J_g m_J'\rangle \\
&\quad \times \langle I m_I' J_g m_J' |\rho| I m_I'' J_g m_J''\rangle \langle I m_I'' J_g m_J'' | d_{-q''} | I m_I J_e m_J\rangle \\
&= \sum (-1)^{q'+q''} \varepsilon_{q'}(\varepsilon^*)_{q''} \langle J_e m_J | d_{-q'} | J_g m_J'\rangle \langle I m_I' J_g m_J' |\rho| I m_I J_g m_J''\rangle \\
&\quad \times \langle J_g m_J'' | d_{-q''} | J_e m_J\rangle.
\end{aligned}$$
(13.20)

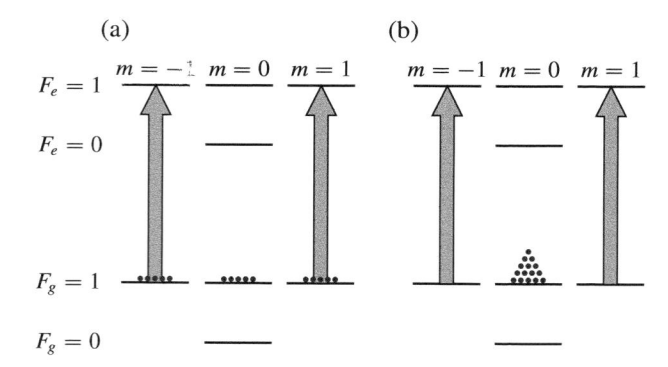

Fig. 13.9 (From Auzinsh et al. 2009a.) $D1$ transition for an atom with $I = 1/2$ subject to linearly polarized light resonant with the $F_g = 1 \rightarrow F_e = 1$ transition. In part (a) the $F_g = 1$ ground state is unpolarized and there is light absorption. In part (b) the $F_g = 1$ state has the same total population, but is aligned, and there is no absorption.

Since the hyperfine interaction has been effectively eliminated, the absorption no longer depends on the nuclear spin: the complete density matrix does not enter, but rather the reduced density matrix

$$\rho^{(J)}_{m'_J, m''_J} = \sum_{m_I} \rho_{m_I m'_J, m_I m''_J} \tag{13.21}$$

that is averaged over the nuclear spin m_I. The reduced density matrix can only support polarization moments up to rank $\kappa = 2J_g$, so any PM in ρ with higher rank cannot affect the absorption. Considering a density matrix that is nonzero only within one ground-state hyperfine level F_g, we see that polarization moments with rank greater than two will not contribute to the signal. Since the other ground-state hyperfine levels are unoccupied, it makes no difference what the ground-state hyperfine splitting is, so we regain the result that, even if the ground-state hfs is resolved, only polarization moments with $\kappa \leq 2J_g$ can affect the absorption of light if the excited-state hfs is unresolved.

There is no corresponding restriction on the polarization moments that can affect absorption when the ground-state hfs is unresolved and the excited-state hfs is resolved. Indeed, we can consider the case in which only one ground-state hyperfine level F_g is populated: the absorption is then exactly as if the transition $F_g \rightarrow F_e$ were completely isolated. For such an isolated transition, the only limit on detectable polarization moments is $\kappa \leq 2$ for the low-power case.

As in the previous subsections, we illustrate this result for a $D1$ transition for an atom with $I = 1/2$ subject to linearly polarized light. In Fig. 13.9 both the ground- and excited-state hfs is resolved, and the light is resonant with the $F_g = 1 \rightarrow F_e = 1$ transition. In part (a) there is no polarization in the $F_g = 1$ ground state: atoms are equally distributed among the Zeeman sublevels. Light is absorbed by atoms in the $|F_g = 1, m = \pm 1\rangle$ sublevels. In part (b) there are the same total number of atoms in the $F_g = 1$ state, but they are collected in the $m = 0$ sublevel. The population is the same, but the $F_g = 1$ state now also has alignment. In this particular case there is no absorption, because the atoms are all in the $m = 0$ dark state. Thus, in this situation, the rank-two polarization moment has a strong effect on the absorption signal.

Figure 13.10 shows the same system, but with unresolved excited-state hfs. In this case there is no dark state; all of the atoms interact with the light. The distribution of the atoms among

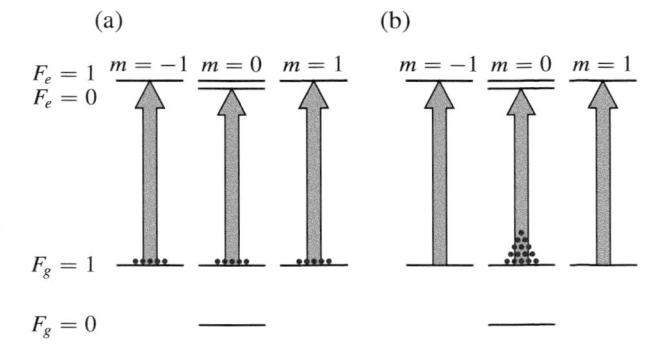

Fig. 13.10 (From Auzinsh et al. 2009a.) As Fig. 13.9, but with unresolved excited-state hfs. In this case there is no difference in the absorption seen for an (a) unpolarized and (b) aligned $F_g = 1$ ground state.

the Zeeman sublevels does not affect the light absorption, and so the rank-two polarization moment is not detectable in the absorption signal.

13.4 Fluorescence

Finally, we consider which excited-state polarization moments can be observed in fluorescence. Assuming broad-band detection, the intensity of fluorescence into a particular polarization $\hat{\varepsilon}$ is given in terms of the excited-state density matrix by (Eq. 12.7)

$$\mathcal{I} \propto \sum_{rsm} \hat{\varepsilon}^* \cdot \mathbf{d}_{mr} \rho_{rs} \hat{\varepsilon} \cdot \mathbf{d}_{sm}. \tag{13.22}$$

Because the sums in r and s go over all excited states, and m runs over all ground states, we can write Eq. (13.22) for our case in terms of the uncoupled-basis states. This gives

$$\mathcal{I} \propto \sum (-1)^{q'+q''} (\varepsilon^*)_{q'} \varepsilon_{q''} \langle I m_I J_g m_J | d_{-q'} | I m_I' J_e m_J' \rangle \\ \times \langle I m_I' J_e m_J' | \rho | I m_I'' J_e m_J'' \rangle \langle I m_I'' J_e m_J'' | d_{-q''} | I m_I J_g m_J \rangle, \tag{13.23}$$

resulting in the restrictions

$$m_J'' = -q'' + m_J, \quad m_J = -q' + m_J', \quad m_I = m_I' = m_I'', \tag{13.24}$$

on the terms that can contribute to the fluorescence. This indicates that the nuclear polarization cannot affect the fluorescence signal, and so only the electronic excited-state polarization of rank $\kappa \leq 2J_e$ can be observed. In addition, only coherences with $|m_J'' - m_J'| = |q' + q''| \leq 2$ can be observed. This rule has appeared earlier as a consequence of the low-light-power assumption; because spontaneous decay is not induced by an incident light field, in this case the rule is exact. This means that no matter the value of J_e, and what polarization moments exist in the excited state, only polarization of rank $\kappa \leq 2$ can be observed in fluorescence.

13.5 Comparison of different cases

In this chapter, we have shown that, when the ground- or excited-state hfs is unresolved, there are restrictions on the rank of the polarization moments that can be created or detected by light. Some of these restrictions may at first seem counter-intuitive, but they can be obtained from very basic considerations. For example, the two facts that nuclear spin can be ignored if the hfs is completely unresolved and that lowest-order depopulation pumping of a given hyperfine level

Table 13.1 Summary of the results of this chapter. For each quantity, the restriction on the rank κ of the polarization that can be created or detected is given in the fourth column. The restriction holds when the ground- or excited-state hfs, as given in the second column, is unresolved with respect to the width given in the third column. For fluorescence with broad-band detection the restriction holds regardless of whether the hfs is resolved.

	unresolved	w.r.t.	maximum κ
Ground-state pol. (depop.)	excited hfs	Doppler	$2J_g$
Ground-state pol. (repop.)	ground hfs	Doppler	$2J_e$
Excited-state pol.	ground hfs	Doppler-free	$2J_e$
Absorption	excited hfs	Doppler	$2J_g$
Fluorescence	—	—	$2J_e$

does not depend on ground-state hyperfine splitting lead directly to the result that polarization moments produced by depopulation pumping are subject to a limit of $\kappa \leq 2J_g$ when the excited-state hfs is unresolved. Various processes of creation and detection of polarization are subject to different restrictions (Table 13.1).

In particular, the two processes that can create ground-state polarization, namely, depopulation and repopulation pumping, are subject to restrictions under different conditions. Consequently, unless the hfs is entirely unresolved, there is always a mechanism for producing polarization limited in rank only by the total angular momentum, rather than the electronic angular momentum.

14
The effect of hyperfine splitting on nonlinear magneto-optical rotation

Now let us examine the more general case of *partially* resolved hyperfine transitions. For this study, we will look at the quantitative dependence on hyperfine splitting of nonlinear optical rotation—rotation of light polarization due to interaction with a $J_g \to J_e$ transition group in the presence of a magnetic field. In this case, the effect of ground-state atomic polarization is brought into starker relief: in the experimental situation that we consider, both the creation and detection of ground-state polarization is required in order to see any signal whatsoever. When linearly polarized light is used, as is supposed here, the lowest-order effect depends on rank-two atomic alignment. Thus, for the alkali atoms, the question of the dependence of the effect on hyperfine structure arises, because, as discussed in the previous section, both the creation and the detection of alignment in the $J_g = 1/2$ ground state can be suppressed due to unresolved hfs. [In fact, a higher-order effect can occur wherein alignment is created, the alignment is converted to orientation, and the orientation is detected (Auzinsh and Ferber 1992; Budker *et al.* 2000a; Auzinsh *et al.* 2006). However, the conversion of alignment to orientation is an effect of tensor AC-Stark shifts, which can be shown by arguments similar to those in Chapter 13 to suffer suppression due to unresolved hfs in the same way as does the direct detection of alignment.]

In the Faraday geometry, linearly polarized light propagates in the direction of an applied magnetic field, and the rotation of the light polarization direction is measured. As we discussed in Chapter 10, a number of magneto-optical effects can contribute to the optical rotation, including the linear Faraday effect, the Bennett-structure effect, and various effects depending on atomic polarization ("coherence effects"). Here we are concerned with optical rotation due to several different forms of the ground-state coherence effect, in which the atomic velocities are treated in three different ways. First we consider the atoms to have no velocity spread, and analyze the Doppler-free "transit effect," as for an atomic beam with negligible transverse velocity distribution (Schuh *et al.* 1993). We then consider the case in which atoms have a Maxwellian distribution, but do not change their velocities in between pumping and probing—this corresponds to the transit effect for buffer-gas-free, dilute atomic vapors (Kanorsky *et al.* 1993). Finally, we treat the case in which atoms undergo velocity-changing collisions between pumping and probing, as for buffer-gas cells (Novikova *et al.* 2001a) or the wall-induced Ramsey effect ("wall effect") in antirelaxation-coated vapor cells (Kanorskii *et al.* 1995). We examine the dependence of these effects on the size of the hyperfine splittings as they vary from much smaller than the natural width to much greater than the Doppler width.

Throughout this chapter we consider formulas for the optical rotation signal valid to lowest order in light power, under the assumption that the ground-state relaxation rate γ is much smaller than both the excited-state natural width Γ and the hyperfine splittings. For the Doppler-free case a single analytic formula can be applied to both resolved and unresolved hfs (i.e., no assumption need be made about the relative size of the hyperfine splittings and the natural width). For the Doppler-broadened cases, analytic results can be obtained in various limits, which together describe the signal over the entire range of hyperfine splittings.

We first focus on the simplest case: the $D1$ line ($J_g = J_e = 1/2$) for an atom with $I = 1/2$. This is a somewhat special case, because one of the two ground-state hyperfine levels has $F_g = 0$, and consequently can neither support atomic alignment nor produce optical rotation. We then consider the differences that arise when considering higher nuclear spin and also the $D2$ line ($J_g = 1/2$ and $J_e = 3/2$). Some details of the calculation and general formulas for arbitrary J_g, J_e, and I are presented in Appendix E. These formulas are generalizations of those first given by Kanorsky et al. (1993).

14.1 Doppler-free transit effect

We consider nonlinear Faraday rotation on a $J_g \rightarrow J_e$ atomic transition for an atom with nuclear spin I. We can limit our attention to the ground-state coherence effects by using a "three-stage" model for Faraday rotation (Kanorsky et al. 1993), in which optical pumping, atomic precession, and optical probing take place sequentially, and the light and magnetic fields are never present at the same time. In this case, the linear and Bennett-structure effects, which require the simultaneous application of light and magnetic fields, do not occur. Such a model can be realized in an atomic-beam experiment, but it is also a good approximation to a vapor-cell experiment that uses low light power and small enough magnetic fields so that the coherence effects are dominant.

The calculation is performed using second-order perturbation theory in the basis of the polarization moments $\rho^{(\kappa q)}(F_1 F_2)$ of the density matrix (Appendix E.1). The three stages of the calculation are as follows. In stage (a), an x-directed light beam linearly polarized along z is applied, and we calculate optical pumping through second order in the optical Rabi frequency. In stage (b), the light field is removed, and a x-directed magnetic field is applied. We calculate the effect of this field on the atomic polarization. Finally, in stage (c), the magnetic field is turned off, and the light field is applied once more to probe the atomic polarization. The nonlinear optical rotation is found to lowest order in the probe-light Rabi frequency (Appendix E.2).

Because the magnetic field is neglected during the optical pumping stage, the atomic ground-state polarization that is produced in this stage is entirely along the light polarization direction, i.e., it has polarization component $q = 0$. Since linearly polarized light has a preferred axis, but no preferred direction, it cannot, in the absence of other fields, produce atomic polarization with a preferred direction, i.e., polarization with odd rank κ. Also, we have seen in Chapter 13 that, to lowest order in the light power, optical pumping cannot produce polarization moments with $\kappa > 2$. Thus the only ground-state polarization moment with rank greater than zero that is produced at lowest order has $\kappa = 2$ and $q = 0$. We first consider the $D1$ line ($J_g = J_e = 1/2$) for an atom with $I = 1/2$. In this case, the only ground-state hyperfine level that can support the $\rho^{(20)}(F_g F_g)$ moment has $F_g = 1$. (Due to the assumption that the hyperfine splittings are much greater than the ground-state relaxation rate, we can

Doppler-free transit effect 273

ignore ground-state hyperfine coherences throughout the discussion.) From Eq. (E.12), the value of this moment is found to be

$$\rho^{(20)}(11) = \frac{\tilde{\kappa}_2}{12\sqrt{6}} \Big(\big[L(\omega'_{0,1}) - L(\omega'_{1,1})\big] + \frac{R}{3} \big[L(\omega'_{1,0}) - L(\omega'_{1,1})\big] \Big), \qquad (14.1)$$

where $\tilde{\kappa}_2 = \langle J_g \| d \| J_e \rangle^2 \mathcal{E}_0^2 / (\Gamma \gamma)$ is the reduced optical-pumping saturation parameter ($\mathcal{E}_0$ is the optical electric field amplitude), R is the branching ratio for the transition $J_e \to J_g$, and $\omega'_{F_e F_g}$ is the transition frequency between excited-state and ground-state hyperfine levels in the frame "rotating" at the Doppler-shifted light frequency ω: $\omega'_{F_e F_g} = \omega_{F_e F_g} - \omega + \mathbf{k} \cdot \mathbf{v}$, where $\omega_{F_e F_g}$ is the transition frequency in the lab frame, ω is the light frequency, $\mathbf{k}$ is the wave vector, and $\mathbf{v}$ is the atomic velocity. We also write $\omega'_{F_e F_g} = -\Delta_{F_e F_g} + \mathbf{k} \cdot \mathbf{v}$, where $\Delta_{F_e F_g}$ is the light detuning from resonance. We have defined the Lorentzian line profile

$$L(\omega') = \frac{(\Gamma/2)^2}{(\Gamma/2)^2 + \omega'^2}. \qquad (14.2)$$

Equation (14.1) is written as the sum of two terms, each surrounded by square brackets. The first term is the contribution to the polarization due to depopulation pumping discussed in Sec. 13.1. This term is itself a sum of contributions due to pumping on the $F_g = 1 \to F_e = 0$ transition and the $F_g = 1 \to F_e = 1$ transition. These two contributions are of opposite sign, as illustrated in Fig. 13.2. Pumping on either transition produces alignment in the $F_g = 1$ ground state; the sign of the corresponding polarization moment depends on whether there is more population in the $m = 0$ sublevel or the $m = \pm 1$ sublevels. We saw in the discussion of Sec. 13.1 that when the excited-state hfs is unresolved, polarization with rank $\kappa > 2J_g$ cannot be created by depopulation pumping (Fig. 13.3). We see here that as $\omega_{0,1}$ approaches $\omega_{1,1}$, i.e., as the excited-state hyperfine splitting goes to zero, the contributions from the two transitions cancel and this term goes to zero. For the Doppler-broadened atomic ensemble discussed in Chapter 13, the hfs was considered unresolved when the hyperfine splittings were smaller than the Doppler width. Since Eq. (14.1) describes a single velocity group, the relevant width here is the natural width Γ.

The second term of Eq. (14.1) is the contribution to the ground-state polarization due to repopulation pumping discussed in Sec. 13.2. This term is also composed of two contributions of opposite sign: one due to pumping on the $F_g = 1 \to F_e = 1$ transition and one due to pumping on the $F_g = 0 \to F_e = 1$ transition. The two contributions are illustrated in Figs. 13.5 and 13.6, which show the origin of the opposite signs. In Sec. 13.2 we found that depopulation pumping cannot create polarization moments with rank $\kappa > 2J_e$ when the ground-state hfs is unresolved (Fig. 13.7). We see here that this term of Eq. (14.1) goes to zero when $\omega_{1,0}$ approaches $\omega_{1,1}$, i.e., as the ground-state hyperfine splitting goes to zero.

In the second and third stages of the model of the coherence effect, the ground-state polarization precesses in a magnetic field and is probed by light with the same polarization as the pump light considered in the first stage. From Eq. (E.19) we find that the normalized optical rotation $d\alpha$ per path length $d\ell$ is proportional to the polarization produced in the first stage and is given by

$$\ell_0 \frac{d\alpha}{d\ell} = \frac{1}{4} \sqrt{\frac{3}{2}} \big[L(\omega'_{0,1}) - L(\omega'_{1,1})\big] x_1 \rho^{(20)}(11), \qquad (14.3)$$

where

$$x_{F_g} = \frac{(\gamma/2)\Omega_{F_g}}{(\gamma/2)^2 + \Omega_{F_g}^2} \qquad (14.4)$$

is the magnetic-resonance lineshape parameter, with $\Omega_{F_g} = g_{F_g}\mu_B B$ the Larmor frequency for the ground-state hyperfine level F_g (g_{F_g} is the Landé factor for the ground state F_g, and μ_B is the Bohr magneton), and

$$\ell_0 = -\left(\frac{1}{\mathcal{I}}\frac{d\mathcal{I}}{d\ell}\right)^{-1} = \frac{2\pi}{Rn\lambda^2}\frac{(2J_g+1)}{(2J_e+1)} \qquad (14.5)$$

is the unsaturated resonant absorption length assuming totally unresolved hyperfine structure, where $\mathcal{I}$ is the light intensity, n is the atomic density, and λ is the light wavelength. The branching ratio R enters here because it factors into the transition strength.

The contributions to the optical rotation signal from the $F_g = 1 \to F_e = 0$ transition and the $F_g = 1 \to F_e = 1$ transition have opposite signs. To understand this, it is helpful to think of the optically polarized medium as a polarizing filter (Kanorsky et al. 1993). When pumping on a $1 \to 0$ or $1 \to 1$ transition, the medium is pumped into a dark (nonabsorbing) state for that transition (Fig. 13.2), corresponding to a polarizing filter with its transmission axis along the input light polarization axis $\hat{\varepsilon}$ (Fig. 14.1a). The Larmor precession induced by the magnetic field causes the transmission axis of the filter to rotate, so that it is no longer along $\hat{\varepsilon}$. This in turn causes the output light polarization axis $\hat{\varepsilon}'$ to rotate. The polarization of light passing through a polarizing filter tends to rotate toward the transmission axis, so that in this case the optical rotation is in the same sense as the Larmor precession (Fig. 14.1b). Now, compare the polarization produced when pumping on a $1 \to 0$ or $1 \to 1$ transition, as shown in Fig. 13.2. We see that the dark state for each transition is a bright (absorbing) state for the other. This means that if we choose one or the other of these states, it will function as just described for one of the transitions, but will function as a polarizing filter with its transmission axis *perpendicular* to $\hat{\varepsilon}$ for the other transition (Fig. 14.1c). When the axis of the filter rotates in this case, the fact that the output light polarization tends to rotate toward the transmission axis means that here the optical rotation is in the other direction, in the opposite sense to the Larmor precession (Fig. 14.1d). In other words, for a particular sign of the rank-two polarization moment, the optical rotation will have one sign when probed on one transition, and the opposite sign when probed on the other, as indicated by Eq. (14.3). Because the observation of optical rotation requires the detection of rank-two polarization moments, we might expect, analogously to the discussion in Sec. 13.3, that it is suppressed when the excited-state hyperfine splitting goes to zero. Equation (14.3) shows that the two contributions indeed cancel when $\omega_{0,1}$ approaches $\omega_{1,1}$.

Equation (14.3) and the two components of Eq. (14.1) are plotted as a function of light detuning from the $F_g = 1 \to F_e = 1$ transition in Fig. 14.2, for particular values of the ground- and excited-state hyperfine coefficients A_g and A_e. (For $J = I = 1/2$, the hyperfine coefficient A is equal to the splitting between the two hyperfine levels.) Here and below numerical values of frequencies are given in units of Γ. As discussed above, each spectrum consists of two peaks of equal magnitude and opposite sign. For the spectrum of alignment due to depopulation and the spectrum of rotation for a given amount of alignment, the peaks are separated by the excited-state hyperfine splitting, so that they cancel as this splitting goes to zero. For the spectrum of alignment due to repopulation, the peaks are separated by the ground-state hyperfine splitting; they cancel as the ground-state splitting goes to zero.

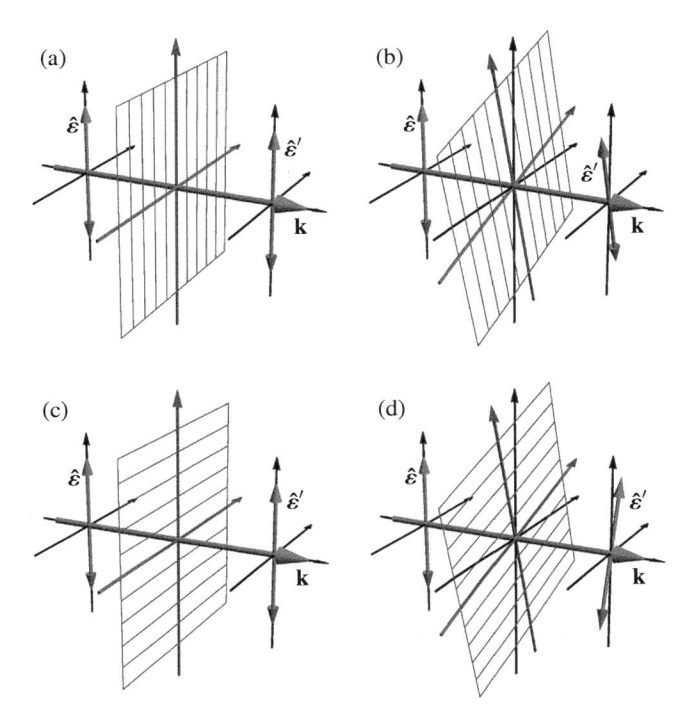

Fig. 14.1 (From Auzinsh *et al.* 2009a.) Illustration of the rotating polarizer model for optical rotation. (a) Optical pumping on a $F_g = 1 \to F_e = 0$ or $F_g = 1 \to F_e = 1$ transition causes the medium to act as a polarizing filter with transmission axis along the input light polarization $\hat{\varepsilon}$. (b) When the transmission axis rotates due to Larmor precession, the output light polarization $\hat{\varepsilon}'$ follows the transmission axis and so rotates in the same sense as the Larmor precession. (c) If polarization produced by pumping on one transition is probed on the other, the polarization functions as a polarizing filter with transmission axis *perpendicular* to the input light polarization. (Attenuation of the light beam is not indicated.) (d) When the medium polarization rotates, the output light polarization tends to rotate toward the transmission axis, in the opposite sense to the Larmor precession in this case.

In this section we are analyzing a Doppler-free system, i.e., we assume that the atoms all have the same velocity, which we take to be zero for simplicity. Then the observed optical rotation signal is found by simply substituting Eq. (14.1) into Eq. (14.3). We first consider the case in which the ground-state hfs is well resolved. The rotation signal is plotted in Fig. 14.3 for large ground-state hyperfine splitting and various excited-state splittings A_e. The components of the rotation signal due to depopulation (dashed) and repopulation (solid) are plotted in the left-hand column, and the total signal is plotted on the right. As the previous discussion indicates, the rotation signal decreases as the excited-state hyperfine splitting A_e becomes smaller, with the component due to depopulation decreasing faster than the component due to repopulation. This is also seen in Fig. 14.4, which shows the maximum magnitude of the rotation spectrum as a function of A_e (for each value of A_e, the signal is optimized with respect to detuning). Thus, for small splittings, the component due to repopulation dominates. To lowest order in A_e, the signal is given by

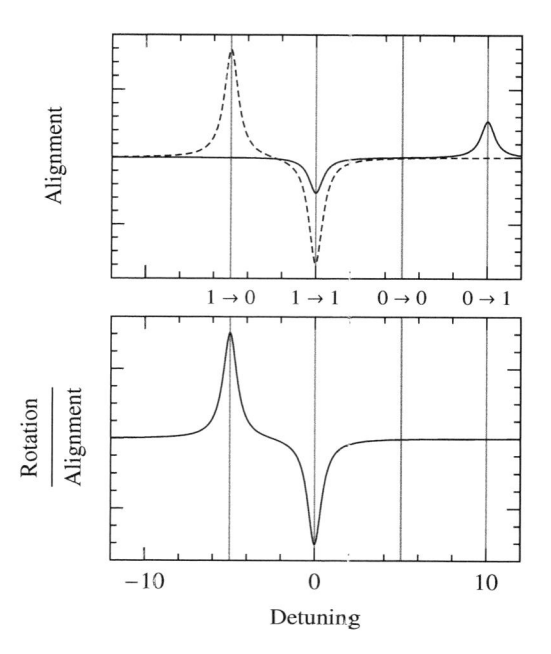

Fig. 14.2 (From Auzinsh et al. 2009a.) Dependence on light detuning from the $F_g = 1 \to F_e = 1$ transition of (top) the components of ground-state alignment due to depopulation (dashed) and repopulation (solid) (Eq. 14.1) and (bottom) optical rotation for a given amount of alignment (Eq. 14.3). Gray vertical lines show $F_g \to F_e$ transition resonance frequencies. Parameter values in units of Γ are $\gamma \ll 1$, $A_g = 10$, $A_e = 5$.

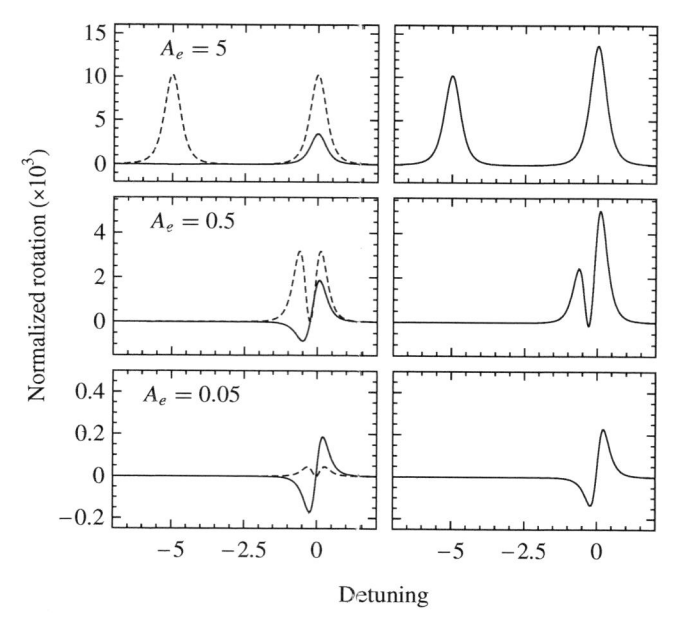

Fig. 14.3 (From Auzinsh et al. 2009a.) Spectra of normalized optical rotation $\ell_0/(\bar{\kappa}_2 x_1)(d\alpha/d\ell)$ for the Doppler-free transit effect. Left column: components due to polarization produced by depopulation (dashed) and repopulation (solid); right column: total signal. Parameter values in units of Γ are $\gamma \ll 1$, $A_g \gg 1$, A_e.

Fig. 14.4 (From Auzinsh et al. 2009a.) Maximum of the spectrum of the Doppler-free nonlinear magneto-optical rotation transit effect as a function of excited-state hyperfine splitting. Plotted are the component due to polarization produced by depopulation (dash-dotted) due to polarization produced by repopulation (dashed) and the total signal (solid). Parameters as in Fig. 14.3.

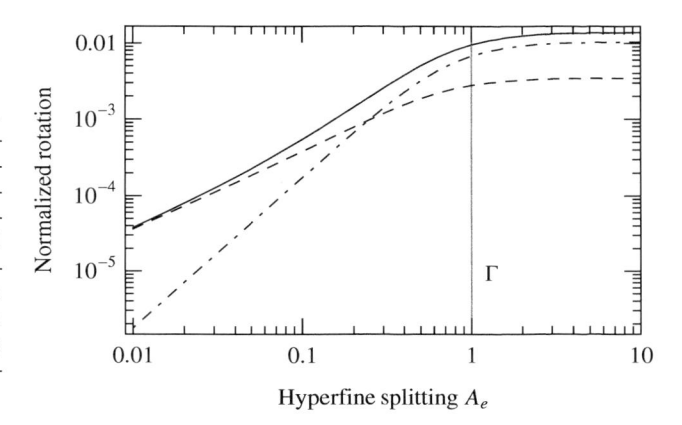

$$\ell_0 \frac{d\alpha}{d\ell} = \frac{A_e \tilde{\kappa}_2 x_1 R(\Gamma/2)^4 \Delta_1}{144 \left[(\Gamma/2)^2 + \Delta_1^2\right]^3}, \tag{14.6}$$

i.e., linear in A_e, with a modified dispersive shape that falls off far from resonance as $1/\Delta_1^5$, where Δ_1 is the detuning from the center of the $F_g = 1 \to F_e$ transition group.

The previous discussion also explains why the two peaks in the component due to depopulation seem to cancel as they overlap, even though they have the same sign: the factors in the signal due to the creation and detection of alignment cancel individually (Fig. 14.2); it is only in their product that the two peaks have the same sign.

We now consider the case in which both the ground- and excited-state hyperfine splittings are small, so that all of the hfs is unresolved. To lowest order in A_g and A_e we have

$$\ell_0 \frac{d\alpha}{d\ell} = A_e \left(A_e - \frac{R}{3} A_g\right) \frac{\tilde{\kappa}_2 x_1 (\Gamma/2)^4 \Delta^2}{24 \left[(\Gamma/2)^2 + \Delta^2\right]^4}, \tag{14.7}$$

where Δ is the light detuning from the line center of the $D1$ transition. As we expect, the component of the signal due to polarization produced by repopulation is proportional to A_g for small hyperfine splitting. The component of the signal resulting from depopulation-induced polarization also enters at this order. The optical rotation spectrum in this case is double-peaked, and falls off as $1/\Delta^6$ (Fig. 14.5).

14.2 Doppler-broadened transit effect

We now consider an atomic ensemble with a Maxwellian velocity distribution, but a low rate of velocity-changing collisions, so that the atomic velocities do not change between optical pumping and probing. This is the case for an atomic-beam experiment, or for the "transit effect" in a dilute-vapor cell. Because the atoms have a fixed velocity, the signal from each velocity group can be found individually and then summed to find the total signal. Thus the signal from the Doppler-broadened transit effect is found by multiplying the Doppler-free signal found in the previous section by a Gaussian weighting function representing the Doppler distribution along the light propagation direction and then integrating over atomic velocity. We can perform this integral analytically in different limiting cases.

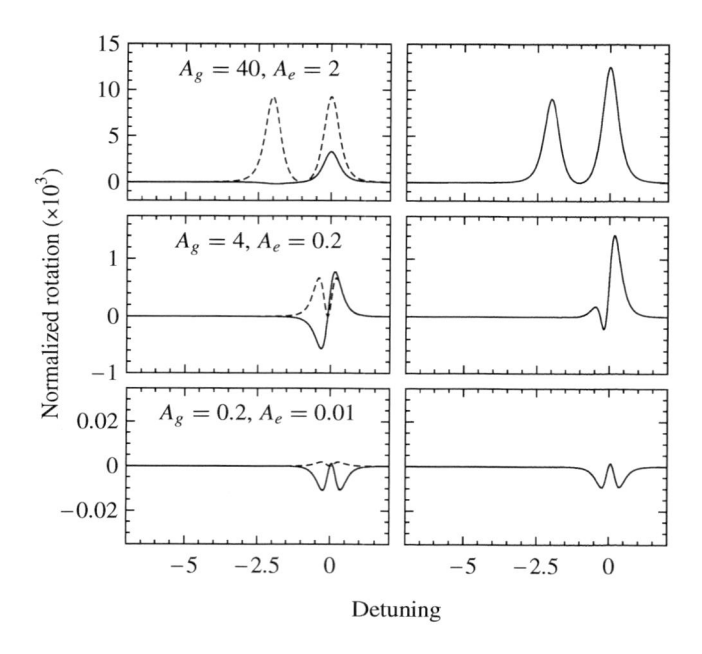

Fig. 14.5 (From Auzinsh et al. 2009a.) As Fig. 14.3, with A_g and A_e varied simultaneously ($A_g = 20 A_e$).

We first consider the commonly encountered experimental case in which the hyperfine splitting is much greater than the natural linewidth of the excited state, i.e., the Doppler-free spectrum is well resolved. In this case, for a given light frequency and atomic velocity, the light acts on at most one transition between hyperfine levels. Thus the excited-state hyperfine coherences can be neglected, and the cancelation effects due to the overlap of resonance lines do not appear. As found in Eq. (E.21), the Doppler-free rotation spectrum then appears as a collection of peaks, one centered at each optical resonance frequency, each with lineshape function $f(\omega'_{F_e F_g}) = L(\omega'_{F_e F_g})^2$, i.e., the square of a Lorentzian lineshape. (One Lorentzian factor is due to optical pumping, the other to probing.)

In this case, the Doppler-broadened signal is found by making the replacement $f \to f_{DB}$, where the velocity integral for f_{DB} takes the form

$$f_{DB}(\Delta_{F_e F_g}) = \int dv_k f(-\Delta_{F_e F_g} + k_B v_k) G(v_k), \tag{14.8}$$

where

$$G(v_k) = \frac{k_B}{\Gamma_D \sqrt{\pi}} e^{-(k_B v_k / \Gamma_D)^2} \tag{14.9}$$

is the normalized distribution of atomic velocities along the light propagation direction $\hat{\mathbf{k}}$, k_B is the Boltzmann constant, and Γ_D is the Doppler width. This integral can be evaluated in terms of the error function. Under the assumption $\Gamma \ll \Gamma_D$ that we will employ here, the integral can be approximated by replacing f with a properly normalized delta function, resulting in

$$f_{DB}(\Delta_{F_e F_g}) \approx \frac{\sqrt{\pi}}{4} \frac{\Gamma}{\Gamma_D} e^{-(\Delta_{F_e F_g} / \Gamma_D)^2}. \tag{14.10}$$

The Doppler-broadened spectrum, given explicitly by Eq. (E.22), thus consists of a collection of resonances, each with Gaussian lineshape. For the $D1$ line with $I = 1/2$, we have

$$\ell_0 \frac{d\alpha}{d\ell} = \frac{\tilde{\kappa}_2 x_1}{576} \left((3+R)e^{-(\Delta_{1,1}/\Gamma_D)^2} + 3e^{-(\Delta_{0,1}/\Gamma_D)^2} \right). \tag{14.11}$$

Here ℓ_0 is the absorption length for the Doppler-broadened case, given by

$$\ell_0 = \frac{4\sqrt{\pi}}{Rn\lambda^2} \frac{\Gamma_D}{\Gamma} \frac{(2J_g+1)}{(2J_e+1)}. \tag{14.12}$$

Equation (14.11) is valid for $A_e, A_g, \Gamma_D \gg \Gamma$. Note that all the terms in this expression have the same sign; thus no cancelation occurs when the resonances overlap. This is because the Doppler-free resonances all have the same sign when the Doppler-free spectrum is well resolved (Fig. 14.3), so when the Doppler-broadened spectrum samples more than one resonance, the contributions from each resonance add.

The same approach can be generalized to describe the case in which some or all of the hyperfine splittings are on the order of or smaller than Γ. In this case, the Doppler-free spectrum is not composed entirely of peaks with a shape given by $f(\omega'_{F_e F_g})$. Nevertheless, as long as each resonance or group of resonances has frequency extent much less than the Doppler width, we can approximate it as a delta function times a coefficient given by the integral of the Doppler-free spectrum over the resonance. For the $D1$ line with $I = 1/2$ and $A_e, \Gamma \ll \Gamma_D \ll A_g$, this procedure yields (Eq. E.23)

$$\ell_0 \frac{d\alpha}{d\ell} = \frac{A_e^2 \tilde{\kappa}_2 x_1 (6+R) e^{-\Delta_{1,1}^2/\Gamma_D^2}}{576 \left(\Gamma^2 + A_e^2 \right)}. \tag{14.13}$$

The rotation in this case goes as A_e^2 for small A_e; the term linear in A_e (Eq. 14.6) is odd in detuning and consequently cancels in the velocity integral.

Since Eq. (14.11) applies when $A_e \gg \Gamma$ and Eq. (14.13) applies when $A_e \ll \Gamma_D$, we have that—if Γ_D is sufficiently larger than Γ—the two formulas together describe the signal over the entire range of A_e to excellent approximation, as verified by a numerical calculation. Figure 14.6 shows the maximum of the rotation spectrum as a function of the excited-state hyperfine splitting. As discussed above, as A_e is reduced, there is no suppression of the optical rotation signal when the Doppler-broadened hfs becomes unresolved. Only when the Doppler-free spectrum for a particular velocity group becomes unresolved is there suppression, as described in the previous subsection.

Spectra for the Doppler-broadened transit effect are shown in Fig. 14.7 for large A_g and various values of A_e, and for A_g and A_e varied together in Fig. 14.8.

In the case in which both A_e and A_g are small, the Doppler-free rotation spectrum is entirely of the same sign [see Eq. (14.7) and the bottom plot of Fig. 14.5]. The Doppler-broadened signal thus behaves similarly to the Doppler-free signal, because no additional cancelation takes place upon integrating over the velocity distribution. The signal for the $D1$ line with $I = 1/2$ and $A_g, A_e, \Gamma \ll \Gamma_D$ is given by

$$\ell_0 \frac{d\alpha}{d\ell} = A_e \left(A_e - \frac{R}{3} A_g \right) \frac{\tilde{\kappa}_2 x_1}{96 \Gamma^2} e^{-\Delta^2/\Gamma_D^2}. \tag{14.14}$$

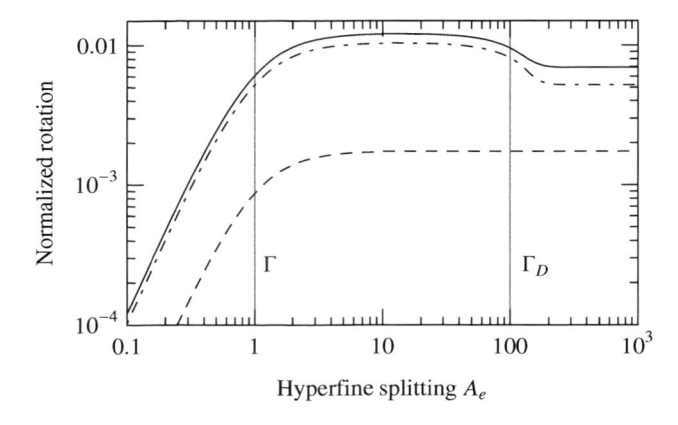

Fig. 14.6 (From Auzinsh et al. 2009a.) Maximum of the spectrum of the Doppler-broadened effect as a function of excited-state hyperfine splitting. Plotted are the component due to polarization produced by depopulation (dash-dotted), due to polarization produced by repopulation (dashed), and the total signal (solid). Equation (14.13) is used for $A_e < 10$, and Eq. (14.11) is used for $A_e > 10$. Parameter values in units of Γ are $\Gamma_D = 100$, $\gamma \ll 1$, $A_g \gg \Gamma_D$.

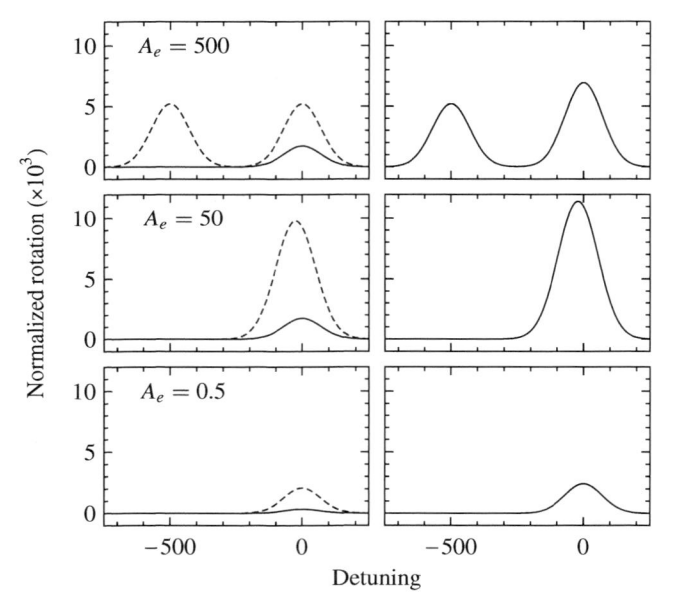

Fig. 14.7 (From Auzinsh et al. 2009a.) Spectra, as Fig. 14.3, but for the Doppler-broadened transit effect. Parameter values in units of Γ are $\Gamma_D = 100$, $\gamma \ll 1$, $A_g \gg \Gamma_D$.

14.3 Wall effect

We now consider systems in which the atomic velocities change in between optical pumping and probing. This is the case for the "wall effect" in antirelaxation-coated vapor cells: atoms are optically pumped as they pass through the light beam, and then retain their polarization through many collisions with the cell walls before returning to the beam and being probed. A similar situation occurs in vapor cells with buffer gas.

We assume that the atomic velocities are completely randomized after optical pumping. Then the density matrix for each velocity group is the same; to lowest order in light power, we can find the velocity-averaged polarization by integrating the perturbative expression (E.12) over velocity with the Gaussian weighting factor (14.9). Since we are now describing the average over all of the atoms in the cell, and not just the illuminated region of the cell, we take

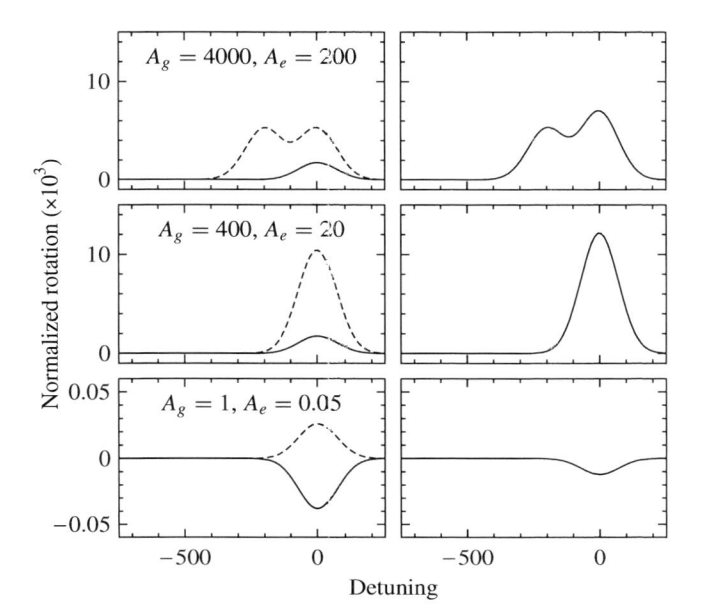

Fig. 14.8 (From Auzinsh et al. 2009a.) As Fig. 14.7, with A_g and A_e varied simultaneously ($A_g = 20 A_e$).

γ to be the average ground-state relaxation rate for an atom in the cell, rather than the transit rate through the light beam. We also multiply the polarization by the illuminated fraction of the cell volume, $V_{\text{illum.}}/V_{\text{cell}}$ (assuming this fraction is small), to account for the fact that the light pumps only some of the atoms at a time.

For the specific case of the $D1$ line for an atom with $I = 1/2$, Eq. (E.12) takes the form, given in Eq. (14.1), of a linear combination of Lorentzian functions $L(\omega'_{F_e F_g})$. This simple form arises because, due to the selection rules for this transition, no coherences are formed between excited-state hyperfine levels. For a general system this is not the case; however, if the excited-state hyperfine splitting is greater than Γ, the excited-state hyperfine coherences are suppressed, and all resonances once again have Lorentzian lineshapes. Thus, assuming that $\Gamma \ll \Gamma_D$, the velocity integral can be accomplished by replacing $L(\omega'_{F_e F_g})$ by

$$\int dv_k L(-\Delta_{F_e F_g} + k v_k) G(v_k) \approx \frac{\sqrt{\pi}}{2} \frac{\Gamma}{\Gamma_D} e^{-(\Delta_{F_e F_g}/\Gamma_D)^2}. \tag{14.15}$$

The polarization in this case is given by (Eq. E.24)

$$\rho^{(20)}(11) = \frac{\tilde{\kappa}_2 \sqrt{\pi}}{24\sqrt{6}} \left(\left[e^{-(\Delta_{0,1}/\Gamma_D)^2} - e^{-(\Delta_{1,1}/\Gamma_D)^2} \right] + \frac{R}{3} \left[e^{-(\Delta_{1,0}/\Gamma_D)^2} - e^{-(\Delta_{1,1}/\Gamma_D)^2} \right] \right), \tag{14.16}$$

where the saturation parameter for the wall effect is defined by

$$\tilde{\kappa}_2 = \frac{\Omega_R^2}{\Gamma \gamma} \frac{\Gamma}{\Gamma_D} \frac{V_{\text{illum.}}}{V_{\text{cell}}}. \tag{14.17}$$

We make this new definition because, in the wall effect, light of a single frequency illuminating just part of the cell effectively pumps all velocity groups in the entire cell.

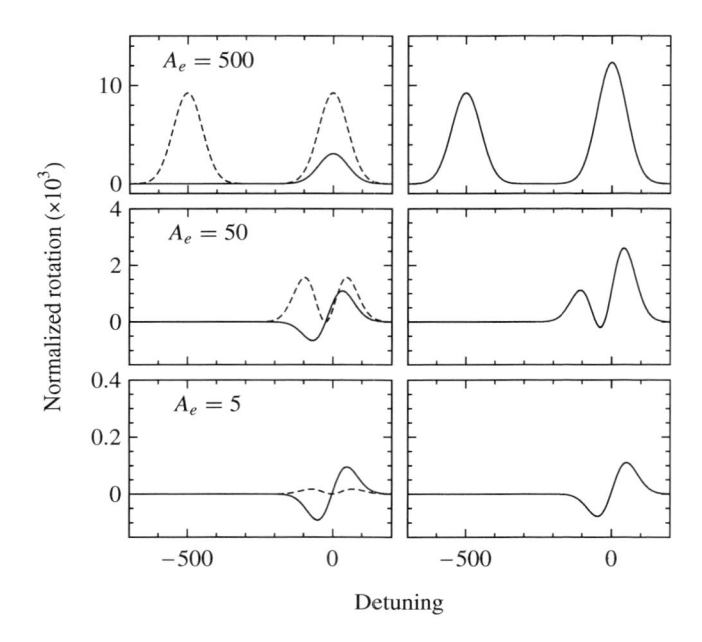

Fig. 14.9 (From Auzinsh et al. 2009a.) Spectra, as Fig. 14.7, but for the wall effect.

The signal due to each velocity group is given in terms of $\rho^{(20)}(1)$ by Eq. (14.3); integrating over velocity to find the total signal, we obtain (Eq. E.25)

$$\ell_0 \frac{d\alpha}{d\ell} = \frac{1}{4}\sqrt{\frac{3}{2}} \left[e^{-(\Delta_{0,1}/\Gamma_D)^2} - e^{-(\Delta_{1,1}/\Gamma_D)^2} \right] x_1 \rho^{(20)}(11). \tag{14.18}$$

The spectrum of the signal due to the wall effect is quite different than the spectrum of the Doppler-broadened transit effect signal, and is in a sense more similar to that of the Doppler-free transit effect (Budker et al. 2000b). Equations (14.16) and (14.18) have the same form as the Doppler-free equations (14.1) and (14.3), with Lorentzians of width Γ replaced by Gaussians of width Γ_D. Thus the rotation signal produced by the wall effect has similar spectra and dependence on hyperfine splitting as the Doppler-free transit effect, but with scale set by the Doppler width rather than the natural width. This is illustrated in Figs. 14.9 and 14.10 for the case of large ground-state hyperfine splitting. Figure 14.9 shows the optical rotation spectrum for various values of A_e, and Fig. 14.10 shows the maximum of the rotation spectrum as a function of A_e. These figures can be compared to Figs. 14.3 and 14.4 for the Doppler-free transit effect. In particular, we see the same phenomenon of two resonance peaks of the same sign appearing to cancel as they overlap [observation of this effect in antirelaxation-coated vapor cells is discussed by Budker et al. (2000b) and in buffer-gas cells by Novikova et al. (2001a)]. The explanation for this is the same as in the Doppler-free case. Also as in the Doppler-free case, the rotation is linear in A_e to lowest order, and this linear term is due to polarization produced by spontaneous emission:

$$\ell_0 \frac{d\alpha}{d\ell} = \frac{\sqrt{\pi}}{288} \tilde{\kappa}_2 x_1 R \frac{A_e \Delta_1}{\Gamma_D^2} e^{-2(\Delta_1/\Gamma_D)^2}. \tag{14.19}$$

Higher nuclear spin and the D2 line **283**

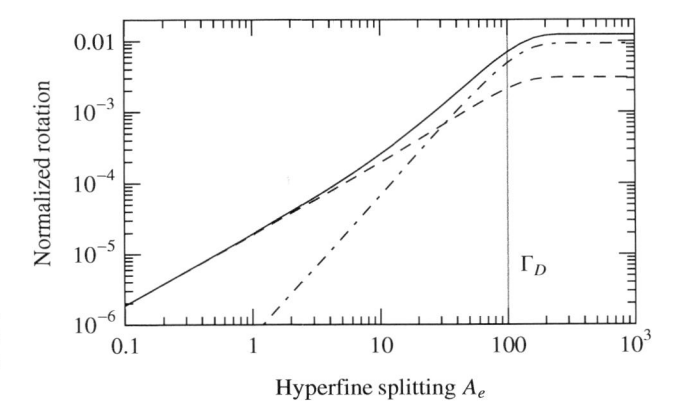

Fig. 14.10 (From Auzinsh *et al.* 2009a.) As Fig. 14.4, but for the wall effect. Parameters are the same as in Fig. 14.6.

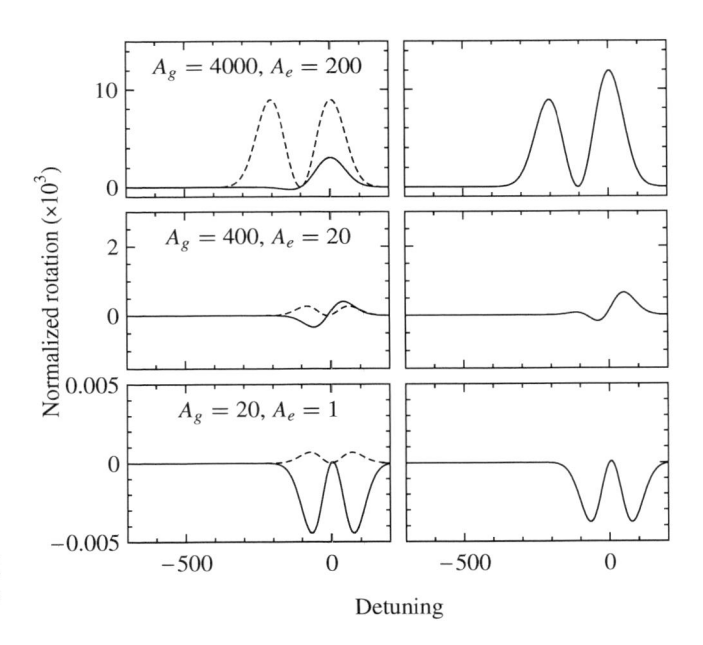

Fig. 14.11 (From Auzinsh *et al.* 2009a.) As Fig. 14.8, but for the wall effect.

Spectra for the case in which A_e and A_g are varied together are shown in Fig. 14.11, and are also similar to the Doppler-free transit effect (Fig. 14.5). When both A_e and A_g are small, the signal to lowest order in these quantities is given by

$$\ell_0 \frac{d\alpha}{d\ell} = A_e \left(A_e - \frac{R}{3} A_g \right) \frac{\tilde{\kappa}_2 x_1 \Delta^2}{48 \Gamma_D^4} e^{-2\Delta^2/\Gamma_D^2}. \tag{14.20}$$

14.4 Higher nuclear spin and the *D*2 line

When nuclear spins $I \geq 1/2$ are considered, several complications arise. The clearest of these is that the two ground states now have angular momenta $F_g = I \pm \frac{1}{2} \geq 1$, so that they can both support atomic alignment and produce optical rotation. A more subtle difference is that, with higher angular momenta in the excited state, coherences between the excited state hyperfine levels can be created when the excited-state hyperfine splitting is on the order of the natural width or smaller. (Ground-state hyperfine coherences can be neglected as long as the ground-state hyperfine splitting is much larger than the ground-state relaxation rate.) This can change the optical rotation spectrum, and also causes the symmetry between the Doppler-free transit and wall effects discussed above to be partially broken, as we see below.

However, many of the results obtained above for the $I = 1/2$ system are a consequence of the general arguments discussed in Chapter 13, and thus hold for any nuclear spin. In particular, the dependence of the optical rotation signal on the hyperfine splitting for large ground-state and small excited state splitting (Eqs. 14.6, 14.13, and 14.19) and for both ground- and excited-state hyperfine splitting small (Eqs. 14.7, 14.14, and 14.20) remains the same. We have, for large A_g and small A_e, and for a particular transition group, the following three expressions. For the Doppler-free transit effect:

$$\ell_0 \frac{d\alpha}{d\ell} \propto A_e \tilde{\kappa}_2 x_{F_g} R \frac{(\Gamma/2)^4 \Delta_{F_g}}{\left[(\Gamma/2)^2 + \Delta_{F_g}^2\right]^3}, \tag{14.21}$$

for the Doppler-broadened transit effect:

$$\ell_0 \frac{d\alpha}{d\ell} \propto A_e^2 \tilde{\kappa}_2 x_{F_g} \frac{e^{-\Delta_{F_g}^2/\Gamma_D^2}}{(\Gamma^2 + A_e^2)}, \tag{14.22}$$

and for the wall effect:

$$\ell_0 \frac{d\alpha}{d\ell} \propto A_e \tilde{\kappa}_2 x_{F_g} R \frac{\Delta_{F_g} e^{-2(\Delta_{F_g}/\Gamma_D)^2}}{\Gamma_D^2}. \tag{14.23}$$

For A_g and A_e both small, we have, for the Doppler-free transit effect:

$$\ell_0 \frac{d\alpha}{d\ell} \propto A_e \left(A_e - \frac{R}{3} A_g \right) \frac{\tilde{\kappa}_2 (\Gamma/2)^4 \Delta^2}{\left[(\Gamma/2)^2 + \Delta^2\right]^4}, \tag{14.24}$$

for the Doppler-broadened effect:

$$\ell_0 \frac{d\alpha}{d\ell} \propto A_e \left(A_e - \frac{R}{3} A_g \right) \frac{\tilde{\kappa}_2}{\Gamma^2} e^{-\Delta^2/\Gamma_D^2}, \tag{14.25}$$

and for the wall effect:

$$\ell_0 \frac{d\alpha}{d\ell} \propto A_e \left(A_e - \frac{R}{3} A_g \right) \frac{\tilde{\kappa}_2 \Delta^2}{\Gamma_D^4} e^{-2\Delta^2/\Gamma_D^2}. \tag{14.26}$$

To illustrate the differences that arise when the nuclear spin is increased, we plot (analogously to Figs. 14.4, 14.6, and 14.10) in Fig. 14.12 the maximum of the rotation spectra for large A_g as a function of A_e for the Doppler-free transit, Doppler-broadened transit, and wall effects. Three values of the nuclear spin are used, $I = 1/2$, $3/2$, and $5/2$, and for $I = 3/2$ and $5/2$ the rotation on the $F_g = I \pm 1/2 \rightarrow F_e$ lines is plotted separately. Rotation due to polarization produced by the depopulation and repopulation mechanisms is plotted, as well

Higher nuclear spin and the D2 line 285

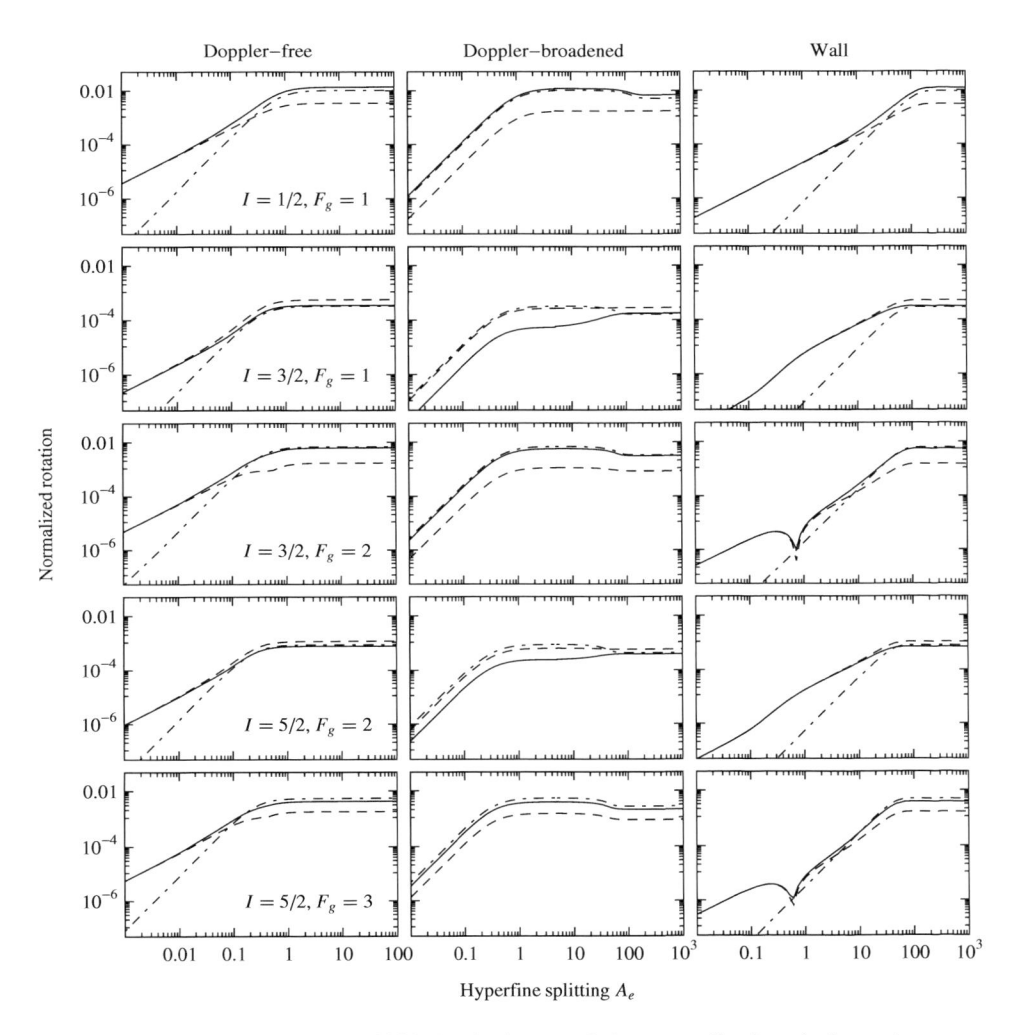

Fig. 14.12 (From Auzinsh *et al.* 2009a.) Maximum of the normalized optical rotation spectra $\ell_0/(\tilde{\kappa}_2 x_{F_g}) d\alpha/d\ell$ for the Doppler-free transit, Doppler-broadened transit, and wall effects on the $D1$ line for $I = 1/2, 3/2$, and $5/2$. We assume $\Gamma_D = 100$, $\gamma \ll 1$, $A_g \gg \Gamma_D$ in units of Γ. The maxima for the $F_g = I \pm 1/2 \rightarrow F_e$ transitions are plotted separately. Each plot shows rotation due to polarization produced by depopulation (dot-dashed line), rotation due to polarization produced by repopulation (dashed line), and total rotation (solid line).

as the total rotation signal. In many cases these two contributions are of opposite sign, so the details of the total signal can depend on how closely the two contributions cancel each other. (The cancelation tends to be more complete for the $F_g = I - 1/2$ lines.) However, the qualitative features of these plots follow, in large part, the pattern exhibited in the $I = 1/2$ case. One exception is the behavior of the wall effect plot for A_e in the neighborhood of the natural width. As mentioned above, when $I > 1/2$, excited-state hyperfine coherences can form when the excited-state hyperfine splitting becomes small. This leads to "interference" effects when the Doppler-free resonance lines overlap that do not occur when the Doppler-broadened

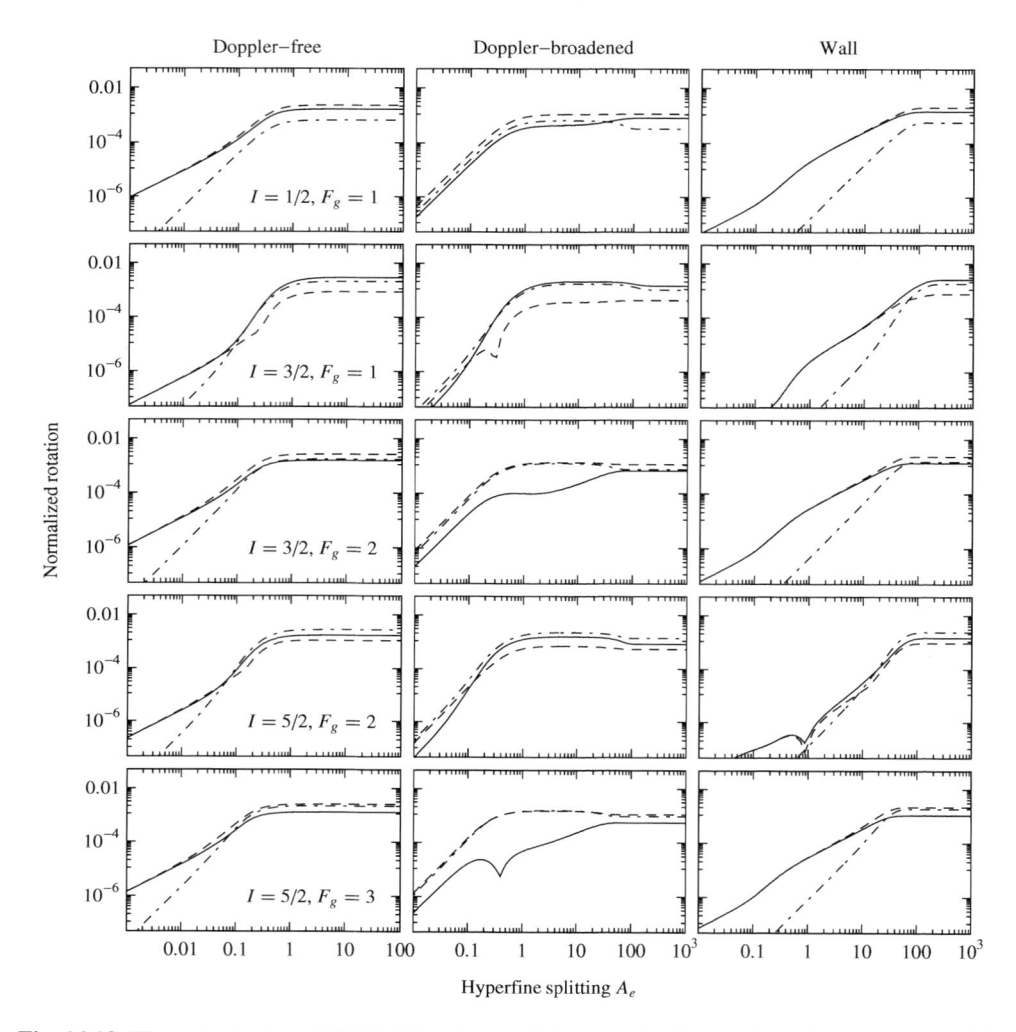

Fig. 14.13 (From Auzinsh *et al.* 2009a.) Maximum of the normalized optical rotation spectra, as in Fig. 14.12, but for the *D*2 line. For the $I = 3/2$, $F_g = 2$ and the $I = 5/2$, $F_g = 3$ systems for the Doppler-broadened transit effect, the two contributions to optical rotation nearly cancel, with the consequence that the approximations used in obtaining the analytic formulas for the total Doppler-broadened signal begin to break down. Numerical convolution is employed in these cases.

resonance lines in the wall effect overlap. This breaks the symmetry between the wall effect and the Doppler-free transit effect that is found in the $I = 1/2$ case.

We now discuss the $J_g = 1/2 \rightarrow J_e = 3/2$ *D*2 transition. The presence of three hyperfine levels in the excited state leads to additional features in the dependence of the signal on the hyperfine splitting (Fig. 14.13). However, the fact that the ground-state electronic momentum is still $J_g = 1/2$ means that the dependence of the signal on the excited-state hyperfine splitting as A_e goes to zero remains the same, for the reasons discussed in Chapter 13. Thus, to lowest order in A_e, the rotation signals on the *D*2 line for large A_g are given by Eqs. (14.21)–(14.23),

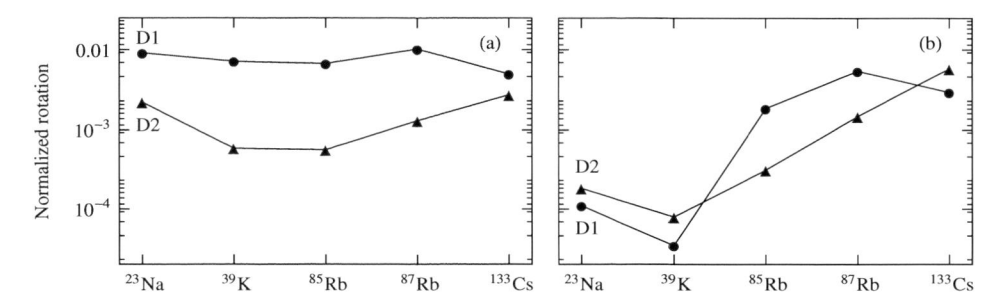

Fig. 14.14 (From Auzinsh *et al.* 2009a.) Maximum of the spectrum of normalized optical rotation for the (a) Doppler-broadened transit effect and (b) wall effect for various alkali atoms. Circles indicate the $D1$ line and triangles indicate the $D2$ line. Room-temperature Maxwellian velocity distributions are assumed. Normalized rotation is defined here as $\ell_0/(\tilde{\kappa}_2 x_{I+J_g})(d\alpha/d\ell)$, where ℓ_0 in this case is the unsaturated absorption length at the detuning that gives maximum absorption. The normalized magnitude of unsuppressed optical rotation is nominally on the order of unity; however, this is to some degree dependent on the normalization convention chosen. For example, if the maximum matrix element of d_z is used in the definition of $\tilde{\kappa}_2$, rather than the reduced matrix element, the values in this plot are increased by a factor of ~ 6.

(We set the hyperfine coefficient B_e to zero for simplicity.)

Considering the signals obtained when both the excited- and ground-state hyperfine splittings are small, we expect somewhat different behavior for the contribution due to polarization produced by repopulation pumping than in the $D1$ case. This is because the excited-state electronic angular momentum is $J_e = 3/2$, so that production of rank $\kappa = 2 < 2J_e$ atomic alignment in the ground state by spontaneous emission is allowed even when the ground-state hfs is unresolved (Sec. 13.2). The lowest order dependence on hyperfine splitting for the $D2$ line is given by

$$\ell_0 \frac{d\alpha}{d\ell} \propto A_e \left[A_e - R\left(2A_e + \frac{1}{3}A_g\right) \right] \quad (14.27)$$

for each of the three effects, with the spectral lineshapes remaining as in Eqs. (14.24)–(14.26). Note that there is now a term that depends on polarization due to repopulation that does not go to zero as A_g goes to zero.

14.5 Comparison of quantitative results for different cases

We now examine the consequences of the preceding discussion for the alkali atoms commonly used in nonlinear magneto-optical experiments. In Fig. 14.14 the maximum of the spectrum of optical rotation is plotted for the $D1$ and $D2$ lines of several alkali atoms. The Doppler-broadened transit effect is shown in Fig. 14.14(a) and the wall effect is shown in Fig. 14.14(b). (Numerical convolution was used to obtain these results, because the alkalis do not all satisfy the conditions under which the analytic formulas were derived.) The nuclear spins, hyperfine splittings, excited-state lifetimes, and Doppler widths all vary between the different alkali atoms. However, focusing our attention on the hyperfine splittings, which have the greatest degree of variation, we can see the correspondence of these results to the preceding discussion. In particular, we have seen that the magnitude of the Doppler-broadened transit effect is largely independent of the hyperfine splitting when the splittings are greater than the natural width of the transition. This is generally the case for the alkalis, leading to the relative constancy of the magnitude of the transit effect among the alkalis. For the wall effect, on the other hand, we have found that the magnitude of the effect diminishes when the hyperfine splitting becomes less than the Doppler width. In the alkalis the excited-state hyperfine splitting is generally on the order of or smaller than the Doppler width, and the general trend is that the ratio of hyperfine splitting to Doppler width increases as the atomic mass number increases. This accounts for the general upward trend in Fig. 14.14(b). The trend is not completely consistent: the hyperfine splitting of K is smaller than that of Na, which is reflected in the plot of the wall effect.

15
Coherence effects revisited

In this chapter we return to the discussion of Chapter 8, which examined various effects resulting from the creation of atomic polarization with linearly polarized light. These examples are valuable in that they represent actual laboratory experiments, they illustrate important concepts concerning atomic coherence, and they are simple enough that it is straightforward to follow all of the steps in the analysis. We therefore repeat some of this discussion using the density matrix, a more general technique than the wave-function approach used in Chapter 8. We will see that this confers certain advantages. We also extend the discussion to cover effects due to polarization produced in the ground state, as well as the excited state.

In this chapter angular-momentum-probability surfaces (AMPS) are used to illustrate the atomic polarization created by the light. These can be compared to the corresponding electron-density plots in Chapter 8 to see the relationship between the two visualization methods. Both methods can be used to illustrate the symmetries and time evolution of the atomic polarization. There are differences between them, however. The AMPS show only the information about the angular momentum of the atomic state, while the electron-density plots also display the radial part of the wave function. This makes the AMPS more general, in a sense, because the same plot applies to all states with a particular angular momentum, while the electron-density plots must assume a specific atomic state (e.g., the 1S or 2P state of "spinless" hydrogen in Chapter 8). A technical difference between the two methods is that the electron-density plot discards phase information, and so is not a complete representation of the atomic state, whereas the AMPS contains all of the information about the angular-momentum polarization of the state, and can be used to reconstruct the density matrix.

15.1 Dark and bright states

We first reexamine the $J = 0 \to J' = 1 \to J'' = 0$ system considered in Sec. 8.1, used to illustrate the concept of dark and bright states. With weak pump light applied to a $J \to J'$ transition in the absence of other fields, the excited-state density-matrix elements $\rho_{m'_1 m'_2}$ are given in terms of the unperturbed ground-state density matrix $\rho^{(0)}$ by Eq. (12.39):

$$\rho_{m'_1 m'_2} = \frac{\pi \mathcal{E}_0^2}{2\Gamma \hbar^2} G(\Delta) \sum_{m_1 m_2} \hat{\boldsymbol{\varepsilon}} \cdot \mathbf{d}_{m'_1 m_1} \rho^{(0)}_{m_1 m_2} \hat{\boldsymbol{\varepsilon}}^* \cdot \mathbf{d}_{m_2 m'_2}, \qquad (15.1)$$

where $\mathcal{E}_0$ is the light electric-field amplitude, $\hat{\boldsymbol{\varepsilon}}$ is the complex light polarization vector, Γ is the excited-state natural linewidth, $G(\Delta)$ is the lineshape function (12.32), and Δ is the light detuning. Expanding the dipole matrix elements using the Wigner-Eckart theorem, we have

290 *Coherence effects revisited*

$$\rho_{m'_1 m'_2} = (-1)^{2J'-m'_1-m'_2} \frac{\pi |\langle J'\|d\|J\rangle|^2 \mathcal{E}_0^2}{2\Gamma \hbar^2} G(\Delta)$$

$$\times \sum_{m_1 m_2} \sum_{q_1 q_2} \varepsilon^{q_1} (\varepsilon^{q_2})^* \begin{pmatrix} J' & 1 & J \\ -m'_1 & q_1 & m_1 \end{pmatrix} \begin{pmatrix} J' & 1 & J \\ -m'_2 & q_2 & m_2 \end{pmatrix} \rho^{(0)}_{m_1 m_2}. \tag{15.2}$$

For a $J = 0 \to J' = 1$ transition the unperturbed ground-state density matrix has only one element, $\rho^{(0)}_{00} = 1$. Assuming x-polarized light, the spherical components of the polarization vector are (Sec. 6.4)

$$\varepsilon^{-1} = \frac{1}{\sqrt{2}}, \quad \varepsilon^0 = 0, \quad \varepsilon^1 = -\frac{1}{\sqrt{2}}. \tag{15.3}$$

Evaluating the $3j$ symbols, we have for the upper-state density matrix under excitation with weak light:

$$\rho' = \frac{\pi}{4} \kappa_1 G(\Delta) \Gamma \begin{pmatrix} 1 & 0 & -1 \\ 0 & 0 & 0 \\ -1 & 0 & 1 \end{pmatrix}, \tag{15.4}$$

where κ_1 is the excited-state optical-pumping saturation parameter (Sec. 9.1)

$$\kappa_1 = \frac{\Omega_R^2}{\Gamma^2} = \frac{|\langle J'\|d\|J\rangle|^2 \mathcal{E}_0^2}{3\hbar^2 \Gamma^2}. \tag{15.5}$$

The angular-momentum probability surface for the $J' = 1$ state corresponding to the density matrix of Eq. (15.4) is shown in Fig. 15.1. This figure illustrates the same angular-momentum state as the electron-density plot of Fig. 8.4. Each figure clearly shows that the x-axis is the alignment axis for the system.

If a second light beam resonant with the $J' = 1 \to J'' = 0$ transition is applied, absorption can be observed due to the atoms excited to the J' state by the first light beam. From Eq. (12.57) we have for the fractional absorption of weak light over a path length $d\ell$:

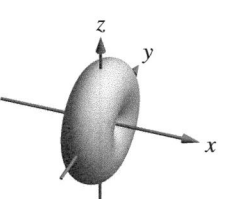

Fig. 15.1 Plot of AMPS for the density matrix (15.4), excited by weak x-polarized light on a $J = 0 \to J' = 1$ transition.

$$\frac{1}{\mathcal{E}_0} \frac{d\mathcal{E}_0}{d\ell} = -\frac{2\pi \omega n}{\hbar c} \sum_{m'' m'_1 m'_2} \operatorname{Im} \left(\frac{\hat{\varepsilon} \cdot \mathbf{d}_{m'' m'_1} \rho'_{m'_1 m'_2} \hat{\varepsilon}^* \cdot \mathbf{d}_{m'_2 m''}}{\widetilde{\omega}_{m'' m'_2}} \right), \tag{15.6}$$

where $\widetilde{\omega}_{m'' m'_2}$ is the complex frequency splitting (12.63). If we assume that the lower and upper states are nearly degenerate compared to the natural width of the upper state, $\widetilde{\omega}_{m'' m'_2}$ becomes independent of m'' and m'_2 and can be taken out of the sum. Further, because $\mathbf{d}$ and ρ' are Hermitian, we see that swapping m'_1 and m'_2 in $\hat{\varepsilon} \cdot \mathbf{d}_{m'' m'_1} \rho'_{m'_1 m'_2} \hat{\varepsilon}^* \cdot \mathbf{d}_{m'_2 m''}$ is equivalent to taking the complex conjugate, so that the imaginary parts of these two terms cancel in the sum. Thus Eq. (15.6) becomes

$$\frac{1}{\mathcal{E}_0} \frac{d\mathcal{E}_0}{d\ell} = -\frac{2\pi \omega n}{\hbar c} \operatorname{Im}\left(\frac{1}{\tilde{\omega}}\right) \sum_{m''m'_1 m'_2} \hat{\boldsymbol{\varepsilon}} \cdot \mathbf{d}_{m''m'_1} \rho'_{m'_1 m'_2} \hat{\boldsymbol{\varepsilon}}^* \cdot \mathbf{d}_{m'_2 m''}$$

$$= -\frac{2\pi^2 \omega n}{\hbar c} G(\Delta) \sum_{m''m'_1 m'_2} \hat{\boldsymbol{\varepsilon}} \cdot \mathbf{d}_{m''m'_1} \rho'_{m'_1 m'_2} \hat{\boldsymbol{\varepsilon}}^* \cdot \mathbf{d}_{m'_2 m''} \qquad (15.7)$$

$$= \sum_{m'_1 m'_2} A_{m'_2 m'_1} \rho'_{m'_1 m'_2}$$

$$= \operatorname{Tr} A\rho'.$$

Here we have written the absorption signal as the expectation value of an *absorption matrix A*, with matrix elements

$$\begin{aligned} A_{m'_2 m'_1} &= -\frac{2\pi^2 \omega n}{\hbar c} G(\Delta) \sum_{m''} \hat{\boldsymbol{\varepsilon}}^* \cdot \mathbf{d}_{m'_2 m''} \hat{\boldsymbol{\varepsilon}} \cdot \mathbf{d}_{m''m'_1} \\ &= -\frac{\lambda^2 n}{8}(2J''+1) G(\Delta) \Gamma \sum_{m'' q_1 q_2} (\varepsilon^{q_2})^* \varepsilon^{q_1} \begin{pmatrix} J'' & 1 & J' \\ -m'' & q_2 & m'_2 \end{pmatrix} \begin{pmatrix} J'' & 1 & J' \\ -m'' & q_1 & m'_1 \end{pmatrix}, \end{aligned} \qquad (15.8)$$

where we have used the Wigner-Eckart theorem and Eq. (12.13) to rewrite the reduced dipole matrix element in terms of the natural width and the wavelength λ.

In the case of a $J' = 1 \to J'' = 0$ transition with x-polarized light on resonance, the absorption matrix becomes

$$A^{(x)} = \frac{n\lambda^2}{24\pi} \begin{pmatrix} 1 & 0 & -1 \\ 0 & 0 & 0 \\ -1 & 0 & 1 \end{pmatrix}, \qquad (15.9)$$

and more generally, for light polarized at an angle α to $\hat{\mathbf{x}}$ in the xy plane, we have

$$A^{(\alpha)} = \frac{n\lambda^2}{24\pi} \begin{pmatrix} 1 & 0 & -e^{-2i\alpha} \\ 0 & 0 & 0 \\ -e^{2i\alpha} & 0 & 1 \end{pmatrix}. \qquad (15.10)$$

The absorption of the probe beam for the $J = 0 \to J' = 1 \to J'' = 0$ system is given by the expectation value of (15.10) with respect to the intermediate-state density matrix (15.4). We have

$$\frac{1}{\mathcal{E}_0} \frac{d\mathcal{E}_0}{d\ell} = \operatorname{Tr} A^{(\alpha)} \rho' = \frac{n\lambda^2 \kappa_1}{12\pi} \cos^2 \alpha, \qquad (15.11)$$

assuming both light fields are on resonance. We see that absorption is maximum for x-polarized light ($\alpha = 0$) and zero for y polarization ($\alpha = \pi/2$), as we found in Chapter 8 using wave functions.

From the dependence of the absorption coefficient on the angle α, we see that atoms in this coherent superposition of states act as a dichroic polarizer (like a Polaroid film). As discussed in Chapters 10 and 14, the dichroic properties of optically polarized atoms are the basis of the nonlinear Faraday rotation and other nonlinear magneto-optical effects.

15.2 Quantum beats

We now turn to the example of quantum beats, previously discussed in Sec. 8.2. We suppose that atoms are initially excited with a short laser pulse on a $J \to J'$ transition. If we assume that the pulse length is shorter than all of the relaxation times in the problem, the laser linewidth is effectively very broad, and we can use the broad-line approximation (Sec. 12.8). We thus begin with the approximate Liouville equations (12.67). Assuming that the pulse length is also shorter than the inverse of the optical pumping rate, $\Gamma_p^{-1} = (\Omega_R^2/\Gamma)^{-1}$, and the Larmor period due to any applied magnetic field, we can neglect the first three terms on the right-hand side of Eq. (12.67), and approximate the density-matrix elements by their initial values. For an initially unpolarized state, the ground-state density-matrix elements before the pulse are given by $\rho_{m_1 m_2}(0) = \delta_{m_1 m_2}/(2J+1)$, and the excited-state density-matrix elements are zero. Equation (12.67) then reduces to

$$\dot{\rho}_{m_1 m_2}(t) = -\frac{2\pi G(\Delta)}{2J+1} \frac{\mathcal{E}_0^2}{4\hbar^2} \sum_{m'} \hat{\varepsilon}^* \cdot \mathbf{d}_{m_1 m'} \hat{\varepsilon} \cdot \mathbf{d}_{m' m_2} \tag{15.12}$$

for the ground state, and

$$\dot{\rho}_{m'_1 m'_2}(t) = \frac{2\pi G(\Delta)}{2J+1} \frac{\mathcal{E}_0^2}{4\hbar^2} \sum_{m} \hat{\varepsilon} \cdot \mathbf{d}_{m'_1 m} \hat{\varepsilon}^* \cdot \mathbf{d}_{m m'_2} \tag{15.13}$$

for the excited state.

After the excitation pulse, the atomic polarization evolves in a static magnetic field. Below, we consider two systems: first a $J = 0 \to J' = 1$ transition, and then a $J = 1 \to J' = 0$ transition.

15.2.1 Excited state

We first consider a $J = 0 \to J' = 1$ transition excited with x-polarized light. Solving Eq. (15.13) for the excited-state density matrix after a short pulse of duration τ_p, we have

$$\rho'(\tau_p) = \frac{1}{2}\Gamma_p(\Delta)\tau_p \begin{pmatrix} 1 & 0 & -1 \\ 0 & 0 & 0 \\ -1 & 0 & 1 \end{pmatrix}, \tag{15.14}$$

where

$$\Gamma_p(\Delta) = \Gamma_p \frac{(\Gamma/2)^2}{\Delta^2 + (\Gamma/2)^2} \tag{15.15}$$

is the detuning-dependent optical pumping rate.

After the excitation pulse, the atomic polarization evolves in a $\hat{z}$-directed magnetic field. We solve the Liouville equation (12.17) to find the upper-state density matrix as a function of time. In the absence of light, we are left with only the first term on the right-hand side of Eq. (12.17):

$$\dot{\rho}_{m'_1 m'_2} = -i\tilde{\omega}_{m'_1 m'_2} \rho_{m'_1 m'_2}, \tag{15.16}$$

where the complex frequency splitting is given in this case by $\tilde{\omega}_{m'_1 m'_2} = \Omega_L(m'_1 - m'_2) - i\Gamma$, with Ω_L the excited-state Larmor frequency. Using Eq. (15.14) as the initial condition, the solution of Eq. (15.16) is

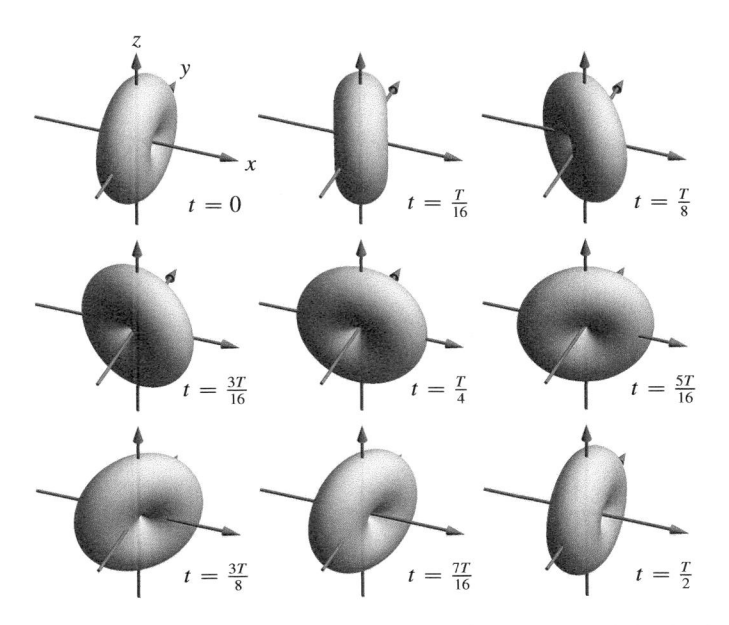

Fig. 15.2 Angular-momentum probability surfaces for the $J' = 1$ state evolving in time after pulsed excitation on a $J = 0 \rightarrow J' = 1$ transition with linearly x-polarized light. T is the period of Larmor precession. The magnetic field is along $\hat{z}$.

$$\rho'(t) = \frac{1}{3}\Gamma_p(\Delta)\tau_p e^{-\Gamma t} \begin{pmatrix} 1 & 0 & -e^{-2i\Omega_L t} \\ 0 & 0 & 0 \\ -e^{2i\Omega_L t} & 0 & 1 \end{pmatrix}. \tag{15.17}$$

Because the excitation-pulse length τ_p is much shorter than the Larmor period, we can make the approximation that this evolution begins at $t = 0$. The evolution is the familiar Larmor precession about the z-axis, illustrated using angular-momentum probability surfaces in Fig. 15.2. This figure can be compared with Fig. 8.5 showing the electron density in the 2P state in a similar situation.

Suppose that we now measure absorption on a transition from the $J' = 1$ state to a higher state with $J'' = 0$. We will assume that the probe light is broad-band, so that the level shifts due to the Zeeman effect can be neglected in calculating the absorption. Then the absorption of light with linear polarization in the xy plane is given by the expectation value of the absorption matrix (15.10):

$$\frac{1}{\mathcal{E}_0}\frac{d\mathcal{E}_0}{d\ell} = \text{Tr } A^{(\alpha)}\rho'(t) = \frac{n\lambda^2\Gamma_p(\Delta)\tau_p}{12\pi}e^{-\Gamma t}\cos^2(\Omega_L t - \alpha). \tag{15.18}$$

The quantum beats are observed as a time-dependent absorption signal with the initial phase dependent on the choice of the angle α for the polarization direction of the probe field.

The quantum beats can also be observed in fluorescence. To find the fluorescence emitted with a particular polarization (measured by placing a polarizing filter in front of a light detector oriented in a chosen direction), we use Eq. (12.7). This equation can be written in terms of a *fluorescence matrix* $\mathcal{F}$: the intensity of fluorescence per unit solid angle for N atoms fluorescing into a mode with polarization vector $\hat{\varepsilon}'$ is given by the expectation value

294 *Coherence effects revisited*

$$\frac{d\mathcal{I}}{d\Omega} = \text{Tr}\,\mathcal{F}\rho', \tag{15.19}$$

where

$$\mathcal{F}_{m'_2 m'_1} = \frac{N\omega^4}{2\pi c^3} \sum_m \hat{\varepsilon}' \cdot \mathbf{d}_{m'_2 m} \left(\hat{\varepsilon}'\right)^* \cdot \mathbf{d}_{mm'_1}. \tag{15.20}$$

Evaluating the fluorescence matrix for a $J' = 1 \to J = 0$ decay and detection of linearly polarized light at an angle α to the x-axis in the xy plane, we find

$$\mathcal{F} = \frac{N\omega^4}{12\pi c^3} \begin{pmatrix} 1 & 0 & -e^{-2i\alpha} \\ 0 & 0 & 0 \\ -e^{2i\alpha} & 0 & 1 \end{pmatrix}. \tag{15.21}$$

Taking the expectation value for pump light on resonance, the measured light intensity is

$$\frac{d\mathcal{I}}{d\Omega} = \frac{N\omega^4 \Gamma_p(\Delta)\tau_p}{6\pi c^3} e^{-\Gamma t} \cos^2(\Omega_L t - \alpha), \tag{15.22}$$

with the same time and angular dependence as the absorption signal (15.18).

15.2.2 Ground state

We now consider quantum beats in a $J = 1 \to J' = 0$ transition, which are associated with ground-state (rather than excited-state) polarization evolving in time. Solving Eq. (15.12) under the conditions used for the $J = 0 \to J' = 1$ case, we find for the ground-state density matrix after an excitation pulse of duration τ_p:

$$\rho(\tau_p) = \frac{1}{3} \begin{pmatrix} 1 - \frac{\Gamma_p \tau_p}{2} & 0 & \frac{\Gamma_p \tau_p}{2} \\ 0 & 1 & 0 \\ \frac{\Gamma_p \tau_p}{2} & 0 & 1 - \frac{\Gamma_p \tau_p}{2} \end{pmatrix}. \tag{15.23}$$

The density matrix then evolves under the action of the $\hat{z}$-directed magnetic field. For the ground state, the repopulation term of the Liouville equation is nonzero, so the evolution equation takes the form

$$\dot{\rho}_{m_1 m_2} = -i\tilde{\omega}_{m_1 m_2} \rho_{m_1 m_2} + \frac{\gamma}{2J+1} \delta_{m_1 m_2}, \tag{15.24}$$

where γ is the ground-state relaxation rate. Solving Eq. (15.16) we find

$$\rho(t) = \frac{1}{3} \begin{pmatrix} 1 - \frac{\Gamma_p \tau_p}{2} e^{-\gamma t} & 0 & \frac{\Gamma_p \tau_p}{2} e^{-(\gamma + 2i\Omega_L)t} \\ 0 & 1 & 0 \\ \frac{\Gamma_p \tau_p}{2} e^{-(\gamma - 2i\Omega_L)t} & 0 & 1 - \frac{\Gamma_p \tau_p}{2} e^{-\gamma t} \end{pmatrix}. \tag{15.25}$$

The evolution of the density matrix (15.25) is shown using AMPS in Fig. 15.3. The figure shows one period of the Larmor precession. As the polarization precesses, it also relaxes at a rate γ toward the unpolarized (isotropic) state.

The quantum beats can be observed either in absorption or fluorescence using a weak cw probe light beam. Assuming that the probe light is polarized at an angle α to $\hat{x}$ in the xy plane,

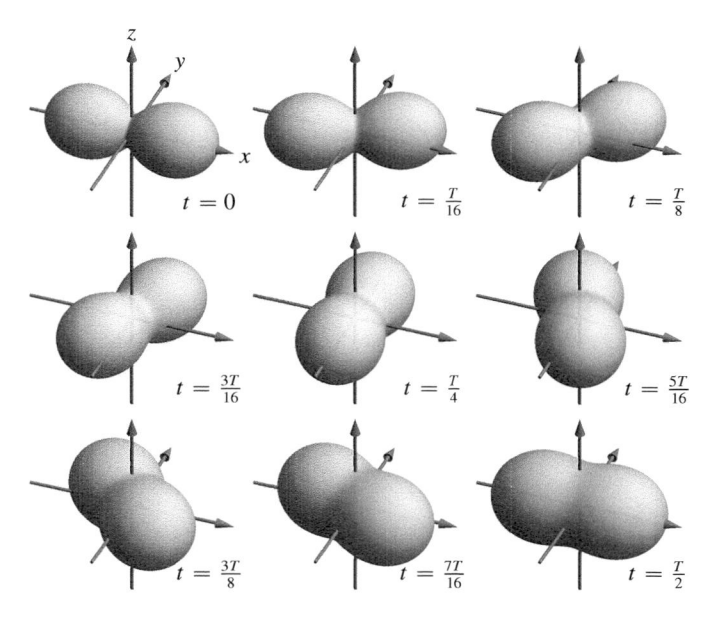

Fig. 15.3 Ground-state quantum beats. Plot of AMPS for the density matrix given by Eq. (15.25) at different times. It is assumed that $\gamma = 0.15\Omega_L$. In order to clearly display the atomic polarization, the maximum amount of the rank-zero polarization moment is subtracted from the density matrix subject to the requirement that the density matrix remains physical for all of the plots. The polarization precesses around the magnetic field and simultaneously relaxes toward the isotropic distribution.

absorption on the $J = 1 \to J' = 0$ transition can be determined using the absorption matrix (15.10). Taking the expectation value with respect to $\rho(t)$ gives

$$\frac{1}{\mathcal{E}_0}\frac{d\mathcal{E}_0}{d\ell} = \text{Tr}\, A^{(\alpha)}\rho(t) = \frac{n\lambda^2}{36\pi}\left[1 - \Gamma_p\tau_p e^{-\gamma t}\cos^2(\Omega_L t - \alpha)\right]. \tag{15.26}$$

We see a signal whose oscillating component decays away at the ground-state relaxation rate γ.

To find the fluorescence signal, we first calculate the excitation of the upper state due to the probe light. Assuming that the Larmor frequency Ω_L and ground-state relaxation rate are much slower than the upper-state relaxation rate Γ, we can consider the ground-state density matrix to be quasi-static for the purposes of calculating the upper state. (Note that this same assumption allows us to neglect the excited-state population generated by the pump pulse, which will rapidly decay due to spontaneous emission.) We can then use the steady-state perturbative formula Eq. (15.2) for the upper-state density matrix elements. For $J' = 0$ there is only one such element, and we have

$$\rho' = \frac{1}{3}\kappa_1\left[1 - \Gamma_p\tau_p e^{-\gamma t}\cos^2(\Omega_L t - \alpha)\right]. \tag{15.27}$$

We now find the fluorescence as before. Note that a $J' = 0$ state is spherically symmetric, so the emitted fluorescence is independent of direction and polarization. Thus the fluorescence signal is just a constant times the upper-state population:

$$\frac{d\mathcal{I}}{d\Omega} = \frac{N\omega^3 3\kappa_1}{18\pi c^3}\left[1 - \Gamma_p\tau_p e^{-\gamma t}\cos^2(\Omega_L t - \alpha)\right]. \tag{15.28}$$

The fluorescence signal is equal to the absorption signal up to an overall constant. This is not surprising, because the atoms that emit fluorescence are precisely those that have absorbed light from the probe beam.

Ground-state Zeeman quantum beats are one example of the dynamic *nonlinear magneto-optical effects* discussed in Chapter 11 and in the review paper by Alexandrov et al. (2005). In addition to Zeeman beats, ground-state quantum beats can occur in other forms, such as ground-state DC- and AC-Stark beats, mixed electric/magnetic-field-induced beats, and hyperfine beats associated with corresponding energy splittings in the ground state.

15.3 The Hanle effect

We now return to the example of the Hanle effect, discussed earlier in Sec. 8.3. The wave-function approach used in Sec. 8.3 is not particularly convenient for this problem: because the effect depends on relaxation processes in the medium, a statistical average must be taken over the wave functions to find the polarization of the ensemble. We will see that when the density-matrix formalism is used, such averaging "by hand" is not necessary, because relaxation is naturally included in the approach.

In this section we will examine the Hanle effect as observed in fluorescence, in contrast to Sec. 8.3 in which absorption was considered. We will first discuss the Hanle effect on a $J = 0 \rightarrow J' = 1$ transition in the presence of a magnetic field, for which the Zeeman effect in the excited state affects the observed fluorescence. We then discuss a $J = 1 \rightarrow J' = 0$ transition, for which the Zeeman effect in the ground state causes the observed dependence of fluorescence on the magnetic field.

We begin with the exact steady-state density-matrix equations for a $J \rightarrow J'$ transition subject to a resonant light field, given by Eq. (12.65). For ground-state density-matrix elements $\rho_{m_1 m_2}$ the equations take the form

$$\begin{aligned}0 = &- i\widetilde{\omega}_{m_1 m_2} \rho_{m_1 m_2} + \frac{\gamma}{2J+1} \delta_{m_1 m_2} + \sum_{m'_1 m'_2} F^{m'_2 m'_1}_{m_1 m_2} \rho_{m'_1 m'_2} \\ &- \frac{i\mathcal{E}_0^2}{4\hbar^2} \sum_{m'_1 m'_2} \left(\frac{1}{\widetilde{\omega}_{m'_1 m_2}} + \frac{1}{\widetilde{\omega}_{m_1 m'_2}} \right) \hat{\varepsilon}^* \cdot \mathbf{d}_{m_1 m'_1} \hat{\varepsilon} \cdot \mathbf{d}_{m'_2 m_2} \rho_{m'_1 m'_2} \\ &+ \frac{i\mathcal{E}_0^2}{4\hbar^2} \sum_{mm'} \left(\frac{\hat{\varepsilon}^* \cdot \mathbf{d}_{m_1 m'} \hat{\varepsilon} \cdot \mathbf{d}_{m'm}}{\widetilde{\omega}_{m'm_2}} \rho_{mm_2} + \frac{\hat{\varepsilon}^* \cdot \mathbf{d}_{mm'} \hat{\varepsilon} \cdot \mathbf{d}_{m'm_2}}{\widetilde{\omega}_{m_1 m'}} \rho_{m_1 m} \right),\end{aligned} \quad (15.29)$$

and for upper-state density-matrix elements $\rho_{m'_1 m'_2}$ they are

$$\begin{aligned}0 = &- i\widetilde{\omega}_{m'_1 m'_2} \rho_{m'_1 m'_2} \\ &- \frac{i\mathcal{E}_0^2}{4\hbar^2} \sum_{m_1 m_2} \left(\frac{1}{\widetilde{\omega}_{m_1 m'_2}} + \frac{1}{\widetilde{\omega}_{m'_1 m_2}} \right) \hat{\varepsilon} \cdot \mathbf{d}_{m'_1 m_1} \hat{\varepsilon}^* \cdot \mathbf{d}_{m_2 m'_2} \rho_{m_1 m_2} \\ &+ \frac{i\mathcal{E}_0^2}{4\hbar^2} \sum_{mm'} \left(\frac{\hat{\varepsilon} \cdot \mathbf{d}_{m'_1 m} \hat{\varepsilon}^* \cdot \mathbf{d}_{mm'}}{\widetilde{\omega}_{mm'_2}} \rho_{m'm'_2} + \frac{\hat{\varepsilon} \cdot \mathbf{d}_{m'm} \hat{\varepsilon}^* \cdot \mathbf{d}_{mm'_2}}{\widetilde{\omega}_{m'_1 m}} \rho_{m'_1 m'} \right).\end{aligned} \quad (15.30)$$

Here it is assumed that any other external fields (in our case, a static magnetic field) are directed along the z-axis.

Note that the complex frequency splitting $\widetilde{\omega}$, which in general depends on the energy-level shifts, light detuning, and relaxation rates, appears in two different contexts in these equations. In the first term on the right-hand side it accounts for the effects of polarization evolution and relaxation, while in the denominators of subsequent terms it describes the resonant character of the interaction with light. For the purpose of describing the Hanle effect, the first role is necessary, while the second is a complication. To avoid unnecessary complexity, we would therefore like to simplify the dependence on $\widetilde{\omega}$ in the light-interaction terms. We can do this by making the assumption that either the natural linewidth Γ of the transition or the bandwidth $\Delta\omega$ of the light field is much larger than both the Zeeman shifts of the sublevels and the detuning of the light from resonance. The appropriate condition to choose depends on the specific effect under study. For magneto-optical effects that depend on ground-state atomic polarization, such as the ground-state Hanle effect, resonances observed as a function of Larmor frequency have a characteristic width equal to the ground-state polarization relaxation rate γ. Because we generally assume that $\gamma \ll \Gamma$, we can, without further assumptions, take the Zeeman shifts to be small in this case. For effects depending on excited-state polarization, on the other hand, the characteristic resonance width is Γ, so that the Zeeman shifts are not necessarily small for the relevant range of magnetic fields. In order to neglect the Zeeman shifts in this case, we must assume a large light bandwidth. As discussed in Sec. 12.8, this causes an increased effective relaxation rate in the complex frequency splittings for optical coherences, $\widetilde{\omega}_{m'm}$.

Using either of these assumptions, we can neglect the real part of the complex frequency splitting for optical coherences, giving $\widetilde{\omega}_{m'm} = \widetilde{\omega}_{mm'} = -i(\Gamma + \Delta\omega)/2$. (Note that the real part of the $\widetilde{\omega}$ for the Zeeman coherences is still relevant.) The equations then become

$$0 = -i\widetilde{\omega}_{m_1 m_2}\rho_{m_1 m_2} + \frac{\gamma}{2J+1}\delta_{m_1 m_2} + \sum_{m'_1 m'_2} F^{m'_2 m'_1}_{m_1 m_2} \rho_{m'_1 m'_2}$$

$$+ \frac{\mathcal{E}_0^2}{(\Gamma + \Delta\omega)\hbar^2} \sum_{m'_1 m'_2} \hat{\varepsilon}^* \cdot \mathbf{d}_{m_1 m'_1} \hat{\varepsilon} \cdot \mathbf{d}_{m'_2 m_2} \rho_{m'_1 m'_2} \qquad (15.31\text{a})$$

$$- \frac{\mathcal{E}_0^2}{2(\Gamma + \Delta\omega)\hbar^2} \sum_{mm'} \left(\hat{\varepsilon}^* \cdot \mathbf{d}_{m_1 m'} \hat{\varepsilon} \cdot \mathbf{d}_{m'm} \rho_{m m_2} + \hat{\varepsilon}^* \cdot \mathbf{d}_{mm'} \hat{\varepsilon} \cdot \mathbf{d}_{m' m_2} \rho_{m_1 m} \right),$$

$$0 = -i\widetilde{\omega}_{m'_1 m'_2}\rho_{m'_1 m'_2} + \frac{\mathcal{E}_0^2}{(\Gamma + \Delta\omega)\hbar^2} \sum_{m_1 m_2} \hat{\varepsilon} \cdot \mathbf{d}_{m'_1 m_1} \hat{\varepsilon}^* \cdot \mathbf{d}_{m_2 m'_2} \rho_{m_1 m_2}$$

$$- \frac{\mathcal{E}_0^2}{2(\Gamma + \Delta\omega)\hbar^2} \sum_{mm'} \left(\hat{\varepsilon} \cdot \mathbf{d}_{m'_1 m} \hat{\varepsilon}^* \cdot \mathbf{d}_{mm'} \rho_{m' m'_2} + \hat{\varepsilon} \cdot \mathbf{d}_{m'm} \hat{\varepsilon}^* \cdot \mathbf{d}_{mm'_2} \rho_{m'_1 m'} \right). \qquad (15.31\text{b})$$

We will use Eqs. (15.31) in the further discussion of the Hanle effect.

15.3.1 Excited state

Consider a closed $J = 0 \rightarrow J' = 1$ transition subject to a $\hat{\mathbf{z}}$-directed magnetic field and x-polarized excitation light. In this chosen geometry the $m = 0$ sublevel of the upper state is not excited. Thus all density-matrix elements can be eliminated from Eqs. (15.31) except for $\rho_{00}, \rho'_{11}, \rho'_{-11}, \rho'_{1-1}$, and ρ'_{-1-1}, where the prime indicates the upper state. The equations for the excited state become

$$\rho'_{11} = \frac{\kappa_1}{2} \left(\frac{1}{2}\rho'_{-11} + \frac{1}{2}\rho'_{1-1} - \rho'_{11} + \rho_{00} \right), \tag{15.32a}$$

$$\rho'_{1-1} = \frac{\kappa_1}{2(1+iu_1)} \left(\frac{1}{2}\rho'_{-1-1} - \rho'_{1-1} + \frac{1}{2}\rho'_{11} - \rho_{00} \right), \tag{15.32b}$$

$$\rho'_{-11} = \frac{\kappa_1}{2(1-iu_1)} \left(\frac{1}{2}\rho'_{-1-1} - \rho'_{-11} + \frac{1}{2}\rho'_{11} - \rho_{00} \right), \tag{15.32c}$$

$$\rho'_{-1-1} = \frac{\kappa_1}{2} \left(-\rho'_{-1-1} + \frac{1}{2}\rho'_{11} + \frac{1}{2}\rho'_{1-1} + \rho_{00} \right), \tag{15.32d}$$

where κ_1 is the saturation parameter for upper-state excitation (Sec. 9.1) and we have written $u_1 = 2\Omega_L/\Gamma$. Because we assume here that the bandwidth $\Delta\omega$ of the light is broad, the saturation parameter is given here by $\kappa_1 = \Omega_R^2/(\Gamma\Delta\omega)$. We have also used the assumption that the ground-state transit width γ is much smaller than Γ.

For a closed transition we can complete the system of equations using the conservation of the trace of the density matrix:

$$1 = \rho'_{-1-1} + \rho_{00} + \rho'_{11}. \tag{15.33}$$

This is useful because we have now eliminated the explicit appearance of the relaxation rates Γ and γ from the system of equations.

Solving for the excited-state density matrix, we find

$$\rho' = \frac{\frac{1}{2}\kappa_1}{(1+\frac{\kappa_1}{2})(1+2\kappa_1) + (1+\frac{3\kappa_1}{2})u_1^2} \begin{pmatrix} 1+\frac{\kappa_1}{2}+u_1^2 & 0 & -1-\frac{\kappa_1}{2}+iu_1 \\ 0 & 0 & 0 \\ -1-\frac{\kappa_1}{2}-iu_1 & 0 & 1+\frac{\kappa_1}{2}+u_1^2 \end{pmatrix}. \tag{15.34}$$

This density matrix exhibits a power-broadened dependence on the magnetic field. In the limit of low light power, the unbroadened dependence is given by

$$\rho' \approx \frac{\kappa_1}{2} \begin{pmatrix} 1 & 0 & -\frac{1}{1+iu_1} \\ 0 & 0 & 0 \\ -\frac{1}{1-iu_1} & 0 & 1 \end{pmatrix}. \tag{15.35}$$

The dependence is displayed in Fig. 15.4 using angular-momentum probability surfaces. (This figure can be compared to Fig. 8.6 showing electron-density plots.) These surfaces show the anisotropy of atomic polarization in the xy plane for zero magnetic field ($\Omega_L = 0$), which is destroyed at high magnetic field ($\Omega_L = 10\Gamma$). An absorption or fluorescence measurement can reflect this change in the amount of polarization transverse to the magnetic field, and the dependence of such a signal on the magnetic field is called the Hanle effect.

In Sec. 8.3 we considered the Hanle effect in absorption—more often, the Hanle effect is observed in fluorescence (see, for example, Moruzzi and Strumia 1991). Traditionally, observation is performed along the magnetic-field direction ($\hat{z}$, in this case) and fluorescence intensity is recorded as a function of the magnetic field strength in two orthogonal polarizations. From the figure, we see that fluorescence emitted along $\hat{z}$ should have a significant degree of linear polarization at zero magnetic field, which should diminish as the field is increased.

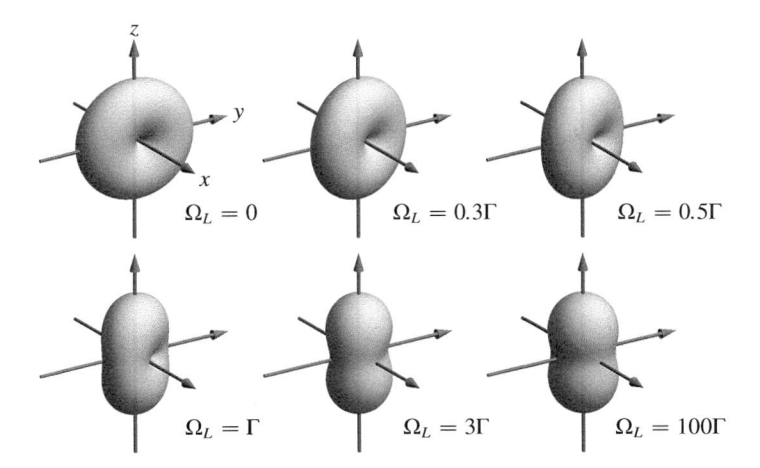

Fig. 15.4 Angular-momentum probability surfaces for various values of the Larmor frequency Ω_L illustrate the essence of the Hanle effect.

To find the intensity of fluorescence propagating in the $\hat{\mathbf{z}}$ direction, we find the expectation value of the fluorescence matrix (15.21) with respect to the density matrix (15.34):

$$\frac{d\mathcal{I}^{(\alpha)}}{d\Omega} = \operatorname{Tr} \mathcal{F} \rho' = \frac{N\omega^4}{12\pi c^3} \frac{\left[1 + \frac{\kappa_1}{2} + u_1^2 + \left(1 + \frac{\kappa_1}{2}\right)\cos 2\alpha + u_1 \sin 2\alpha\right] \kappa_1}{\left(1 + \frac{\kappa_1}{2}\right)(1 + 2\kappa_1) + \left(1 + \frac{3\kappa_1}{2}\right)u_1^2}. \quad (15.36)$$

We observe both the intensity $\mathcal{I}_{\parallel}$ of light polarized parallel to the polarization of the excitation light (along $\hat{\mathbf{x}}$ in our case), and the intensity $\mathcal{I}_{\perp}$ of orthogonally polarized light (along $\hat{\mathbf{y}}$). The Hanle effect is characterized by the normalized Stokes parameter (see Sec. 6.1)

$$S_1 = \frac{\mathcal{I}_{\parallel} - \mathcal{I}_{\perp}}{\mathcal{I}_{\parallel} + \mathcal{I}_{\perp}} = \frac{\mathcal{I}^{(0)} - \mathcal{I}^{(\frac{\pi}{2})}}{\mathcal{I}^{(0)} + \mathcal{I}^{(\frac{\pi}{2})}} = \frac{1 + \frac{\kappa_1}{2}}{1 + \frac{\kappa_1}{2} + u_1^2}. \quad (15.37)$$

The magnetic-field dependence of the signal takes the form of a Lorentzian centered at zero magnetic field (Fig. 15.5). We see that, in this particular case, $S_1 = 1$ at zero field, so that the light is 100% polarized. This does not hold true in general, but only for $J = 0 \to J' = 1$ transitions. (In the case of an open transition, it applies when the fluorescence is emitted in a decay to another $J = 0$ state.) For other transition schemes, the zero-field polarization can be calculated in a similar manner, and is generally lower. A summary of the values for both linearly and circularly polarized excitation and observation can be found in the book by Auzinsh and Ferber (1995).

In the opposite limit of $u_1 \to \infty$, both fluorescence components $\mathcal{I}_{\parallel}$ and $\mathcal{I}_{\perp}$ are equal for all possible transition schemes and the Stokes parameter vanishes: $S_1 \to 0$.

The Hanle effect is a useful tool for measuring the properties of atomic states. For example, if the dependence of the Larmor frequency $\Omega_L = g\mu_B B$ on the magnetic field is known (or, in other words, if we know the Landé factor g, which can be calculated for many atoms quite accurately), we can determine Γ from the S_1 dependence on the magnetic-field strength, which gives the lifetime $\tau = 1/\Gamma$ of the excited state. For molecules, the Landé factors are often not well known, and, in this case, from the known lifetime of a molecular state, one can determine the Landé factor. If both quantities are unknown, one can at least determine their product.

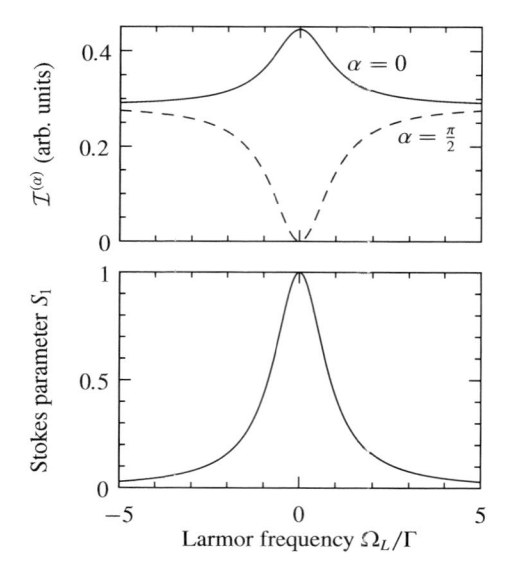

Fig. 15.5 Magnetic-field dependence of the excited-state Hanle effect. The top plot shows the fluorescence intensities observed for polarizations at $\alpha = 0$ (solid line) and $\alpha = \pi/2$ (dashed line). The bottom plot shows the observed Stokes parameter S_1. We assume that the saturation parameter $\kappa_1 = 4$.

15.3.2 Ground state

We now consider the Hanle effect on an open $J = 1 \rightarrow J' = 0$ transition. In this case it is the creation of ground-state polarization that affects the observed signal, so the Zeeman shifts can be neglected compared to the spectral linewidth whether or not the spectral profile of the light is broad. To simplify the results, we assume that the light power is weak enough so that the upper-state saturation parameter $\kappa_1 = \Omega_R^2/\Gamma^2 \ll 1$. Due to this assumption, we can neglect the term in the Liouville equation (12.65) describing stimulated emission. Note that there is no restriction on the optical-pumping saturation parameter $\kappa_2 = \Omega_R^2/(\Gamma\gamma)$, so that ground-state polarization saturation effects can still be observed. The steady-state equations for the ground state reduce to

$$\rho_{11} = \frac{1}{3} + \frac{\kappa_2}{2}\left(\frac{\rho_{-11}}{2} + \rho_{00} + \frac{\rho_{1-1}}{2} - \rho_{11}\right), \quad (15.38a)$$

$$\rho_{1-1} = \frac{\kappa_2}{2(1+iu)}\left(\frac{\rho_{-1-1}}{2} - \rho_{00} - \rho_{1-1} + \frac{\rho_{11}}{2}\right), \quad (15.38b)$$

$$\rho_{-11} = \frac{\kappa_2}{2(1-iu)}\left(\frac{\rho_{-1-1}}{2} - \rho_{-11} - \rho_{00} + \frac{\rho_{11}}{2}\right), \quad (15.38c)$$

$$\rho_{-1-1} = \frac{1}{3} + \frac{\kappa_2}{2}\left(-\rho_{-1-1} + \frac{\rho_{-11}}{2} + \rho_{00} + \frac{\rho_{1-1}}{2}\right), \quad (15.38d)$$

where $u = 2\Omega_L/\gamma$. Solving them, we find for the ground-state density matrix

$$\rho = \frac{1}{3}\begin{pmatrix} \frac{1}{1+\frac{\kappa_2}{2}}\left(1 + \frac{\kappa_2^2}{4(1+\kappa_2+u^2)}\right) & 0 & \frac{1}{1+\frac{\kappa_2}{2}}\frac{\kappa_2(1+\frac{\kappa_2}{2}-iu)}{2(1+\kappa_2+u^2)} \\ 0 & 1 & 0 \\ \frac{1}{1+\frac{\kappa_2}{2}}\frac{\kappa_2(1+\frac{\kappa_2}{2}+iu)}{2(1+\kappa_2+u^2)} & 0 & \frac{1}{1+\frac{\kappa_2}{2}}\left(1 + \frac{\kappa_2^2}{4(1+\kappa_2+u^2)}\right) \end{pmatrix}. \quad (15.39)$$

We again see a power-broadened magnetic-field dependence. In the limit of low light power, the ground-state density matrix becomes

The Hanle effect

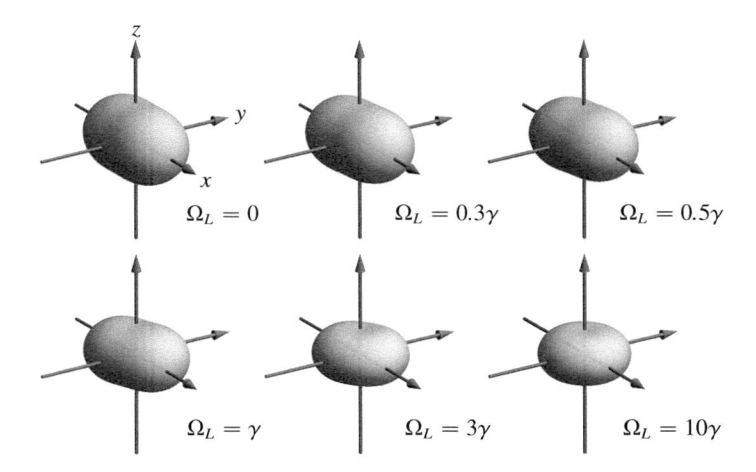

Fig. 15.6 The ground-state Hanle effect. Depicted are the angular-momentum probability surfaces corresponding to the density matrix of Eq. (15.39) for $\kappa_2 = 3$ and various values of the Larmor frequency Ω_L. The shape of the surfaces arises from the concurrent action of three processes: continuous optical pumping, Larmor precession of the atomic polarization, and relaxation.

$$\rho = \frac{1}{3}\begin{pmatrix} 1 - \frac{\kappa_2}{2} & 0 & \frac{\kappa_2}{2}\frac{1}{1+iu} \\ 0 & 1 & 0 \\ \frac{\kappa_2}{2}\frac{1}{1-iu} & 0 & 1 - \frac{\kappa_2}{2} \end{pmatrix}. \tag{15.40}$$

As for the excited-state case, there is a Lorentzian dependence, but the characteristic width in Larmor frequency here is γ, the ground-state polarization relaxation rate, rather than Γ as for the excited state.

Angular-momentum-probability surfaces corresponding to the density matrix (15.39) are shown in Fig. 15.6 for saturation parameter $\kappa_2 = 3$ and various values of the Larmor frequency. The dependence of the polarization in the xy plane on the magnetic field once again leads to a zero-field resonance that can be observed, for example, in fluorescence (Fig. 15.7).

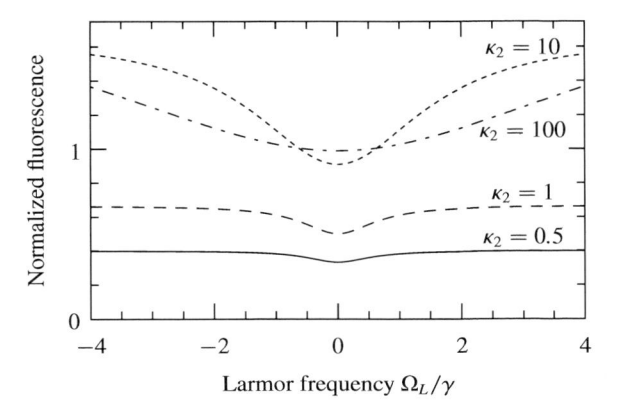

Fig. 15.7 Magnetic-field dependence of the ground-state Hanle effect for various values of the saturation parameter κ_2. Plotted is the normalized fluorescence $[18\pi c^3 \Gamma/(N\omega^4 \gamma)]d\mathcal{I}/d\Omega$.

Fluorescence is emitted by atoms excited by the pump light to the upper state. In this way the same light field serves as both pump and probe. The equation for the upper-state population is

$$\rho'_{00} = \frac{\kappa_1}{2} \left(\rho_{-1-1} - \rho_{-11} - \rho_{1-1} + \rho_{11} - 2\rho'_{00} \right). \quad (15.41)$$

This gives

$$\rho'_{00} = \frac{\kappa_1}{3 \left(1 + \frac{\kappa_2}{2}\right)} \left(1 - \frac{\kappa_2}{2 \left(1 + \kappa_2 + u^2\right)} \right). \quad (15.42)$$

As discussed above, fluorescence from a $J' = 0$ state is independent of direction and polarization, and is simply proportional to the upper-state population. The Hanle effect is observed in the overall intensity of the light:

$$\frac{dI}{d\Omega} = \frac{N\omega^4}{18\pi c^3} \frac{\kappa_1}{\left(1 + \frac{\kappa_2}{2}\right)} \left(1 - \frac{\kappa_2}{2 \left(1 + \kappa_2 + u^2\right)} \right). \quad (15.43)$$

In this case the Lorentzian resonance is superimposed on a constant fluorescence background (Fig. 15.7). The minimum fluorescence is observed when the magnetic field is zero.

16
Collapse and revival in quantum beats

As discussed in earlier chapters, *quantum beats* is the general term for the time evolution of a coherent superposition of nondegenerate energy eigenstates. At $t = 0$, a state $|\psi\rangle$ can be written in terms of eigenstates $|n\rangle$ with energies E_n as

$$|\psi\rangle = \sum_n c_n |n\rangle, \tag{16.1}$$

where $c_n = \langle n|\psi\rangle$ are the component amplitudes. The time evolution of a particular eigenstate is given by $|n(t)\rangle = |n\rangle \exp(-i\omega_n t)$, where $\omega_n = E_n/\hbar$, so we have for the time evolution of the superposition state

$$|\psi(t)\rangle = \sum_n c_n |n\rangle e^{-i\omega_n t}. \tag{16.2}$$

As time passes, the different frequencies ω_n beat against each other, similar to the beats heard when two slightly detuned guitar strings are plucked simultaneously. In general, however, there are more than two component states in a coherent superposition, allowing for more complicated evolution. We will analyze this process in the context of atomic polarization evolution in this chapter.

> In the general case, a localized superposition of eigenstates is known as a *wave packet*. The behavior of various wave packets is an active research topic in atomic, molecular, and optical physics. For familiarization with the concept of wave packets and their collapse and revival in atomic systems, we recommend the paper by Bluhm *et al.* (1996).

We consider a superposition of Zeeman sublevels $|m\rangle$ of a state with total angular momentum F, i.e.,

$$|\psi(t)\rangle = \sum_{m=-F}^{F} c_m |m\rangle e^{-i\omega_m t}. \tag{16.3}$$

In the absence of external fields, the sublevels are degenerate. If external fields are applied along the quantization axis, the Zeeman sublevels are split, but remain energy eigenstates. We can think of ω_m as a continuous function, which, when supplied with integer (or half-integer) values of m, returns the eigenfrequencies determining the evolution. For example, if the sublevels are split by the linear Zeeman effect, ω_m is a linear function in m. In general, if the dependence on m is smooth enough, ω_m can be approximated by a Taylor series

$$\omega_m = \omega_0 + \left.\frac{d\omega_m}{dm}\right|_{m=0} m + \frac{1}{2!}\left.\frac{d^2\omega_m}{dm^2}\right|_{m=0} m^2 + \frac{1}{3!}\left.\frac{d^3\omega_m}{dm^3}\right|_{m=0} m^3 + \ldots, \tag{16.4}$$

keeping only the lowest-order terms, so that we have, for example

$$\omega_m \approx \omega_0 + \Omega_1 m + \Omega_2 m^2 + \Omega_3 m^3, \qquad (16.5)$$

with

$$\Omega_j = \frac{1}{j!} \frac{d^j \omega_m}{dm^j}\bigg|_{m=0}. \qquad (16.6)$$

Substituting Eq. (16.5) into Eq. (16.3) allows us to analyze the quantum-beat evolution in terms of oscillation at the frequencies Ω_j:

$$|\psi(t)\rangle = \sum_{m=-F}^{F} c_m |m\rangle e^{-im\Omega_1 t} e^{-im^2 \Omega_2 t} e^{-im^3 \Omega_3 t}, \qquad (16.7)$$

where we have dropped the overall phase factor.

Each of the frequencies Ω_j corresponds to an oscillation period $\tau_j = 2\pi \Omega_j^{-1}$. Note that the time evolution described by Eq. (16.7) is not truly periodic unless all of the frequencies Ω_j are integer multiples of some "fundamental" frequency. This is generally not the case in physical systems. However, if the time scales τ_j are widely separated, we can analyze the oscillation at each frequency essentially independently. We therefore assume $\tau_1 \ll \tau_2 \ll \tau_3$, as is common for quantum systems. In the following, when we talk about a period of oscillation, we are actually referring to a quasiperiod determined by considering only one of the evolution frequencies of the system over a corresponding timescale.

To see this in more detail, consider Eq. (16.7) for $t \ll \tau_2, \tau_3$. Then the second and third exponentials can be neglected. The remaining time dependence is just due to linear shifts, i.e., it corresponds to precession about the quantization axis with a period τ_1. This evolution is referred to as classical motion.

At longer time scales, on the order of τ_2, the precession at Ω_1 is modulated at a frequency Ω_2 due to the influence of the second exponential. Since this corresponds to the effect of quadratic shifts, the evolution is not simple precession, but instead involves a change in the magnitudes of different polarization moments—the simplest case of this is alignment-to-orientation conversion (Sec. 5.6). Thus the polarization state at $t = 0$ is altered during the period τ_2, only to re-form at $t = \tau_2$. (In the discussion of a wave packet, the packet is said to *collapse* and then *revive*.) Thus τ_2 called the *revival period*. Similarly, over longer periods, this evolution is modulated due to the cubic shifts, resulting in *super-revival* with a period τ_3.

> A peculiar property of atoms in an electric field under the conditions of the quadratic Stark effect is that there is no classical evolution of the wave packets in this case (because of the absence of linear energy shifts), and the classical oscillation period τ_1 does not exist (Auzinsh 1999).

As a specific example, we consider the nonlinear Zeeman shifts (Sec. 4.2) that occur when atoms with hyperfine structure are subjected to a sufficiently large magnetic field. For states with total electron angular momentum $J = 1/2$, such as in the alkali atoms, the shifted frequencies ω_m are given by the Breit–Rabi formula, which can be derived along the lines of our discussion in Sec. 4.2:

$$\omega_m = -\frac{\Delta_{\text{hf}}}{2(2I+1)\hbar} - g_I \mu_B mB/\hbar \pm \frac{\Delta_{\text{hf}}}{2\hbar}\left(1 + \frac{4m\xi}{2I+1} + \xi^2\right)^{1/2}, \qquad (16.8)$$

where $\xi = (g_J + g_I)\mu_B B/\Delta_{\text{hf}}$, g_J and g_I are the electronic and nuclear Landé factors, respectively, B is the magnetic field strength, μ_B is the Bohr magneton, Δ_{hf} is the hyperfine-structure interval, I is the nuclear spin, and the $\pm$ sign refers to the upper and the lower hyperfine level, respectively.

Consider an atomic sample of cesium ($I = 7/2$) in the presence of an $\hat{\mathbf{x}}$-directed magnetic field. If the quantization axis is chosen to be along $\hat{\mathbf{x}}$, the Zeeman sublevels are energy eigenstates, with time evolution given by $c_m|Fm\rangle \exp(-i\omega_m t)$, as described above. Suppose that the atoms are initially in a stretched state along $\hat{\mathbf{z}}$, $|Fm_z = 4\rangle$. Written in terms of the Zeeman sublevels in the x basis, the initial state is seen to be a superposition of the nondegenerate energy eigenstates, and so quantum beats are seen in the evolution of the system.

For moderate field strengths such that the parameter ξ is small, the energy shifts deviate only slightly from linearity. Thus, expanding Eq. (16.8) in powers of m as described above, we see that over time scales comparable to the Larmor period, the evolution (to first order) is just Larmor precession with the period of $\tau_1 \simeq 8[g_J\mu_B B/(2\pi\hbar)]^{-1}$ (neglecting here g_I compared to g_J). The evolution due to the second-order quadratic shifts is also periodic, but with a much longer period that from the expansion of Eq. (16.8) is $\tau_2 \simeq 64\hbar\Delta_{\rm hf}(g_J\mu B)^{-2}/(2\pi)$. (For $B = 0.5$ G, $\tau_1 \simeq 6$ μs and $\tau_2 \simeq 0.3$ s.) One way to illustrate these quantum beats is to produce graphs of the spatial distribution of angular momentum at a given time. To do this, we plot the angular-momentum probability surfaces (Sec. 5.3). The angular-momentum probability surface illustrates the symmetries of the polarization state, indicating which polarization moments are present. A collection of surfaces showing the polarization over half of a period τ_2 of the second-order evolution is shown in Fig. 16.1. The first plot represents the initial stretched state—the surface is literally stretched in the $\hat{\mathbf{z}}$-direction. This state undergoes rapid precession around $\hat{\mathbf{x}}$ with period τ_1. At the same time, the slower second-order evolution results in changes in the shape of the probability surface. By "stroboscopically" drawing successive surfaces each at the same phase of the fast Larmor precession (i.e., at integer multiples of τ_1), the polarization can be seen to evolve into states with higher-order symmetry before becoming stretched along $-\hat{\mathbf{z}}$ at $t = \tau_2/2$. In particular, at $t = \tau_2/4$ the state is symmetric with respect to the xy plane, a characteristic of the even orders in the decomposition of the polarization state into irreducible tensor moments ρ^κ (see Sec. 5.7 and Fig. 11.11).

In order to explore the decomposition into polarization moments further, it is useful to plot the norms of the polarization moments. We define the norm of a rank κ polarization moment as the square root of the scalar product with itself:

$$W_\kappa = \left(\sum_q \rho_q^\kappa \rho_{-q}^\kappa\right)^{1/2}. \tag{16.9}$$

The norms are plotted as a function of time in Fig. 16.2. The figure shows that initially the lowest-order moments predominate. At $t = \tau_2/4$ the odd-order moments are zero and the state is comprised of even-order moments only.

We can now connect these pictures of the atomic polarization state to an experimentally observable signal, e.g., the absorption of weak, circularly polarized light propagating along $\hat{\mathbf{z}}$. To find the absorption coefficient, assuming that the upper-state hyperfine structure is not resolved, we transform to the $|J, m_J\rangle|I, m - m_J\rangle$ basis with quantization axis along $\hat{\mathbf{z}}$ and sum over the transition rates for the Zeeman sublevels. At short time scales (Fig. 16.3a) we see absorption modulated at the Larmor frequency, due to the precession of the atomic polarization about the x-axis. The absorption is minimal when the state is oriented along $\hat{\mathbf{z}}$, and maximal

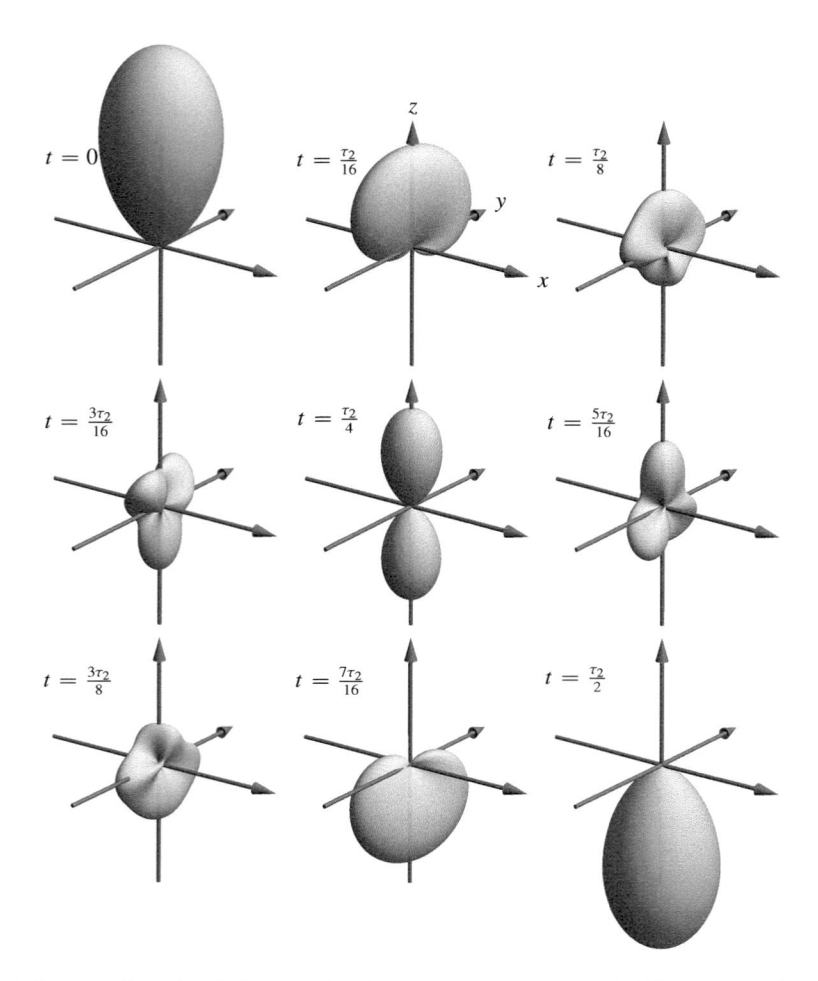

Fig. 16.1 Quantum beats in Cs illustrated with angular-momentum probability surfaces (Sec. 5.3). This sequence is "stroboscopic" in the sense that the surfaces correspond to times chosen to have the same phase of the fast Larmor precession around the direction of the magnetic field ($\hat{x}$). From the symmetry of the plots one clearly sees that orientation present in the initial state collapses and revives in the process of the temporal evolution. Temporal variation of higher polarization moments gives rise to higher-order-symmetry contributions to the probability surface (see also Fig. 16.2).

when it is oriented along $-\hat{z}$. Looking at the envelope of this modulation at longer time scales, we see "collapse and revival" with period $\tau_2/2$ of the absorption oscillation amplitude (Fig. 16.3b). The maxima of the envelope are associated with the stretched states shown in Fig. 16.1 and the minima with the states that are symmetric with respect to the xy plane. This can also be seen by comparison to Fig. 16.2; the envelope of the signal plotted in Fig. 16.3(b) is proportional to the norm of the $\kappa = 1$ moment (orientation), and does not have any of the time-dependent behavior exhibited by the higher-order moments in Fig. 16.2. While it is true in general that weak probe light is not coupled to atomic polarization moments of rank greater than two (Chapter 12), the fact that the absorption is insensitive to the $\kappa = 2$ moment

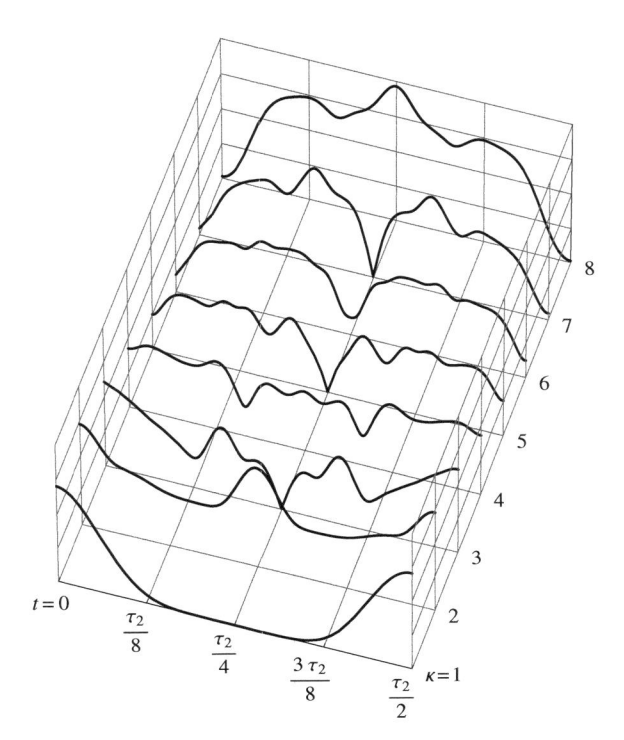

Fig. 16.2 (From Alexandrov *et al.* 2005.) Temporal evolution of the norms of various-rank polarization moments of the $F = 4$ ground state of Cs corresponding to the case of Figs. 16.1 and 16.3. The initial stretched state is dominated by the lowest-order moments; at $t = \tau_2/4$ the state is composed only of even-order moments.

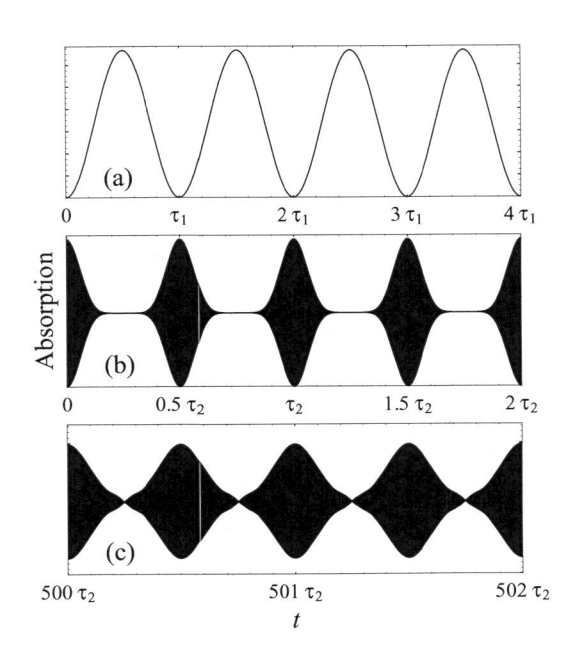

Fig. 16.3 (From Alexandrov *et al.* 2005.) Collapse and revival beats arising in optically pumped Cs atoms due to nonlinearity of Zeeman shifts. (a) Time-dependent absorption of the probe light (see text) observed on a short time scale reveals an oscillation at the Larmor frequency. (b) Observation on a longer time scale reveals the characteristic collapse and revival (beating) behavior. Note the period of essentially complete collapse of the oscillation pattern. (c) At even longer time scales, the beat pattern is modified due to third-order nonlinearity.

(alignment) is a consequence of our assumption that the upper-state hyperfine structure is unresolved (Chapter 13). In Fig. 16.3(c) one can see the effect of the third-order terms in the expansion of Eq. (16.8), reducing the contrast of the envelope function.

Apart from being interesting phenomena in their own right, the collapse and revival of Zeeman beats due to the nonlinear Zeeman effect can be utilized in sensitive atomic magnetometers operating in the geophysical field range of ∼0.5 G, increasing the resolution and reducing the *heading error*, i.e., the dependence of the reading of a nominally scalar sensor to its orientation with respect to the direction of the magnetic field (Seltzer *et al.* 2007). Optical atomic magnetometers were reviewed by Budker and Romalis (2007).

Collapse and revival phenomena similar to the effect described here have been observed in nuclear precession by Majumder *et al.* (1990). In their work, spin precession of an $I = 3/2$ system, ^{201}Hg, was studied and the slight deviations from linearity in the Zeeman shifts responsible for the collapse and revival beats were due to quadrupole-interaction shifts arising from the interaction of the atoms with the walls of a rectangular vapor cell. Related effects were also studied by Donley *et al.* (2009) with another $I = 3/2$ system, ^{131}Xe. In their experiment, the cubic vapor cell was very small (millimeter-sized), and the quadrupolar shifts were significantly enhanced.

Collapse and revival phenomena in molecules with large angular momenta are considered in the tutorial paper by Auzinsh (1999).

17
Nuclear quadrupole resonance and alignment-to-orientation conversion

The concepts and ideas we have encountered in the context of light interactions with free atoms and molecules can be applied in other areas. They are frequently applicable to light interactions with liquids and solids, and also in situations involving electromagnetic fields other than light fields. Here we present a striking illustration of this from the field of condensed-matter nuclear magnetic resonance.

We have previously described the effect on a polarized system of quantum beats that do not correspond to simple precession in Secs. 5.6, 11.3, and in Chapter 16. We have seen that when external fields produce nonlinear energy shifts, polarization moments can transform into moments of other ranks. An important case of this is alignment-to-orientation conversion (AOC). In a simple example of atomic AOC, optical pumping by linearly polarized light produces alignment in the initially unpolarized atomic ground state. Then a static electric field, pointed in any direction other than along or perpendicular to the atomic alignment axis, induces quantum beats that result in an oriented state.

In this chapter, which is based on the work of Budker *et al.* (2003), we will describe how AOC is also the mechanism for *nuclear quadrupole resonance* (NQR) (Das and Hahn 1958; Kopfermann 1958; Abragam 1962), which finds applications, for example, in biochemistry (Edmonds and Summers 1973), and in explosives, land mine, and narcotics detection (Garroway *et al.* 2001; Fraissard 2009).

In pulsed NQR experiments, a radio-frequency (RF) magnetic-field excitation pulse is applied to a *crystalline solid*, resulting in AC magnetization of the nuclei in the sample that is detected by a pick-up coil after the completion of the RF pulse. For this to occur, there must be nuclear polarization present in the sample prior to the RF pulse, since the pulse does not generate polarization but simply rotates the polarization direction. In addition, there must be a field (external or internal) present after the pulse in order to induce the quantum beats that provide the signal. In contrast to the more common technique of nuclear magnetic resonance (NMR), which supplies these conditions with an external magnetic field, NQR takes advantage of *electric field gradients (EFG)* produced by the crystalline lattice of the sample itself that interact with the nuclear quadrupole moment. In both the usual NMR and NQR, the function of the RF pulse is to rotate the polarization axis of symmetry away from the static field symmetry axis, so that quantum beats will occur.

Consider a nucleus with a nonzero quadrupole moment (i.e., possessing angular momentum

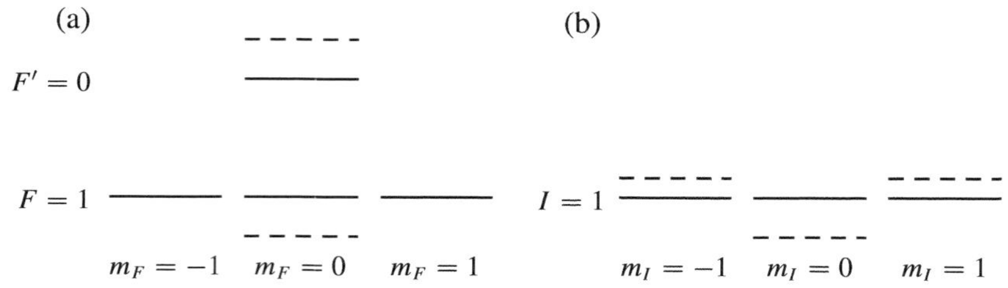

310 *Nuclear quadrupole resonance and alignment-to-orientation conversion*

(a)
$F' = 0$ ─────

(b)

$F = 1$ ─── ─── ─── $I = 1$ ───── ───── ─────
    $m_F = -1$ $m_F = 0$ $m_F = 1$ $m_I = -1$ $m_I = 0$ $m_I = 1$

Fig. 17.1 An energy splitting between levels with different absolute values of the magnetic quantum number m can arise due to the interaction of the atomic system with a uniform electric field (the Stark effect) or due to the interaction of a quadrupole moment with electric field gradients. In both the $F = 1$ atomic system and the $I = 1$ nuclei with axially symmetric electric field gradients, the splitting between the $m = 0$ and $m = \pm 1$ levels can result in quantum beats that convert alignment to orientation.

$I \geq 1$). While the average electric field seen by the nucleus is zero, there are nonzero electric field gradients that interact with the quadrupole moment according to the Hamiltonian (expressed in the Cartesian basis x_1, x_2, x_3)

$$H_{E2} = -\frac{1}{6} \sum_{i,j} Q_{ij} \frac{\partial \mathcal{E}_j}{\partial x_i}. \tag{17.1}$$

Here Q_{ij} is the quadrupole moment tensor and $\mathcal{E}$ is the local electric field in the vicinity of the nucleus. As a simple example of NQR, we consider an $I = 1$ nucleus (such as the ^{14}N or ^{2}H nuclei important in many applications) and assume that the local field gradients at each nucleus have cylindrical symmetry. In this case, the quadrupole interaction Hamiltonian written in the I, m basis with the symmetry axis as the quantization axis reduces to the diagonal operator

$$(H_{E2})_{mm} \propto 3m^2 - I(I+1) = 3m^2 - 2. \tag{17.2}$$

We can see this by noting that because the quadrupole tensor is a rank-two irreducible tensor, only the rank-two part of the gradient tensor $\partial \mathcal{E}_j / \partial x_i$ contributes to the sum in Eq. (17.1), (In order to produce a scalar, two tensors contracted together must have the same rank.) The only component of a second-rank tensor that is symmetric about the z-axis is the $\kappa = 2, q = 0$ component, which according to Eq. (3.114c) is proportional to $\partial \mathcal{E}_z / \partial x_z$. Thus, H_{E2} is proportional to Q_{zz}. Now recall that, by the Wigner-Eckart theorem, any two irreducible tensor operators of equal rank defined on a state with a particular angular momentum are proportional to each other. Consider then the second-rank irreducible tensor component of the Cartesian tensor operator $I_i I_j$. The Hamiltonian must be proportional to the rank-two irreducible component of this tensor; from Eq. (3.109) the rank-two part of $I_z I_z$ is $I_z^2 - \mathbf{I}^2/3$, giving the result above.

Note that the energy splitting caused by the quadrupole interaction is akin to the quadratic (second order) Stark splitting of $F = 1$ atomic energy levels relevant to AOC (Fig. 17.1).

The interaction (17.1) lifts the degeneracy between sublevels corresponding to different magnetic quantum numbers $|m|$ of the nucleus. Typical values of the sublevel frequency splittings are between 100 kHz and 10 MHz. In a sample at thermal equilibrium, the energy splitting gives rise to nuclear polarization because, according to the Boltzmann law, there is a higher probability of finding a nucleus in a lower energy state. At room temperature, a representative value of the relative population difference is $\sim 10^{-7}$. Although each nucleus

Nuclear quadrupole resonance and alignment-to-orientation conversion **311**

(a)   (b)

$F' = 0$

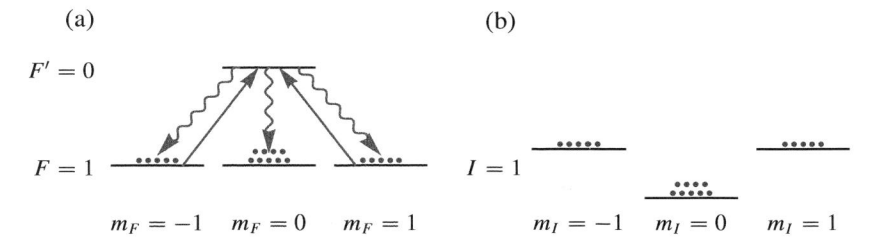

$F = 1$ $I = 1$

    $m_F = -1$ $m_F = 0$ $m_F = 1$ $m_I = -1$ $m_I = 0$ $m_I = 1$

Fig. 17.2 A difference in the population of magnetic sublevels with different absolute values of the magnetic quantum number m can be created via optical pumping of an atomic system by linearly polarized light, or, when the energy levels are split by an electric field gradient, as a result of thermal distribution, as in nuclear systems. The presence of such a population difference gives rise to a quadrupole moment (alignment) along the quantization axis. Optical excitation by linearly polarized light transfers atoms from a particular (aligned) ground state to the upper state, while spontaneous decay repopulates all the ground states equally, since spontaneous emission can occur with any polarization. More atoms are left in the aligned state, determined by the light polarization, that does not interact with light. In the figure, solid lines represent the excitation light, and the wavy lines represent spontaneous decay. In the case of quadrupole splitting and thermal population, the alignment axis is determined by the direction of the electric field gradients in the crystal.

is in an aligned state, with a preferred axis (although no preferred direction), in a disordered medium, such as a powder, there is no macroscopic polarization of the sample because the crystallites are randomly oriented with respect to each other. However, remarkably, NQR signals can still be observed in such media, as discussed below.

Figure 17.2 illustrates the different ways in which initial alignment is achieved in the case of optical pumping and in NQR. Note that in the former case, initial alignment is determined by the polarization of the pumping light, and not by static fields or gradients.

The initial nuclear alignment of several crystallites with different orientations of the local field gradients is illustrated in the first column of Fig. 17.3 using angular-momentum probability surfaces (Sec. 5.3). The distance to such a surface from the origin in a given direction is proportional to the probability of finding the projection $m = I$ along this direction. For clarity, we assume complete polarization, i.e., that all the nuclei are in the lowest energy state; we can see that the aligned states have a preferred axis (the EFG axis of symmetry) but no preferred direction.

The excitation pulse consists of a resonant RF magnetic-field pulse, given by $\mathbf{B}(t) = \mathbf{B}_1 \cos(\omega t + \phi)$, applied for a time τ. Here $\mathbf{B}_1$ is the magnetic-field amplitude (the field is directed at an angle β to the EFG axis of symmetry), ω is the frequency (equal to the quadrupolar splitting frequency), and ϕ is the phase. Assuming that $\gamma B_1 \ll \omega$, where γ is the gyromagnetic ratio, and decomposing this field into components along and perpendicular to the EFG axis of symmetry, we see that the longitudinal component causes fast-oscillating level shifts that have no effect on atomic polarization, whereas the transverse component consists of two circular components, each of which is resonant with one transition from $m = 0$ to $m' = \pm 1$. We can neglect the nonresonant component for each transition; the resonant components, of amplitude $B_1 \sin(\beta)/2$, cause rotation of the nuclear polarization by an angle $\gamma B_1 \tau \sin(\beta)/2$ around the direction of the transverse component of the magnetic field. Since in a typical NQR experiment the pulse length is much longer than the quantum-beat period $T = 2\pi/\omega$, quantum beats begin to occur during the RF pulse. However, at the end of the pulse, the net result, shown in the

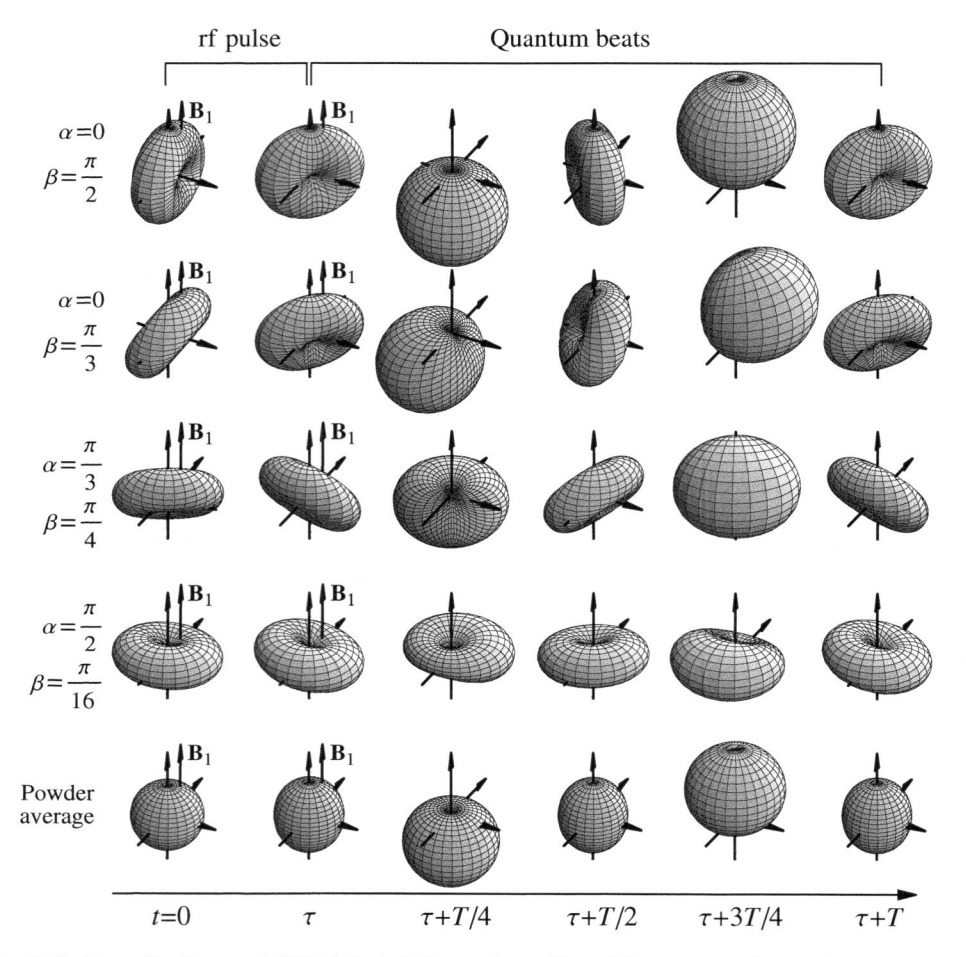

312 *Nuclear quadrupole resonance and alignment-to-orientation conversion*

Fig. 17.3 (From Budker *et al.* 2003.) Probability surfaces (Sec. 5.3) corresponding to the evolution of the nuclear polarization in several crystallites with different representative orientations of the (axially symmetric) local field gradients (given by the Euler angles α and β of the symmetry axes of the local electric field gradients with respect to the fixed lab frame). Surfaces corresponding to a given crystallite are shown in a row. The last row is an average over crystallites with all possible orientations. The columns correspond to different times. The first column represents the local nuclear polarizations prior to the excitation pulse, while the second column shows these polarizations at the end of the resonant RF excitation pulse (with magnetic field along $\mathbf{B}_1$). The excitation is simply a rotation of the polarization by a certain angle. For these plots, the parameters of the excitation pulse were chosen so the rotation is by $\pi/4$ for $\beta = \pi/2$; it is also assumed that the pulse length τ is an integer multiple of the quantum-beat period T, so that the phase of the quantum beats is zero (full alignment) at time τ. The last three columns show the probability surfaces at times $T/4$, $T/2$, and $3T/4$ after the end of the pulse. As is shown in the bottom row, macroscopic oscillating orientation appears along the direction of $\mathbf{B}_1$. These plots are produced by performing an averaged-Hamiltonian calculation in the quadrupolar interaction frame (assuming $\gamma B_1 \ll \omega$) as described, for example, by Lee (2002) to find the density matrix as a function of time, and then plotting the polarization as described in Sec. 5.3. The powder average is found by integrating analytically over the Euler angles.

second column of Fig. 17.3, corresponds to simple rotation. (We plot only the effect of the rotation, and not of the fast quantum-beat oscillation, by assuming that the pulse length τ is an integer number of quantum-beat periods, but this assumption is not important for any of the mechanisms described here.) We have chosen the parameters of the excitation pulse such that the rotation is by $\pi/4$ for $\beta = \pi/2$.

> According to the common NMR/NQR terminology, the pulse that accomplishes such a rotation is called a $\pi/2$ pulse. The terminology stems from the two-level spin-1/2 system, where if one starts, for example, with "spin-down" and applies a pulse creating a coherent superposition of "spin-down" and "spin-up" (with equal amplitudes of the two components) this corresponds to rotating the orientation direction by $\pi/2$. Similarly, a π pulse transfers all atoms from "spin-down" to "spin-up," and rotates the orientation by π. In the present case of a spin-one system, if the excitation pulse transfers the initial $m = 0$ population into a superposition of the $m = \pm 1$ sublevels, this actually corresponds to a physical rotation of the alignment by $\pi/2$ (not by π!). Unfortunately, there appears to be some confusion in the literature about this point.

After the excitation pulse is over, nuclear polarization undergoes evolution in the presence of the quadrupolar interaction. Because the excitation pulse has rotated the polarization of the nuclei, these nuclei are now in coherent superpositions of eigenstates of different energies—the condition for quantum beats. These quantum beats correspond to a cycle of alignment-to-orientation conversion, as shown in the last four columns of Fig. 17.3. In one period of the cycle, alignment is converted into orientation (angular momentum biased in one direction), then into alignment at an angle of $\pi/2$ with respect to the original alignment, followed by conversion to the opposite orientation, and back to the original state. This illustrates that the evolution of the nuclear system in the presence of an axially symmetric EFG is the same as the evolution of an aligned atomic system in the presence of an electric field not directed along the axis of alignment (see, for example, Rochester and Budker 2001).

Since, as illustrated by Fig. 17.3, the orientation produced in each crystallite is perpendicular to both the EFG axis and the axis of the alignment prepared by the excitation pulse, all crystallites contribute coherently to the orientation along $\mathbf{B}_1$, which leads to a net orientation of the entire sample (Fig. 17.3, bottom row). Such AOC-induced orientation corresponds to a net sample AC magnetization that is the source of the NQR signal detected by Faraday induction in the pick-up coil.

Finally, we mention that various techniques for converting nuclear alignment into orientation have been developed for the studies of nuclear moments of short-lived nuclides (Matsuta *et al.* 1998; Coulier *et al.* 1999).

As we see, alignment-to-orientation conversion plays a prominent role in the phenomenon of nuclear quadrupole resonance, converting local nuclear alignment into global orientation, causing the appearance of a macroscopic oscillating magnetic moment. The method of angular-momentum probability surfaces has provided us with a convenient way to visualize this, and to make apparent the relationship of this mechanism to that of AOC in atomic physics.

18
Selective addressing of high-rank polarization moments

In our discussion of atomic polarization moments (PM), we have been primarily concerned with rank-one orientation and rank-two alignment. As discussed in Chapter 13, these are the moments that can be created and detected via single-photon interactions. Higher-rank polarization moments and associated high-order coherences have recently attracted considerable attention due to the role of these moments in optical nonlinearities. Such nonlinearities are important in many applications, for example, the creation of nonclassical atomic and photonic states (Parkins *et al.* 1993) and atomic magnetometry (Alexandrov *et al.* 1997; Okunevich 2001; Yashchuk *et al.* 2003; Acosta *et al.* 2008).

Signatures of high-order atomic PM were detected in several early experiments; see, for example, the review by Alexandrov *et al.* (2005). Aleksandrov *et al.* (1999) studied four-quantum radio-frequency resonances in potassium in the context of precision magnetometry. The production by radio-frequency techniques of $\Delta m = 8$ coherences in laser-cooled cesium, where m is the spin projection quantum number, was demonstrated by Xu and Heinzen (1999). Chin *et al.* (2001) have proposed an alternative method of creating $\Delta m = 4$ coherences using a pulse from a pair of laser beams applied to an $m = 0$ initial state in ultracold cesium, which are thus induced to undergo a *Raman transition* to the desired coherent superposition of the Zeeman sublevels.

Yashchuk *et al.* (2003) described a method, based on nonlinear magneto-optical rotation with frequency-modulated light (*FM NMOR*; see Sec. 11.7.2 and Budker *et al.* 2002c), by which one can selectively induce, control, and study any possible rank multipole moment. This technique is discussed in Sec. 18.1. Applying the method to ^{87}Rb atoms in a paraffin-coated cell, they verified the expected power and spectral dependences of the resonant signals and obtained a quantitative comparison of relaxation rates for the even-rank moments. A similar method, discussed in Sec. 18.2, was used by Pustelny *et al.* (2006) to produce and detect the $\kappa = 6$ polarization moment. Extension of the technique to the geomagnetic-field range (as required in applications such as Earth-field magnetometry) presents challenges that have been addressed by Acosta *et al.* (2008), as discussed in Sec. 18.3.

18.1 General technique and production and detection of the $\kappa = 2$ and $\kappa = 4$ moments

As discussed in Sec. 5.7, $|q| \neq 0$ components of the PM are related to coherences between Zeeman sublevels for which $\Delta m = m - m' = q$, while $q = 0$ components depend on sublevel populations. Thus, coherences $\rho_{m,m'}$ generally contribute to all PM with $\kappa \geq |\Delta m|$; for a

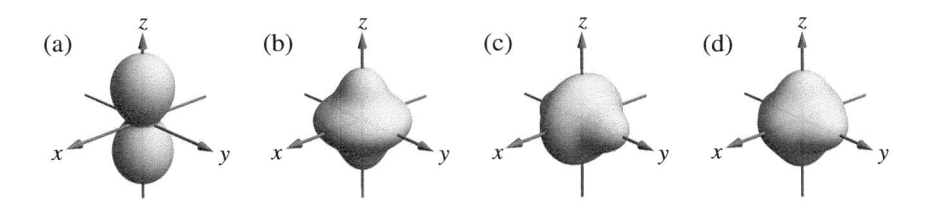

Fig. 18.1 Polarization moments visualized by angular-momentum probability surfaces. (a): "pure" quadrupole $\kappa = 2, q = 0$; (b): $\kappa = 4, q = 0$ hexadecapole; (c): same as in (b), but rotated by $\pi/2$ around the x-axis; (d): the average of (b) and (c), which has a four-fold symmetry with respect to rotations around $\hat{x}$. In all cases, the minimum necessary amount of population was added to ensure that all sublevel populations are nonnegative (Sec. 5.7). Probability surfaces (a) and (d) rotating around an $\hat{x}$-directed magnetic field with the Larmor frequency correspond to the polarization states produced in the experiment of Yashchuk *et al.* (2003).

state with angular momentum F, a $|\Delta m| = 2F$ coherence is uniquely associated with the highest possible PM, e.g., the quadrupole moment ($\kappa = 2$) for $F = 1$, or the hexadecapole ($\kappa = 4$) for $F = 2$. The method introduced by Yashchuk *et al.* (2003) exploits the different axial symmetries of the PM (two-fold and four-fold for the quadrupole and hexadecapole, respectively; Fig. 18.1) to selectively create and detect them.[1]

While multipole moments of rank $\kappa \leq 2$ can easily be generated and detected with weak light (since a photon has spin one), higher-rank moments require multi-photon interactions for both production and detection. In their method, Yashchuk *et al.* (2003) used a single laser beam for the nonlinear interactions required to pump and probe the high-multipole moments. (Although these effects are dependent on high-order interactions with the light, the required laser power was still below a milliwatt).

Under the conditions of the experiment of Yashchuk *et al.* (2003), in which an antirelaxation-coated vapor cell with ^{87}Rb and a single laser beam were used, FM NMOR can be understood as a three-stage (pump, precession, probe) process: atoms are polarized in an interaction with the laser beam (whose diameter is much smaller than the vapor cell dimensions), then leave the beam and bounce around the cell while undergoing Larmor precession, and finally return into the laser beam region and undergo the "probe" interaction. The laser light is frequency modulated, causing the optical pumping and probing to acquire a periodic time dependence. When the pumping rate is synchronized with the precession of atomic polarization, a resonance occurs and the atomic medium is pumped into a polarized state which rotates around the direction of the magnetic field at the Larmor frequency Ω_L. The optical properties of the medium are modulated at the frequency $\kappa\Omega_L$, due to the symmetry of atomic polarization with rank κ. For example, for the quadrupole moment the modulation is at $2\Omega_L$, and for the hexadecapole it is at $4\Omega_L$ (Fig. 18.1a,d). This periodic change of the optical properties of the atomic vapor modulated the angle of the light polarization, leading to the FM NMOR resonances. When the time-dependent optical rotation was measured at the first harmonic of Ω_m, a resonance is seen when Ω_m coincides with $\kappa\Omega_L$ (Fig. 18.2).

With linearly polarized light, only even-rank PMs are excited. At the resonance for a PM of rank κ (which should be absent for states with $2F \leq \kappa$), the signal amplitude is expected to go

---

[1] Directly relevant earlier work includes that of Suter *et al.* (1993); Suter and Marty (1994); and Xu *et al.* (1997a;b).

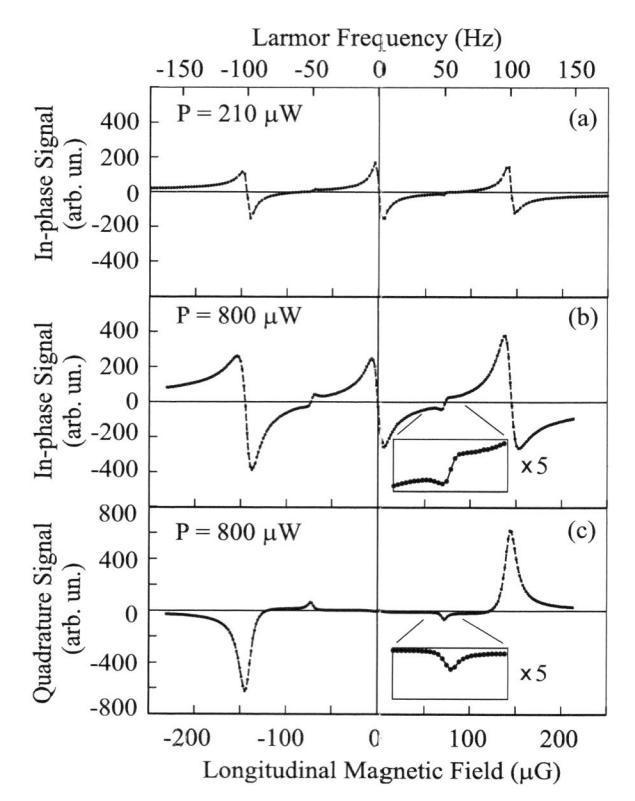

Fig. 18.2 An example of the magnetic-field dependence of the FM NMOR signals observed by Yashchuk *et al.* (2003), showing quadrupole resonances at $B = \pm 143.0$ μG, and the hexadecapole resonances at ± 71.5 μG. The laser modulation frequency was 200 Hz, modulation amplitude was 40 MHz peak-to-peak; the central frequency was tuned to the low-frequency slope of the $F = 2 \to F' = 1$ absorption line. Plots (a,b) show the in-phase component of the signal at two different light powers; plot (c) shows the quadrature component. Note the increase in the relative size of the hexadecapole signals at the higher power. The insets show zooms on hexadecapole resonances.

as the κ-th power of the light intensity at low intensities, as discussed below. These predictions were indeed verified in the experiment.

Yashchuk *et al.* (2003) used the FM NMOR technique (Budker *et al.* 2002c) with ^{87}Rb atoms. The central laser frequency was tuned near various hfs components of the $D1$ line. The typical light power was a few hundred μW and the laser beam diameter was ∼3 mm. The laser frequency was modulated at $\Omega_m/(2\pi)$ from 50 Hz to 1 kHz, and the frequency modulation amplitude was approximately 40 MHz (peak-to-peak). The vapor cell, with isotopically enriched ^{87}Rb, was 10 cm in diameter and had an antirelaxation coating and no buffer gas (Alexandrov *et al.* 1996; 2002). The cell was surrounded with four layers of magnetic shielding. A system of coils inside the innermost shield was used to compensate the residual fields (at a level $\lesssim 0.1$ μG) and first-order gradients, and to apply a well-controlled, arbitrarily directed magnetic field to the atoms (Yashchuk *et al.* 1999). This allowed observation of FM NMOR resonances with magnetic-field widths of about 1 μG in the low-light-intensity limit.

Figure 18.2 shows the magnetic-field dependence of the observed FM NMOR signals. The central laser frequency was tuned to the low-frequency slope of the $F = 2 \to F' = 1$ absorption line as shown in Fig. 18.3. At relatively low light power (Fig. 18.2a), there are three prominent resonances: one at $B = 0$, and two corresponding to $2\Omega_L = \Omega_m$. Much smaller signals, whose relative amplitudes rapidly grow with light power, are seen at $4\Omega_L = \Omega_m$, the expected positions of the hexadecapole resonances.

The spectral dependences for both types of resonance signals with fixed magnetic field

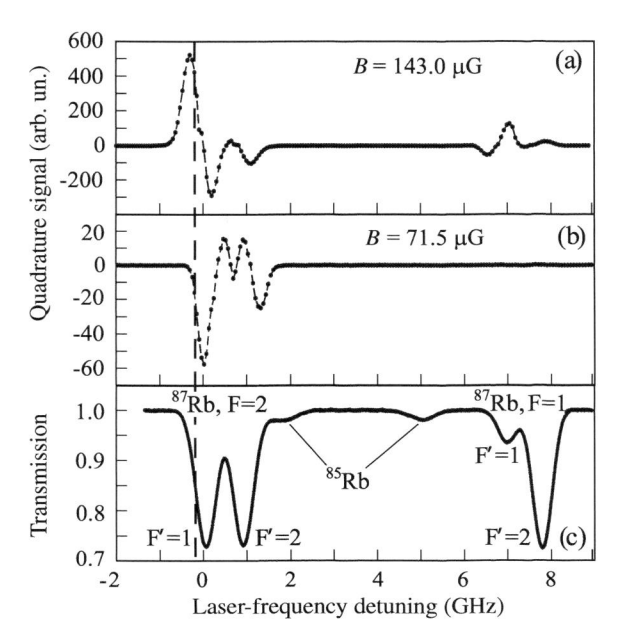

Fig. 18.3 (From Yashchuk et al. 2003.) Spectral dependences of the quadrupole (a) and the hexadecapole (b) signals measured at light power of 800 μW, and the low-power (0.4 μW) transmission spectrum (no FM) (c). Note the different spectral dependences for the FM NMOR signals in (a) and (b), and in particular, the absence of the signal at the $F = 1 \to F'$ transitions in (b). The vertical line indicates the central laser frequency where the measurements represented in Figs. 18.2, 18.4, and 18.5 were taken.

and modulation frequency are shown in Fig. 18.3. While the quadrupole resonance signals ($\Omega_m = 2\Omega_L$; Fig. 18.3a) are observed for both ground-state hyperfine components, no signals are observed for the hexadecapole resonances ($\Omega_m = 4\Omega_L$; Fig. 18.3b) near the lines involving the $F = 1$ ground state, which cannot support a hexadecapole moment.

> The spectral dependences shown in Fig. 18.3(a,b) were obtained by subtracting the average of spectra of the quadrature component taken approximately 15 μG above and below the resonance from the spectrum recorded with magnetic field set at the resonance value. This procedure removes the small residual contribution from the zero-field resonance arising from the "transit" effect (see Chapter 11 and the review by Budker et al., 2002c).

The dependences of the resonance amplitudes on the light intensity I are shown in Fig. 18.4. The observed low-intensity asymptotics of these curves (which show saturation at higher intensities) scale approximately as I^2 and I^4. These power dependences and many other salient features of the observed resonances can be understood with the help of a simple model of an $F = 2 \to F' = 1$ transition with independent pumping, evolution in a magnetic field, and probing. In this model, to lowest-order the creation of the quadrupole and hexadecapole moments are first ($\rho^{(2)} \propto I_{\text{pump}}$) and second order ($\rho^{(4)} \propto I_{\text{pump}}^2$) in light intensity, respectively. The component of the signal due to the induced optical rotation in the probe beam proportional to the quadrupole moment goes as $I_{\text{probe}}\rho^{(2)}$, while that proportional to the hexadecapole moment goes as $I_{\text{probe}}^2 \rho^{(4)}$. The intensity dependences predicted by this model, as well as the periodicity of the signals with respect to Larmor precession and the relative widths of the resonances, match the observations from the experiment, in which a single beam serves as pump and probe.

Figure 18.5 shows the dependence of the resonance widths on power at a fixed magnetic field. While both the quadrupole and hexadecapole resonances exhibit power broadening, it is much less pronounced in the latter case. This is important for applications such as magnetometry

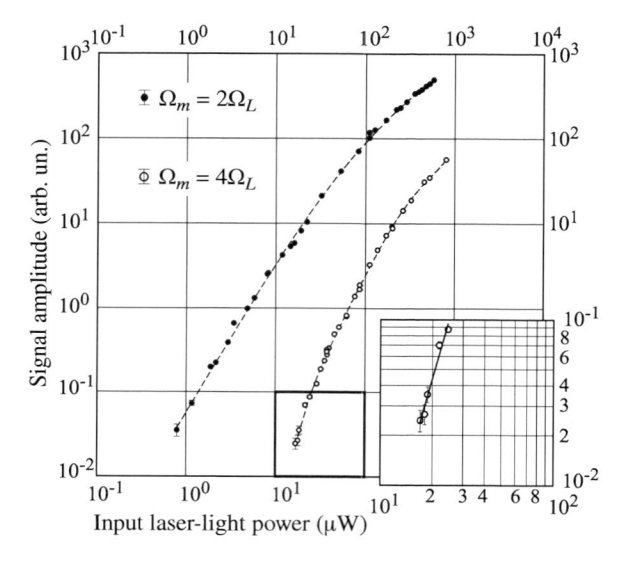

318 *Selective addressing of high-rank polarization moments*

Fig. 18.4 Signal amplitude vs. the input laser power as measured by Yashchuk *et al.* (2003). Filled circles: the quadrupole resonance; open circles: the hexadecapole resonance. The inset shows the expanded low-power region for the latter. From these data, the authors determined the initial linear slope on these log–log plots (corresponding to the exponent in the power dependence of the signal) as 1.96(6) and 3.75(37) for the quadrupole and hexadecapole cases, respectively. The corresponding expected exponents are 2 and 4.

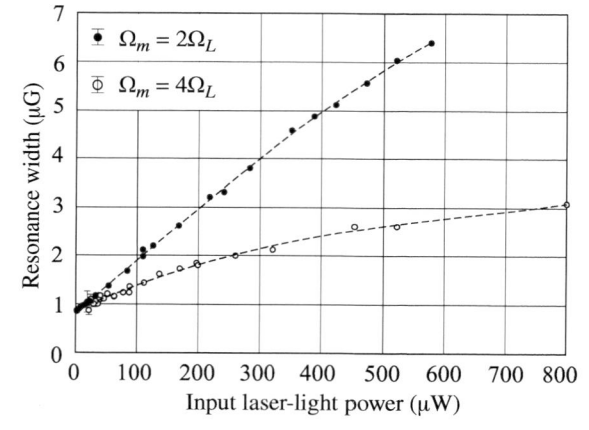

Fig. 18.5 (From Yashchuk *et al.* 2003.) Resonance width vs. the input laser power. Filled circles: the quadrupole resonance; open circles: the hexadecapole resonance. The widths extrapolated to zero light power were found to be $\Delta B_{\text{quad}} = 0.848(4)$ µG and $\Delta B_{\text{hex}} = 0.904(33)$ µG for the quadrupole and hexadecapole cases, respectively.

because it allows operation at higher light powers with better statistical sensitivity.

The resonance widths in the limit of zero light power are also of interest. In general, even in the absence of light, polarization moments of different rank can relax at different rates. The relative rates at which the PM of different rank decay depends on the mechanism of the relaxation. For example, if the relaxation is dominated by the loss of atoms from the vapor into the reservoir containing the supply of the alkali metal, then all PMs relax at the same rate. On the other hand, atoms may lose some of their polarization due to collisions, but not be lost entirely from the system. The *electron-randomization collision model* (see, for example, Knize *et al.* 1988), predicts that the quadrupole moment relaxes at a rate 9/16 that of the hexadecapole moment (Pustelny *et al.* 2006), which relaxes at the electron-randomization rate.

> The electron-randomization collision model assumes that the electron spin of an atom is completely randomized in each collision. On the other hand, the nuclear spin is not affected by the collision. In polarized alkali atoms, only a fraction of the total angular momentum is carried by electrons. Therefore, it takes many electron-randomization collisions to depolarize such atoms (see Prob. 5.6 in the book

by Budker *et al.*, 2008). *Spin-exchange collisions* between alkali atoms are electron-randomization collisions subject to a condition that the total electron angular momentum of two colliding atoms is conserved; see, for example, Prob. 5.4 in the book by Budker *et al.* (2008).

Another factor affecting the observations is that the relationship between relaxation rate and resonance width is different for different polarization moments. It can be shown that the width of an FM NMOR resonance corresponding to a PM of rank κ goes as $\Delta B_\kappa \propto \gamma_\kappa/(\kappa g \mu_B)$, where γ_κ is the rate of light-independent relaxation for a PM of rank κ, g is the gyromagnetic ratio, and μ_B is the Bohr magneton.

In the experiment of Yashchuk *et al.* (2003), the resonance widths ΔB_2 and ΔB_4 for the two polarization moments were found to tend to values near 1 μG in the zero-power limit (Fig. 18.5), with ratio $\Delta B_2/\Delta B_4 = 0.94(4)$. Thus we can deduce the ratio of the relaxation rates γ_2 and γ_4 to be $\gamma_2/\gamma_4 = (1/2)\Delta B_2/\Delta B_4 = 0.47(2)$. This is close to the value 9/16 predicted by the electron-randomization model, indicating that relaxation of the polarization was dominated by residual relaxation on the paraffin-coated cell walls and spin-exchange collisions between Rb atoms.

An interesting question that has not as yet been fully addressed in the literature is the role of conversion of the high-rank moments into alignment and orientation under the combined action of the magnetic field and the light shifts, which can be of crucial importance in nonlinear optical rotation (see, for example, the review by Budker *et al.* 2002a). Another extension is into the domains of nuclear physics and applications of nuclear magnetic resonance where high-order *nuclear* PM and coherences are of great importance (see Jerschow *et al.* 2001, and Chapter 17).

18.2 Production and observation of the $\kappa = 6$ hexacontatetrapole moment

The highest-rank polarization moment that has been isolated using techniques similar to that discussed in the previous section is the rank-six hexacontatetrapole moment (Pustelny *et al.* 2006).

Using separated pump and probe laser beams, Pustelny *et al.* (2006) took data with [85]Rb. This atom has an $F = 3$ ground state, which is the smallest angular momentum that can support a rank $\kappa = 6$ polarization moment. Due to the fact that the averaged hexacontatetrapole moment has six-fold symmetry, the FM NMOR signals related to this multipole were observed at $\Omega_m = 6\Omega_L$.

The quadrupole, hexadecapole, and hexacontatetrapole signal amplitudes are plotted as a function of pump- and probe-beam intensities in Figs. 18.6 and 18.7, respectively. The pump-intensity dependence of the FM NMOR resonance related to the hexacontatetrapole moment is stronger than those of the quadrupole and hexadecapole resonances and is consistent with the expected I_{pump}^3 dependence at low intensity. (Three photons are needed for the creation or detection of the hexacontatetrapole moment.) The amplitudes of the quadrupole, hexadecapole, and hexacontatetrapole resonances show, respectively, linear, quadratic, and cubic dependence on I_{probe} (Fig. 18.7).

The resonance-width dependences for the three multipoles are shown in Figs. 18.8 and 18.9. Although all resonances recorded as a function of pump-beam intensity (Fig. 18.8) broaden with the light intensity, the difference in the broadening is significant only in the first part of the measured dependences. The difference is due to the different number of photons needed to create the various multipoles. However, for higher pump-light intensities the saturation

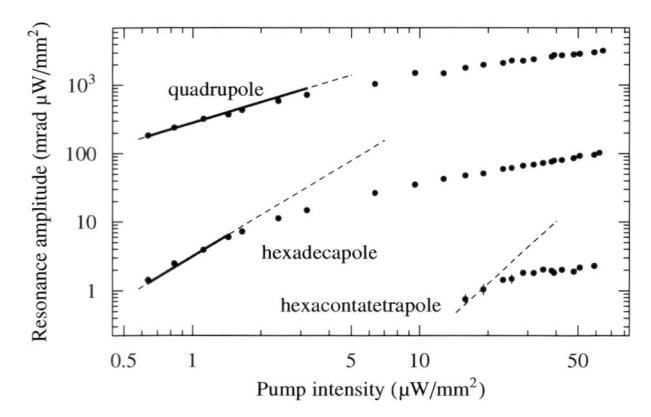

Fig. 18.6 (From Pustelny *et al.* 2006.) The amplitudes of the FM NMOR resonances related to the quadrupole, hexadecapole, and hexacontatetrapole moments in ^{85}Rb vs. pump-beam intensity. The dashed lines show linear, quadratic, and cubic slopes while the solid segments indicate the region in which recorded dependences obey theoretical predictions. Under these experimental conditions ($I_{\text{probe}} = 36$ μW/mm^2) the measured amplitude dependence of the hexacontatetrapole signal is weaker than cubic, which is a result of saturation. The pump beam central frequency was tuned to the center of the $F = 3 \to F'$ transition group while the probe beam was tuned to the high-frequency wing of the $F = 3 \to F''$ transition group of ^{85}Rb.

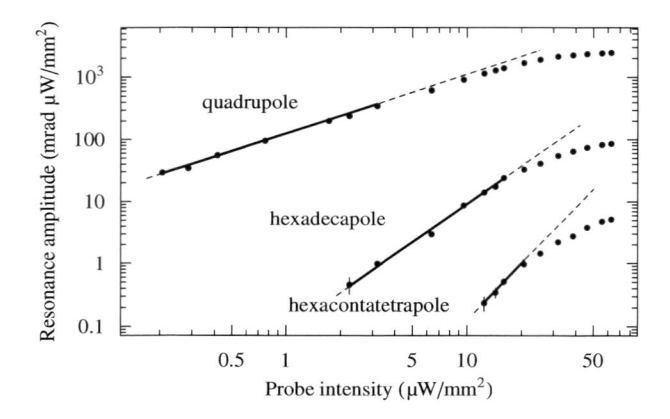

Fig. 18.7 (From Pustelny *et al.* 2006.) The amplitudes of the FM NMOR signals related to the quadrupole, hexadecapole, and hexacontatetrapole moments in ^{85}Rb as a function of probe-light intensity. The dashed and solid lines have the same meaning as in Fig. 18.6. The tuning of the pump beam was the same as in the previous case and its intensity was 52 μW/mm^2. The probe beam tuning was slightly different than in the previous case and was locked to the center of the $F = 3 \to F''$ transition group of ^{85}Rb. This difference in tuning of the probe-beam is a contributor to the fact that the amplitudes of the FM NMOR signals presented in this plot are smaller than those in Fig. 18.6.

behavior starts to play an important role and all resonances exhibit similar broadening with the pump-light intensity, i.e., the slopes of the width dependences related to a given multipole are almost the same.

The width dependences of the FM NMOR signals on pump-beam intensity also show a very interesting feature. For a range of pump intensities the width of the hexadecapole resonance

Fig. 18.8 (From Pustelny et al. 2006.) The widths of the ^{85}Rb quadrupole, hexadecapole, and hexacontatetrapole (inset) FM NMOR resonances vs. pump-light intensity. The central frequency of the pump beam was tuned near the center of the $F = 3 \rightarrow F'$ transition group. The probe beam was tuned to the high-frequency wing of the $F = 3 \rightarrow F''$ transition group and its intensity was 36 μW/mm^2.

Fig. 18.9 (From Pustelny et al. 2006.) The widths of the FM NMOR signals related to the quadrupole, hexadecapole, and hexacontatetrapole moments vs. probe-light intensity. The pump beam central frequency was tuned to the center of the $F = 3 \rightarrow F'$ transition group and its intensity was 52 μW/mm^2. The probe beam was tuned to the center of the $F = 3 \rightarrow F''$ transition group.

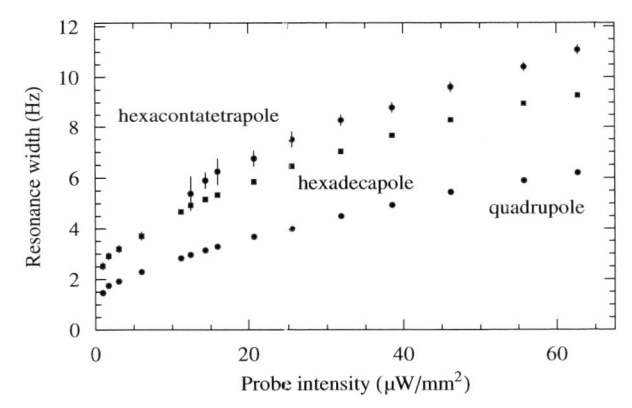

is broader than that of the hexacontatetrapole resonance. A more detailed analysis revealed that the ratio between these widths changes with the tuning of the pump and probe lasers. For instance, for the probe beam tuned toward higher frequencies the hexadecapole resonance is broader than the hexacontatetrapole resonance. The different tuning of the probe beam is the reason why similar behavior was not observed when signals were recorded as a function of probe-light intensity (Fig. 18.9).

In Fig. 18.10 the hexacontatetrapole signal is shown as a function of pump-light central frequency. The probe light was tuned to the high-frequency wing of the $F = 3 \rightarrow F''$ transition group while the pump beam was scanned over all ^{85}Rb hyperfine components of the $D1$ line. Under these experimental conditions, a strong hexacontatetrapole signal was measured at the $F = 3 \rightarrow F'$ and $F = 2 \rightarrow F'$ transition group. The signal recorded at the $F = 2 \rightarrow F'$ transition group is a result of the hexacontatetrapole-moment transfer from the $F' = 3$ excited state to the $F = 3$ ground state via spontaneous emission. The comparable amplitudes of the two spectral contributions show that the coherence-transfer mechanism could be very efficient. The amplitude of the hexacontatetrapole signal recorded when tuned to the $F = 2 \rightarrow F'$ transition group is comparable with the signal recorded when tuned to the $F = 3 \rightarrow F'$ transition group that allows direct creation of this type of coherence in the ground state. This means that there could be potential applications of this mechanism in the transfer of the Zeeman coherences between different atomic states.

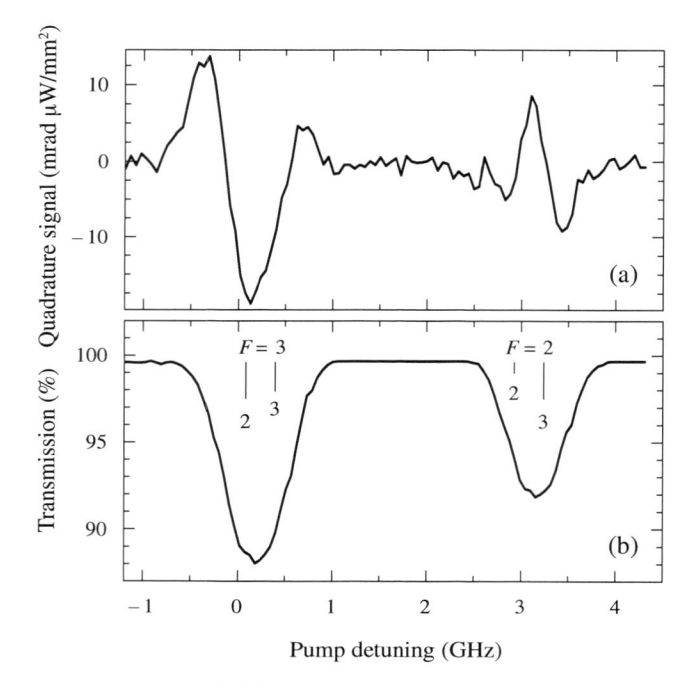

Fig. 18.10 (From Pustelny et al. 2006.) (a) Hexacontatetrapole signal and (b) a reference absorption spectrum taken with low light intensity (2.5 μW/mm²) vs. pump-beam central frequency recorded with ^{85}Rb. The strong hexacontatetrapole signal is measured at the $F = 3 \rightarrow F'$ transition group which supports generation of this multipole in the ground state. The hexacontatetrapole signal is also observed for the pump-light tuned to the $F = 2 \rightarrow F'$ transition group which does not support creation of the quadrupole in the ground state. This signal is due to the transfer of the coherences from the excited state $F' = 3$ to the ground state $F = 3$ via spontaneous emission. The probe laser was tuned to the center of the $F = 3 \rightarrow F''$ transition group, while the pump-light central frequency was scanned over all hyperfine components of ^{85}Rb $D1$ transition. For the measurements of the signal related to the hexacontatetrapole moment the pump-beam intensity was 64 μW/mm² and the probe-beam intensity was 65 μW/mm².

In Fig. 18.11 the hexacontatetrapole signal is shown as a function of probe-beam frequency. The spectrum was recorded for the pump-light central frequency tuned to the high-frequency wing of the $F = 3 \rightarrow F'$ transition group. The probe beam was scanned over all hyperfine components of the ^{85}Rb $D2$ line. The pump- and probe-beam intensities were the same as for the previous case.

The hexacontatetrapole signal is seen only for the probe beam tuned to the $F = 3 \rightarrow F''$ transition group, as expected.

18.3 Production and detection of the hexadecapole moment in the Earth's magnetic field

A complication in the manipulation of high-order coherences arises when the energies of spin states are nonuniformly spaced. Such nonuniform splittings arise in a variety of contexts, including nonlinear Zeeman (NLZ) splitting of alkali ground-state levels in the Earth's magnetic field (Sec. 4.2), splittings of nuclear-spin energy levels under the influence of electric-field

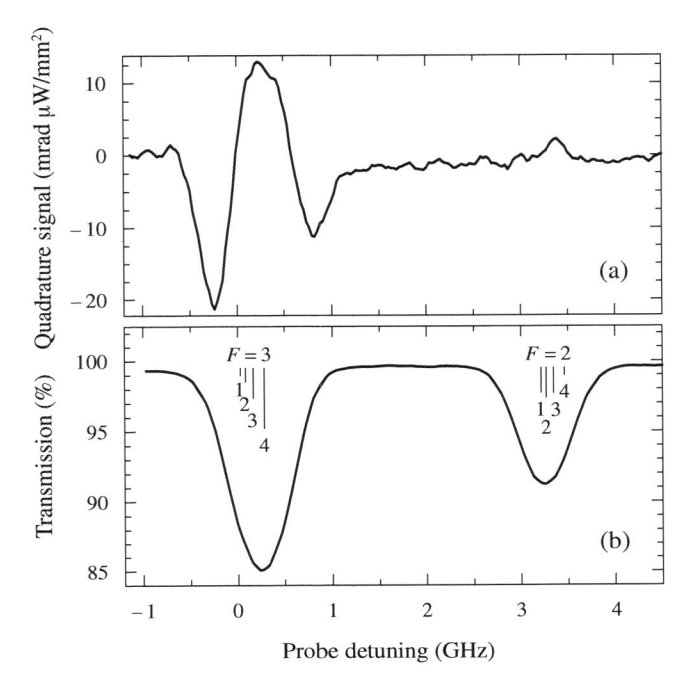

Fig. 18.11 (From Pustelny *et al.* 2006.) The signal related to the hexacontatetrapole moment (a) and the reference spectrum taken for low light intensity (b) vs. probe-light frequency. The central frequency of the pump beam was tuned to the center of the $F = 3 \rightarrow F'$ transition group, while the probe-beam frequency was scanned over all hyperfine components of the ^{85}Rb $D2$ line. The pump- and probe-beam intensities were chosen to be the same as in Fig. 18.10.

gradients in a crystal (Chapter 17), and Stark shifts of atomic states by static (Sec. 4.3) or oscillating (Sec. 4.5) electric fields.

In Earth-field atomic magnetometry, for example, the NLZ splitting leads to split or poorly resolved resonances, resulting in reduced signals and heading errors, i.e., errors in the magnetometer reading that depend on its orientation. Such errors are typically on the tens of microgauss scale, far larger than the picogauss-scale sensitivity achievable with atomic magnetometers (see, for example, the review by Budker and Romalis 2007).

In an attempt to address the significant technological problem of eliminating heading error in optical or atomic magnetometers, Acosta *et al.* (2008) have developed a method that allows selective creation of a long-lived ground-state $|\Delta m| = 4$ coherence in the Earth's field. The energy levels and optical transitions involved are illustrated in Fig. 18.12(a). Because the hexadecapole moment involves states whose energy as a function of magnetic field is strictly linear (see Sec. 4.2 and Fig. 18.12b), it is expected to show no NLZ shifts (Alexandrov *et al.* 1997; Okunevich 2001). As a consequence of having only a single resonant frequency, a practical hexadecapole-based magnetometer would be a better *scalar magnetometer*—it would register the same magnetic field regardless of its spatial orientation. Acosta *et al.* (2008) experimentally demonstrated the immunity of the hexadecapole signal to NLZ splitting using total spin $F = 2$ ^{87}Rb atoms.

The atomic hexadecapole ($\kappa = 4$) has nine components, of which two (those with $q = \pm 4$) are of particular importance for the present discussion. These components are given in terms of matrix elements $\rho_{mm'}$ of the density matrix in the Zeeman basis by

$$\rho^4_{+4} = \rho_{2,-2}, \quad \rho^4_{-4} = \rho_{-2,2}. \tag{18.1}$$

The time evolution of these components is given by phase factors $e^{\pm 4 i \Omega_L t}$; physically measurable

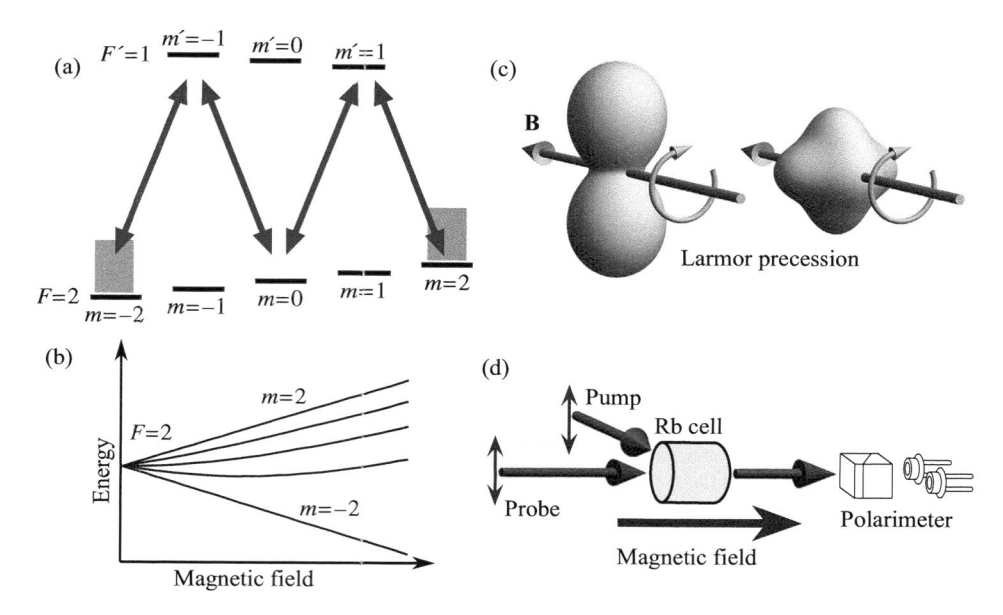

324 *Selective addressing of high-rank polarization moments*

Fig. 18.12 (From Acosta *et al.* 2008.) States, energies, and layout of the experiment. Part (a) shows the states involved in the four-quantum coherence. Part (b) shows the linearity of the $m = \pm 2$ states' energies as a function of magnetic field (shown for purposes of illustration over a much larger range of fields than are experimentally relevant). Part (c) shows angular-momentum probability surfaces for quadrupole (left) and hexadecapole (right) for $F = 2$. The atomic polarizations are transverse to and precess around the magnetic-field quantization axis. Part (d) shows the respective directions of laser polarizations (two-headed arrows), propagation directions, and magnetic field.

signals are then proportional to the real-valued sum and difference of these components, i.e., to $\rho^4_{+4} + \rho^4_{-4}$ and to $i\rho^4_{+4} - i\rho^4_{-4}$. Such superpositions of the two extremal states of the hexadecapole were produced and detected by Acosta *et al.* (2008).

Angular-momentum probability surfaces illustrating the symmetries of the polarization moments under study are shown in Fig. 18.12(c). In a magnetic field, the atomic angular momentum precesses at the Larmor frequency Ω_L, which is proportional to the applied field. Consider the "peanut"-shaped quadrupole moment of Fig. 18.12(c). When the peanut has rotated by an angle π, it is impossible to differentiate it from its initial state, i.e., it has a two-fold symmetry. Consequently, efficient pumping of the quadrupole is achieved with light modulated at an angular frequency $2\Omega_L$. Precession of the quadrupole results in optical rotation of an incident probe beam oscillating at $2\Omega_L$. Similarly, precession of the hexadecapole results in a rotation signal at $4\Omega_L$. More generally, a PM with a component q has $|q|$-fold symmetry and can therefore be pumped with light harmonically modulated at $|q|\Omega_L$ or with pulses at a repetition rate of $|q|\Omega_L/(2\pi n)$, where n is an integer. The polarization moments prepared in this way produce an optical rotation signal at $|q|\Omega_L$.

The experimental geometry for the measurements is pictured in Fig. 18.12(d). A diode laser tuned to the ^{87}Rb $F = 2 \to F' = 2$ transition on the $D1$ (795 nm) line was separated into two beams: a 4-mW linearly polarized pump beam that was amplitude-modulated and a continuous probe beam with the same initial polarization. The probe power was 25 µW for the data shown in Figs. 18.13 and 18.14. The laser frequency was fine-tuned to maximize the hexadecapole

Production and detection of the hexadecapole moment in the Earth's magnetic field 325

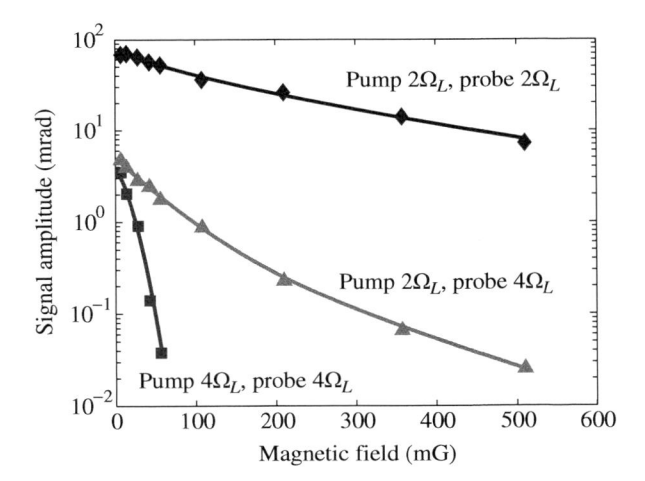

Fig. 18.13 (From Acosta *et al.* 2008.) Magnetic-field dependence of optical rotation amplitudes for quadrupole (diamonds), hexadecapole pumped with light modulated at $4\Omega_L$ (squares), and hexadecapole pumped at $2\Omega_L$ (triangles). The quadrupole and hexadecapole, pumped at $2\Omega_L$ decrease much more slowly with magnetic field compared to hexadecapole pumped at $4\Omega_L$. The solid lines are fits by ad hoc functions.

signal. Additional data were taken with a single laser beam serving as both pump and probe, tuned to the $F = 2 \rightarrow F' = 1$ transition. The Rb atoms were contained in an evacuated glass cell with paraffin antirelaxation coating. A stable magnetic field of 510 mG was maintained in the experiment. The angle of polarization of the outgoing probe beam was measured.

Atomic magnetometers operate by producing transverse atomic polarization and optically detecting its precession in a magnetic field (Budker and Romalis 2007). In the experiments of Acosta *et al.* (2008), the $F = 2$ ground-state atoms were pumped using a sequence of short pulses of linearly polarized light with a repetition rate determined by the Larmor precession frequency of the atomic angular momentum. At the end of the sequence, the pump light was blocked. The evolution of the atomic polarization was observed by measuring the angle of optical rotation of an unmodulated probe beam whose initial polarization was the same as that of the pump. Acosta *et al.* (2008) modulated the pumping light at a frequency which approximately matched the Larmor frequency at the applied field, and this field was then determined precisely by measuring the frequency of the optical-rotation signal.

Pumping at a rate $4\Omega_L$ creates a hexadecapole moment without an accompanying transverse quadrupole moment. This is desirable, as the presence of the typically much larger quadrupole signal makes it difficult to observe the hexadecapole. Unfortunately, the amount of hexadecapole pumped in this way rapidly decreases with the magnetic field strength, as seen in Fig. 18.13 (filled squares). This can be understood by noting that angular-momentum conservation requires participation of two photons to produce the atomic hexadecapole. When pumping at $4\Omega_L$, both photons must interact with the atoms within the same pulse in order to create the hexadecapole, since separate single-photon processes occurring during successive pumping cycles create quadrupole moments that are orthogonal to each other. This orthogonality is a result of Larmor precession by $\pi/2$ radians in one pumping period. Thus, the quadrupole polarizations from successive cycles cancel the transverse quadrupole, and the net result is

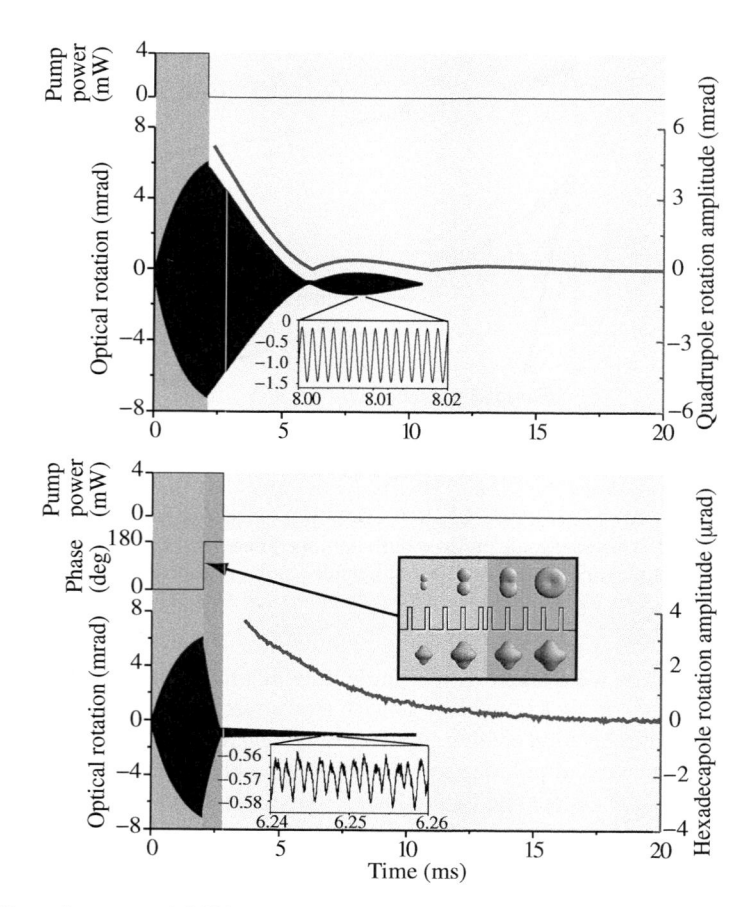

Fig. 18.14 (From Acosta et al. 2008.) Comparison of the demodulated quadrupole signal (top) without phase flip (see text) and of the hexadecapole signal obtained with phase flip (bottom). For the case of pumping without phase flip (top), the signal demodulated at $2\Omega_L$ is plotted alongside the raw optical rotation signal. The inset on the top plot is a magnification of the raw optical-rotation signal during a revival stage of the quadrupole-signal beats occurring due to the nonlinear Zeeman effect (NLZ). This signal on the bottom plot (with phase flip) is demodulated at $4\Omega_L$, and the resulting curve shows the absence of the beats related to NLZ. The inset on the right shows the details of the pumping pulses near the phase flip, along with the AMPS characterizing the ensemble at different stages of pumping. The magnetic field around which these surfaces precess is normal to the page.

a longitudinal quadrupole, seen as a "doughnut" in the inset of Fig. 18.14, which does not aid in the creation of transverse hexadecapole. As the field increases, the pumping period $[T = 2\pi/(4\Omega_L)]$ decreases, reducing the probability of a process occurring which involves interactions with two photons in a single pulse. Even using the strongest pump power available in the experiment (4 mW), the hexadecapole signals could not be distinguished above noise in the Earth's field ($4\Omega_L/2\pi \approx 1.4$ MHz) in this scheme.

An alternative method for efficiently pumping the hexadecapole moment is to use light modulated at $2\Omega_L$, producing both quadrupole and hexadecapole moments. In this scheme, the requirement of pumping the hexadecapole in a single pulse is alleviated as the hexadecapole can be obtained by "promoting" the quadrupole polarization with just a single-photon interaction. This method allows one to obtain hexadecapole signals that, while still decreasing with magnetic field in the experiment, nevertheless remain observable in the Earth's field (Fig. 18.13). Both hexadecapole and quadrupole signals decrease as a function of magnetic field, the hexadecapole because it requires a two-photon probing step to occur in an ever-decreasing time interval, and the quadrupole because Acosta *et al.* (2008) used the same high probe-laser power for both signals in order to compare them directly. This decrease of the hexadecapole signal is qualitatively reproduced by a model density-matrix calculation.

Figure 18.14 (top) shows the optical rotation signal obtained with 1500 square pump pulses with repetition rate of $2\Omega_L/(2\pi) \approx 714$ kHz and 1/8 duty cycle. Significant beating of the signal is observed after the pulse sequence ends, resulting from NLZ (the effect of NLZ on nonlinear magneto-optical rotation is discussed by Acosta *et al.*, 2006). These beats are the time-domain manifestation of three closely spaced frequencies in the optical rotation signal. Fits of the demodulated quadrupole signal indicate a splitting between adjacent frequencies of ≈ 72 Hz close to the calculated value of the NLZ splitting, $\delta_{\text{NLZ}} = 74.65$ Hz for this field. The overall exponential decay time ($\tau \approx 4.7$ ms) is determined by the relaxation of the PMs due to dephasing from collisions and magnetic-field inhomogeneities, and by residual probe-power broadening. Buried under the much larger quadrupole signal is also a hexadecapole signal producing modulation of the optical rotation at $4\Omega_L$. Unfortunately, due to nonlinearities in the detection electronics as well as those due to the interaction of atoms with a strong probe light (Series 1966), the large quadrupole signal leads to the presence of a "false hexadecapole" signal at $4\Omega_L$, and it is hard to distinguish the two contributions.

The solution implemented by Acosta *et al.* (2008) was to eliminate the quadrupole just before probing. To accomplish this, the pumping was separated into two stages. The first stage is the same as in the pumping scheme above: the atoms are pumped at $2\Omega_L$ for ≈ 1500 cycles. Then, the phase of the pumping is flipped by π radians (see the inset in the bottom plot of Fig. 18.14). The quadrupole is now pumped orthogonally to its previous alignment and the resulting sum of orthogonal quadrupole moments leaves no net transverse alignment (the resulting doughnut shape, corresponding to longitudinal alignment that causes no optical rotation of the probe light, is shown in the inset at the bottom plot in Fig. 18.14). However, the hexadecapole produced before and after the phase flip is identical, so it continues to be pumped even after the phase flip. After ≈ 500 cycles, the quadrupole reaches a minimum and the pump is shut off. Figure 18.14 (bottom) demonstrates the phase-flip pumping scheme and the resulting signal (see also Fig. 18.15). The quadrupole signal (not shown) is reduced by about a factor of 40 in this scheme, allowing for the reliable recovery of the signal due to the hexadecapole moment. The demodulated hexadecapole signal is a simple exponential decay ($\tau = 4.2$ ms), clearly demonstrating the absence of NLZ-induced beating.

Thus, Acosta *et al.* (2008) demonstrated a way to create and detect macroscopic hexadecapole polarization (corresponding to a four-quantum Zeeman coherence) in the geomagnetic field range. The amount of hexadecapole created at these fields was dramatically enhanced by pumping at $2\Omega_L$, the frequency associated with efficient production of the (lower-rank) quadrupole moment. Phase flipping allowed the elimination of the large quadrupole signal, unmasking the smaller hexadecapole signal. The resulting hexadecapole signal demonstrates the absence of beating associated with the nonlinear Zeeman effect. The NLZ-free hexadecapole signals are attractive for applications in optical magnetometry because the linear relation between the magnetic field and the spin-precession frequency is maintained over all magnetic

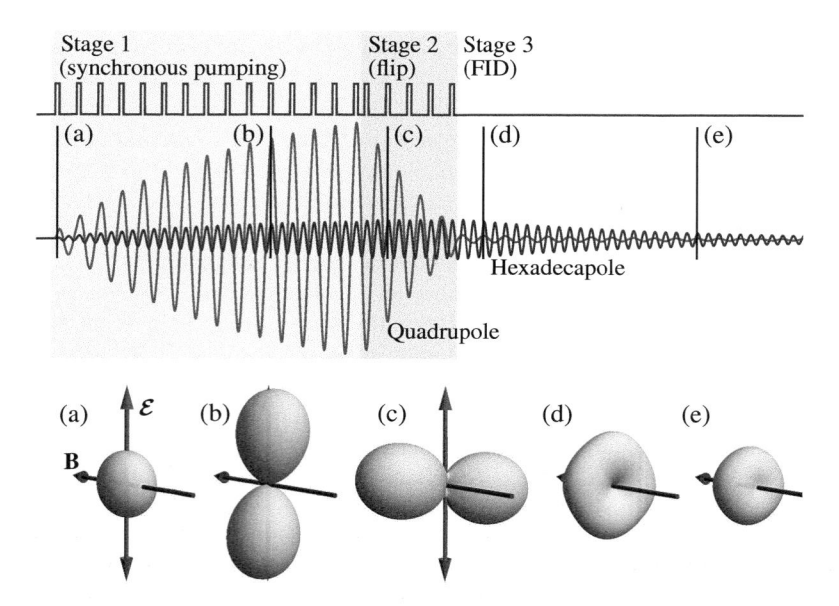

Fig. 18.15 The results of a density-matrix calculation of the phase-flip pumping scheme applied to a $F = 2 \rightarrow F' = 1$ transition. Plotted are the pulse sequence, the real parts of the ρ_2^2 quadrupole component and ρ_4^4 hexadecapole component of the ground state as a function of time, and angular-momentum probability surfaces corresponding to the ground-state polarization at various times during the procedure, marked on the plot with vertical bars. For (a) and (b) optical pumping with the light field $\mathcal{E}$ acts to increase the quadrupole moment, while in (c) the pulse is out of phase with the precession about the magnetic field **B** and the quadrupole moment is reduced. In (d) and (e) the pump light is turned off and the hexadecapole moment decays away. The magnetic-field strength used for the calculation is much smaller than the experimental value in order for the Larmor precession to be visible on the optical-pumping time scale.

fields. The demonstrated technique eliminates the major mechanism for heading error in atomic magnetometers, however the degradation of the hexadecapole signal at geomagnetic fields remains a significant obstacle in the use of this technique in practical high-sensitivity magnetometry. There are expectations still awaiting experimental verification that it may be possible to augment the signal-to-noise ratio of the hexadecapole signal by detecting fluorescence rather than optical rotation (Ducloy *et al.* 1973; Auzinsh *et al.* 1986; Okunevich 2001).

> We should also mention an altogether different solution of the NLZ problem analyzed theoretically and demonstrated experimentally by Jensen *et al.* (2009). It involves canceling the NLZ level shifts by application of an auxiliary light field, whose intensity and detuning from the resonance are chosen to match the tensor AC-Stark (Sec. 4.5) light shifts to the NLZ shifts.

19
Tensor structure of the DC- and AC-Stark polarizabilities

In this chapter, we give another example of the utility of thinking of a physical process in terms of irreducible tensor operators. In Sec. 4.3.5 we considered the effect of a static electric field on a manifold of Zeeman sublevels, and showed explicitly how the energy shifts may be described in terms of the scalar and tensor polarizabilities, α_0 and α_2. Here, we will re-derive this result by considering irreducible tensors, and, following Stalnaker et al. (2006), we will generalize it to the case of AC electric fields and the *AC-Stark polarizabilities*. One immediate result of considering irreducible tensor operators is the observation that, because any two operators of the same rank defined on a particular angular-momentum state are proportional to each other, it is possible to model the AC-Stark shifts using fictitious static fields. The AC-Stark shifts resulting from rank-zero and rank-two operators can be thought of as corresponding to the scalar and tensor shifts due to DC electric fields (Sec. 4.3.5), while the rank-one AC-Stark operator, being proportional to the rank-one magnetic moment operator, produces shifts analogous to those of an effective magnetic field. For more details on this approach see articles by Happer (1970); Cohen-Tannoudji and Dupont-Roc (1972); Cho et al. (1997); Park et al. (2002).

Understanding the energy-level shifts and splittings due to oscillating electric fields, particularly light fields, is important in many atomic and molecular experiments, such as atomic parity-violation experiments (see, for example, Stalnaker et al. 2006, and references therein), in *laser trapping and cooling*, in *optical atomic clocks*, and in *microwave clocks* using laser trapped atoms (see, for example, Beloy et al. 2009, and references therein).

Consider an atom with states $|\xi J m\rangle$, where ξ represents quantum numbers in addition to J and m, subject to an oscillating electric field (e.g., a light field) $\mathcal{E}(t) = \mathcal{E}_0 \operatorname{Re}(\hat{\varepsilon} e^{-i\omega t})$. Recall that, because there is no permanent atomic electric-dipole moment, there is no first-order Stark shift (Sec. 4.3). Thus we can use second-order perturbation theory, along with the rotating-wave approximation (as described in Sec. 4.5.3) to find the average energy shifts. Note that under the RWA the negative-frequency component of the electric field is neglected. As was done in Sec. 4.5.3, a second term is added with $\omega \to -\omega$ to recover the effect of this component. The average Stark shift of the state $|\xi J m\rangle$ is found to be

$$\Delta E_{\xi J m} = \frac{\mathcal{E}_0^2}{4} \sum_{\xi' J' m'} \langle \xi J m | d_i | \xi' J' m' \rangle \langle \xi' J' m' | d_j | \xi J m \rangle$$
$$\times \left(\frac{\varepsilon_i \varepsilon_j^*}{E_{\xi J m} - E_{\xi' J' m'} - \hbar\omega} + \frac{\varepsilon_i^* \varepsilon_j}{E_{\xi J m} - E_{\xi' J' m'} + \hbar\omega} \right), \quad (19.1)$$

where **d** is the dipole operator, $E_{\xi Jm}$ is the unperturbed energy of state $|\xi Jm\rangle$, and summation over the repeated indices i, j is assumed (here and henceforth). The explicitly written sum in Eq. (19.1) is taken over all of the opposite-parity atomic energy eigenstates, including the continuum. We have assumed that the light is not resonant with any of the transition frequencies $|E_{\xi Jm} - E_{\xi' J'm'}|/\hbar$, so that we can ignore the natural widths of the atomic states.

Without referring to specific intermediate energy levels, we may say that the energy shift of a given atomic state, being a scalar quantity, must result from a contraction of irreducible tensors of the same rank describing the light and the atom, respectively. The light tensors are bilinear in the components of the light polarization vector $\hat{\varepsilon}$ and its complex conjugate $\hat{\varepsilon}^*$. The three irreducible tensor components that can be built from $\hat{\varepsilon}$ and $\hat{\varepsilon}^*$ are (Sec. 3.7)

$$\frac{1}{3}\delta_{ij} \qquad \text{scalar,} \qquad (19.2a)$$

$$\frac{1}{2}\left(\varepsilon_i\varepsilon_j^* - \varepsilon_j\varepsilon_i^*\right) \qquad \text{vector,} \qquad (19.2b)$$

$$\frac{1}{2}\left(\varepsilon_i\varepsilon_j^* + \varepsilon_j\varepsilon_i^*\right) - \frac{1}{3}\delta_{ij} \qquad \text{second-rank tensor.} \qquad (19.2c)$$

These three tensor components correspond to the normalized intensity, orientation, and alignment of the light field, respectively. The vector part of the light tensor, Eq. (19.2b), can be also written in terms of the dual normalized "circular-intensity" vector V_k:

$$\frac{1}{2}\left(\varepsilon_i\varepsilon_j^* - \varepsilon_j\varepsilon_i^*\right) = \frac{1}{2}\epsilon_{ijk}V_k, \qquad (19.3)$$

where

$$V_k = \frac{1}{2}\epsilon_{lmk}\left(\varepsilon_l\varepsilon_m^* - \varepsilon_m\varepsilon_l^*\right), \qquad (19.4)$$

and ϵ_{ijk} is the totally antisymmetric tensor.

> The validity of this manipulation can be checked by substituting the definition (19.4) of V_k back into Eq. (19.3) and noting that $\epsilon_{ijk}\epsilon_{lmk} = \delta_{il}\delta_{jm} - \delta_{im}\delta_{jl}$.

Note that for the circular intensity vector to be nonzero, the light-polarization-vector components cannot all be of the same phase. Consequently, the circular-intensity vector vanishes in the limit of a linearly polarized field, of which a static field is a particular case. As an example of a nonvanishing circular-intensity vector, for left-circularly polarized light propagating along $\hat{\mathbf{z}}$,

$$\hat{\varepsilon} = -\frac{\hat{\mathbf{x}} + i\hat{\mathbf{y}}}{\sqrt{2}}, \qquad (19.5a)$$

$$\mathbf{V} = -i\hat{\mathbf{z}}. \qquad (19.5b)$$

To obtain a shift of an atomic state, three atomic tensor operators T^κ of ranks $\kappa = 0, 1$, and 2, respectively, must be contracted with the light-field tensor components given in Eqs. (19.2), i.e., the AC-Stark shift operator can be written as

$$H_{\text{AC}} = \mathcal{E}_0^2 \left\{ \frac{1}{3}\delta_{ij}T_{ji}^0 + \frac{1}{2}\epsilon_{ijk}V_k T_{ji}^1 + \left[\frac{1}{2}\left(\varepsilon_i\varepsilon_j^* + \varepsilon_j\varepsilon_i^*\right) - \frac{1}{3}\delta_{ij}\right]T_{ji}^2 \right\}. \qquad (19.6)$$

Generally, the eigenstates of the Hamiltonian (19.6) are superpositions of the magnetic sublevels. However, if mixing of the different magnetic sublevels is negligible, we can find the shifts of the magnetic sublevels by simply evaluating the diagonal elements of the above Hamiltonian. This would be the case if the magnetic sublevels are split by a sufficiently strong DC magnetic field that defines the quantization axis, or if linearly polarized light is applied in the absence of other fields, so that the light field defines the quantization axis. In this case, only the tensor operators with the tensor index $q = 0$ contribute to the matrix elements. From Eqs. (3.114) we see that these are T^1_{xy} (and T^1_{yx}) and T^2_{zz} for the rank-one and rank-two tensors, in addition to the diagonal elements of the rank-zero tensor. These contract with the corresponding components of the electric-field tensors.

To evaluate the components of the operators $T^{(\kappa)}$, we use the fact, noted above, that by the Wigner–Eckart theorem, $T^{(\kappa)}$ is proportional to any other rank-κ tensor operator. In particular, the total-angular-momentum operator $\mathbf{J}$ is a rank-one tensor operator, so $T^1_0 \propto J_z$, and, from Eq. (3.109), T^2_{ij} is proportional to $J_i J_j - \delta_{ij} \mathbf{J}^2/3$. Thus we have

$$T^0_{xx} = T^0_{yy} = T^0_{zz} = C_0, \tag{19.7a}$$

$$T^1_{yx} = -T^1_{xy} = C_1 J_z/\hbar, \tag{19.7b}$$

$$T^2_{zz} = C_2 (J_z^2 - \mathbf{J}^2/3)/\hbar^2, \tag{19.7c}$$

where the constants C_κ are determined by the specifics of the atomic energy levels. Taking the matrix element $\langle \xi J m | H_{\text{AC}} | \xi J m \rangle$, we find the AC-Stark shift for the $|\xi J m \rangle$ state:

$$\Delta E_{\xi J m} = \mathcal{E}_0^2 \left[C_0 + C_1 m V_z + C_2 \left(m^2 - \frac{J(J+1)}{3} \right) \left(|\varepsilon_z|^2 - \frac{1}{3} \right) \right]. \tag{19.8}$$

The parameters C_κ are often written in terms of the AC-Stark polarizabilities α_κ, commonly defined so that the scalar and tensor AC polarizabilities, α_0 and α_2, become equal to the corresponding static polarizabilities in the limit of a very-low-frequency linearly polarized field. In making this correspondence, we must remember that the AC-Stark shift is the average shift over an oscillation cycle of the electric field. Thus a low-frequency oscillating field of amplitude $\mathcal{E}_0$ will produce an average AC shift that is one-half as large as the DC shift due to a static field $\mathcal{E}_{\text{static}} = \mathcal{E}_0$, since the average of the square of the oscillating field is $\overline{\mathcal{E}^2} = \mathcal{E}_0^2/2$.

Comparing with the definitions for the static electric polarizabilities as given by Eq. (4.65), and including the factor of one-half, we find that the AC scalar shift should be given by $-\alpha_0 \mathcal{E}_0^2/4$, while the value of α_2 can be specified by fixing the AC tensor shift as $-\alpha_2 \mathcal{E}_0^2/4$ for the stretched states $m = \pm J$ in the special case of a field linearly polarized along $\hat{\mathbf{z}}$. Since there is no vector polarizability in the limit of a static field, its definition is somewhat arbitrary; a commonly used convention is that the vector shift of the $m = J$ state for left circularly polarized light propagating along $\hat{\mathbf{z}}$ is given by $-\alpha_1 \mathcal{E}_0^2/8$. (Note that some variation in the literature may be encountered in the definitions of α_0 and α_2, as well as in the definition of α_1.) Using these definitions, we have the expression:

$$\Delta E_{\xi J m} = -\frac{\mathcal{E}_0^2}{4} \left(\alpha_0 + i \alpha_1 \frac{m}{2J} V_z + \alpha_2 \frac{3m^2 - J(J+1)}{J(2J-1)} \frac{3|\varepsilon_z|^2 - 1}{2} \right). \tag{19.9}$$

Note that we are not in danger of dividing by zero for $J = 0$ or $1/2$, because from Eq. (19.8) we see that the vector shifts vanish for $J = 0$ states, while the second-rank tensor shifts vanish for $J = 0$ and $J = 1/2$ states. This dictates selection rules for the corresponding polarizabilities.

The form of Eq. (19.9) suggests that, indeed, the effect of an off-resonant oscillating electric field is analogous to that of static fields. For example, a linearly polarized oscillating electric field causes scalar and tensor shifts similar to a static electric field. A circularly polarized electric field, on the other hand, produces a combination of a scalar shift and a splitting of Zeeman sublevels linear in m analogous to that produced by a static magnetic field applied along the quantization axis.

In the case of an atomic state with hyperfine structure, the analysis above has to be modified, as in the treatment of static fields (Sec. 4.3.6).

20
Photoionization of polarized atoms with polarized light

In this chapter, we will discuss an application of atomic polarization moments in a context somewhat different from those encountered so far: *photoionization* of an atom prepared in a polarized state. To photoionize an atom, a photon must have an energy exceeding the *ionization potential* of the initial state, i.e., the energy gap between the initial state and the continuum of energy states with at least one of the electrons detached from the original atom. Following a paper by Li and Budker (2006), we will show that there are, in general, only three independent parameters that determine the total photoionization cross-section for ionization of a polarized atom with polarized light.

With tunable lasers, it is possible to obtain high populations and polarizations of selected excited states even if these states have short lifetimes. The measurements of photoionization cross-sections of atoms in excited states are valuable for testing atomic theory, and are important for understanding of processes in plasmas, including those in stellar atmospheres, lighting devices, etc.

In the work of Li and Budker (2006), photoionization of some even-parity states of Ba excited by a sequence of two laser pulses was investigated (Fig. 20.1). The dependence of the photoionization cross-section on the relative polarizations of the atoms and that of the third, ionizing laser beam was studied. To produce different polarizations of the atoms in the probed

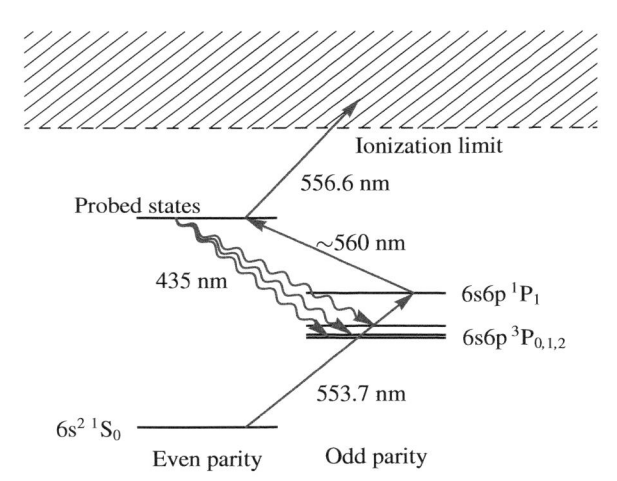

Fig. 20.1 The excitation-detection scheme. The probed states are the $5d6d\,^3D_1$ and $6s7d\,^3D_2$ states. Solid arrows indicate laser excitation; the squiggle arrows indicate fluorescence.

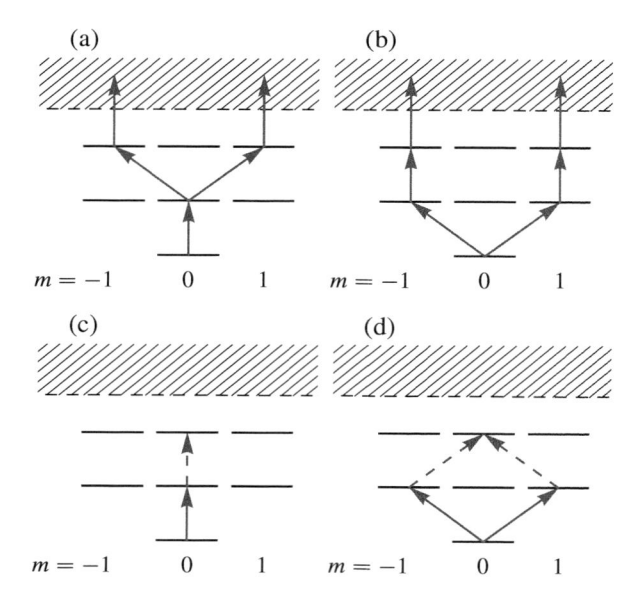

Fig. 20.2 Three-step ionization on a $0 \to 1 \to 1$ transition. Dashed arrows indicate forbidden transitions. When the polarizations of the two excitation lasers are orthogonal, as in plots (a) and (b), excitation can occur. Excitation cannot occur when the polarizations of the excitation lasers are parallel, as in plots (c) and (d), because the transition amplitude for the second step is zero. Case (d) is physically equivalent to case (c); however, in this basis the suppression shows up as a cancelation of coherent excitation paths via the $m = 1$ and $m = -1$ sublevels of the intermediate state.

state, the polarizations of the two excitation-laser beams were varied with half-wave plates.

Suppose that linearly polarized ionization light is used, and the atoms are initially in a $J = 0$ ground state, as in the case of atomic barium. Let us define the quantization axis z along the ionization-light polarization axis. Suppose also that the excitation and ionization laser beams propagate along the same axis, defined as the x-axis. Since light is transverse, the polarization of the excitation laser beams can only be in the yz plane. For a $J = 1$ probed state, only $m = \pm 1$ Zeeman sublevels can be coherently excited (Fig. 20.2). The $m = 0$ sublevel cannot be excited because the corresponding Clebsch–Gordan coefficient for a $J = 1 \to J' = 1$ transition is zero (see the discussion in Sec. 7.3.2). Various different polarizations that can be obtained for a $J = 2$ probed state are depicted in Fig. 20.3. When both excitation lasers are polarized along the z-axis, the $m = 0$ sublevel is populated in this case. When one of the excitation lasers is polarized along the z-axis and the other is polarized along the y-axis, $m = \pm 1$ sublevels are coherently excited. When the polarizations of both lasers are along the y-axis, $m = \pm 2$ and $m = 0$ sublevels can be coherently excited.

20.1 Photoionization cross-section

In this subsection, we show, using the polarization moments of the density matrix, that there are at most three parameters in the total photoionization cross-section. We denote the atomic density matrix by ρ and the photon density matrix by Φ, each with unit normalization:

$$\mathrm{Tr}\,\rho = \mathrm{Tr}\,\Phi = 1. \tag{20.1}$$

Each density matrix can be decomposed into polarization moments, given in terms of the density-matrix elements in the Zeeman basis by Eq. (5.68). The atomic density matrix, with angular momentum J, can have polarization moments up to rank $\kappa = 2J$. Under the electric-dipole approximation, photons have total angular momentum $J = 1$, so the photon density matrix can have polarization moments up to rank two.

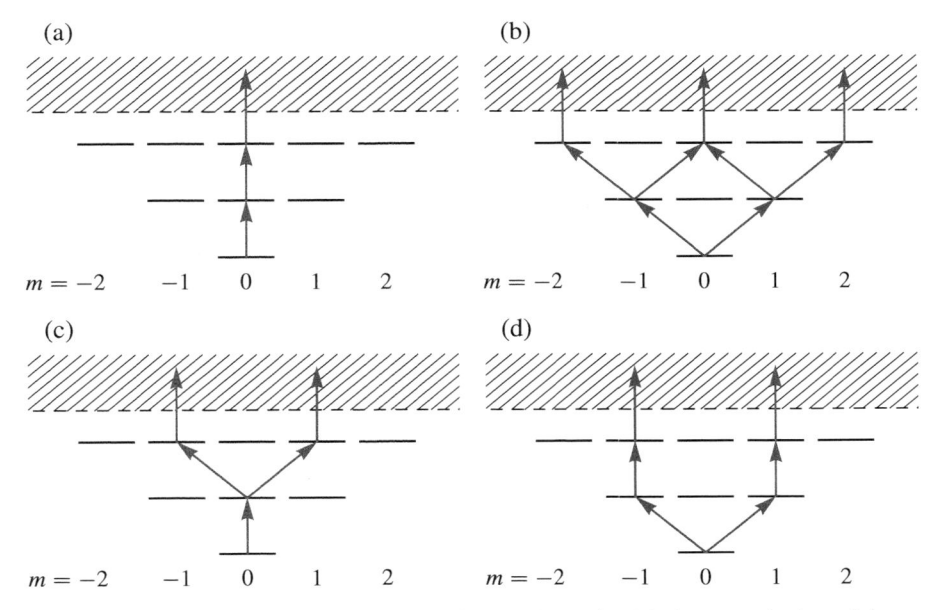

Fig. 20.3 Three-step ionization on a $0 \to 1 \to 2$ transition. Plot (a) shows excitation of the $m = 0$ sublevels with both excitation lasers polarized along the z-axis. Plot (b) shows excitation of a coherent superposition of Zeeman sublevels with both laser beams polarized along the y-axis. Plots (c) and (d) show excitation with one laser beam polarized along the z-axis and the other polarized along the y-axis.

The photoionization process is related to the density matrices of the ionizing photons and the probed state. Because the total photoionization cross-section is a scalar (since here we are not concerned with the angular distribution of the electrons and ions), the irreducible tensors of the density matrix of the photons should be contracted with those of the atoms of the same ranks. The photoionization cross-section can be expressed as:

$$\sigma = \sqrt{3(2J+1)} \left(\sigma_0 \Phi_0^0 \rho_0^0 + \sigma_1 \sum_{q=-1}^{1} (-1)^q \Phi_q^1 \rho_{-q}^1 + \sigma_2 \sum_{q=-2}^{2} (-1)^q \Phi_q^2 \rho_{-q}^2 \right), \quad (20.2)$$

where $\sigma_{0,1,2}$ are coefficients determined by the total angular momenta of the initial and final (continuum) states. The normalization factor $\sqrt{3(2J+1)}$ is chosen because according to Eq. (5.68),

$$\rho_0^0 = \frac{1}{\sqrt{2J+1}}, \quad \Phi_0^0 = \frac{1}{\sqrt{3}}. \quad (20.3)$$

Therefore, we conclude that in general there are at most three parameters in the photoionization cross-section. (There is only one parameter for a $J = 0$ state and there are two parameters for a $J = 1/2$ state.) For an unpolarized initial atomic state and/or an unpolarized ionization light source, the photoionization cross-section is σ_0. As we discussed in Chapter 6, if the light source is completely unpolarized, the diagonal elements of the density matrix of the light are equal and the off-diagonal elements are all zero. A directional light beam cannot be completely unpolarized under this definition because the electric field has no component along the propagation direction of the light. Even if the light is "unpolarized" according to the commonly used definition (i.e., the Stokes parameters $S_{1,2,3}$ are all zero) it in fact possesses alignment along the propagation direction.

If both the atoms and the light are polarized, the cross-section may be different depending on their relative orientation ($\rho^{(1)}$, $\Phi^{(1)}$) and their relative alignment ($\rho^{(2)}$, $\Phi^{(2)}$).

20.2 Formulas for $\sigma_{0,1,2}$

In this section, we derive general formulas for $\sigma_{0,1,2}$ for states of arbitrary angular momenta. Note that these coefficients are independent of the polarizations of the atoms and light. Therefore, we are free to choose any polarizations that are convenient for performing the calculation. Consider, then, an ensemble of atoms prepared in a particular Zeeman sublevel $|aJ\bar{m}\rangle$ of the probed state, where a represents all other quantum numbers of the state. We also assume that the ionizing light is left-circularly polarized. The density matrix for the photons is

$$\Phi = \begin{pmatrix} 1 & 0 & 0 \\ 0 & 0 & 0 \\ 0 & 0 & 0 \end{pmatrix}. \tag{20.4}$$

Using Eq. (5.68), we can decompose it into irreducible tensors with components Φ_q^κ:

$$\Phi_0^0 = \frac{1}{\sqrt{3}}, \quad \Phi_0^1 = \frac{1}{\sqrt{2}}, \quad \Phi_0^2 = \frac{1}{\sqrt{6}}, \tag{20.5}$$

with all other components equal to zero. All elements of the density matrix of the probed state are zero except $\rho_{\bar{m}\bar{m}} = 1$. This matrix can also be decomposed according to Eq. (5.68):

$$\rho_0^0 = (-1)^{J-\bar{m}} \begin{pmatrix} J & J & 0 \\ \bar{m} & -\bar{m} & 0 \end{pmatrix}, \tag{20.6a}$$

$$\rho_0^1 = (-1)^{J-\bar{m}} \sqrt{3} \begin{pmatrix} J & J & 1 \\ \bar{m} & -\bar{m} & 0 \end{pmatrix}, \tag{20.6b}$$

$$\rho_0^2 = (-1)^{J-\bar{m}} \sqrt{5} \begin{pmatrix} J & J & 2 \\ \bar{m} & -\bar{m} & 0 \end{pmatrix}, \tag{20.6c}$$

with all other components equal to zero. Substituting these values into Eq. (20.2), we find for the photoionization cross-section in this case

$$\sigma = \sqrt{3(2J+1)} \left[\sigma_0 \frac{(-1)^{J-\bar{m}}}{\sqrt{3}} \begin{pmatrix} J & J & 0 \\ \bar{m} & -\bar{m} & 0 \end{pmatrix} + \sigma_1 (-1)^{J-\bar{m}} \sqrt{\frac{3}{2}} \begin{pmatrix} J & J & 1 \\ \bar{m} & -\bar{m} & 0 \end{pmatrix} \right.$$
$$\left. + \sigma_2 (-1)^{J-\bar{m}} \sqrt{\frac{5}{6}} \begin{pmatrix} J & J & 2 \\ \bar{m} & -\bar{m} & 0 \end{pmatrix} \right]. \tag{20.7}$$

We now use another expression for σ to eliminate it and find formulas for $\sigma_{0,1,2}$. As derived, for example, in the book by Sobelman (1992), the photoionization cross-section can be written in terms of a sum over transition probabilities for transitions from the initial state $|\psi_a\rangle$ to the continuum states $|\psi_n\rangle$:

$$\sigma = \frac{4\pi^2 m_e}{\hbar^2} \frac{k}{p} \sum_n |\langle \psi_n | \hat{\varepsilon} \cdot \mathbf{d} | \psi_a \rangle|^2, \tag{20.8}$$

where m_e is the mass of the electron, k is the momentum of an ionizing photon, p is the momentum of an ionized electron, and $\hat{\varepsilon}$ is the light-polarization vector.

For left-circularly polarized light, $\hat{\varepsilon} \cdot \mathbf{d} = d_1$. Using the relation

$$|\langle nJ_n, m+1|d_1|aJm\rangle|^2 = -\begin{pmatrix} J_n & 1 & J \\ -m-1 & 1 & m \end{pmatrix}\begin{pmatrix} J & 1 & J_n \\ -m & -1 & m+1 \end{pmatrix}|\langle nJ_n\|d\|aJ\rangle|^2, \quad (20.9)$$

the photoionization cross-section in the case we are considering can be written as

$$\sigma = -\sum_n A_n \begin{pmatrix} J_n & 1 & J \\ -m-1 & 1 & m \end{pmatrix}\begin{pmatrix} J & 1 & J_n \\ -m & -1 & m+1 \end{pmatrix}, \quad (20.10)$$

where

$$A_n = \frac{4\pi^2 m_e k}{\hbar^2 p}|\langle nJ_n\|d\|aJ\rangle|^2. \quad (20.11)$$

Rather than substitute Eq. (20.10) into Eq. (20.7) directly, it is convenient to introduce an auxiliary function Z defined as

$$Z(l) = \sum_{\overline{m}=-J}^{J} (-1)^{J+\overline{m}} \begin{pmatrix} J & J & l \\ m & -m & 0 \end{pmatrix} \sigma. \quad (20.12)$$

This allows us to use Clebsch–Gordan sum rules to obtain independent equations for $\sigma_{0,1,2}$. Substituting Eq. (20.7) into Eq. (20.12) and using the identity

$$\sum_{m=-J}^{J} \begin{pmatrix} J & J & \kappa \\ m & -m & 0 \end{pmatrix}\begin{pmatrix} J & J & l \\ m & -m & 0 \end{pmatrix} = \frac{1}{2l+1}\delta_{\kappa l}, \quad (20.13)$$

we find

$$Z(0) = \sqrt{2J+1}\,\sigma_0, \quad Z(1) = \sqrt{\frac{2J+1}{2}}\,\sigma_1, \quad Z(2) = \sqrt{\frac{2J+1}{10}}\,\sigma_2. \quad (20.14)$$

On the other hand, using Eq. (20.10) in Eq. (20.12) yields

$$Z(l) = \sum_n A_n \sum_{\overline{m}} (-1)^{J+\overline{m}+1} \begin{pmatrix} J_n & 1 & J \\ -m-1 & 1 & m \end{pmatrix}\begin{pmatrix} J & 1 & J_n \\ -m & -1 & m+1 \end{pmatrix}\begin{pmatrix} J & J & l \\ m & -m & 0 \end{pmatrix}$$
$$= \sum_n A_n (-1)^{J-J_n} \begin{pmatrix} 1 & 1 & l \\ 1 & -1 & 0 \end{pmatrix} \begin{Bmatrix} 1 & 1 & l \\ J & J & J_n \end{Bmatrix}, \quad (20.15)$$

where, in the last line, we have used the identity

$$\sum_{m_4 m_5 m_6} (-1)^{j_4+j_5+j_6-m_4-m_5-m_6} \begin{pmatrix} j_1 & j_5 & j_6 \\ m_1 & -m_5 & m_6 \end{pmatrix}\begin{pmatrix} j_4 & j_2 & j_6 \\ m_4 & m_2 & -m_6 \end{pmatrix}\begin{pmatrix} j_4 & j_5 & j_3 \\ -m_4 & m_5 & m_3 \end{pmatrix}$$
$$= \begin{pmatrix} j_1 & j_2 & j_3 \\ m_1 & m_2 & m_3 \end{pmatrix}\begin{Bmatrix} j_1 & j_2 & j_3 \\ j_4 & j_5 & j_6 \end{Bmatrix}. \quad (20.16)$$

Comparing Eq. (20.15) with Eqs. (20.14), we find the formulas for $\sigma_{0,1,2}$:

Table 20.1 Ratios between photoionization cross-sections. The ratios are calculated assuming that a probed state with total angular momentum J is dominantly coupled to continuum states with total angular momentum J_c.

J	J_c	σ_0 :	σ_1 :	σ_2
1	0	1 :	-1 :	1
	1	1 :	$-\frac{1}{2}$:	$-\frac{1}{2}$
	2	1 :	$\frac{1}{2}$:	$\frac{1}{10}$
2	1	1 :	$-\sqrt{\frac{3}{4}}$:	$\sqrt{\frac{7}{20}}$
	2	1 :	$-\sqrt{\frac{1}{12}}$:	$-\sqrt{\frac{7}{20}}$
	3	1 :	$\sqrt{\frac{1}{3}}$:	$\sqrt{\frac{1}{35}}$

$$\sigma_0 = \sum_n A_n \frac{(-1)^{1-2J_n}}{3(2J+1)}, \tag{20.17a}$$

$$\sigma_1 = \sum_n A_n \frac{(-1)^{J-J_n}}{\sqrt{3(2J+1)}} \begin{Bmatrix} 1 & 1 & 1 \\ J & J & J_n \end{Bmatrix}, \tag{20.17b}$$

$$\sigma_2 = \sum_n A_n \frac{(-1)^{J+J_n}}{\sqrt{3(2J+1)}} \begin{Bmatrix} 1 & 1 & 2 \\ J & J & J_n \end{Bmatrix}. \tag{20.17c}$$

If the probed state is dominantly coupled to continuum states with a specific total angular momentum $J_n = J_c$, the ratios between photoionization cross-sections are

$$\frac{\sigma_1}{\sigma_0} = (-1)^{J+J_c-1} \sqrt{3(2J+1)} \begin{Bmatrix} 1 & 1 & 1 \\ J & J & J_c \end{Bmatrix}, \tag{20.18a}$$

$$\frac{\sigma_2}{\sigma_0} = (-1)^{J-J_c-1} \sqrt{3(2J+1)} \begin{Bmatrix} 1 & 1 & 2 \\ J & J & J_c \end{Bmatrix}. \tag{20.18b}$$

In Table 20.1, we list the ratios between $\sigma_{0,1,2}$ of a probed state with total angular momentum $J = 1, 2$ if it is dominantly coupled to continuum states with total angular momentum J_c.

Thus we can see that measuring the dependence of the photoionization cross-section on the relative polarization of the atomic state and the ionization light offers a powerful tool for investigating the structure of the continuum—specifically, for determining the angular momentum of the continuum states including the so-called *autoionization states* (see, for example, the book by Letokhov, 1987), which in certain cases can be almost as narrow as the usual atomic bound states. Autoionization states arise when two or more atomic electrons are excited in such a away that each single-electron excitation energy is below the ionization limit; however, the sum of the excitation energies exceeds the limit. An electron is ejected from such a system only when the excitation energy can somehow be transferred to a single electron, a process that can take a long time on the atomic scale.

A
Constants, units, and notations

We have mostly used Gaussian units in this book (as we do in our research) because we find that it is both convenient and aesthetically appealing to have the symmetry between electric and magnetic quantities "built-in" into the system of units. Of course, it is straightforward to convert between various systems of units, and it had better be that all physics results are invariant with respect to unit conversion!

Many workers in atomic, molecular, and optical physics choose to use *atomic units*, in which the magnitude of the electron charge e, the electron mass m, and Planck's constant $\hbar$ are all set to unity. Because e, $\hbar$, and the speed of light, c, form a dimensionless combination of Eq. (2.23)

$$\alpha = \frac{e^2}{\hbar c} \approx \frac{1}{137}, \tag{A.1}$$

the speed of light in atomic units is $c \approx 137$. All other units can be derived from these basic atomic units. For example, the unit of distance turns out to be the Bohr radius

$$a_0 = \frac{\hbar^2}{m_e e^2}, \tag{A.2}$$

while the unit of the electric-dipole moment is ea_0 as found by multiplying the unit of charge by the unit of distance.

One needs to make a distinction between an atomic unit and the characteristic value of some quantity. For example, electric and magnetic dipole moments have the same units, so ea_0 is also the unit of the magnetic dipole moment. However, the natural scale of magnetic moment in an atom (see Chapter 2) is $\mu_B = (\alpha/2)ea_0$.

Table A.1 lists the values of physical constants in the Gaussian system of units. Numerical values of physical constants are taken from http://physics.nist.gov/constants; see also Mohr et al. (2008). Some common notations used throughout this book are summarized in Table A.2.

Table A.1 Numerical values of physical constants in CGS (Gaussian) units.

Symbol	Meaning	Value
m_e	electron mass	$9.10938215(45) \times 10^{-28}$ g $0.510998910(13)$ MeV/c^2
m_p	proton mass	$1.672621637(83) \times 10^{-24}$ g $938.272013(23)$ MeV/c^2
m_n	neutron mass	$1.674927211(84) \times 10^{-24}$ g $939.565346(23)$ MeV/c^2
m_n/m_p	nucleon mass ratio	$1.00137841918(46)$
e	electron charge magnitude	$4.80321284(12) \times 10^{-10}$ esu
h $\hbar = h/(2\pi)$	Planck's constant	$6.62606896(33) \times 10^{-27}$ erg s $1.054571628(53) \times 10^{-27}$ erg s
$\alpha = e^2/(\hbar c)$	fine-structure constant	$1/137.035999679(94)$
$a_0 = \hbar^2/(m_e e^2)$	Bohr radius	$0.52917720859(36) \times 10^{-8}$ cm
$\mu_B = e\hbar/(2m_e c)$	Bohr magneton	$0.927400915(23) \times 10^{-20}$ erg/G $1.399624604(35)$ MHz/G
$\mu_N = e\hbar/(2m_p c)$	nuclear magneton	$5.05078324(13) \times 10^{-24}$ erg/G $762.259384(19)$ Hz/G
$R_\infty = m_e e^4/(4\pi\hbar^3 c)$	Rydberg constant	$109737.31568527(73)$ cm^{-1}
k_B	Boltzmann's constant	$1.3806504(24) \times 10^{-16}$ erg/K $8.617343(15) \times 10^{-5}$ eV/K

Table A.2 Some notations used throughout this book.

Symbol	Meaning
L, **l**	total (individual-particle) orbital-angular-momentum operator
L, l	total (individual-particle) orbital-angular-momentum quantum number
S, **s**, S, s	electron spin
J, **j**, J, j	total electronic angular momentum
I, I	nuclear spin
F, F	total atomic angular momentum
m	angular-momentum projection
d	electric-dipole operator
$\alpha_{0,1,2}$	scalar, vector, tensor Stark polarizabilities
μ	magnetic-moment operator
g	Landé factor
H	Hamiltonian
E	energy

Table A.2 Some notations, continued.

Symbol	Meaning
κ	tensor rank
q	tensor projection index
$\mathcal{T}_q^\kappa$	polarization operator
Y_{lm}	spherical harmonic
κ_1	excitation saturation parameter
κ_2	optical-pumping saturation parameter
$\hat{\mathbf{x}}_{1,2,3}$	Cartesian unit vectors
$\hat{\mathbf{e}}_{1,2}$	real orthogonal unit vectors
$\hat{\boldsymbol{\varepsilon}}_{1,0,-1}$	spherical-basis unit vectors
$\hat{\varepsilon}$	light-electric-field unit vector
ω	light frequency
$\mathbf{k}$	light wave vector
$\boldsymbol{\mathcal{E}}$	electric-field vector
$\mathbf{B}$	magnetic-field vector
$\mathcal{E}_0$	light-electric-field amplitude
φ	overall phase factor for light field
α	light polarization angle
ϵ	light ellipticity
$\mathbf{P}$	ensemble polarization
$P_{1,2,3,4}$	ensemble polarization components
$S_{1,2,3}$	normalized Stokes parameters
ℓ	optical path length
R	classical rotation operator
R	classical rotation matrix
$\mathscr{D}$	quantum-mechanical rotation operator
$D_{m'm}^{(J)}$	Wigner D-function
ρ	density matrix
Φ	density matrix for light
γ	atomic ground-state relaxation rate
Γ	upper-state natural width
$\widehat{\Gamma}$	relaxation matrix
Λ	repopulation matrix
Γ_D	Doppler width
Ω_L	Larmor frequency
Ω_R	Rabi frequency
Ω_m	modulation frequency
$\mathcal{A}$	absorption matrix
$\mathcal{F}$	fluorescence matrix
$\mathcal{I}$	fluorescence intensity
$F_{m_1 m_2}^{m'_1 m'_2}$	spontaneous-emission operator

B
Units of energy, frequency, and wavelength

It is useful in practice to be able to convert quantities between various units of energy, between units of energy and frequency, as well as between photon energy and inverse wavelength of radiation. Here we present the conversion factors for reference.

$$1\text{eV} \approx 1.60 \times 10^{-19}\text{ J} \approx 1.60 \times 10^{-12}\text{ erg}$$
$$\approx 8066\text{ cm}^{-1} \times hc \approx 2.41 \times 10^{14}\text{ Hz} \times h. \tag{B.1a}$$
$$1\text{ cm}^{-1} \times c \approx 30\text{ GHz}. \tag{B.1b}$$

The conversion (B.1b) is derived from the relations between the light frequency ν, and wavelength λ: $\nu = c/\lambda$.

The temperature corresponding to an energy E of 1 eV is

$$T = E/k_B \approx 11,600\text{ K}. \tag{B.2}$$

Here $k_B \approx 1.38 \times 10^{-16}$ erg/K is the *Boltzmann constant*.

When talking about things like the energy of a molecular bond, dissociation or ionization energies, etc., people sometimes use the unit of energy kcal/mole, originating from chemistry. We can relate this unit to the units more commonly used in atomic physics by recalling that a calorie is the energy needed to heat a gram of water by 1 K, which is ≈ 4.185 J. Since the number of molecules in a mole (given by the *Avogadro number*) is (Mohr *et al.* 2008)

$$N_A = 6.02214179(30) \times 10^{23}, \tag{B.3}$$

one can easily calculate that 1 kcal/mole corresponds to $\approx 4.3 \times 10^{-2}$ eV/molecule.

In spectroscopy, wavelengths are frequently specified using the *ångström* (Å) and the nanometer, which are related according to

$$1\text{ Å} = 0.1\text{ nm} = 10^{-8}\text{ cm}. \tag{B.4}$$

C
Reference data for hydrogen and the alkali atoms

Table C.1 Parameters of the lowest-energy resonance transitions from the ground state for hydrogen ($1s \rightarrow 2p_{1/2,3/2}$) and the alkali atoms [the $D1(2)$ transitions: $ns \rightarrow np_{1/2(3/2)}$]. Wavelengths are given in vacuum; $\|d_J\|$ is the reduced matrix element in the J-basis. High-precision experimental data for the matrix elements are taken from a compilation by Safronova *et al.* (1999).

Atom	Upper state	Energy (cm^{-1} × hc)	Wavelength (nm)	Lifetime (ns)	$\|d_J\|$ (ea_0)
H	$2\,^2P_{1/2}$	82258.91	121.5674	1.60	1.05
	$2\,^2P_{3/2}$	82259.27	121.5668	1.60	1.49
Li	$2\,^2P_{1/2}$	14903.66	670.976	27.1	3.33
	$2\,^2P_{3/2}$	14904.00	670.961	27.1	4.71
Na	$3\,^2P_{1/2}$	16956.18	589.755	16.3	3.5246(23)
	$3\,^2P_{3/2}$	16973.38	589.158	16.2	4.9838(34)
K	$4\,^2P_{1/2}$	12985.17	770.109	26.2	4.102(5)
	$4\,^2P_{3/2}$	13042.89	766.701	26.1	5.800(8)
Rb	$5\,^2P_{1/2}$	12578.96	794.978	27.7	4.231(3)
	$5\,^2P_{3/2}$	12816.56	780.241	26.2	5.977(4)
Cs	$6\,^2P_{1/2}$	11178.24	894.595	34.8	4.4890(65)
	$6\,^2P_{3/2}$	11732.35	852.344	30.4	6.3238(73)
Fr	$7\,^2P_{1/2}$	12236.66	817.216	29.5	4.277(8)
	$7\,^2P_{3/2}$	13923.20	718.226	21.0	5.898(15)

D
Classical rotations

In this Appendix we fill in some details in the discussion of Sec. 3.1 regarding classical rotations in the Cartesian and spherical bases. For additional information regarding co- and contravariance, see the authoritative treatment by Fano and Racah (1959).

D.1 Rotations in the Cartesian basis

Suppose we have a vector **v** in three-dimensional space and have chosen a Cartesian basis $\hat{\mathbf{x}}_i$, so that we can write **v** in terms of its Cartesian components v_i:

$$\mathbf{v} = \hat{\mathbf{x}}_i v_i, \quad \text{with} \quad v_i = \hat{\mathbf{x}}_i \cdot \mathbf{v}, \tag{D.1}$$

where we use the orthonormality of the Cartesian basis: $\hat{\mathbf{x}}_i \cdot \hat{\mathbf{x}}_j = \delta_{ij}$.

Now consider a rotation of **v** by an angle θ about a direction $\hat{\mathbf{n}}$ to obtain a rotated vector $\mathbf{v}' = R_{\hat{\mathbf{n}}}(\theta)\mathbf{v}$, where $R_{\hat{\mathbf{n}}}(\theta)$ (abbreviated as R) is the rotation operator. The right-hand rule is used to fix the rotation direction: if the thumb of the right hand points along $\hat{\mathbf{n}}$, the fingers curl in the direction of a positive rotation. We can expand the rotated vector in the Cartesian basis to obtain

$$\mathbf{v}' = R\mathbf{v} = R\hat{\mathbf{x}}_i v_i = \hat{\mathbf{x}}_j \hat{\mathbf{x}}_j \cdot R\hat{\mathbf{x}}_i v_i = \hat{\mathbf{x}}_j R_{ji} v_i, \tag{D.2}$$

where $R_{ji} = \hat{\mathbf{x}}_j \cdot R\hat{\mathbf{x}}_i$ is the rotation matrix corresponding to R. If we consider R to act on the components v_i, as indicated in Eq. (D.2), we see that they are transformed to a new set of components v'_i according to $v'_i = R_{ij} v_j$, and the rotation of **v** can be written $\mathbf{v}' = \hat{\mathbf{x}}_i v'_i$. A rotation thought of in this way, in which the basis vectors are held fixed and the vector components are rotated, is called an *active rotation*. An example of an active rotation by an angle θ about the $\hat{\mathbf{z}}$ direction shown in Fig. D.1. From trigonometric arguments, we can find that the rotation matrix in this case is given by

$$R_{\hat{\mathbf{z}}}(\theta) = \begin{pmatrix} \cos\theta & -\sin\theta & 0 \\ \sin\theta & \cos\theta & 0 \\ 0 & 0 & 1 \end{pmatrix}. \tag{D.3}$$

Writing the expressions in matrix notation, and using the case of Fig. D.1(a),

$$\mathbf{v} = \hat{\mathbf{x}} = \begin{pmatrix} \hat{\mathbf{x}} & \hat{\mathbf{y}} & \hat{\mathbf{z}} \end{pmatrix} \begin{pmatrix} 1 \\ 0 \\ 0 \end{pmatrix}, \tag{D.4}$$

as an example, we have for the rotated vector components

Rotations in the Cartesian basis

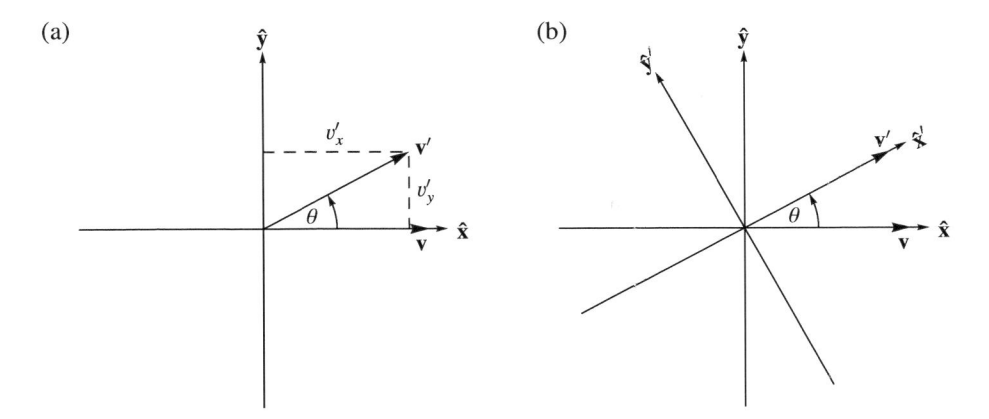

Fig. D.1 A rotation of a vector **v**, originally along the $\hat{\mathbf{x}}$ direction, by an angle θ about the $\hat{\mathbf{z}}$ direction. By inspection, we find that $v'_x = |\mathbf{v}| \cos\theta$ and $v'_y = |\mathbf{v}| \sin\theta$.

$$\begin{pmatrix} v'_x \\ v'_y \\ v'_z \end{pmatrix} = \begin{pmatrix} \cos\theta & -\sin\theta & 0 \\ \sin\theta & \cos\theta & 0 \\ 0 & 0 & 1 \end{pmatrix} \begin{pmatrix} 1 \\ 0 \\ 0 \end{pmatrix} = \begin{pmatrix} \cos\theta \\ \sin\theta \\ 0 \end{pmatrix}, \tag{D.5}$$

and for the vector itself,

$$\mathbf{v}' = \begin{pmatrix} \hat{\mathbf{x}} & \hat{\mathbf{y}} & \hat{\mathbf{z}} \end{pmatrix} \begin{pmatrix} \cos\theta \\ \sin\theta \\ 0 \end{pmatrix} = \hat{\mathbf{x}}\cos\theta + \hat{\mathbf{y}}\sin\theta. \tag{D.6}$$

It is also possible to think of the rotation matrix in Eq. (D.2) as acting on the basis vectors $\hat{\mathbf{x}}_i$. A general property of rotation matrices in the Cartesian basis is that they are *orthogonal*, meaning that their transpose equals their inverse: $\mathsf{R}^t = \mathsf{R}^{-1}$.

This results from the fact that a rotation preserves the length of the vector, i.e.,

$$\mathbf{v} \cdot \mathbf{v} = \mathbf{v}' \cdot \mathbf{v}' = \mathsf{R}_{ki} v_i \mathsf{R}_{kj} v_j = v_i \mathsf{R}^t_{ik} \mathsf{R}_{kj} v_j, \tag{D.7}$$

so that $v_i \mathsf{R}^t_{ik} \mathsf{R}_{kj} v_j = v_i \delta_{ij} v_j$. Since this is true for any vector **v**, we have $\mathsf{R}^t \mathsf{R} = \mathbb{1}$, where $\mathbb{1}$ is the identity matrix.

Using this property, we can write Eq. (D.2) as

$$\mathbf{v}' = \mathsf{R}\mathbf{v} = \hat{\mathbf{x}}_j \mathsf{R}_{ji} v_i = \mathsf{R}^t_{ij} \hat{\mathbf{x}}_j v_i = \mathsf{R}^{-1}_{ij} \hat{\mathbf{x}}_j v_i. \tag{D.8}$$

In this form, the same rotation is due to a transformation of the basis vectors, rather than the vector components: $\mathbf{v}' = \hat{\mathbf{x}}'_i v_i$, where the rotated basis vectors are given by $\hat{\mathbf{x}}'_i = \mathsf{R}^{-1}_{ij} \hat{\mathbf{x}}_j$. The effect on the vector **v** is the same as for the active rotation, but here the basis vectors move along with **v**, so that the relationship between **v** and the basis vectors (represented by the vector components) is unchanged [Fig. D.1(b)]. This type of rotation is not normally given a specific name, but we can call it a *joint rotation* because the vector and the basis vectors are rotated jointly. As the vector does not move in relation to the basis vectors, this kind of rotation is often thought of as leaving the system unchanged—it is equivalent, in a sense, to simply tilting one's head! Thus, two rotations are generally thought of as equivalent if they differ only by a given joint rotation.

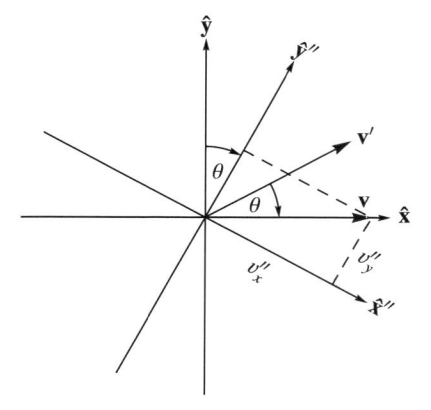

Fig. D.2 The inverse rotation to that shown in Fig. D.1, returning the vector **v**′ back to **v**. The combination of the two rotations can be written as a passive rotation, with the original vector **v** being written in terms of the rotated basis vectors $\hat{\mathbf{x}}_i''$.

Note that applying R to the vector components has the same effect on **v** as applying R^{-1} to the basis vectors. This shows that the rotation matrix has the opposite effect on the components as it does on the basis vectors.

By applying a joint rotation to the rotated vector **v**′, we can find a rotation that is equivalent to the active rotation (D.2), but leaves the vector **v** unchanged. In this type of rotation, called a *passive rotation*, we think of the basis being rotated and the vector being held fixed, as opposed to the vector rotating and the basis being held fixed, as for an active rotation. In a passive rotation the basis vectors and the vector components must rotate in a complementary way so that the vector itself does not rotate. This relationship is expressed by saying that the components and the basis vectors are *relatively contravariant*. Applying the inverse rotation R^{-1} to **v**′ gives back the original vector **v** (Fig. D.2). Thus we can write

$$\mathbf{v} = R^{-1}\mathbf{v}' = R^{-1}\hat{\mathbf{x}}_j \mathsf{R}_{ji} v_i = \hat{\mathbf{x}}_k \mathsf{R}^{-1}_{kj} \mathsf{R}_{ji} v_i = \mathsf{R}_{jk} \hat{\mathbf{x}}_k \mathsf{R}_{ji} v_i. \tag{D.9}$$

We can therefore write the passive rotation as $\mathbf{v} = \hat{\mathbf{x}}_j'' v_j''$, with $\hat{\mathbf{x}}_j'' = \mathsf{R}_{jk}\hat{\mathbf{x}}_k$ and $v_j'' = \mathsf{R}_{ji} v_i = v_j'$. In the example of Figs. D.1 and D.2, we have

$$\begin{pmatrix} \hat{\mathbf{x}}'' \\ \hat{\mathbf{y}}'' \\ \hat{\mathbf{z}}'' \end{pmatrix} = \begin{pmatrix} \cos\theta & -\sin\theta & 0 \\ \sin\theta & \cos\theta & 0 \\ 0 & 0 & 1 \end{pmatrix} \begin{pmatrix} \hat{\mathbf{x}} \\ \hat{\mathbf{y}} \\ \hat{\mathbf{z}} \end{pmatrix} = \begin{pmatrix} \hat{\mathbf{x}}\cos\theta - \hat{\mathbf{y}}\sin\theta \\ \hat{\mathbf{x}}\sin\theta + \hat{\mathbf{y}}\cos\theta \\ 0 \end{pmatrix}, \tag{D.10}$$

and $v_j'' = v_j'$ given by Eq. (D.5). Then

$$\mathbf{v} = \begin{pmatrix} \hat{\mathbf{x}}\cos\theta - \hat{\mathbf{y}}\sin\theta & \hat{\mathbf{x}}\sin\theta + \hat{\mathbf{y}}\cos\theta & 0 \end{pmatrix} \begin{pmatrix} \cos\theta \\ \sin\theta \\ 0 \end{pmatrix} = \hat{\mathbf{x}}. \tag{D.11}$$

Here we see that because R has the opposite effect on the basis vectors as it does on the components, applying it to each has the desired effect of leaving **v** constant. Thus, even though they are relatively contravariant, in the Cartesian basis the vector components and the basis vectors transform according to the same rule.

We have found this expression by performing an active rotation of the components by an angle θ, and then a joint rotation on the basis vectors by $-\theta$. Accordingly, an active rotation on the vector components by θ is equivalent to a passive rotation of the basis vectors by $-\theta$. In our discussion, we have parameterized the rotation matrix in terms of active rotations: $\mathsf{R}_{\hat{\mathbf{n}}}(\theta)$ rotates the vector by θ when applied to the vector components. We can make a note of this by writing $\mathsf{R}_{\hat{\mathbf{n}}}^{(a)}(\theta)$. However, we normally describe a passive rotation in terms of the rotation of the basis vectors. Since $R_{\hat{n}}^{(a)}(\theta)$ rotates the basis vectors by $-\theta$, we can write the rotation matrix for passive rotations as

$$R_{\hat{n}}^{(p)}(\theta) = R_{\hat{n}}^{(a)}(-\theta) = \left(R_{\hat{n}}^{(a)}(\theta)\right)^{-1}. \tag{D.12}$$

Then an active rotation of $\mathbf{v}$ by an angle θ is given by

$$\mathbf{v}' = \hat{\mathbf{x}}_i R_{ij}^{(a)}(\theta) v_j, \tag{D.13}$$

a passive rotation of the basis vectors by an angle θ is described by

$$\mathbf{v} = R_{jk}^{(p)}(\theta) \hat{\mathbf{x}}_k R_{ji}^{(p)}(\theta) v_i, \tag{D.14}$$

and the active rotation equivalent to this passive rotation is given by

$$\mathbf{v}'' = \hat{\mathbf{x}}_j R_{ji}^{(p)}(\theta) v_i = \hat{\mathbf{x}}_j R_{ji}^{(a)}(-\theta) v_i. \tag{D.15}$$

When we omit the superscript label on the rotation operator, we will continue to mean an active rotation.

D.2 The spherical basis

Let us introduce a new set of basis vectors, defined by

$$\hat{\boldsymbol{\epsilon}}_1 = -\frac{1}{\sqrt{2}}(\hat{\mathbf{x}} + i\hat{\mathbf{y}}), \tag{D.16a}$$

$$\hat{\boldsymbol{\epsilon}}_0 = \hat{\mathbf{z}}, \tag{D.16b}$$

$$\hat{\boldsymbol{\epsilon}}_{-1} = \frac{1}{\sqrt{2}}(\hat{\mathbf{x}} - i\hat{\mathbf{y}}). \tag{D.16c}$$

The fact that these unit vectors, describing the *spherical basis*, are complex has consequences for the algebra of rotations.

The transformation between the Cartesian and spherical bases can be written in terms of a matrix U, given by

$$U = \begin{pmatrix} -\frac{1}{\sqrt{2}} & -\frac{i}{\sqrt{2}} & 0 \\ 0 & 0 & 1 \\ \frac{1}{\sqrt{2}} & -\frac{i}{\sqrt{2}} & 0 \end{pmatrix}, \tag{D.17}$$

so that

$$\begin{pmatrix} \hat{\boldsymbol{\epsilon}}_1 \\ \hat{\boldsymbol{\epsilon}}_0 \\ \hat{\boldsymbol{\epsilon}}_{-1} \end{pmatrix} = U \begin{pmatrix} \hat{\mathbf{x}}_1 \\ \hat{\mathbf{x}}_2 \\ \hat{\mathbf{x}}_3 \end{pmatrix}. \tag{D.18}$$

This can also be notated as

$$\hat{\boldsymbol{\epsilon}}_q = U_{qj} \hat{\mathbf{x}}_j, \tag{D.19}$$

keeping track of the fact that q runs from 1 to -1, while j runs from 1 to 3. Here the matrix U is *unitary*, meaning that its inverse is equal to its conjugate transpose:

$$U^{-1} = (U^*)^t = U^\dagger. \tag{D.20}$$

348 *Classical rotations*

This means that the inverse transformation is given by
$$\hat{\mathbf{x}}_j = U_{jq}^{-1}\hat{\boldsymbol{\epsilon}}_q = U_{jq}^{\dagger}\hat{\boldsymbol{\epsilon}}_q. \tag{D.21}$$

We now consider the components of **v** in the spherical basis, v^q, chosen so that they satisfy
$$\mathbf{v} = \hat{\boldsymbol{\epsilon}}_q v^q. \tag{D.22}$$

Here we have written the component index q as a superscript, for reasons to be explained below. We can treat the spherical components using a procedure similar to that used in Eq. (D.1) for the Cartesian components. Note, however, that the tensor $\hat{\boldsymbol{\epsilon}}_{q'} \cdot \hat{\boldsymbol{\epsilon}}_q$ is not equivalent to the identity tensor. Rather,
$$\hat{\boldsymbol{\epsilon}}_q^* \cdot \hat{\boldsymbol{\epsilon}}_{q'} = U_{qk}^*\hat{\mathbf{x}}_k \cdot U_{q'j}\hat{\mathbf{x}}_j = \hat{\mathbf{x}}_k \cdot \hat{\mathbf{x}}_j U_{qk}^* U_{jq'}^t = \left(U_{qj}U_{jq'}^{\dagger}\right)^* = \delta_{qq'}, \tag{D.23}$$
where we have used the fact that $\hat{\mathbf{x}}_j$ is real and U is unitary. This means that we can find the spherical components by taking the dot product between $\hat{\boldsymbol{\epsilon}}_{q'}^*$ and **v**:
$$\hat{\boldsymbol{\epsilon}}_q^* \cdot \mathbf{v} = \hat{\boldsymbol{\epsilon}}_q^* \cdot \hat{\boldsymbol{\epsilon}}_{q'} v^{q'} = v^q. \tag{D.24}$$

How are the components v^q related to the Cartesian-basis components v_j? We have
$$\mathbf{v} = \hat{\mathbf{x}}_j v_j = U_{jq}^{\dagger}\hat{\boldsymbol{\epsilon}}_q v_j = \left(U_{jq}^*\right)^t \hat{\boldsymbol{\epsilon}}_q v_j = \hat{\boldsymbol{\epsilon}}_q U_{qj}^* v_j. \tag{D.25}$$

Using $\mathbf{v} = \hat{\boldsymbol{\epsilon}}_q v^q$, we find that the components in the new basis are given by
$$v^q = U_{qj}^* v_j, \tag{D.26}$$
or, explicitly,
$$v^1 = -\frac{1}{\sqrt{2}}(v_x - iv_y), \tag{D.27a}$$
$$v^0 = v_z, \tag{D.27b}$$
$$v^{-1} = \frac{1}{\sqrt{2}}(v_x + iv_y). \tag{D.27c}$$

The inverse transformation is given by
$$v_j = U_{jq}^{\dagger*} v^q = U_{jq}^t v^q. \tag{D.28}$$

We see that the transformation that takes the vector components into the spherical basis is the complex conjugate of the one given by Eq. (D.19) that acts on the basis vectors.

Now let us consider rotations in the spherical basis. We first look at passive rotations. Following the procedure of Eq. (D.9), we can write a passive rotation as
$$\mathbf{v} = R^{-1}R\mathbf{v} = \hat{\boldsymbol{\epsilon}}_{q_2}\tilde{R}_{q_2q_1}^{-1}\tilde{R}_{q_1q}v^q, \tag{D.29}$$
where $\tilde{R}$ is the matrix for rotation in the spherical basis. In the Cartesian basis, R is orthogonal. This is not necessarily the case for $\tilde{R}$, because it may be complex. However, it is guaranteed to be unitary, i.e., $\tilde{R}^{\dagger} = \tilde{R}^{-1}$. Thus, we can write
$$\mathbf{v} = \hat{\boldsymbol{\epsilon}}_{q_2}\tilde{R}_{q_2q_1}^{-1}\tilde{R}_{q_1q}v^q = \hat{\boldsymbol{\epsilon}}_{q_2}\tilde{R}_{q_2q_1}^{\dagger}\tilde{R}_{q_1q}v^q = \tilde{R}_{q_1q_2}^*\hat{\boldsymbol{\epsilon}}_{q_2}\tilde{R}_{q_1q}v^q. \tag{D.30}$$

The spherical basis vectors and vector coefficients must be relatively contravariant in order that **v** remain unchanged under a passive rotation. In contrast to the Cartesian case, here the relatively contravariant quantities have different transformation properties—the rotation matrix for the basis vectors, $\tilde{R}^*$, is the complex conjugate of that for the vector components, $\tilde{R}$.

Note that it is not true that basis vectors always transform one way and vector components another. Taking the complex conjugate of Eq. (D.30), we have

$$\mathbf{v} = \tilde{R}_{q_1 q_2} \hat{\boldsymbol{\epsilon}}^*_{q_2} \tilde{R}^*_{q_1 q} v^{q*}, \tag{D.31}$$

where we have assumed that $\mathbf{v}$ is real for simplicity. This shows that the alternate set of unit vectors $\hat{\boldsymbol{\epsilon}}^*_q$ transforms the same way that v^q does, whereas v^{q*} transforms like $\hat{\boldsymbol{\epsilon}}_q$. In addition, from Eqs. (D.19) and (D.26) we find that $\hat{\boldsymbol{\epsilon}}^*_q$ and v^q transform to the Cartesian basis in the same way as each other, and likewise $\hat{\boldsymbol{\epsilon}}_q$ and v^{q*}. To keep track of these different types of quantities, we will call sets of quantities that transform like $\hat{\boldsymbol{\epsilon}}_q$ under passive rotations *covariant*, and sets that transform like v^q *contravariant*. As may be already clear, this distinction is indicated in the notation by writing the index on contravariant quantities as a superscript. We will discuss the relationship between covariant and contravariant quantities further below.

The matrix $\tilde{R}_{\hat{n}}(\theta)$ produces a transformation of the spherical components equivalent to an active rotation by an angle θ about $\hat{n}$. To find an expression for $\tilde{R}$, we consider an active rotation in the Cartesian basis:

$$v'_j = R_{jk} v_k. \tag{D.32}$$

Multiplying on the left by U^* and inserting $(U^{-1} U)^*$ before v_k, we have

$$(v^q)' = U^*_{qj} v'_j = U^*_{qj} R_{jk} U^{\dagger *}_{kq_1} U^*_{q_1 l} v_l = U^*_{qj} R_{jk} U^{\dagger *}_{kq_1} v^{q_1} = \tilde{R}_{qq_1} v^{q_1}, \tag{D.33}$$

with $\tilde{R} = (URU^\dagger)^*$. For example, for the rotation by an angle θ about $\hat{z}$ specified by Eq. (D.3), the corresponding rotation matrix for contravariant quantities in the spherical basis is

$$\tilde{R} = \begin{pmatrix} -\frac{1}{\sqrt{2}} & \frac{i}{\sqrt{2}} & 0 \\ 0 & 0 & 1 \\ \frac{1}{\sqrt{2}} & \frac{i}{\sqrt{2}} & 0 \end{pmatrix} \begin{pmatrix} \cos\theta & -\sin\theta & 0 \\ \sin\theta & \cos\theta & 0 \\ 0 & 0 & 1 \end{pmatrix} \begin{pmatrix} -\frac{1}{\sqrt{2}} & 0 & \frac{1}{\sqrt{2}} \\ -\frac{i}{\sqrt{2}} & 0 & -\frac{i}{\sqrt{2}} \\ 0 & 1 & 0 \end{pmatrix}$$

$$= \begin{pmatrix} e^{-i\theta} & 0 & 0 \\ 0 & 1 & 0 \\ 0 & 0 & e^{i\theta} \end{pmatrix}. \tag{D.34}$$

This shows that a rotation about the z-axis induces phase shifts in the spherical components, but does not mix them. This is an important feature of the spherical basis.

The matrix $\tilde{R}$ applies to active rotations of the contravariant components, and can be notated in full as $\tilde{R}^{(a)}_{\hat{n}}(\theta)$. In a passive rotation of the basis vectors by an angle θ, the components undergo a transformation equivalent to an active rotation of $-\theta$, as discussed in the previous section. The matrix that describes this equivalent rotation is then given by $\tilde{R}^{(p)}_{\hat{n}}(\theta) = \tilde{R}^{(a)}_{\hat{n}}(-\theta) = [\tilde{R}^{(a)}_{\hat{n}}(\theta)]^{-1}$.

We have seen above that taking the complex conjugate evidently converts covariant quantities into contravariant and vice versa. Therefore we can define the set of contravariant basis vectors

$$\hat{\boldsymbol{\epsilon}}^q = \hat{\boldsymbol{\epsilon}}^*_q = U^*_{qj} \hat{\mathbf{x}}_j, \tag{D.35}$$

which transform under passive rotations as $(\hat{\boldsymbol{\epsilon}}^q)' = \tilde{R}^{(p)}_{q'q} \hat{\boldsymbol{\epsilon}}^q$. Note, however, that we have used here the fact that the Cartesian basis vectors $\hat{\mathbf{x}}_j$ are real. If we attempt to use a similar scheme to define a set of covariant vector components, we run into problems if the vector is complex (i.e., if its Cartesian components are complex). Taking the complex conjugate of Eq. (D.22), we have

$$\mathbf{v}^* = \hat{\boldsymbol{\epsilon}}_q^* v^{q*} = \hat{\boldsymbol{\epsilon}}^q v^{q*}. \tag{D.36}$$

This shows that taking the complex conjugate of the contravariant spherical components of $\mathbf{v}$ gives a set of covariant spherical components, but they are the components of $\mathbf{v}^*$, not $\mathbf{v}$. Another problem with the conjugation method is that we will wish to generalize the discussion to include co- and contravariant components of quantities that do not possess complex conjugates.

There is another, more general, way to convert between covariant and contravariant components. Inspection of the form (D.17) of the transformation matrix U shows that its matrix elements obey the relation

$$U_{qj}^* = (-1)^q U_{-q,j}. \tag{D.37}$$

Consider a rotation applied to a contravariant set y^q, where y could represent either a vector or a basis. Because they are contravariant, the components y^q transform according to

$$(y^q)' = \tilde{R}_{qq_1} y^{q_1}, \tag{D.38}$$

where the transformation matrix $\tilde{R}$ can be written using Eq. (D.37) as

$$\tilde{R}_{qq_1} = U_{qj}^* R_{jk} U_{kq_1}^{\dagger*} = (-1)^q U_{-q,j} R_{jk} U_{k,-q_1}^{\dagger} (-1)^{q_1} = (-1)^{q+q_1} \tilde{R}_{-q,-q_1}^*. \tag{D.39}$$

Substituting back into Eq. (D.38) and flipping the signs of the indices, we have

$$[(-1)^q y^{-q}]' = \tilde{R}_{qq_1}^* (-1)^{q_1} y^{-q_1}. \tag{D.40}$$

This shows that $(-1)^q y^{-q}$ transforms as a covariant set. Likewise, if y_q is a covariant set of quantities, $(-1)^q y_{-q}$ is contravariant. For the spherical basis vectors, the contravariant set obtained in this way coincides with the above definition of $\hat{\boldsymbol{\epsilon}}^q$:

$$(-1)^q \hat{\boldsymbol{\epsilon}}_{-q} = (-1)^q U_{-q,j} \hat{\mathbf{x}}_j = U_{q,j}^* \hat{\mathbf{x}}_j = \hat{\boldsymbol{\epsilon}}_q^* = \hat{\boldsymbol{\epsilon}}^q. \tag{D.41}$$

For the spherical components, $(-1)^q v^{-q}$ is equal to v^{q*} only if $\mathbf{v}$ is real. It is clear that this set comprises the covariant components corresponding to $\hat{\boldsymbol{\epsilon}}^q$:

$$\mathbf{v} = \hat{\boldsymbol{\epsilon}}_q v^q = (-1)^q \hat{\boldsymbol{\epsilon}}_{-q} (-1)^q v^{-q} = \hat{\boldsymbol{\epsilon}}^q (-1)^q v^{-q}. \tag{D.42}$$

Therefore, we make the definition $v_q = (-1)^q v^{-q}$.

The transformation converting between covariant and contravariant sets can be written in terms of a matrix Y given by

$$Y = \begin{pmatrix} 0 & 0 & -1 \\ 0 & 1 & 0 \\ -1 & 0 & 0 \end{pmatrix}, \tag{D.43}$$

so that $U^* = YU$, $\hat{\boldsymbol{\epsilon}}^q = Y_{qq'} \hat{\boldsymbol{\epsilon}}_{q'}$, and $v_q = Y_{qq'} v^{q'}$. The transformation Y is equivalent up to an overall sign to a rotation by π about the y-axis.

Using the definition of $\hat{\boldsymbol{\epsilon}}^q$, we can write the spherical-basis orthonormality relation (D.23) as

$$\hat{\boldsymbol{\epsilon}}^q \cdot \hat{\boldsymbol{\epsilon}}_{q'} = \delta_{qq'}. \tag{D.44}$$

The scalar product in this relation is between a covariant and a contravariant vector. This is a general requirement for scalar products—because a scalar is invariant under rotations, the effect of rotating one of the vectors in the scalar product must cancel that of the other. Thus the scalar product must be between relatively contravariant quantities.

To write the scalar product of two vectors **v** and **w** in terms of their spherical components, we convert from the Cartesian components:

$$\mathbf{v} \cdot \mathbf{w} = v_j w_j = v_j U^t_{jq} w^q = U_{qj} v_j w^q = v_q w^q. \tag{D.45}$$

Employing the transformation between covariant and contravariant components, the scalar product can be written in the alternate forms:

$$\mathbf{v} \cdot \mathbf{w} = v_q w^q = v_q Y_{qq'} w_{q'} = (-1)^q v_q w_{-q} = v^q w_q = (-1)^q v^q w^{-q}. \tag{D.46}$$

E

Nonlinear magneto-optical rotation with hyperfine structure

E.1 Perturbation theory with polarization moments

The time evolution of the atomic density matrix ρ subject to the action of a time-independent Hamiltonian H is given by the Liouville equation (Eq. 12.16):

$$\dot{\rho} = \frac{1}{i\hbar}[H,\rho] - \frac{1}{2}\{\widehat{\Gamma},\rho\} + \Lambda + \text{Tr}\,(F\rho), \tag{E.1}$$

where $\widehat{\Gamma}$ is the relaxation matrix, Λ accounts for repopulation, and F is the spontaneous emission operator (Eq. 12.2). (We neglect other relaxation and repopulation mechanisms, such as spin-exchanging collisions, which may require the inclusion of additional terms.)

We will now rewrite Eq. (E.1) in terms of the polarization moments. To do this, we expand the operators appearing in the Liouville equation in terms of polarization operators. In order to describe coherences between two states, as well as polarization moments of an atomic state with a particular value of F, we must extend the definition of the polarization operators given in Sec. 5.7. Thus we define the generalized polarization operators $\mathcal{T}_q^{(\kappa)}(F_1 F_2)$ connecting states F_1 and F_2 via the orthonormality condition

$$\text{Tr}\,\mathcal{T}_q^\kappa(F_1 F_2)\mathcal{T}_{q'}^{\kappa'}(F_1' F_2')^\dagger = \delta_{\kappa\kappa'}\delta_{qq'}\delta_{F_1 F_1'}\delta_{F_2 F_2'} \tag{E.2}$$

and the phase convention

$$\mathcal{T}_q^\kappa(F_1 F_2)^\dagger = (-1)^{F_1-F_2+q}\mathcal{T}_{-q}^\kappa(F_2 F_1) = \mathcal{T}^{\kappa q}(F_1 F_2). \tag{E.3}$$

Here F is understood to represent the total angular momentum quantum number as well as any additional quantum numbers necessary to distinguish between two states with the same total angular momentum.

The reduced matrix element of $\mathcal{T}^\kappa(F_1' F_2')$ is then given by

$$\langle F_1 \| \mathcal{T}^\kappa(F_1' F_2') \| F_2 \rangle = \sqrt{2\kappa+1}\,\delta_{F_1 F_1'}\delta_{F_2 F_2'} \tag{E.4}$$

and the matrix elements of the polarization operators are given by

$$\langle F_1 m_1 | \mathcal{T}_q^\kappa(F_1' F_2') | F_2 m_2 \rangle = \sqrt{\frac{2\kappa+1}{2F_1+1}}\langle F_2 m_2 \kappa q | F_1 m_1 \rangle \delta_{F_1 F_1'}\delta_{F_2 F_2'}$$
$$= (-1)^{F_2-m_2}\langle F_1 m_1 F_2, -m_2 | \kappa q \rangle \delta_{F_1 F_1'}\delta_{F_2 F_2'}. \tag{E.5}$$

An arbitrary operator A can now be expanded according to

$$A = \sum A^{\kappa q}(F_1 F_2) T_q^{\kappa}(F_1 F_2), \qquad (E.6)$$

where $F_{1,2}$ runs over all pairs of states in the system. Using Eq. (E.5) and the orthonormality condition (E.2), we find the expansion coefficients $A^{\kappa q}(F_1 F_2)$ in terms of the Zeeman-basis matrix elements of A:

$$\begin{aligned} A^{\kappa q}(F_1 F_2) &= \text{Tr}\left[A T_q^{\kappa}(F_1 F_2)^{\dagger}\right] \\ &= \sum_{m_1 m_2} A_{F_1 m_1, F_2 m_2}\left[T_q^{\kappa}(F_1 F_2)\right]_{F_1 m_1 F_2 m_2}^* \\ &= \sum_{m_1 m_2} (-1)^{F_2 - m_2} \langle F_1 m_1 F_2, -m_2 | \kappa q \rangle A_{F_1 m_1, F_2 m_2}. \end{aligned} \qquad (E.7)$$

Performing the expansion of each operator, and using appropriate tensor product and sum rules, the equation of motion for the polarization moments is found from the Liouville equation to be

$$\begin{aligned} \dot{\rho}^{\kappa q}(F_1 F_2) = &- i(-1)^{F_1 + F_2 + \kappa} \sum \sqrt{[\kappa'][\kappa'']} \langle \kappa' q' \kappa'' q'' | \kappa q \rangle \begin{Bmatrix} \kappa' & \kappa'' & \kappa \\ F_2 & F_1 & F_3 \end{Bmatrix} \\ &\times \Bigg[\left(\frac{1}{\hbar} H^{\kappa' q'}(F_1 F_3) - \frac{i}{2}\Gamma^{\kappa' q'}(F_1 F_3)\right) \rho^{\kappa'' q''}(F_3 F_2) \\ &\quad - \rho^{\kappa' q'}(F_1 F_3)\left(\frac{1}{\hbar} H^{\kappa'' q''}(F_3 F_2) + \frac{i}{2}\widehat{\Gamma}^{\kappa'' q''}(F_3 F_2)\right) \Bigg] \\ &+ \Lambda^{\kappa q}(F_1 F_2) \\ &+ \frac{4\omega^3}{3\hbar c^3} \sum \langle F_1 \|d\| F_e \rangle \rho^{\kappa q}(F_e F_e') \langle F_e' \|d\| F_2 \rangle (-1)^{F_e + F_e' + \kappa + 1} \begin{Bmatrix} \kappa & F_2 & F_1 \\ 1 & F_e & F_e' \end{Bmatrix}, \end{aligned} \qquad (E.8)$$

where all variables not appearing on the left-hand side are summed over (the variables F_e and F_e' appearing in the last term relate to spontaneous emission and run over only those states of higher energy than $F_{1,2}$). Here the arrays enclosed in curly brackets are the $6j$ symbols, and we use the notation $[x] = 2x + 1$.

We now suppose that the total Hamiltonian is $H = H_0 + \hbar V$, where H_0 is diagonal and V is a time-independent perturbation. We also assume that Γ and Λ are diagonal. More precisely, we assume that only $\Gamma^{00}(FF)$, $\Lambda^{00}(FF)$, and $H^{00}(FF)$ are nonzero (for arbitrary F). Taking the steady-state limit in Eq. (E.8) and expanding to second order in the perturbation V, we find for a ground-state polarization moment

$$\rho^{\kappa q}(F_g F_g) = \frac{\gamma}{i\widetilde{\omega}_{FF} N_g} \Bigg[\delta_{\kappa 0} \delta_{q 0} \sqrt{[F_g]} - (-1)^{2F_g + \kappa' + \kappa''} \sqrt{[\kappa'][\kappa'']}$$

$$\times \langle \kappa' q' \kappa'' q'' | \kappa q \rangle \begin{Bmatrix} \kappa'' & \kappa' & \kappa \\ F_g & F_g & F' \end{Bmatrix} \frac{\widetilde{\omega}_{F_g F'} + \widetilde{\omega}_{F' F_g}}{\widetilde{\omega}_{F_g F_g} \widetilde{\omega}_{F_g F'} \widetilde{\omega}_{F' F_g}}$$

$$\times V^{\kappa' q'}(F' F_g) V^{\kappa'' q''}(F_g F')$$

$$- i \frac{4\omega^3}{3\hbar c^3} (-1)^{2F_1' + 2F_2' + \kappa + \kappa' + \kappa''} \sqrt{[\kappa'][\kappa'']} \langle \kappa' q' \kappa'' q'' | \kappa q \rangle$$

$$\times \begin{Bmatrix} \kappa'' & \kappa' & \kappa \\ F_2' & F_1' & F' \end{Bmatrix} \begin{Bmatrix} \kappa & F_g & F_g \\ 1 & F_1' & F_2' \end{Bmatrix} \langle F_2' \| d \| F_g \rangle \langle F_g \| d \| F_1' \rangle$$

$$\times \frac{\widetilde{\omega}_{F_1' F'} + \widetilde{\omega}_{F' F_2'}}{\widetilde{\omega}_{F_1' F'} \widetilde{\omega}_{F_1' F_2'} \widetilde{\omega}_{F' F_1'} \widetilde{\omega}_{F' F_2'}} V^{\kappa'' q''}(F_1' F') V^{\kappa' q'}(F' F_2') \Bigg]. \tag{E.9}$$

Here we have neglected the possibility of cascade decays and assumed that V does not couple a state to itself. We have also assumed that Λ repopulates all ground-state sublevels equally; i.e. $\Lambda_{Fm,Fm} = \gamma/N_g$, where γ is the ground-state relaxation rate and N_g is the total number of ground-state sublevels. The complex frequency splitting $\widetilde{\omega}_{F_1 F_2}$ is given by

$$\widetilde{\omega}_{F_1 F_2} = \frac{1}{\hbar}(E_{F_1} - E_{F_2}) - \frac{i}{2}(\Gamma_{F_1} + \Gamma_{F_2}), \tag{E.10}$$

where $E_F = (2F+1)^{-1/2} H_0^{00}(FF)$ is the unperturbed energy and

$$\Gamma_F = (2F+1)^{-1/2} \Gamma^{00}(FF) \tag{E.11}$$

is the total relaxation rate of a state F.

E.2 Doppler-free transit effect

We now apply the results obtained in Appendix E.1 to the three-stage calculation described in Sec. 14.1. In stage (a), we consider a z-polarized light field $\mathcal{E} = \mathcal{E}_0 \, \text{Re}\left(\hat{\varepsilon} e^{i(\mathbf{k}\cdot\mathbf{r}-\omega t)}\right)$ with $\hat{\varepsilon} = \hat{z}$. We let V represent the electric-dipole Hamiltonian in the rotating-wave approximation: $V' = -\frac{1}{2} d_z \mathcal{E}_0$. (Here the prime refers to the rotating frame.) We assume that the magnetic field is absent in this stage. From Eq. (E.9) we find

$$\rho_a^{(20)}(F_g F_g) = -\sqrt{\frac{2}{3}} \sum_{F_e} (-1)^{F_g - F_e} \widetilde{\kappa}_2 \frac{[F_e][F_g]}{[I][J_g]}$$

$$\times \Bigg((-1)^{2I + 2J_g} \begin{Bmatrix} 1 & 1 & 2 \\ F_g & F_g & F_e \end{Bmatrix} \begin{Bmatrix} J_e & F_e & I \\ F_g & J_g & 1 \end{Bmatrix}^2 L(\omega_{F_e F_g}')$$

$$+ R \sum_{F_g' F_e'} (-1)^{F_g' - F_e'} [J_e][F_g'][F_e'] \begin{Bmatrix} 1 & 1 & 2 \\ F_e & F_e' & F_g' \end{Bmatrix} \begin{Bmatrix} F_g & F_g & 2 \\ F_e & F_e' & 1 \end{Bmatrix} \begin{Bmatrix} J_e & F_e & I \\ F_g & J_g & 1 \end{Bmatrix}$$

$$\times \begin{Bmatrix} J_e & F_e & I \\ F_g' & J_g & 1 \end{Bmatrix} \begin{Bmatrix} J_e & F_e' & I \\ F_g & J_g & 1 \end{Bmatrix} \begin{Bmatrix} J_e & F_e' & I \\ F_g' & J_g & 1 \end{Bmatrix} \frac{L(\omega_{F_e F_g'}') L(\omega_{F_e' F_g'}')}{L\left(\sqrt{\omega_{F_e F_g'}' \omega_{F_e' F_g}'}\right)} \Bigg), \tag{E.12}$$

where all variables are as defined in Sec. 14.1. We have evaluated matrix elements using the Wigner–Eckart theorem and have used the relation (see, for example, Sobelman 1992)

$$R\Gamma = \frac{4\omega^3}{3\hbar c^3} \frac{1}{2J_e+1} |\langle J_g \| d \| J_e \rangle|^2. \quad \text{(E.13)}$$

The unperturbed energies can be evaluated with

$$E_{JFM} = E_J + \frac{1}{2} K_{IJF} A_J + \frac{3}{8} \frac{K_{IJF}(K_{IJF}+1) - \frac{4}{3} I(I+1) J(J+1)}{I(2I-1) J(2J-1)} B_J, \quad \text{(E.14)}$$

where $K_{IJF} = F(F+1) - I(I+1) - J(J+1)$ and A_J and B_J are the hyperfine coefficients. The last term is zero for $J \leq 1/2$ or $I \leq 1/2$.

In the case in which the excited-state hfs is well resolved in the Doppler-free spectrum ($\omega_{F_e F_e'} \gg \Gamma$), Eq. (E.12) reduces to

$$\rho_a^{(20)}(F_g F_g) = -\sqrt{\frac{2}{3}} \sum_{F_e} (-1)^{F_g - F_e} \tilde{\kappa}_2 \frac{[F_e][F_g]}{[I][J_g]}$$

$$\times \left((-1)^{2I+2J_g} \begin{Bmatrix} 1 & 1 & 2 \\ F_g & F_g & F_e \end{Bmatrix} \begin{Bmatrix} J_e & F_e & I \\ F_g & J_g & 1 \end{Bmatrix}^2 L(\omega'_{F_e F_g}) \right.$$

$$+ \sum_{F_g'} R \, (-1)^{F_g' - F_e} [J_e][F_g'][F_e] \begin{Bmatrix} 1 & 1 & 2 \\ F_e & F_e & F_g' \end{Bmatrix} \quad \text{(E.15)}$$

$$\left. \times \begin{Bmatrix} F_g & F_g & 2 \\ F_e & F_e & 1 \end{Bmatrix} \begin{Bmatrix} J_e & F_e & I \\ F_g & J_g & 1 \end{Bmatrix}^2 \begin{Bmatrix} J_e & F_e & I \\ F_g' & J_g & 1 \end{Bmatrix}^2 L(\omega'_{F_e F_g'}) \right).$$

In stage (b), the ground-state density matrix, which is initially in the state found in stage (a), evolves under the influence of a magnetic field $B\hat{\mathbf{x}}$. We will require only the value of the polarization moment $\rho_b^{(21)}(F_g F_g)$. Using the Hamiltonian $H_B = -\boldsymbol{\mu} \cdot \mathbf{B}$ in Eq. (E.8) and solving for the steady state, we find

$$\rho_b^{(21)}(F_g F_g) = i \frac{\sqrt{3}}{2\sqrt{2}} x_{F_g} \rho_a^{(20)}(F_g F_g), \quad \text{(E.16)}$$

where the magnetic-resonance lineshape parameter x_{F_g} is defined in Eq. (14.4).

In stage (c) the ground-state polarization is probed. The optical rotation for weak linearly polarized probe light is given by (Eq. 12.57)

$$\frac{d\mathcal{O}}{d\ell} = -\frac{2\pi \omega n}{\hbar c} \text{Im} \left[\hat{\boldsymbol{\varepsilon}} \cdot \overset{\leftrightarrow}{\boldsymbol{\beta}} \cdot (\hat{\mathbf{k}} \times \hat{\boldsymbol{\varepsilon}}) \right], \quad \text{(E.17)}$$

where $\overset{\leftrightarrow}{\boldsymbol{\beta}}$ is a tensor defined in Eq. (12.58) that depends on the ground-state density matrix. Expanding $\overset{\leftrightarrow}{\boldsymbol{\beta}}$ in terms of the ground-state polarization moments, we obtain

$$\overset{\leftrightarrow}{\boldsymbol{\beta}} = \sum_{F_g F_e \kappa q' q''} \frac{(-1)^{F_g + F_e + \kappa}}{\tilde{\omega}'_{F_e F_g}} \hat{\boldsymbol{\varepsilon}}_{-q'} \hat{\boldsymbol{\varepsilon}}_{-q''} \langle 1 q' 1 q'' | \kappa, q' + q'' \rangle \begin{Bmatrix} 1 & 1 & \kappa \\ F_g & F_g & F_e \end{Bmatrix}$$

$$\times |\langle F_g \| d \| F_e \rangle|^2 \rho^{(\kappa, q' + q'')}(F_g F_g), \quad \text{(E.18)}$$

where $\hat{\epsilon}_q$ are the spherical basis vectors. Evaluating (E.17) for z-polarized light and using Eq. (E.16) gives

$$\ell_0 \frac{d\alpha}{d\ell} = -\frac{3\sqrt{3}}{4\sqrt{2}} \sum_{F_g F_e} (-1)^{F_g+F_e} [F_g][F_e][J_g] \begin{Bmatrix} 1 & 1 & 2 \\ F_g & F_g & F_e \end{Bmatrix} \begin{Bmatrix} J_e & F_e & I \\ F_g & J_g & 1 \end{Bmatrix}^2 \qquad \text{(E.19)}$$
$$\times L(\omega'_{F_e F_g}) x_{F_g} \rho_a^{(20)}(F_g F_g),$$

where the unsaturated absorption length for the $J_g \to J_e$ transition is defined in Eq. (14.5). Substituting in Eq. (E.12) results in the full expression for optical rotation due to the Doppler-free transit effect:

$$\ell_0 \frac{d\alpha}{d\ell} = \frac{3}{4} \tilde{\kappa}_2 \sum_{F_g F_e F_e''} (-1)^{2F_g + F_e'' - F_e} \frac{[F_e][F_e''][F_g]^2}{[I]} \begin{Bmatrix} 1 & 1 & 2 \\ F_g & F_g & F_e'' \end{Bmatrix} \begin{Bmatrix} J_e & F_e'' & I \\ F_g & J_g & 1 \end{Bmatrix}^2 x_{F_g}$$

$$\times \left((-1)^{2I+2J_g} \begin{Bmatrix} 1 & 1 & 2 \\ F_g & F_g & F_e \end{Bmatrix} \begin{Bmatrix} J_e & F_e & I \\ F_g & J_g & 1 \end{Bmatrix}^2 L(\omega'_{F_e F_g}) L(\omega'_{F_e'' F_g}) \right.$$

$$+ R \sum_{F_g' F_e'} (-1)^{F_g' - F_e'} [J_e][F_g'][F_e'] \begin{Bmatrix} 1 & 1 & 2 \\ F_e & F_e' & F_g' \end{Bmatrix} \begin{Bmatrix} F_g & F_g & 2 \\ F_e & F_e' & 1 \end{Bmatrix} \begin{Bmatrix} J_e & F_e & I \\ F_g & J_g & 1 \end{Bmatrix}$$

$$\left. \times \begin{Bmatrix} J_e & F_e & I \\ F_g' & J_g & 1 \end{Bmatrix} \begin{Bmatrix} J_e & F_e' & I \\ F_g & J_g & 1 \end{Bmatrix} \begin{Bmatrix} J_e & F_e' & I \\ F_g' & J_g & 1 \end{Bmatrix} \frac{L(\omega'_{F_e F_g'}) L(\omega'_{F_e' F_g'}) L(\omega'_{F_e'' F_g})}{L\left(\sqrt{\omega'_{F_e F_g'} \omega'_{F_e' F_g'}}\right)} \right). \qquad \text{(E.20)}$$

For completely resolved hfs ($\omega_{F_e F_e'}, \omega_{F_g F_g'} \gg \Gamma$), this reduces to

$$\ell_0 \frac{d\alpha}{d\ell} = \frac{3}{4} \tilde{\kappa}_2 \sum_{F_g F_e} (-1)^{2F_g} \frac{[F_e]^3 [F_g]^3}{[I]} \begin{Bmatrix} 1 & 1 & 2 \\ F_g & F_g & F_e \end{Bmatrix} \begin{Bmatrix} J_e & F_e & I \\ F_g & J_g & 1 \end{Bmatrix}^4 x_{F_g} [L(\omega'_{F_e F_g})]^2$$

$$\times \left(\frac{(-1)^{2I+2J_g}}{[F_e][F_g]} \begin{Bmatrix} 1 & 1 & 2 \\ F_g & F_g & F_e \end{Bmatrix} \right. \qquad \text{(E.21)}$$

$$\left. + R(-1)^{F_g - F_e} (2J_e + 1) \begin{Bmatrix} 1 & 1 & 2 \\ F_e & F_e & F_g \end{Bmatrix} \begin{Bmatrix} F_g & F_g & 2 \\ F_e & F_e & 1 \end{Bmatrix} \begin{Bmatrix} J_e & F_e & I \\ F_g & J_g & 1 \end{Bmatrix}^2 \right).$$

E.3 Doppler-broadened transit effect

The procedure used to obtain the optical rotation signal in the Doppler-broadened case is described in Sec. 14.2. When the ground- and excited-state hyperfine splittings are all much greater than the natural width ($\omega_{F_e F_e'}, \omega_{F_g F_g'}, \Gamma_D \gg \Gamma$) we have, applying the integration procedure to Eq. (E.21),

$$\ell_0 \frac{d\alpha}{d\ell} = \frac{3}{8} \tilde{\kappa}_2 \sum_{F_g F_e} (-1)^{2F_g} \frac{[F_e]^3 [F_g]^3}{[I]} \begin{Bmatrix} 1 & 1 & 2 \\ F_g & F_g & F_e \end{Bmatrix} \begin{Bmatrix} J_e & F_e & I \\ F_g & J_g & 1 \end{Bmatrix}^4 e^{-(\Delta_{F_e F_g}/\Gamma_D)^2} x_{F_g}$$

$$\times \left(\frac{(-1)^{2I+2J_g}}{[F_e][F_g]} \begin{Bmatrix} 1 & 1 & 2 \\ F_g & F_g & F_e \end{Bmatrix} + R(-1)^{F_g - F_e} [J_e] \begin{Bmatrix} 1 & 1 & 2 \\ F_e & F_e & F_g \end{Bmatrix} \begin{Bmatrix} F_g & F_g & 2 \\ F_e & F_e & 1 \end{Bmatrix} \begin{Bmatrix} J_e & F_e & I \\ F_g & J_g & 1 \end{Bmatrix}^2 \right), \qquad \text{(E.22)}$$

where the unsaturated absorption length for the Doppler-broadened case is given by Eq. (14.12).

In a different limit in which the ground-state hyperfine splittings are much greater than the natural width and the excited-state splittings are much smaller than the Doppler width ($\omega_{F_e F'_e}, \Gamma \ll \Gamma_D, \Gamma \ll \omega_{F_g F'_g}$), we have

$$\ell_0 \frac{d\alpha}{d\ell} = \frac{3}{8}\tilde{\kappa}_2 \sum_{F_g F_e F''_e} (-1)^{2F_g + F''_e - F_e} \frac{[F_e][F''_e][F_g]^2}{[I]} \begin{Bmatrix} 1 & 1 & 2 \\ F_g & F_g & F''_e \end{Bmatrix} \begin{Bmatrix} J_e & F_e & I \\ F_g & J_g & 1 \end{Bmatrix}^2 \begin{Bmatrix} J_e & F''_e & I \\ F_g & J_g & 1 \end{Bmatrix}^2$$

$$\times x_{F_g} \left((-1)^{2I+2J_g} \begin{Bmatrix} 1 & 1 & 2 \\ F_g & F_g & F_e \end{Bmatrix} + R \sum_{F'_e} (-1)^{F_g - F'_e} [J_e][F_g][F'_e] \begin{Bmatrix} 1 & 1 & 2 \\ F_e & F'_e & F_g \end{Bmatrix} \right.$$

$$\left. \times \begin{Bmatrix} F_g & F_g & 2 \\ F_e & F'_e & 1 \end{Bmatrix} \begin{Bmatrix} J_e & F'_e & I \\ F_g & J_g & 1 \end{Bmatrix}^2 \frac{2\Gamma^4 + (2\Gamma^2 + \omega^2_{F_e F'_e})\omega_{F_e F''_e}\omega_{F'_e F''_e}}{2(\Gamma^2 + \omega^2_{F_e F'_e})(\Gamma^2 + \omega^2_{F'_e F''_e})} \right) \frac{e^{-(\Delta_{F_g}/\Gamma_D)^2}\Gamma^2}{\Gamma^2 + \omega^2_{F_e F''_e}}.$$

(E.23)

E.4 Wall effect

The procedure for obtaining the signal in the wall-effect case is described in Sec. 14.3. For excited-state hyperfine splittings much greater than the natural width ($\Gamma \ll \omega_{F_e F'_e}, \Gamma_D$), we have for the ground-state polarization

$$\rho^{(20)}_a(F_g F_g) = -\sqrt{\frac{\pi}{6}} \sum_{F_e} (-1)^{F_g - F_e} \tilde{\kappa}_2 \frac{[F_e][F_g]}{[I][J_g]} \begin{Bmatrix} J_e & F_e & I \\ F_g & J_g & 1 \end{Bmatrix}^2$$

$$\times \left((-1)^{2I+2J_g} \begin{Bmatrix} 1 & 1 & 2 \\ F_g & F_g & F_e \end{Bmatrix} e^{-(\Delta_{F_e F_g}/\Gamma_D)^2} \right.$$

(E.24)

$$\left. + R \sum_{F'_g} (-1)^{F'_g - F_e} [J_e][F'_g][F_e] \begin{Bmatrix} 1 & 1 & 2 \\ F_e & F_e & F'_g \end{Bmatrix} \begin{Bmatrix} F_g & F_g & 2 \\ F_e & F_e & 1 \end{Bmatrix} \begin{Bmatrix} J_e & F_e & I \\ F'_g & J_g & 1 \end{Bmatrix}^2 \right),$$

where the saturation parameter for the wall effect is defined by Eq. (14.17). The optical rotation signal is then given by

$$\ell_0 \frac{d\alpha}{d\ell} = -\frac{3\sqrt{3}}{4\sqrt{2}} \sum_{F_g F'_e} (-1)^{F_g + F'_e} [F_g][F'_e][J_g] \begin{Bmatrix} 1 & 1 & 2 \\ F_g & F_g & F'_e \end{Bmatrix} \begin{Bmatrix} J_g & F_g & I \\ F'_e & J_e & 1 \end{Bmatrix}^2 e^{-(\Delta_{F'_e F_g}/\Gamma_D)^2}$$

$$\times x_{F_g} \rho^{(20)}_a(F_g F_g)$$

$$= \frac{3\sqrt{\pi}}{8}\tilde{\kappa}_2 \sum_{F_g F_e F'_e} (-1)^{2F_g + F'_e - F_e} \frac{[F_e][F'_e][F_g]^2}{[I]} \begin{Bmatrix} 1 & 1 & 2 \\ F_g & F_g & F'_e \end{Bmatrix} \begin{Bmatrix} J_e & F_e & I \\ F_g & J_g & 1 \end{Bmatrix}^2 \begin{Bmatrix} J_e & F'_e & I \\ F_g & J_g & 1 \end{Bmatrix}^2$$

$$\times x_{F_g} e^{-(\Delta_{F'_e F_g}/\Gamma_D)^2} \left((-1)^{2I+2J_g} \begin{Bmatrix} 1 & 1 & 2 \\ F_g & F_g & F_e \end{Bmatrix} e^{-(\Delta_{F_e F_g}/\Gamma_D)^2} \right.$$

$$\left. + R \sum_{F'_g} (-1)^{F'_g - F_e} [J_e][F_e][F'_g] \begin{Bmatrix} 1 & 1 & 2 \\ F_e & F_e & F'_g \end{Bmatrix} \begin{Bmatrix} F_g & F_g & 2 \\ F_e & F_e & 1 \end{Bmatrix} \begin{Bmatrix} J_e & F_e & I \\ F'_g & J_g & 1 \end{Bmatrix}^2 e^{-(\Delta_{F_e F'_g}/\Gamma_D)^2} \right).$$

(E.25)

F
The Atomic Density Matrix software package

For many calculations and figures presented in this book, we used a *Mathematica* package written by one of the authors (S.M.R.). Although use of *Mathematica* is by no means necessary for using the book, the readers who are familiar with this software system may find the package useful for performing practical calculations and/or working out specific examples. The package is available at the book web site (http://ukcatalogue.oup.com/product/9780199565122.do, also linked from the Oxford web page http://www.oup.co.uk/). The book web site and the Berkeley authors' web site http://budker.berkeley.edu/Tutorials also contain a number of tutorials and examples that can be used with *Mathematica* or viewed using free *Mathematica* Player software available from Wolfram Research (http://www.wolfram.com/products/player/).

In this Appendix, we provide a brief description of the capabilities and the underlying principles behind the *Atomic Density Matrix* (*ADM*) package. A more detailed description is accessible via the documentation supplied with the package.

The package is designed to facilitate analytic and numerical simulations of atoms interacting with light and other external fields. The first step in using the package is to define a list of the atomic states making up the system, with all relevant angular-momentum quantum numbers specified. (The package automatically generates hyperfine and Zeeman sublevels of the specified states.) This list of states is then provided to various functions that generate the atomic density matrix, the Hamiltonian, and the system of Liouville equations. The evolution equations can then be solved using built-in *Mathematica* routines, or specialized methods built into the *ADM* package. Once the density matrix is known, functions can be used to find observed signals in transmitted or fluorescent light.

Calculations can be performed in either the Zeeman basis or the basis of polarization moments, and functions are provided to translate between these two representations. Various aspects of tensor algebra are implemented, including inner (scalar) and outer (direct) products, tensor products, and rotations.

The package includes visualization routines, including a function that draws angular-momentum probability surfaces, and one that automatically draws a level diagram for a specified atomic system.

It is also possible to address more complicated experimental systems, in which experimental conditions vary as a function of position or atomic velocity, and atoms travel between the different regions. A list of experimental regions can be provided to *ADM* functions along with the list of atomic states to generate evolution equations for multiple coupled density matrices describing the entire system. This can be used, for example, to do nonperturbative calculations for antirelaxation-coated vapor cells, in which atoms in the light beam are described by a different density matrix than those in the unilluminated region of the cell, and the density matrices for atoms in different velocity groups are coupled due to velocity-changing collisions.

The package is designed, as much as possible, to avoid the use of specific formulas derived for particular purposes, and to rather make calls to a function that uses the Wigner-Eckart theorem to return matrix elements of an appropriate irreducible tensor operator. This approach has two benefits: it lessens the chance of errors appearing as formulas are entered into the code, and it makes the package more easily extensible. For example, it is possible to define a new Hamiltonian simply by specifying the reduced matrix element of the corresponding operator.

Bibliography

Abragam, A. (1962). *The Principles of Nuclear Magnetism*. International Series of Monographs on Physics, Claredon, Oxford.

Acosta, V., Ledbetter, M. P., Rochester, S. M., Budker, D., Jackson Kimball, D. F., Hovde, D. C., Gawlik, W., Pustelny, S., Zachorowski, J. and Yashchuk, V. V. (2006). *Physical Review A*, **73**, 053404.

Acosta, V. M., Auzinsh, M., Gawlik, W., Grisins, P., Higbie, J. M., Kimball, D. F. J., Krzemien, L., Ledbetter, M. P., Pustelny, S., Rochester, S. M., Yashchuk, V. V. and Budker, D. (2008). *Optics Express*, **16**, 11423.

Aleksandrov, E. (1964). *Optics and Spectroscopy (USSR)*, **17**, 522.

Aleksandrov, E. B., Balabas, M. V., Vershovskii, A. K. and Pazgalev, A. S. (1999). *Technical Physics*, **44**, 1025.

Aleksandrov, E. B., Chaika, M. P. and Khvostenko, G. I. (1993). *Interference of Atomic States*. Springer Series on Atoms and Plasmas; 7, Springer-Verlag, Berlin.

Alexandrov, E. B., Auzinsh, M., Budker, D., Kimball, D. F., Rochester, S. M. and Yashchuk, V. V. (2005). *Journal of the Optical Society of America B*, **22**, 7.

Alexandrov, E. B., Balabas, M. V., Budker, D., English, D., Kimball, D. F., Li, C. H. and Yashchuk, V. V. (2002). *Physical Review A*, **66**, 042903.

Alexandrov, E. B., Balabas, M. V., Pasgalev, A. S., Vershovskii, A. K. and Yakobson, N. N. (1996). *Laser Physics*, **6**, 244.

Alexandrov, E. B. and Bonch-Bruevich, V. A. (1992). *Optical Engineering*, **31**, 711.

Alexandrov, E. B., Pazgalev, A. S. and Rasson, J. L. (1997). *Optics and Spectroscopy (USSR)*, **82**, 14.

Allen, L., Beijersbergen, M. W., Spreeuw, R. J. C. and Woerdman, J. P. (1992). *Physical Review A*, **45**, 8185.

Allen, L. and Eberly, J. H. (1987). *Optical Resonance and Two-Level Atoms*. Dover, New York.

Allen, L. and Padgett, M. (2002). *American Journal of Physics*, **70**, 567.

Arfken, G. B. and Weber, H.-J. (2005). *Mathematical Methods for Physicists*. Elsevier, Boston, 6th ed.

Arimondo, E. (1996). E. Wolf (ed.), *Progess in Optics*, Elsevier Science B.V., New York, vol. XXXV, pp. 259–354.

Armstrong, L. (1971). *Theory of the Hyperfine Structure of Free Atoms*. Wiley-Interscience, New York.

Atkins, P. W. and Friedman, R. S. (1996). *Molecular Quantum Mechanics*. Oxford University Press, New York, 3rd ed.

Audoin, C. and Guinot, B. (2001). *The Measurement of Time: Time, Frequency, and the Atomic Clock*. Cambridge University Press, Cambridge.

Auzinsh, M. (1997). *Canadian Journal of Physics*, **75**, 853.

——— (1999). *Canadian Journal of Physics*, **77**, 491.

Auzinsh, M., Blushs, K., Ferber, R., Gahbauer, F., Jarmola, A. and Tamanis, M. (2006). *Physical Review Letters*, **97**, 043002.

Auzinsh, M., Budker, D. and Rochester, S. M. (2009a). *Physical Review A*, **80**, 053406.

Auzinsh, M. and Ferber, R. (1992). *Physical Review Letters*, **69**, 3464.

——— (1995). *Optical Polarization of Molecules*. Cambridge University Press, Cambridge.

Auzinsh, M., Ferber, R., Gahbauer, F., Jarmola, A. and Kalvans, L. (2009b). *Physical Review A*, **79**, 053404.

Auzinsh, M. P. (2004). A. Lagana and G. Lendvay (eds.), *Theory of Chemical Reaction Dynamics*, Kluwer Academic/Plenum Publishers, New York, pp. 447–466.

Auzinsh, M. P. and Ferber, R. S. (1991). *Physical Review A*, **43**, 2374.

Auzinsh, M. P., Tamanis, M. Y. and Ferber, R. S. (1986). *Zhurnal Eksperimentalnoi I Teoreticheskoi Fiziki*, **90**, 1182.

Baird, P. E. G., Irie, M. and Wolfenden, T. D. (1989). *Journal of Physics B*, **22**, 1733.

Balabas, M. V., Karaulanov, T., Ledbetter, M. P. and Budker, D. (2010). *Physical Review Letters*, **105**, 070801.

Barkov, L. M., Melik-Pashayev, D. A. and Zolotorev, M. S. (1989). *Opt. Comm.*, **70**, 467.

Barrat, J. P. and Cohen-Tannoudji, C. (1961a). *Journal de Physique et Le Radium*, **22**, 329.

——— (1961b). *Journal de Physique et Le Radium*, **22**, 443.

Barrow, J. D. (2002). *The Constants of Nature: From Alpha to Omega–the Numbers That Encode the Deepest Secrets of the Universe*. Pantheon Books, New York.

Beijersbergen, M. W., Allen, L., van der Veen, H. and Woerdman, J. P. (1993). *Optics Communications*, **96**, 123.

Bell, W. and Bloom, A. (1961). *Physical Review Letters*, **6**, 280.

Beloy, K., Derevianko, A., Dzuba, V. A. and Flambaum, V. V. (2009). *Physical Review Letters*, **102**, 120801.

Bergmann, K., Theuer, H. and Shore, B. W. (1998). *Reviews of Modern Physics*, **70**, 1003.

Berry, M. V. (1984). *Proceedings of the Royal Society of London, Series A*, **392**, 45.

Bethe, H. A. and Salpeter, E. E. (1977). *Quantum Mechanics of One-and Two-Electron Atoms*. Plenum, New York.

Bluhm, R., Kostelecky, V. A. and Porter, J. A. (1996). *American Journal of Physics*, **64**, 944.

Blum, K. (1996). *Density Matrix Theory and Applications*. Physics of Atoms and Molecules, Plenum Press, New York, 2nd ed.

Blushs, K. and Auzinsh, M. (2004). *Physical Review A*, **69**, 063806.

Born, M. and Wolf, E. (1999). *Principles of Optics: Electromagnetic Theory of Propagation, Interference and Diffraction of Light*. Cambridge University Press, Cambridge, 7th ed.

Bouchiat, M. A. and Brossel, J. (1966). *Physical Review*, **147**, 41.

Boyd, R. W. (2008). *Nonlinear Optics*. Academic Press, Amsterdam, 3rd ed.

Brandt, S. (1999). *Data Analysis: Statistical and Computational Methods for Scientists and Engineers*. Springer, New York, 3rd ed.

Bransden, B. H. and Joachain, C. J. (2003). *Physics of Atoms and Molecules*. Prentice Hall, Harlow, 2nd ed.

Brink, D. M. and Satchler, G. R. (1993). *Angular Momentum*. Clarendon Press, Oxford.

Budker, D., Gawlik, W., Kimball, D. F., Rochester, S. M., Yashchuk, V. V. and Weis, A. (2002a). *Reviews of Modern Physics*, **74**, 1153.

Budker, D., Kimball, D. F. and DeMille, D. (2008). *Atomic Physics. An Exploration through Problems and Solutions*. Oxford University Press, Oxford, 2nd ed.

Budker, D., Kimball, D. F., Rochester, S. M. and Urban, J. T. (2003). *Chemical Physics Letters*, **378**, 440.

Budker, D., Kimball, D. F., Rochester, S. M. and Yashchuk, V. V. (2000a). *Physical Review Letters*, **85**, 2088.

——— (2002b). *Physical Review A*, **65**, 033401.

Budker, D., Kimball, D. F., Rochester, S. M., Yashchuk, V. V. and Zolotorev, M. (2000b). *Physical Review A*, **62**, 043403.

Budker, D., Kimball, D. F., Yashchuk, V. V. and Zolotorev, M. (2002c). *Physical Review A*, **65**, 055403.

Budker, D., Orlando, D. J. and Yashchuk, V. (1999). *American Journal of Physics*, **67**, 584.

Budker, D. and Rochester, S. M. (2004). *Physical Review A*, **70**, 25804.

Budker, D. and Romalis, M. V. (2007). *Nature Physics*, **3**, 227.

Budker, D., Yashchuk, V. and Zolotorev, M. (1998). *Physical Review Letters*, **81**, 5788.

Carruthers, P. and Nieto, M. M. (1968). *Reviews of Modern Physics*, **40**, 411.

Chen, X., Telegdi, V. L. and Weis, A. (1990). *Optics Communications*, **74**, 301.

Childs, W. J. (1992). *Physics Reports*, **211**, 113.

Chin, C., Leiber, V., Vuletic, V., Kerman, A. J. and Chu, S. (2001). *Physical Review A*, **63**, 033401.

Cho, D., Wood, C. S., Bennett, S. C., Roberts, J. L. and Wieman, C. E. (1997). *Physical Review A*, **55**, 1007.

Cingöz, A., Lapierre, A., Nguyen, A.-T., Leefer, N., Budker, D., Lamoreaux, S. K. and Torgerson, J. R. (2007). *Physical Review Letters*, **98**, 040801.

Cohen-Tannoudji, C. (1962a). *Annales de Physique*, **7**, 423.

——— (1962b). *Annales de Physique*, **7**, 469.

——— (1994). *Atoms in Electromagnetic Fields*. World Scientific, Singapore.

Cohen-Tannoudji, C. and Dupont-Roc, J. (1969). *Optics Communications*, **1**, 184.

——— (1972). *Physical Review A*, **5**, 968.

Cohen-Tannoudji, C., Dupont-Roc, J. and Grynberg, G. (1989). *Photons and Atoms : Introduction to Quantum Electrodynamics*. Wiley, New York.

Condon, E. U. and Shortley, G. (1951). *The Theory of Atomic Spectra*. Cambridge University Press, Cambridge.

Corney, A. (1988). *Atomic and Laser Spectroscopy*. Clarendon, Oxford.

Coulier, N., Neyens, G., Teughels, S., Balabanski, D. L., Coussement, R., Georgiev, G., Ternier, S., Vyvey, K. and Rogers, W. F. (1999). *Physical Review C (Nuclear Physics)*, **59**, 1935.

Das, T. P. and Hahn, E. L. (1958). *Nuclear Quadrupole Resonance Spectroscopy*. Solid State Physics. Supplement 1, Academic Press, New York.

Davies, I. O. G., Baird, P. E. G. and Nicol, J. L. (1987). *Journal of Physics B*, **20**, 5371.

DeBenedetti, S. (1964). *Nuclear Interactions*. John Wiley, New York.

Dehmelt, H. (1990). *American Journal of Physics*, **58**, 17.

DeMille, D., Budker, D., Derr, N. and Deveney, E. (2000). R. C. Hilborn and G. Tino (eds.), *Spin-Statistics Connection and Commutation Relations*, American Institute of Physics Conference Proceedings, vol. 545, pp. 227–40.

Demtröder, W. (1996). *Laser spectroscopy: basic concepts and instrumentation*. Springer, Berlin, 2nd ed.

Dodd, J. N., Kaul, R. and Warrington, D. (1964). *Proceedings of the Physical Society*, **84**, 176.

Donley, E. A., Long, J. L., Liebisch, T. C., Hodby, E. R., Fisher, T. A. and Kitching, J. (2009). *Physical Review A*, **79**, 013420.

Dovator, N. A. and Okunevich, A. I. (2001). *Optics and Spectroscopy (USSR)*, **90**, 23.

Dowling, J. P., Agarwal, G. S. and Schleich, W. P. (1994). *Physical Review A*, **49**, 4101.

Drake, G. W. F. (2006). *Springer Handbook of Atomic, Molecular, and Optical Physics*. Springer, New York.

Ducloy, M. (1976). *Journal of Physics B*, **9**, 357.

Ducloy, M., Gorza, M. P. and Decomps, B. (1973). *Optics Communications*, **8**, 21.

Dyakonov, M. I. (1964). *Zhurnal Eksperimentalnoi I Teoreticheskoi Fiziki*, **47**, 2213.

——— (1965). *Soviet Physics JETP*, **20**, 1484.

Edmonds, A. R. (1974). *Angular Momentum in Quantum Mechanics*. Investigations in Physics, 4, Princeton University Press, Princeton, N.J.

Edmonds, D. T. and Summers, C. P. (1973). *Journal of Magnetic Resonance*, **12**, 134.

Ewen, H. I. and Purcell, E. M. (1951a). *Nature*, **168**, 356.

——— (1951b). *Physical Review*, **83**, 881.

Fano, U. and Racah, G. (1959). *Irreducible Tensorial Sets*. Academic Press, New York.

Feynman, R. P., Leighton, R. B. and Sands, M. L. (1989). *The Feynman Lectures on Physics*. Addison-Wesley, Redwood City, Calif.

Feynman, R. P. and Weinberg, S. (1987). *Elementary Particles and the Laws of Physics: The 1986 Dirac Memorial Lectures*. Cambridge University Press, Cambridge.

Fleischhauer, M., Imamoglu, A. and Marangos, J. P. (2005). *Reviews of Modern Physics*, **77**, 633.

Foot, C. J. (2005). *Atomic Physics*. Oxford University Press, Oxford.

Fraissard, J. (2009). *Explosives Detection Using Magnetic and Nuclear Resonance Techniques*. Springer, New York.

Friese, M. E. J., Enger, J., Rubinsztein-Dunlop, H. and Heckenberg, N. R. (1996). *Physical Review A*, **54**, 1593.

Frish, S. E. (1963). *Opticheskie Spektry Atomov*. Gosizdat Fizmatlit, Moscow.

Frois, B. and Bouchiat, M.-A. (1999). *Parity Violation in Atoms and Polarized Electron Scattering*. World Scientific, Singapore.

Garroway, A. N., Buess, M. L., Miller, J. B., Suits, B. H., Hibbs, A. D., Barrall, G. A., Matthews, R. and Burnett, L. J. (2001). *IEEE Transactions on Geoscience and Remote Sensing*, **39**, 1108.

Ginges, J. S. M. and Flambaum, V. V. (2004). *Physics Reports*, **397**, 63.

Gordon, R. J. and Rice, S. A. (1997). *Annual Review of Physical Chemistry*, **48**, 601.

Griffiths, D. J. (2005). *Introduction to Quantum Mechanics*. Pearson Prentice Hall, Upper Saddle River, NJ, 2nd ed.

Grossetête, F. (1964). *Journal de Physique*, **25**, 383.

Hanle, W. (1924). *Zeitschrift für Physik*, **30**, 93.

Hanneke, D., Fogwell, S. and Gabrielse, G. (2008). *Physical Review Letters*, **100**, 120801.

Happer, W. (1970). *Progress in Quantum Electronics*, **1**, 51.

────── (1972). *Reviews of Modern Physics*, **44**, 169.

Happer, W., Jau, Y.-Y. and Walker, T. (2010). *Optically Pumped Atoms*. John Wiley-VCH.

Happer, W. and Mathur, B. (1967). *Physical Review*, **163**, 12.

Haroche, S. and Raimond, J. M. (2006). *Exploring the Quantum: Atoms, Cavities and Photons*. Oxford University Press, Oxford.

Harris, S. E. (1997). *Physics Today*, **50**, 36.

Hartemann, F. V. (2002). *High-Field Electrodynamics*. CRC Press, Boca Raton.

Heitler, W. (1984). *The Quantum Theory of Radiation*. Dover Publications, New York, 3rd ed.

Hilborn, R., Hunter, L., Johnson, K., Peck, S., Spencer, A. and Watson, J. (1994). *Physical Review*, **50**, 2467.

Hilborn, R. C. and Tino, G. M. (2000). *Spin-Statistics Connection and Commutation Relations: Experimental Tests and Theoretical Implications: Anacapri, Capri Island, Italy, 31 May-3 June 2000*. American Institute of Physics, Melville, N.Y.

Huard, S. (1997). *Polarization of Light*. John Wiley, Chichester.

Jackson, J. D. (1975). *Classical Electrodynamics*. John Wiley, New York, 2nd ed.

Jensen, K., Acosta, V. M., Higbie, J. M., Ledbetter, M. P., Rochester, S. M. and Budker, D. (2009). *Physical Review A*, **79**, 023406.

Jerschow, A., Logan, J. W. and Pines, A. (2001). *Journal of Magnetic Resonance*, **149**, 268.

Judd, B. R. (1998). *Operator Techniques in Atomic Spectroscopy*. Princeton University Press, Princeton, N.J.

Kanorskii, S. I., Weis, A. and Skalla, J. (1995). *Applied Physics B: Laser Optics*, **60**, S165.

Kanorsky, S. I., Weis, A., Wurster, J. and Hänsch, T. W. (1993). *Physical Review A*, **47**, 1220.

Karshenboim, S. G. and Peik, E. (2004). *Astrophysics, Clocks and Fundamental Constants*. Lecture Notes in Physics, 648, Springer, Berlin.

Kastler, A. (1967). *Science*, **158**, 214.

Kazantsev, A., Smirnov, V., Tumaikin, A. and Yagovarov, I. (1984). *Optics and Spectroscopy (USSR)*, **57**, 116.

Khrapko, R. (2001). *American Journal of Physics*, **69**, 405.

Khriplovich, I. B. (1991). *Parity Nonconservation in Atomic Phenomena*. Gordon and Breach Science Publishers, Philadelphia.

Khriplovich, I. B. and Lamoreaux, S. K. (1997). *CP Violation without Strangeness: Electric Dipole Moments of Particles, Atoms, and Molecules*. Springer-Verlag, Berlin.

Kleppner, D., Ramsey, N. F. and Fjelstadt, P. (1958). *Physical Review Letters*, **1**, 232.

Knize, R. J., Wu, Z. and Happer, W. (1988). *Advances in Atomic, Molecular, and Optical Physics*, **24**, 223.

Kocharovskaya, O. (1992). *XXth Solvay Conference on Physics; Physics Reports (Netherlands)*, Brussels, Belgium, pp. 175–90.

Kocharovskaya, O. A. and Khanin Ya, I. (1988). *JETP Letters*, **48**, 630.

Kopfermann, H. (1958). *Nuclear Moments*. Academic Press, New York.

Kozlov, M. G. (1989). *Optics and Spectroscopy (USSR)*, **67**, 1342.

Kuntz, M. C., Hilborn, R. and Spencer, A. M. (2002). *Physical Review A.*, **65**, 023411.

Landau, L. D. and Lifshitz, E. M. (1976). *Mechanics*. Pergamon Press, Oxford, 3d ed.

────── (1977). *Quantum Mechanics: Non-Relativistic Theory*. Pergamon Press, Oxford, 3d ed.

Lee, Y. K. (2002). *Concepts in Magnetic Resonance*, **14**, 155.

Lehmann, J. (1967). *Annales de Physique*, **2**, 345.

Leone, M., Paoletti, A. and Robotti, N. (2004). *Physics in Perspective*, **6**, 271.

Letokhov, V. S. (1987). *Laser Photoionization Spectroscopy*. Academic Press, Orlando.

Lezama, A., Barreiro, S. and Akulshin, A. M. (1999). *Physical Review A*, **59**, 4732.

Li, C. H. and Budker, D. (2006). *Physical Review A*, **74**, 12512.

Li, C.-H., Rochester, S. M., Kozlov, M. G. and Budker, D. (2004). *Physical Review A*, **69**, 042507.

Lombardi, M. (1969). *Journal de Physique*, **30**, 631.

Loudon, R. (2000). *The Quantum Theory of Light*. Oxford University Press, Oxford, 3d ed.

Macaluso, D. and Corbino, O. M. (1898). *Nuovo Cimento*, **8**, 257.

────── (1899). *Nuovo Cimento*, **9**, 384.

Majumder, P. K., Venema, B. J., Lamoreaux, S. K., Heckel, B. R. and Fortson, E. N. (1990). *Physical Review Letters*, **65**, 2931.

Margolis, H. S. (2009). *Journal of Physics B*, **42**, 16.

Martin, W. C. and Wiese, W. L. (2007). *National Institute of Standards and Technology: Atomic Spectroscopy Database*, http://physics.nist.gov/pubs/atspec/index.html.

Matsko, A. B., Kocharovskaya, O., Rostovtsev, Y., Welch, G. R., Zibrov, A. S. and Scully, M. O. (2001). B. Bederson (ed.), *Advances in Atomic, Molecular and Optical Physics*, Academic Press, New York, vol. 46, pp. 191 – 242.

Matsuta, K., Minamisono, T., Nojiri, Y., Fukuda, M., Onishi, T. and Minamisono, K. (1998). *Nuclear Instruments and Methods in Physics Research, Section A (Accelerators, Spectrometers, Detectors and Associated Equipment)*, **402**, 229.

Migdal, A. B. (2000). *Qualitative Methods in Quantum Theory*. Advanced Book Program, Perseus, Cambridge, Mass.

Milner, V. and Prior, Y. (1998). *Physical Review Letters*, **80**, 940.

Milonni, P. W. (1994). *The Quantum Vacuum: An Introduction to Quantum Electrodynamics*. Academic Press, Boston.

────── (2005). *Fast Light, Slow Light, and Left-Handed Light*. Institute of Physics, Bristol.

Mohr, P. J., Taylor, B. N. and Newell, D. B. (2008). *Reviews of Modern Physics*, **80**, 633.

Moruzzi, G. and Strumia, F. (1991). *The Hanle Effect and Level-Crossing Spectroscopy*. Physics of Atoms and Molecules, Plenum Press, New York.

Nasyrov, K., Cartaleva, S., Petrov, N., Biancalana, V., Dancheva, Y., Mariotti, E. and Moi, L. (2006). *Physical Review A*, **74**, 13811.

Novikova, I., Matsko, A. B., Velichansky, V. L., Scully, M. O. and Welch, G. R. (2001a). *Physical Review A*, **63**, 063802.

Novikova, I., Matsko, A. B. and Welch, G. R. (2001b). *Optics Letters*, **26**, 1016.

Okunevich, A. I. (1994). *Optics and Spectroscopy (USSR)*, **77**, 178.

——— (1995). *Optics and Spectroscopy (USSR)*, **79**, 12.

——— (2001). *Optics and Spectroscopy (USSR)*, **91**, 193.

Padgett, M., Courtial, J. and Allen, L. (2004). *Physics Today*, **57**, 35.

Pais, A. (1991). *Niels Bohr's Times: In Physics, Philosophy, and Polity*. Clarendon Press, Oxford.

Papoyan, A. V., Auzinsh, M. and Bergmann, K. (2002). *European Physical Journal D*, **21**, 63.

Park, B. K., Pustelny, S., Family, A., Gawlik, W. and Budker, D. (2010). (In preparation).

Park, C. Y., Ji Young, K., Jong Min, S. and Cho, D. (2002). *Physical Review A*, **65**, 033410.

Parkins, A. S., Marte, P., Zoller, P. and Kimble, H. J. (1993). *Physical Review Letters*, **71**, 3095.

Particle Data Group (2008). *Physics Letters B*, **667**, 1.

Pinard, M. and Aminoff, C. (1982). *Journal de Physique*, **43**, 1327.

Pustelny, S., Jackson Kimball, D. F., Rochester, S. M., Yashchuk, V. V., Gawlik, W. and Budker, D. (2006). *Physical Review A*, **73**, 023817.

Ramsey, N. F. (1990). *Reviews of Modern Physics*, **62**, 541.

Rauch, H. and Werner, S. A. (2000). *Neutron Interferometry: Lessons in Experimental Quantum Mechanics*. Clarendon Press, New York.

Robinson, H., Ensberg, E. and Dehmelt, H. (1958). *Bulletin of the American Physical Society*, **3**, 9.

Rochester, S., Bowers, C. J., Budker, D., DeMille, D. and Zolotorev, M. (1999). *Physical Review A*, **59**, 3480.

Rochester, S. M. and Budker, D. (2001). *American Journal of Physics*, **69**, 450.

——— (2002). *Journal of Modern Optics*, **49**, 2543.

Rochester, S. M., Hsiung, D. S., Budker, D., Chiao, R. Y., Kimball, D. F. and Yashchuk, V. V. (2001). *Physical Review A*, **63**, 043814.

Safronova, M. S., Johnson, W. R. and Derevianko, A. (1999). *Physical Review A*, **60**, 4476.

Sakurai, J. J. and Tuan, S. F. (1994). *Modern Quantum Mechanics*. Addison-Wesley, Reading, Mass.

Schuh, B., Kanorsky, S. I., Weis, A. and Hänsch, T. W. (1993). *Optics Communications*, **100**, 451.

Scully, M. O., Shi-Yao, Z. and Gavrielides, A. (1989). *Physical Review Letters*, **62**, 2813.

Scully, M. O. and Zubairy, M. S. (1997). *Quantum Optics*. Cambridge University Press, Cambridge.

Seltzer, S., Meares, P. and Romalis, M. (2007). *Physical Review A*, **75**, 51407.

Series, G. W. (1966). *Proceedings of the Physical Society*, **88**, 957.

Shen, Y. R. (1984). *The Principles of Nonlinear Optics*. John Wiley, New York.

Shore, B. W. (1990). *The Theory of Coherent Atomic Excitation*. John Wiley, New York.

——— (1995). *Contemporary Physics*, **36**, 15.

Siegman, A. E. (1986). *Lasers*. University Science Books, Mill Valley.

Skalla, J. and Waeckerle, G. (1997). *Applied Physics B (Lasers and Optics)*, **B64**, 459.
Slichter, C. P. (1996). *Principles of Magnetic Resonance*. Springer, Berlin, 3rd ed.
Sobelman, I. I. (1992). *Atomic Spectra and Radiative Transitions*. Springer, Berlin.
Sommerfeld, A. and Brose, H. L. (1934). *Atomic Structure and Spectral Lines*. Methuen & Co., London,, 3d ed.
Stalnaker, J. E., Budker, D., Freedman, S. J., Guzman, J. S., Rochester, S. M. and Yashchuk, V. V. (2006). *Physical Review A*, **73**, 43416.
Stenholm, S. (2005). *Foundations of Laser Spectroscopy*. Dover Publications, Mineola, N.Y.
Suter, D. (1997). *The Physics of Laser-Atom Interactions*. Cambridge University Press, Cambridge.
Suter, D. and Marty, T. (1994). *Journal of the Optical Society of America B*, **11**, 242.
Suter, D., Marty, T. and Klepel, H. (1993). *Optics Letters*, **18**, 531.
Townes, C. H. and Schawlow, A. L. (1975). *Microwave Spectroscopy*. Dover Publications, New York.
Vallés, J. A. and Alvarez, J. M. (1994). *Physical Review A*, **50**, 2490.
—— (1996). *Physical Review A*, **54**, 977.
Varshalovich, D. A., Moskalev, A. N. and Khersonskii, V. K. (1988). *Quantum Theory of Angular Momentum: Irreducible Tensors, Spherical Harmonics, Vector Coupling Coefficients, 3j Symbols*. World Scientific, Singapore.
Walker, T. G. (1989). *Physical Review A*, **40**, 4959.
Walker, T. G., Thywissen, J. H. and Happer, W. (1997). *Physical Review A*, **56**, 2090.
Warren, W. S., Rabitz, H. and Dahleh, M. (1993). *Science*, **259**, 1581.
Weis, A. and Ulzega, S. (2007). *Proceedings of SPIE*, **6604**, 660408.
Weis, A., Wurster, J. and Kanorsky, S. I. (1993). *Journal of the Optical Society of America B*, **10**, 716.
White, H. E. (1934). *Introduction to Atomic Spectra*. McGraw-Hill, New York.
Xu, G. and Heinzen, D. J. (1999). *Physical Review A*, **59**, R922.
Xu, J. D., Wäckerle, G. and Mehring, M. (1997a). *Physical Review A*, **55**, 206.
—— (1997b). *Zeitschrift fur Physik D (Atoms, Molecules and Clusters)*, **42**, 5.
Yashchuk, V., Budker, D. and Zolotorev, M. (1999). D. Dubin and D. Schneider (eds.), *Trapped Charged Particles and Fundamental Physics*, American Institute of Physics, Asilomar, CA, USA, *AIP Conference Proceedings*, vol. 457, pp. 177–81.
Yashchuk, V. V., Budker, D., Gawlik, W., Kimball, D. F., Malakyan, Y. P. and Rochester, S. M. (2003). *Physical Review Letters*, **90**, 253001.
Zare, R. (1988). *Angular Momentum: Understanding Spatial Aspects in Chemistry and Physics*. John Wiley, New York.
Zetie, K. P., Warrington, R. B., Macpherson, M. J. D., Stacey, D. N. and Schuller, F. (1992). *Optics Communications*, **91**, 210.

Index

3j symbol, 46, 47, 49, 137–139, 147, 180, 181
6j symbol, 48, 49, 141–143, 353
21-cm transition in hydrogen, 145

absorption
   coefficient, 169, 170, 182, 229–231, 291, 305
   length, 279
      resonant, 202
      unsaturated, 201, 274
   matrix, 291
AC-Stark polarizability, 329
adiabatic elimination, 255
alignment-to-orientation conversion (AOC), 72, 96, 209, 219, 222–225, 309, 310, 313
alkali atom, 343
amplitude
   quantum mechanical, 7
   transition, 126, 130, 131, 137, 144, 145, 148, 149, 152, 157, 163
angular momentum
   lowering operator, 52
   orbital of the light, 117
   raising operator, 52
angular-momentum probability surface (AMPS), 89–92, 101–103, 122, 123, 175, 177–180, 182, 184, 185, 223, 290, 293, 294, 299, 301, 305, 306, 311, 324, 326, 328
anticommutator, 243
aperture, 116
atomic clocks, 70, 132, 145
   microwave, 329
   optical, 329

atomic magnetometers, 210, 228, 234, 257, 308, 323, 325, 328
   based on NMOR, 223, 228, 234, 238
   opical-pumping, 146
   scalar, 323
atomic units, 66, 339
autoionization, 338
Avogadro number, 342
axis-angle
   rotation, 35

basis
   Cartesian, 310
   change of, 87
   complete, 6, 42, 82, 209, 267
   coupled, 48, 49, 53, 60–62, 71, 260
   judicious choice of, 87, 120, 159, 173–175, 241, 257
   of polarization moments, 272
   orthonormal, 6
   spherical, 31–34, 44, 49–51, 118, 119, 137, 148, 160, 174, 242, 347–350, 356
   transformation of, 195
   uncoupled, 47, 49, 60, 61, 70, 71, 260, 264, 267, 269
   vectors, 119
   with no coherences, 86
   Zeeman, 72, 82, 173, 261
beat frequencies, 19
Bessel functions, 77, 78
   spherical, 150
birefringence, circular, 203, 208, 209, 223
body-fixed
   axes, 35, 36
   rotations, 36
Bohr

magneton, 14, 26, 58, 59, 132, 144, 164, 229, 274, 304, 319, 340
model, 9, 15
radius, 11, 63, 64, 133, 148, 339, 340
Bohr, Niels, 9, 16
Boltzmann
constant, 87, 210, 278, 340, 342
law, 310
boson, 153
branching ratio, 142, 180, 181, 200, 241, 242, 265, 266, 273, 274
Breit–Rabi
diagram, 61, 66
formula, 304
broad-line approximation (BLA), 248, 252, 292
buffer gas, 191, 211, 224, 229, 245, 259, 271, 280, 282

central field, 11, 20–23, 28
charge conjugation, 28
chirality, 110
classical radius of the electron, 13
Clebsch–Gordan coefficient, 46, 47, 52, 53, 71, 100, 241, 260, 334
closed transition, 172, 174, 183–185, 220, 221
co/contravariance, 31, 33, 34, 43, 44, 99, 346, 348–350
coherent
control, 164
population trapping, xi, 164, 257
commutation relations, 36–38, 40
commutator, 36, 52, 94, 136, 214
completeness relation, 65
Compton wavelength, 12
configuration
electronic, 20–25, 130
configuration, electronic, 20
conservation law, 28, 29
continuous transformation, 28
continuum, 330, 333, 335, 336, 338
contravariant
components, 147
spherical, 119
convolution, 211, 212, 227, 286, 287

Corbino, Orso M., 204
correlation
matrix, 87
time, 115
correspondence principle, 7
covariant
basis
vectors, 119
CP violation, 29
CPT symmetry, 29
cross-section
for depolarizing collisions, 224
for ionization, 333
on resonance, 202
photon-absorption, 202

decoherence, 255
degeneracy
of atomic states, 9, 16, 20, 21, 23, 28, 63, 66, 68, 159, 196, 245, 248, 259, 262
of photons, 153
density matrix, 82, 85, 86
and AMPS, 89–92, 102, 122, 175, 177, 179
and atomic velocity distribution, 227, 245, 246, 280
and dark states, 178
and irreducible tensors, 261, 335
and optical properties of the medium, 187, 189, 213, 216, 229
and polarization moments, 98–101, 268, 272, 352
and spin-exchange collisions, 228
and the wave function, 7, 87
and two-photon transitions, 153
coherences in, 85, 86, 91, 176
eigenvalues of, 86
Hermiticity of, 86, 100, 199
light polarization, 121–123, 125
multipole expansion of, 101, 257
nonphysical, 101
of photons, 334
of single atom, 86
perturbative expansion of, 215
properties of, 86

repopulation and depopulation, 246
rotating-frame, 201
rotation of, 88
steady-state, 180, 244
temporal evolution of, 92, 94, 95, 243, 292, 312, 352
density operator, 84, 85
dichroism
  circular, 205
  linear, 204, 208, 209, 222
Dirac
  equation, 14
  formula, 15, 16
  notation, 6
    bra, 7
    ket, 6
direct product, 51
discrete
  symmetry, 29
    violation, 29
  transformation, 28, 29
displacement, electric, 187
Doppler
  averaging, 219
  broadening, 183, 191, 204, 211–213, 219, 220, 224, 227, 229–233, 245, 246, 258–260, 264, 272, 273, 277–280, 282, 284–288, 356, 357
  distribution, 212, 220, 221, 231, 232, 252, 258, 259, 277
  profile, 211
  shift, 191, 211, 220, 226, 227, 245, 246, 273
Doppler-free, 202, 211, 212, 224, 229, 231–233, 260, 270–272, 275–279, 282–286, 354–356
dyadic, 50, 147
dynamic range, 234, 238

eccentricity, 9
eigenfunction, 6, 8, 28, 61, 69, 73
eigenstate, 6, 7, 10, 15, 18, 19, 22, 23, 28, 39–43, 45, 53, 60, 62, 86, 95, 126, 128, 132, 154, 157, 164, 243, 260, 303, 313, 330

eigenvalue, 6, 8, 28, 39–42, 61, 71, 72, 79, 86, 87
electric field gradient (EFG), 26, 147–149, 309–312, 323
electric-dipole
  moment, 7, 27, 63, 132, 144, 145, 154–157, 199, 339
    induced, 63
    instantaneous, 155, 156
    permanent, xi, 63, 329
    time dependent, 154
  selection rules, 66
electro-optical effects, 215
electromagnetically induced
  absorption (EIA), 257
  transparency (EIT), 164, 257
electron
  charge, 340
  mass, 340
elliptical dark states, 174
elliptically polarized light, 110, 111, 115, 122, 123, 153, 174, 176
ellipticity, 112, 186, 189, 190, 202, 205, 230, 249
entropy, von Neumann, 87
Euler angles, 35, 36, 39, 43, 121, 175, 176, 312
exchange interaction, 21, 22
expansion, 48, 70, 71, 82, 85, 148–150, 213, 214, 216, 305, 353
  Clebsch–Gordan, 60
  irreducible-tensor, 261
  multipole, 101, 144, 149, 257
  perturbative, 214, 215, 251, 308
  series, 243

Faraday
  effect, 55, 204, 215
    linear, 215, 271
    nonlinear, 208, 210, 215, 220, 228, 236, 272, 291
  geometry, 229, 271
  induction, 313
  rotation, 112
Faraday, Michael, 55, 204
Fermi's golden rule, 171

fermion, 21
Feynman
    diagram, 152, 153
        trunk, 152
        vertex, 152
Feynman, Richard, 13
fictitious
    field, 78, 79
    force, 195
fine structure, 8, 22, 60
fine-structure constant α, 15, 66, 145, 340
fluorescence matrix, 293
FM NMOR, xi, 236, 238, 314–317, 319–321
four-wave mixing, 215

Gaussian (CGS) system of units, xi, 339, 340
generator of infinitesimal rotations, 39
gross structure, 8, 16, 22

Hamiltonian, 6, 7, 55, 60, 61, 72, 75, 192
    and relaxation, 197, 199
    Coulomb-gauge, 148
    coupled and uncoupled bases, 60, 71
    effective, 70, 71, 95, 195
    electric interaction, 62, 64, 70, 73, 80, 95, 128, 330
    electric-dipole, 128, 214, 244, 354
    electric-quadrupole, 147
    for NQR, 310, 312
    hyperfine, 60, 71
    light–atom interaction, 192
    magnetic interaction, 60, 72, 193, 355
        time dependent, 79
    magnetic-dipole, 144
    non-Hermitian, 93
    of atom in the presence of light, 133
    perturbation, 127
    under RWA, 79, 193, 195, 196, 244, 354
Hanle effect, 166, 168, 296, 300
Hartree, 8
heading error, 308, 323, 328
Hermite–Gaussian mode, 118
Hertzian
    coherences, 146

    resonances, 146
hexacontatetrapole, 100, 319–323
hexadecapole, 27, 100, 236, 315, 317, 318, 320, 323–328
    resonance, 316–321
    signal, 316, 317, 319, 321, 323, 325–328
        false, 327
hydrogen, 59–61, 67, 343
hyperfine
    interactions, 26, 132
    polarization, 82

infinitesimal rotation, 39
    generator of, 39
intermediate coupling, 24
ionization potential, 333
irreducible tensor
    operators, 52, 329

jj coupling, 23, 130

Lamb shift, 16, 67
Landé g-factor, 14, 56, 58, 59, 222
    nuclear, 26, 304
Landau, Lev D., 5, 86, 153
Landau-Yang theorem, 153
laser
    cooling, 75, 81, 191, 210, 314, 329
    diode, 324
    femtosecond, 133
    ionization, 333, 334
    modulation, 316
    physics, 115
    pulse, 292, 333
    trapping, 75, 81, 191, 210, 329
    tunable, 333
lasing without inversion, 164, 175
level crossing, 215
    avoided, 73, 74
lifetime, 115, 142, 154, 166, 168, 170, 171, 287, 299, 333
    ground state, 173, 224–226, 228, 259
light shifts, 81, 173, 223, 319, 328
linear Faraday rotation, 215
linear optical process, 169

Liouville equation, 93–95, 101, 180, 186, 198, 214, 239, 243, 253, 254, 352, 353
Lo Surdo, Antonino, 62
lock-in
    amplifier, 235
    detection, 235, 237
Lorentz
    law, 109
    theory, 56
Lorentz, Hendrik, 55, 56, 109
Lorentzian, 115, 203, 206, 208, 246, 255, 273, 278, 281, 282, 301
    absorptive, 204, 207, 208
    dispersive, 206, 207, 210, 234
lowering operator, 52
LS coupling, 20, 22–24, 56, 130, 141, 142

Macaluso, Damiano, 204
magnetic
    decoupling, 60
    moment, anomalous, 14
    quantum number, 9
magneto-optical
    effects, 72, 160, 257, 271
        linear, 215
        nonlinear, vi, 101, 183, 204, 213, 225, 262, 291
    experiments, 258
    rotation, 205, 212, 219, 234
        linear, 208, 212
        nonlinear, xi, 182, 206–208, 212, 216, 228, 235, 257, 277, 296, 314, 327
matrix
    element, 7
    notation, 6
Maxwell equations, 4, 63, 116, 187
Maxwell–Boltzmann distribution, 210, 211, 220
Maxwellian velocity distribution, 231, 246, 252, 260, 271, 277, 287
modulation index, 77, 78
molecular
    beam, 226
    bond, 342
    physics, 4

spectroscopy, 5, 101, 166
molecules, 5, 29, 66, 149, 177, 183, 299, 308, 342
    diatomic, 101, 183
monopole moment, 100, 236
multiplicity, 22, 25, 246

neutron
    interferometry, 43
    mass, 340
Nobel prize, 146, 226, 228
Noether's theorem, 28
noncommutativity, 36
noninertial frame, 195
nonlinear
    Faraday rotation, 210, 220, 228, 236, 272, 291
    light–atom interaction, 169
    optics, 3, 216
    polarization, 215
    wave mixing, 215
nuclear magneton, 26, 132, 340
nuclear quadrupole resonance (NQR), 309–311, 313

octupole, 100, 149, 236
    electric, 146, 149
    magnetic, 27
open transition, 183, 185, 221, 300
operator, 38
    in quantum mechanics, 6
    vector, 9, 53
optical coherence, 253
optical gain, 186
optical pumping, 4, 163, 169, 170, 172–185, 208, 209, 220, 221, 223, 227, 236, 238, 239, 247, 258, 259, 261, 270, 272, 274, 275, 277, 278, 280, 301, 309, 311, 315
    cycle, 209
    depopulation, 185, 209, 246, 247, 259, 262, 264, 269, 270, 273
    early work on, 228
    into dark states, 183
    magnetometers, 146
    relaxation due to, 209

repopulation, 185, 247, 263, 264, 273, 287
saturation parameter, 200, 229, 273
synchronous, 235
optically
  thick medium, 189, 228, 230
  thin medium, 189, 190, 229
optically driven spin precession, 235
orbital quantum number, 9
oscillation theorem, 20

parity, 28, 63, 64, 66, 69, 70, 131, 145
  conservation, 27, 29
  intrinsic, 150
  photon, 150
  selection rules, 64, 131, 145, 146, 150, 153, 214
  violation, 29, 63, 72, 131, 145, 329
partially
  polarized light, 114, 115, 121, 123
  resolved hyperfine structure, 3, 258, 271
Paschen–Back effect, 60
Pauli
  exclusion principle, 21
  matrices, 43, 79, 83
  spin operator, 42
periodic table, 21, 59
perturbation theory
  degenerate, 66
  first-order, 135
  higher-order, 258
  nondegenerate, 66
  second-order, 64, 215, 272, 329
  time-dependent, 126, 329
phase diffusion, 115, 255
phase-sensitive detection, 235, 237
photoionization, 333, 335
  cross section, 333–338
Poincaré sphere, 114, 123
polarizability, 74
  AC-Stark, 329
  atomic unit of, 64
  electric, 63, 64
    of barium, 66
    of conducting sphere, 64
    of hydrogen ground state, 65

  of Rydberg states, 66
  scalar, 69–71, 73, 95, 329
  static, 331
  tensor, 69–71, 73, 74, 95, 329
  large, 66
polarization
  degree of, 123, 298
  eigenstate, 209
  ellipse, 110–112, 174, 186
  longitudinal, 100, 326, 327
  moment (PM), 94, 96–103, 177, 227, 234, 237, 252, 257, 258, 260–265, 267–270, 272–274, 305–307, 314, 319, 324, 333, 353, 355
  operator, 98
  rotation, 186, 204, 219, 228, 252
  transverse, 100, 240
power broadening, 173, 209, 228, 233, 234, 300, 317, 327
probability
  amplitude, 5, 131
  density, 6, 163
projection operator, 97
proton
  magnetic moment, 62
  mass, 340

QED, xi, 14, 16, 110, 116, 128, 133, 239
quadrature component, 188, 237, 316, 317
quadrupole, 100, 149, 236, 311, 315, 317, 320, 322, 324–328
  electric, 26, 27, 144, 146, 147, 149, 150
  interaction, 27, 308, 310
  longitudinal, 326
  magnetic, 146, 149
  nuclear, 309, 310
  operator, 146, 149
  pumping of, 324, 327
  relaxation, 318
  resonance, 316–321
  signal, 317, 319, 321, 325–327
  splitting, 311
  tensor, 310
  transverse, 325

quantum beat, 5, 19, 165–168, 173, 215, 221, 222, 292, 303, 305, 309–313, 326, 327
  collapse and revival, 307, 308
  excited-state, 293
  frequencies, 19, 234, 237
  ground-state, 294, 295
  hyperfine, 296
  illustrated with AMPS, 306
  in absorption, 293, 294
  in fluorescence, 293, 294
  related to NLZ, 326, 327
  resonances, 235, 238
  Stark, 95, 96, 296
  Zeeman, 296, 308
quantum number
  good, 20, 22–24, 28, 68, 69, 131, 132
  principal, 8, 16, 20, 25, 64, 66, 145
quarter-wave plate, 117

Rabi
  frequency, 127–129, 170, 171, 181, 193, 199, 229, 272
  oscillations, 5, 128, 171
  period, 127
radiation trapping, 229
radiophysics, 115
raising operator, 48, 52
Raman transition, 314
Ramsey, Norman, 226, 228
rate equations, 180, 181, 253, 255
reduced matrix element, 52, 53, 68, 137, 139, 141–144, 181, 193, 229, 268, 287, 343
reducible
  tensor, 51
relaxation matrix, 94, 197, 243
repopulation, 92–94, 185, 197–199, 209, 214, 239, 243, 247, 265, 270, 274–277, 280, 284, 285, 287, 352
residual Coulomb interaction, 20, 23
revival
  period, 306
  stage, 326
  super, 304
rotating

frame, 78–80, 126, 127, 194–198, 200, 245, 250, 273, 354
polarizer model, 208, 209, 275
rotating-wave approximation (RWA), 78–80, 128, 191, 196, 222, 243, 244, 255, 354
rotation matrix, 75
Rydberg
  constant, 8, 148, 340
  states, 66

saturated absorption, 170, 230
saturation, 171, 230, 317, 319, 320
  intensity, 169, 170, 172
  parameter, 170–174, 184, 185, 200, 202, 205, 206, 210, 212, 229, 230, 232, 233, 273, 281, 357
scalar product, 6, 7, 50, 51, 137, 147, 305
Schrödinger equation, 7, 9, 15, 19, 20, 55, 60, 93
  hydrogenic, 20
  radial, 20
  relativistic corrections to, 15
  time independent, 8
  time-dependent, 92
secular equation, 61
self-rotation, 174
"slow" light, 164
space-fixed
  axes, 35, 36
  rotations, 36
spatial inversion, 27–29, 131
spectroscopic notation, 24
spherical harmonics, 11, 102, 137, 177, 261
spin
  nuclear, 23, 26, 27, 58, 59, 61, 129, 132, 258, 259, 261–263, 268, 272, 283, 284, 287, 305, 318, 340
  of the light, 117
spin–orbit interaction, 14–16, 20–23, 25
spinor representation, 79, 126
spontaneous emission, 94, 128, 133, 136, 137, 139, 151, 157, 163, 170, 229, 239, 240, 243, 246, 263–265, 282, 287, 311, 321, 322, 352, 353
  operator, 239, 242

Stark
  effect, 62, 69, 70, 73, 75, 310
    AC, 75, 78, 80, 81, 209, 222, 223, 254, 271, 296, 328–331
    linear, 63, 66, 67
    quadratic, 63, 68, 70, 304
  Hamiltonian, 71, 80
  quantum beats, 95, 96, 296
  shift, 67, 70, 71, 76, 223, 323
    first-order, 329
    quadratic, 74
  splitting, 95, 310
Stark, Johannes, 62
Stark-induced transitions, 70
state multipoles, 98
stationary state, *see* eigenstate
statistics
  Bose–Einstein, 153
  Fermi–Dirac, 21
stimulated
  absorption, 128
  emission, 128, 133, 137, 157, 171, 300
  Raman adiabatic passage (STIRAP), 164, 257
Stokes parameters, 112–114, 123–125, 188, 299, 336
stretched state, 46, 62, 88, 90, 96, 305–307, 331
sum rules, 138, 353
  for $3j$ symbols, 47, 138
  for Clebsch–Gordan coefficients, 46
  for line strengths, 143
  for transition probabilities, 138
summation
  over repeated indices, 52, 330
"superluminal" light, 164
susceptibility tensor, 189, 216

tensor
  Cartesian, 51
  product, 50, 51, 261, 267, 353
  rank, 26, 27, 50–52, 68, 74, 101, 103, 121, 138, 146, 147, 150, 177, 227, 252, 257, 258, 261–263, 265–274, 287, 305–307, 309, 314, 315, 318, 319, 327, 330, 335, 341

  spherical, 4, 68
term
  electronic, 22, 24, 25
  mixing, 130
time reversal, 28
  invariance, 27, 63
transit
  effect, 225–227, 245, 271, 277, 279, 280, 282–284, 286–288, 354, 356
  rate, 94, 184, 191, 199, 209, 211, 225, 229, 233, 281
  relaxation, 94, 198, 199, 205, 209, 220, 229, 233
transition
  forbidden, 70, 130–133, 138, 140, 144–146, 151, 153, 157, 334
  probability, 126, 131, 152, 171, 240
  strength, 139, 140, 147, 180, 262
triacontadipole, 100
triangle
  inequality, 48
  rule, 45, 46, 129, 131, 137, 150

uncertainty relations, 10, 11, 17
unitary
  matrix, 195, 196
  operator, 37
  transformation, 244
unpolarized
  atoms, 94, 184, 197, 244
  density matrix, 261
  ensemble, 82–86, 88–90
  light, 115, 116, 123, 138, 139, 335
  radiation, 115
  state, 82, 84, 89, 101, 182, 244, 248, 257, 268, 269, 309, 335
unsaturated absorption, 287, 356

vacuum, 128, 343
  fluctuations, 16, 128, 139, 157, 240, 242
  permittivity, 4
  quantum, 93, 128, 136
vector model, 9, 18, 22, 26, 39, 53, 56, 177, 178
vector-addition coefficient, *see* Clebsch–Gordan coefficient

velocity mixing, 211, 225–227, 229, 246
  complete, 245, 246, 252, 259
von Neumann entropy, 87
von Neumann, John, 86

wave equation, 186–189, 248
wave function, 5–7, 27, 36, 43, 55, 82–84, 86, 130, 140, 154, 155, 160, 161, 164
  and the density matrix, 87, 92, 291
  angular, 12, 145
  atom-light, 135
  hydrogen, 8, 11, 65
  identical particles, 21
  multi-electron, 21, 28
  parity of, 28, 131
  photon, 150
  radial, 11, 12, 20, 145
  Rydberg atoms, 66
  time dependence of, 164, 166, 167
wave packet, 303
Wigner
  $3j$ symbol, *see* $3j$ symbol
  $6j$ symbol, *see* $6j$ symbol
  D-functions, 43, 44
  D-matrices, 120, 121, 125
  function, 103
Wigner's formula, 45
Wigner–Eckart theorem, 52–54, 57, 68, 119, 137, 160, 181, 192, 331, 355

Zeeman
  basis, 176, 261
  coherence, 88, 100, 159, 174, 175, 214, 314, 321, 327
  effect, 56, 58–61, 73, 146, 204
    AC, 75, 78, 80, 81
    anomalous, 55, 58
    linear, 101
    nonlinear, xi, 101, 308, 326, 327
    normal, 55, 56
  polarization, 82, 215
  quantum beats, 296, 308
  shift, 60, 164, 221
    linear, 332
    nonlinear, 304, 307, 308

splitting, 59, 222
  nonlinear, 322
structure, 129
sublevel, 55, 56, 69, 73, 82, 89–91, 100, 146, 159, 160, 163, 174, 175, 177, 194, 222, 263, 268, 269, 305, 329, 334, 336
Zeeman, Pieter, 55
zero-field resonance, 234, 237, 238, 299, 317